AF785826

Reviews in Modern Astronomy

3

Gerhard Klare (Ed.)

Reviews in Modern Astronomy 3

Accretion and Winds

With 175 Figures

Springer-Verlag

Berlin Heidelberg New York
London Paris Tokyo
Hong Kong Barcelona

Edited on behalf of the *Astronomische Gesellschaft* by

Dr. *Gerhard Klare*

Landessternwarte, Königstuhl,
D-6900 Heidelberg 1, Fed. Rep. of Germany

ISBN-13: 978-3-642-76240-6 e-ISBN-13: 978-3-642-76238-3
DOI: 10.1007/978-3-642-76238-3

Softcover reprint of the hardcover 1st edition 1990

56/3140-543210 – Printed on acid-free paper

Preface

In order to bring the scientific events of the meetings of the Astronomische Gesellschaft (AG) to the attention of the worldwide astronomical community, an annual publication, *Reviews in Modern Astronomy*, was established. It is devoted exclusively to the invited reviews, the Karl Schwarzschild lectures, and the highlight contributions from leading scientists reporting on recent progress and scientific achievements at their respective institutes.

This third volume continues the yearbook series of publications of the society. It comprises the complete set of contributions presented during the spring meeting of the AG at Berlin in March 1990, which was dedicated to the topic "Accretion and Winds". In addition four latecomers (two review and two highlight papers) delivered at the fall meeting at Graz, Austria in September 1989 close this volume.

Heidelberg, September 1990 *G. Klare*

Contents

Some New Elements in Accretion Disk Theory

F. Meyer

Max-Planck-Institut für Physik und Astrophysik,
Institut für Astrophysik, Karl-Schwarzschild-Str. 1
D-8046 Garching, Fed. Rep. of Germany

Accretion on compact bodies provides the energy for some of the highest energy sources in astrophysics. Due to the always present angular momentum it almost always takes the form of an accretion disk. We mention only active galactic nuclei, quasars and giant radio galaxies, protostellar disks and T Tauri stars, symbiotic stars, cataclysmic variables, X-ray binaries, SS433, γ-burst sources and the TeV-γ-ray sources as examples where an accretion disk is thought to play an essential role.

The theory of accretion disks has been quite far developed in recent years. In particular the disk instability responsible for the dwarf nova outbursts has been found, and it has proved of important element for deriving theoretical clues from observations. Though much knowledge is now available, some fundamental elements could not have been totally clarified. One of the central issues is the parameter α, the ratio of frictional stress to pressure. We do still not yet know which mechanism is responsible of the friction observed to be present nor which value of α it would provide.

Every new element that can be introduced and shown to be operative by successful modeling is therefore of high interest since it may allow further insight into the basic physics.

We discuss here two new features (out of possible many others) which promise new conclusions to be drawn from them. One is the evaporation of cold accretion disks through a hot accreting corona in dwarf novae systems, the other is the removal of a basic stumbling stone in understanding FU Orionis outbursts as disk instabilities.

Two long standing problems in the theory of accretion disks seem to have been:

a) In outbursts of dwarf novae the UV light appears substantially later than the optical light ("UV-lag").
b) Theory predicts a very cool and practically inactive accretion disk after the end of a dwarf nova outburst, while observation indicates substantial UV and X-ray light far into the quiescence period.

These two problems appear to be solved by a basic feature of such cool disks. In the potential gravitational field of the central white dwarf star a coronal "siphon flow" sets in that drives coronal mass accretion onto the central star which powers the hard X-rays and UV light. It also can evaporate the innermost accretion disk. As a consequence during the next outburst this region must first

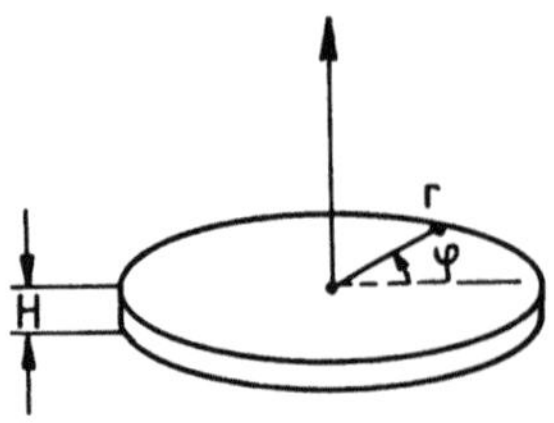

Fig. 1. Thin disk coordinates

be filled in again by mass diffusion before it can radiate at UV temperatures. This accounts easily for the observed UV-time lag.

The outbursts of FU Orionis stars appear to be accretion events in which a large amount of mass ($\Delta M \simeq 10^{-3} M_\odot$) is accreting over a short time ($\approx$ 100 years) Though in appearance similar to disk instability in dwarf novae, theory could not account for it. The solution appears to be an unusually small value of the standard frictional parameter $\alpha \simeq 10^{-4}$. Such a small value is still not in contradiction to convectively driven friction. A proper treatment of the reduction that the Coriolis force introduces can account for such small values if other sources of friction are absent or irrelevant.

These new elements establish important features in accretion disk theory:

Hot and extended coronae are now also shown to be subject to accretion flow with sizeable values of $\alpha(\simeq 1/3)$. Winds transport a significant fraction ($\frac{1}{3}$ to $\frac{1}{5}$) of the accretion rate away into space. The values of the α parameter cover a wide range of now 10^{-4} to 1. In FU Ori outbursts there appears to be a sudden (possibly temperature dependent) transition of α by about a factor 10. This points to the appearance of an additional and effective friction once some critical parameter is passed.

In the follwing we first give a short derivation of accretion disk physics. We then discuss as an example the mechanism of the dwarf nova instability. Two apparently unrelated observational features mentioned in the beginning have in this context withstood theoretical explanation and thereby cast some doubt on our understanding of the basic process. We show how these are resolved by a model that relies an fairly basic physics only.

We then address the question whether the FU Orionis outburst phenomenon in protostellar accretion disks (and T Tauri stars) cannot be naturally derived from instability in a disk of appropriate low mass accretion rate. We show that this long-standing difficulty is resolvable, and that several independent observational features point to very low α-values in such disks.

The article ends with a few concluding remarks.

1 Thin disk approximation

Important features of disk accretion can already be discussed in the simple model of a rotationally symmetric, non-selfgravitating, and thin accretion disk.

Introducing cylindrical coordinates r, φ, z, one has $\frac{\partial}{\partial\varphi} = 0$. The component of gravity perpendicular to the plane of the disk,

$$g_{\text{eff}} \approx \frac{GM}{r^2}\frac{z}{r} \quad (M \text{ central mass}) \tag{1}$$

is small near $z = 0$. This yields the hydrostatic equilibrium of a locally isothermal gas of temperature T and pressure p,

$$p = p_0(r)\exp\left[-\left(\frac{z}{H}\right)^2\right] \quad , \tag{2}$$

where the scaleheight H is obtained as

$$H = \left(\frac{2\frac{\Re}{\mu}Tr^3}{GM}\right)^{1/2} \approx \frac{V_s}{v_K} r \ll r \quad . \tag{3}$$

Here $\Re$ is the gas constant, μ the molecular weight, V_s is the isothermal sound speed and $v_K = (GM/r)^{1/2}$ is the Kepler velocity. This value of H gives a measure of the disk thickness when we take for T the midplane temperature.

Since the sound velocity is small compared to the Kepler velocity the radial pressure gradiant cannot balance the gravitational attraction and matter must circulate in nearly Keplerian motion, $v_\varphi = v_K$.

The Kepler frequency

$$\Omega(r) = \frac{v_\varphi}{r} = \left(\frac{GM}{r^3}\right)^{1/2} \tag{4}$$

thus introduces shear, $d\Omega/dr \neq 0$, and viscosity enters.

Any viscosity μ present in an accretion disk leads to a frictional stress

$$\tau_{r\varphi} = \mu r \frac{d\Omega}{dr} \quad . \tag{5}$$

This can remove angular momentum radially and allow matter to spiral inwards. Estimates of the value of μ can be obtained by relating observed time and spatial scales of accretion flows. They yield values which are larger than the molecular viscosity by a huge factor ($\approx 10^{12}$). The frictional stress can best be parametrized in terms of the pressure,

$$\tau_{r\varphi} = \alpha P \tag{6}$$

(Shakura and Sunyaev, 1973), approximately equivalent to

$$\mu \approx \alpha \varrho H V_s \quad . \tag{7}$$

This tells one that the α parameter measures the "Mach number" of a "turbulent exchange process" which takes place on the scale of the disk thickness itself. One therefore can expect validity of the viscosity description only in the mean over sufficiently large disk areas and/or sufficiently long timescales. Values deduced from observations yield $10^{-3} \leq \alpha \leq 1$.

Here lies one of the fundamental difficulties for a consistent theory of friction: The processes responsible are clearly macroscopic and non-linear. As models for such processes small scale magnetic dynamo fields, unstable wave modes in the accretion disk differential rotation and spiral shock waves have been discussed.

For thin disks the vertical structure is locally homogeneous in horizontal planes and can conveniently be separated from the radial structure. The latter is obtained by integrating across the disk in vertical direction. The resulting equations of conservation of mass and angular momentum are

$$\frac{\partial}{\partial t}(2\pi r \Sigma) - \frac{\partial \dot{M}}{\partial r} = 0 \quad , \tag{8}$$

$$\frac{\partial}{\partial t}(2\pi \Sigma r^2 \Omega) - \frac{\partial}{\partial r}\left[\dot{M} r^2 \Omega - 3\pi f r^2 \Omega\right] \tag{9}$$

with $\Sigma \equiv \int_{-\infty}^{\infty} \varrho dz$, the surface density and $f = \int_{-\infty}^{\infty} \mu dz = \frac{2\alpha}{3\Omega}\int_{-\infty}^{\infty} P dz$ a viscosity integral. $\dot{M}$ is the inward directed mass accretion rate.

Elimination of $\dot{M}$ yields the standard form for viscous diffusion of the surface density,

$$\frac{\partial \Sigma}{\partial t} = \frac{3}{r}\frac{\partial}{\partial r}\left[r^{1/2}\frac{\partial}{\partial r}(r^{1/2} f)\right] \quad . \tag{10}$$

This equation contains a diffusion coefficient

$$D = 3\frac{\partial f}{\partial \Sigma} \tag{11}$$

where the functional dependence $f = f(\Sigma, r)$ is obtained from the vertical structure of the disk.

The vertical structure is determined by the equations of hydrostatic equilibrium, heat transport, and frictional generation of heat. For optically deep structures these are

$$\frac{\partial P}{\partial z} = -\frac{GM}{r^3} z \varrho \quad , \tag{12}$$

$$F = \begin{cases} -\frac{4acT^3}{3\kappa\varrho}\frac{\partial T}{\partial z} \quad , & \nabla \equiv \frac{d\ln T}{d\ln P} < \nabla_{\mathrm{ad}} \quad , \\ F_{\mathrm{conv}} \quad , & \nabla > \nabla_{\mathrm{ad}} \quad , \end{cases} \tag{13}$$

$$\frac{\partial F}{\partial z} = \mu\left(r\frac{d\Omega}{dr}\right)^2 = \alpha P r \frac{-d\Omega}{dr} \quad . \tag{14}$$

κ is the opacity. If the temperature gradiant ∇ becomes overadiabatic the flux F_{conv} contains a convective contribution and is usually determined by a mixing length model for convective flow. The density ϱ is related to P and T by an equation of state, e.g.

$$P = \varrho\frac{\Re}{\mu}T \tag{15}$$

if radiation pressure is negligible. At the midplane and the surface one has to add boundary conditions, for example an optically thin atmosphere at the top.

From these equations one may derive a number of elementary relations of accretion disk physics, of which we list a few:

$F_0 = \sigma T_{\text{eff}}^4 = \frac{9}{8}\frac{GMf}{r^3}$	flux from surface ,	(16a)
$\dot{M} = 3\pi f$	stationary mass flow rate ,	(16b)
$T \approx T_{\text{eff}}\left(\frac{3}{8}\kappa\Sigma\right)^{1/4}$	midplane temperature ,	(16c)
$\dot{M} \approx 2\mu r\Sigma\alpha\frac{V_s^2}{v_{\text{K}}}$	stationary mass flow rate ,	(16d)
$\tau_{\text{th}} = \frac{1}{\alpha\Omega}$	thermal relaxation time ,	(16e)
$\tau_d \approx \left(\frac{r}{H}\right)^2 \tau_{\text{th}}$	radial diffusion time .	(16f)

Equation (16b) holds when the net angular momentum flow (advective plus frictional) is zero, and is a good approximation far from the inner boundary. Often one assumes that matter accretes at the inner radius R with its Keppler angular momentum. Then a factor $1-\sqrt{R/r}$ should be added on the left side of (16b). Equation (16c) holds if the energy transport is (mainly) by radiation. It shows that the opacity κ essentially influences the midplane temperature. (16d) is a result of (16b) when the pressure integral is replaced by V_s^2 times the density integral over z. It contains a stationary "radial diffusive velocity" $\alpha V_s^2/v_{\text{K}}$. The small value of this velocity is responsible for the large difference between the thermal and radial diffusive relaxation times of equations 16e and f, an essential feature of the thin accretion disks.

2 The dwarf nova instability

The equations of vertical structure (12), (13), and (14) allow to determine the relation between the surface density Σ and the viscosity f (equivalently the stationary mass flow rate $\dot{M}$ or the effective temperature T_{eff}). Realistic computations have led to the discovery of the cause for the dwarf nova oubursts (Meyer and Meyer-Hofmeister, 1981, see also Osaki 1974, Hoshi 1979, Cannizzo *et al.* 1982, Cannizzo and Wheeler 1984, Faulkner *et al.* 1983, Mineshige and Osaki 1983, Smak 1983).

In the range of partial ionization of hydrogen and helium the $f(\Sigma)$ relation shows an unusual behaviour. Here $df/d\Sigma$ is negative. This results from the strong temperature dependence of the opacity in this region. One may investigate how differential changes in $\dot{M}$, f, Σ, T, ϱ, and κ are related in disk equilibrium by way of equations (16a, b, c, d) and (3) using $\varrho \approx \Sigma/2H$. This yields

$$\frac{d\log f}{d\log\Sigma} = \frac{d\log\dot{M}}{d\log\Sigma} = \frac{5+\frac{3}{2}\kappa_\varrho - \kappa_T}{3+\frac{1}{2}\kappa_\varrho - \kappa_T} \quad . \tag{17}$$

Here κ_ϱ and κ_T are the logarithmic derivatives $\frac{\partial\log\kappa}{\partial\log\varrho}$, $\frac{\partial\log\kappa}{\partial\log T}$, and dominant radiative heat transport was assumed.

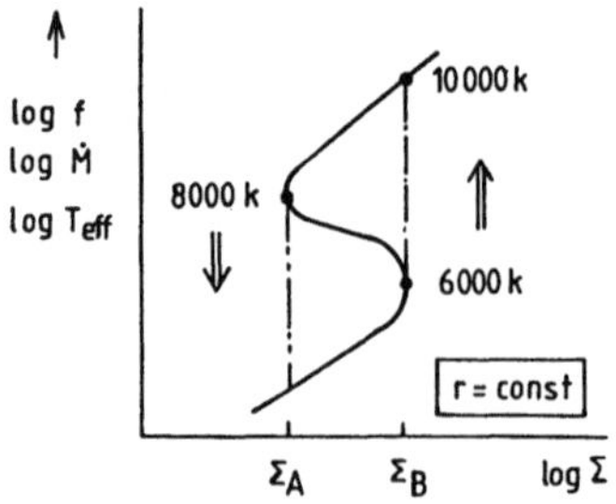

Fig. 2. Viscosity surface density relation with approximate surface temperatures. Single arrows: Path of an outburst cycle. Double arrows: Direction of thermal relaxation.

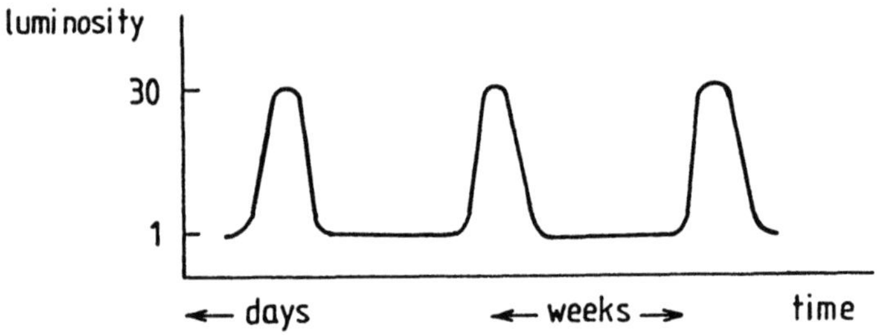

Fig. 3. Dwarf nova light curve. Luminosity in relative units.

For $\frac{\partial \log \kappa}{\partial \log T} > 3 + \frac{1}{2}\frac{\partial \log \kappa}{\partial \log \varrho}$ the denominator becomes negative and $d \log f / d \log \Sigma$ changes sign.

Such disk equilibria are unstable and therefore do not exist: $df/d\Sigma < 0$ implies a negative diffusion in the radial equation (10). The smooth $\Sigma(r)$ distribution would break up into rings of increasingly uneven surface density. These disk equilibria are however also thermally unstable. On the equilibrium curve of figure 2 heating and cooling are in balance, to the right of it the disk heats up, to the left of it the disk cools down. For a given surface density Σ the disk must therefore lie on either the hot or the cold stable branch with $\frac{df}{d\Sigma} > 0$.

In dwarf nova systems the mass accretion is determined by the rate at which the secondary star sheds mass onto the accretion disk around the white dwarf. When this mass accretion rate falls into the unstable region of figure 2 no stationary flow is possible. The disk must instead perform a limit cycle indicated by the arrowed path in figure 2. On the cool branch it passes less mass inward than it receives from the outside. The surface density increases until above Σ_B only hot branch equilibria are available to which the disk has to transit. On this branch the disk region looses mass faster inwards than it receives from the outside, and the surface density decreases. At a value Σ_A it must transit back to the cool branch. This behaviour explains the dwarf nova light curves, schematically given in figure 3. The short outburst phase and the long quiescent phase result from the long and short diffusion timescales on the cool and hot branches, respectively.

Quantitative analysis and fit to observed light curves indicates that α on the hot branch is larger by a factor of ≈ 3 than that on the cool branch.

The spatial coherence of the outburst is provided by transition waves. Starting from a radius where the critical density Σ_B is first reached and a transition from the cool to the hot state is inforced, they sweep over the whole disk region whose surface density lies between Σ_A and Σ_B and transforms it to the hot outburst stage (heating wave). Conversely, the transition at some radius where Σ has first dropped below Σ_A spreads likewise to the whole region with Σ between Σ_A and Σ_B (cooling wave, end of outburst). These waves are of a diffusive type (like combustion waves) and move at characteristic speeds

$$V_F \approx \alpha V_s \quad . \tag{18}$$

3 Two problems of outburst models

Two features of observed dwarf novae outbursts have until now withstood theoretical understanding, and the inability of accretion disk theory to account for them has cast doubt on whether this theory is basically correct. It is therefore of fundamental importance to resolve these problems.

3.1 The UV-delay problem

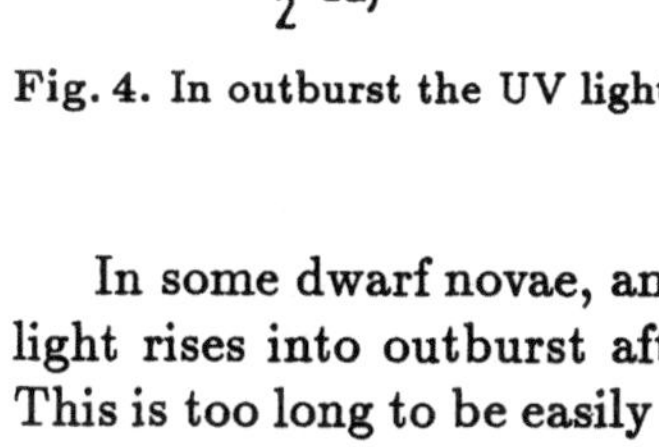

Fig. 4. In outburst the UV light lags the rise in optical light by $\geq 1/2$ day

In some dwarf novae, among them VW Hyi, WX Hyi and SU Uma, the UV light rises into outburst after a time delay of sometimes more than 1/2 day. This is too long to be easily accounted for by outburst modelling (Cannizzo and Kenyon, 1987). The difficulty is the following. When the outer disk is going into outburst transition waves rapidly move inward and in a few hours have brought the inner disk to a high mass flow stage resulting in hot UV temperatures. This has led to the suggestion that the disk transition may linger unusually long near transition temperature of 6000 to 7000 K at Σ_B. This could be caused by the increase of the thermal time scale due to the heat consumed for ionization and by a still approximate equilibrium between heat gain and loss due to radial heat diffusion (Mineshige, 1989). Alternatively, the α parameter might be much reduced near this point if friction is mainly due to convective motion that tends to disappear at these temperatures. This would extend the equilibrium curve of Fig. 2 to larger Σ values near the 6000 K effective temperature point and produce a standstill at this temperature until the required surface density has been reached (Duschl, 1989).

These suggestions however will not suffice. The observations for VW Hyi show that the optical light is already near maximum before the UV light starts rising. From this one has to conclude that the transition to the hot branch is already completed and has passed the 6000 K level before the UV light appears.

3.2 The post-outburst light problem

The standard model predicts very low mass accretion rates of order $10^{-12.5}$ to $10^{-13} M_\odot/\mathrm{yr}$ near the white dwarf when the outburst is ended and the disk has returned to the cool state. Observations however show that UV light and X-rays are present after the outburst, declining gradually into quiescence, at an energy rate equivalent to $\dot{M} \approx 10^{-11} M_\odot/\mathrm{yr}$ (Verbunt *et al.*, 1987, van der Woerd and Heise, 1987).

The UV light might perhaps be explained by cooling of a white dwarf that was heated during the outburst. It is however not clear whether the outburst will provide high enough temperatures ($\approx 10^6$K) in deep enough surface layers for this to be a valid proposal (Pringle, 1988). In any case this cannot explain the X-rays, and one must conclude that unaccounted for accretion is going on in quiescence.

We report here recent progress which seems to resolve both of these apparently unrelated difficulties and might open the way to further interesting questions (Meyer and Meyer-Hofmeister, 1990).

4 Disk evaporation model

The suggestion is that a coronal "siphon flow" evaporates matter from the cool disk which then accretes onto the white dwarf. This flow is driven by the gravitational potential difference between the disk and the white dwarf which provides enough energy to keep the evaporation going and power the radiation of X-rays from the hot corona and of UV light from the white dwarf surface.

The matter evaporating into the hot corona still has to loose it's angular momentum in order to accrete. This can best be modelled with frictional stresses $\tau_{r\varphi} = \alpha P$. The elements entering into the theory are then:

coronal heating by friction,
radiative losses in an optically thin plasma,
hear conduction,
mass evaporation from the disk and accretion on the white dwarf,
wind losses.

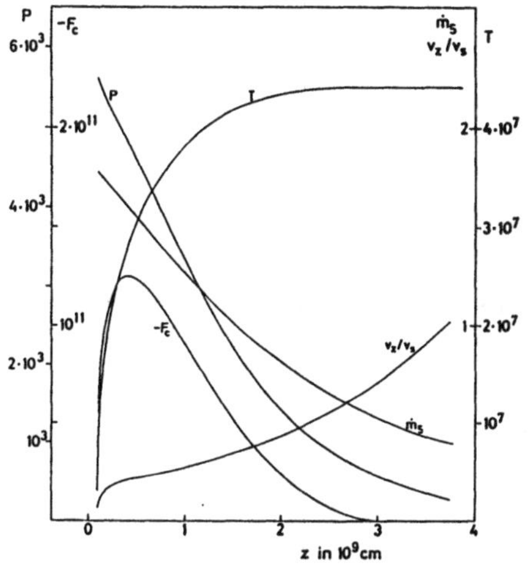

Fig. 5. Temperature T (in K), pressure P, Mach number v_z/v_s, downward conductive flux F_c, and specific mass flux rate $\dot{m}_5$ (in 10^{-5} g/cm^{2}s) as function of height z above the cool disk. Distance from white dwarf $r = 10^{9.5}$ cm, $M_{WD} = 1M_\odot$, $\alpha = 1/3$. cgs units.

The problem is two-dimensional but one can derive approximate results from a simplified one-zone model. In such a model one computes the vertical dependence of all quantities from the conservation laws and incorporates the sidewise losses in radial direction by distributing them smoothly as sink terms.

This leads to the following equations

$$\frac{d\dot{m}}{dz} = -2\varrho\frac{|v_r|}{r} - 2\frac{\dot{m}z}{(r^2+z^2)} \qquad \text{mass conservation} \quad , \tag{19}$$

$$\varrho v_z \frac{dv_z}{dz} = -\frac{dP}{dz} - \varrho\frac{GMz}{(r^2+z^2)^{3/2}} \qquad \text{vertical dynamical equilibrium} \quad , \tag{20}$$

$$\begin{aligned}
&\frac{d}{dz}\left\{(F_c)_z + \dot{m}\left[\frac{v^2}{2} + \frac{\gamma}{\gamma-1}\frac{P}{\varrho} - \frac{GM}{(r^2+z^2)^{12}}\right]\right\} \\
&= \frac{3}{2}\alpha P\Omega - AP^2\frac{\Lambda(T)}{T^2} \\
&-2\varrho\frac{|v_r|}{r}\left[\frac{v^2}{2} + \frac{\gamma}{\gamma-1}\frac{P}{\varrho} - \frac{GM}{(r^2+z^2)^{1/2}}\right] - 2\frac{\{\;\}z}{(r^2+z^2)} \qquad \text{energy equation} \quad ,
\end{aligned} \tag{21}$$

Here $\dot{m} = \varrho v_z$ is the density of the vertical mass flow, $v_r = -\alpha V_s^2/(r\Omega)$ is the radial flow velocity. The z component of the conductive heat flux is

$$(F_c)_z = -\chi_0 T^{5/2}\frac{dT}{dz} \tag{22}$$

where χ_0 is a constant of order 10^{-6} in cgs units. The first two terms on the right hand side of the energy equation are the frictional heating and the radiative losses (with the radiative loss function $\Lambda(T)$, Raymond *et al.*, 1976). The last terms in equations (19) and (21) approximately take into account the flaring of the vertical flux tubes when z becomes comparable to r and the geometry changes from cylindrical to spherical.

Since there is no confinement at large values of $(r^2+z^2)^{1/2}$, both P and F_c must go to zero. This excludes "hydrostatic" solutions.

Figure 5 shows the solution of these equations for $r = 10^{9.5}$cm and $\alpha = 1/3$, which leads to a coronal temperature $T \approx 10^{7.7}$K, a pressure $P \approx 10^{3.7}$dyn/cm^2, a density $n \approx 10^{12}$/cm^3, and a value of $\dot{m} = 10^{-4.5}$g/cm^{2}sec, equivalent to a mass accretion rate $\dot{M} \approx \pi r^2 \dot{m} \simeq 10^{-11} M_\odot$/yr. About one quarter of this is lost through the wind with speeds of ≈ 1000 km/s.

The values of T, n, and $\dot{M}$ are in very close agreement with those inferred from Exosat observations of the dwarf nova VW Hyi (van der Woerd and Heise, 1987). Evaluation of the boundary layer of this corona at the white dwarf yields an effective temperature of the white dwarf of

$$T_{\text{eff}} = 16\,000\text{K} \frac{(M/M_\odot)^{1/4}}{R_q^{3/4}} \dot{M}_{-11}^{1/4} \tag{23}$$

where R_q is the white dwarf radius in 10^9cm and $\dot{M}_{-11}$ is the mass accretion rate in $10^{-11} M_\odot$/yr. This is again in good agreement with the line profile analysis by Mateo and Szkody (1984) which gave $T = 18\,000 \pm 2\,000$K.

Dimensional analysis of the equations allows to scale the solution with r and α. In the case considered the radiative loss in the corona is small compared to conductive energy losses toward the boundaries and the wind losses. This yields

$$T = 10^{7.7}\text{K} \; \frac{M/M_\odot}{r_{9.5}} \tag{24a}$$

$$P = \frac{10^{3.2}}{\alpha} \frac{(M/M_\odot)^3}{r_{9.5}^4} \frac{\text{g}}{\text{cm s}^2} \tag{24b}$$

$$\dot{m} = 10^{-4.5} \; \frac{(M/M_\odot)^{5/2}}{r_{9.5}^{7/2}} \frac{\text{g}}{\text{cm}^2\text{s}} \tag{24c}$$

$$\dot{M} = 10^{-4} \; \frac{(M/M_\odot)^{5/2}}{r_{9.5}^{3/2}} \frac{M_\odot}{\text{yr}} \tag{24d}$$

This shows that the innermost disk regions are the most important contributor and justifies the one-zone model approach. We note that α enters only into the pressure, thereby into the density, and from there into the X-ray radiation. Observations of this latter can therefore in principal serve to determine a coronal value for α. Our good agreement with the observations for $\alpha \simeq 1/3$ appears to indicate that coronal accretion operates with not very small values of α.

We now note that the coronal mass flow can lead to the complete evaporation of the innermost accretion disk. For a typical post-outburst value of $\log \Sigma \approx 0.9$ and the obtained specific evaporation $\log \dot{m} \approx -4.5$ this occurs on a time

$$t_{\text{ev}} = \frac{\Sigma}{\dot{m}} = 10^{5.4} \text{ s} \approx 3 \text{ days} \quad . \tag{25}$$

Progressively, by this process the inner radius r_i of the disk moves outward in time, the coronal accretion decreases accordingly (equation 24d). Standard outburst models yields a scaling of the post-outburst surface density with r, e.g. $\Sigma \propto r$. Using equation (24c) this predicts $r_i \propto t^{2/9}$ and $\dot{M} \propto t^{-1/3}$ which is again in good agreement with the observed UV light decay during quiescent (Verbunt *et al.*, 1987). Thus the post-outburst light problem seems to be solved.

The disappearance of the innermost disk region also will solve the UV-delay problem. This problem resulted from the spread of the outburst to the inner disk region by way of the fast moving transition wave. If however the innermost regions have evaporated during quiescence they must first be reformed by the slower radial diffusion from the outer hot region. The time scale for this is again in good agreement with the observed lag of 1/2 day.

The model thus removes two important difficulties and at the same time establishes a strong observational support for theoretically expected coronal accretion flows with a not small value of the parameter α.

5 FU Orionis outbursts in protostellar disks

Substantial spectral and photometric evidence indicates that these outbursts are mass accretion events in protostellar disks around T Tauri stars (Hartmann and Kenyon, 1987; Kenyon, Hartmann, and Hewitt, 1988). Inferred mass accretion rates are $\geq 10^{-4} M_\odot/\mathrm{yr}$. For outburst durations of order 10 years the total mass accreted in one such outburst is at least of order $10^{-3} M_\odot$. It is however not understood how such mass accumulation could naturally arise from disks whose outer steady state accretion rate is limited by the observed infrared excess to about $10^{-7} M_\odot/\mathrm{yr}$. Likewise the cause for such sudden outbursts is not known. It was therefore speculated that some outside disturbance would set up the disk such that the observed high mass rates will result. Clarke, Lin, and Pringle (1990) investigated what such conditions would have to be and found, that about $10^{-3} M_\odot$ had to be laid down at a distance of $10^{12.4}$cm from the star, spread out over a range of similar size, $10^{12.2}$cm, and heated to more than the temperature required for ionization, $T > 10^4$K. Starting from this initial condition the subsequent development of the accretion disk gave light curves similar to those of FU Orionis outburst for α values of order 10^{-3}. It remains however an open question what might lead to such an initial condition. Passage of a massive free-falling body or disruption of an infalling gas cloud were suggested as possible causes (Clarke *et al.*, 1990).

Herbig (1977) has estimated that every T Tauri star would undergo a FU Orionis outburst once every 10 000 years. This time between outburst is just what would be required to store $\Delta M = 10^{-3} M_\odot$ for the next outburst from steady outside accretion flow of $\dot{M} \approx 10^{-7} M_\odot/\mathrm{yr}$ as observed. This is strong evidence that the FU Orionis phenomenon should be a natural consequence of disk accretion flow like the dwarf nova outburst.

We have therefore reinvestigated this instability for the case of a protostellar accretion disk and asked which value of α would be required to produce such an instability. For this one may combine equations (16) to obtain the following relation

$$\frac{T^3}{\kappa(\varrho, T)} = \frac{9}{32} \frac{\alpha}{\sigma} \frac{\Re}{\mu} \Sigma^2 \Omega \quad . \tag{26}$$

$\varrho \approx \Sigma/2H$ is itself a function of Σ and T by equation (3). Equation (26) shows that for α constant the maximum value Σ_B of Σ on the cool branch of Figure 2 is obtained when T^3/κ reaches it's maximum value. This determines the midplane temperature at the point of outburst $\Sigma = \Sigma_B$. Now for $\Sigma = \Sigma_B$ one should have

$$\Sigma \approx \frac{\Delta M}{\pi r^2} = 10^5 \frac{\Delta M_{-3}}{r_{12.4}^2} \frac{\mathrm{g}}{\mathrm{cm}^2} \tag{27}$$

where ΔM is the mass required to be stored before outburst, ΔM_{-3} the same in units of $10^{-3} M_\odot$, and $r_{12.4}$ is the distance from the central star in $10^{12.4}$cm. This yields

$$\varrho \approx 10^{-6.7} \frac{\Delta M_{-3} (M/M_\odot)^{1/2}}{T_{3.5}^{1/2} r_{12.4}^{7/2}} \tag{28}$$

where the temperature is in units of $10^{3.5}$K. Using this and opacity values computed by Alexander (1975) and graphically displayed by Duschl (1985), one determines the critical midplane temperature for maximal Σ as $T \approx 10^{3.45}$K, and $\kappa(\varrho, T) \approx 10^{-2.1}$. All this put into equation (26) allows to derive a required value of α,

$$\alpha = 10^{-3.8} \frac{(r_{12.4})^{11/2}}{\Delta M_{-3} (M/M_\odot)^{1/2}} \quad . \tag{29}$$

For this value the disk outburst will be triggered at $r \simeq 10^{12.4}$cm when $\Delta M \simeq 10^{-3} M_\odot$ is accumulated.

There is however an independent condition on α, namely that these values of T, Σ, and r should correspond to the given low mass accretion rate $\dot{M}$. Equation (16d) yields

$$\alpha = \frac{\dot{M}\Omega}{2\pi\Sigma \frac{\Re}{\mu} T} = 10^{-3.9} \frac{\dot{M}_{-7} (M/M_\odot)^{1/2} r_{12.4}^{1/2}}{\Delta M_{-3}} \tag{30}$$

where $\dot{M}_{-7}$ is the mass accretion rate in units of $10^{-7} M_\odot$/yr.

The coincidence of these two estimates, equation (29) and (30) strongly suggests that the FU Orionis phenomenon is indeed a dwarf nova type instability with, however, an unexpected low value of α. It is often thought that such low values of α might not be obtainable in a cool, convective accretion disk where convective turbulence will already sustain a minimum value of turbulent friction. One can however show that a proper evaluation of the reduction by Coriolis force of radial turbulent exchange can yield convective α values in accretion disks even smaller than this.

The simulation by Clarke *et al.* (1990) indicates that α values in outbursts and thus on the hot branch of protostellar accretion disks might be about 10^{-3} and thus by a factor 10 larger than on the cool branch. A similar phenomenon is encountered in the dwarf nova case proper.

6 Concluding remarks

It is interesting to note a differnece between the range of α values discussed here and those derived for accretion disks in symbiotic stars by Duschl (1986), $\alpha \approx 10^{-1.3}$. Mass accretion rates, central masses and radial distances for disks in both systems are comparable. The apparent difference in α might therefore point to a tidal effect on the disk by the companion star in the symbiotic case. Such a companion is not present in the FU Orionis accretion disk.

In this article we have discussed how some of the crucial difficulties encountered by accretion disk theory can be resolved. This has led to the introduction of evaporating accretion disks which feed a hot corona through which the mass accretes onto the central star while a fraction escapes via a coronal wind. The evaluation of observations indicate that angular momentum transport in these accreting coronae is described by α-values that are not small.

It is further shown that the difficulty to understand FU Orionis outburst in protostellar accretion disks can possibly be removed by considering very small values of α. This would be able to account both for the instability mechanism as one of dwarf nova type and for the accumulation of $10^{-3} M_\odot$ before outburst. It also naturally relates the low mean mass accretion rate, the outburst magnitude, and the repetition time inferred from observations. One may hope that further evaluation of observations by such models will also shed more light on the processes that are responsible for friction in these accretion disks.

Alexander, D.R.: 1975 *Astrophys. J. Suppl.* **29**, 363
Cannizzo, J.K., and Wheeler, J.C.: 1984 *Astrophys. J. Suppl.* **55**, 367
Cannizzo, J.K., Gosh, P., and Wheeler, J.C.: 1982 *Astrophys. J. (Letters)* **260**, L83
Cannizzo, J.K., and Kenyon, S.J.: 1987 *Astrophys. J.* **320**, 319
Clarke, C.J., Lin, D.N.C., and Pringle, J.E.: 1990 *Mon. Not. R. Astr. Soc.* **242**, 439
Duschl, W.J.: 1985, Thesis, Ludwig-Maximilians-University, Munich
Duschl, W.J.: 1985, in Proc. 23rd ESLAB Symp. on Two Topics in X-ray Astronomy, ESA
Duschl, W.J.: 1986 *Astron. Astrophys.* **163**, 61
Faulkner, J., Lin, D.N.C., and Papaloizou: 1983 *Mon. Not. R. Astr. Soc.* **205**, 359
Hartmann, L., and Kenyon, S.J.: 1987 *Astrophys. J.* **312**, 243
Herbig, G.H.: 1977 *Astrophys. J.* **217**, 693
Hoshi, R.: 1979 *Prog. Theor. Phys.* **61**, 1307
Kenyon, S.J., Hartmann, L., and Hewitt, R.: 1988 *Astrophys. J.* **325**, 231
Mateo, M., and Skyody, P.: 1984 *Astrophys. J.* **89**, 863
Meyer, F., and Meyer-Hofmeister, E.: 1981 *Astron. Astrophys.* **104**, L10
Meyer, F., and Meyer-Hofmeister, E.: 1982 *Astron. Astrophys.* **106**, 34
Meyer, F., and Meyer-Hofmeister, E.: 1990, to be published
Mineshige, S.: 1988 *Astron. Astrophys.* **104**, L10
Osaki, Y.: 1974 *Publ. Astron. Soc. Japan* **26**, 429
Raymond, J.C., Cox, D.P., and Smith, B.W.: 1976 *Astrophys. J.* **204**, 290
Shakura, N.I., and Sunyaev, R.A.: 1973 *Astron. Astrophys.* **24**, 337
Verbunt, F., Hassall, B.J.H., Pringle, J.E., Warner, B., and Marang, F.: 1987 *Mon. Not. R. Astr. Soc.* **225**, 113
van der Woerd, H., and Heise, J.: 1987 *Mon. Not. R. Astr. Soc.* **225**, 141

Mass Transfer and Evolution in Close Binaries

A.R. King

Astronomy Group, University of Leicester,
Leicester LE1 7RH, UK

DAEC and DARC, Observatoire de Paris, Section de Meudon,
Université de Paris VII, France

1 Introduction

This article discusses close binary systems with orbital periods of a few hours, in which a compact primary star accretes mass from a low–mass companion. If the primary is a white dwarf these are called cataclysmic variables (CVs); if it is a neutron star or black hole the systems are known as low mass X–ray binaries (LMXBs). The evolution of these systems has been reviewed by many authors, and the interested reader is referred to these papers (e.g. King, 1988) for further details.

The precise condition for mass transfer to occur is that the secondary star should fill its Roche lobe: more intuitively, this amounts to requiring the orbital period to be short enough to be comparable with this star's dynamical time. Transferring mass from the less massive secondary to the primary would cause the system to expand if its angular momentum were conserved (matter is brought closer to the centre of mass, so the secondary must move in a slightly wider orbit). Accordingly, long-lasting mass transfer in close binaries requires a steady drain of orbital angular momentum to drive it. At periods $P \lesssim 3$ hr gravitational radiation is a plausible source of this; at longer periods the mechanism is less clear, but may involve the loss of spin angular momentum from the secondary in a magnetic stellar wind. As this star is locked to the orbital rotation by tides, this removes angular momentum from the system also. As a result of the loss of orbital angular momentum, the binary period and separation progressively decrease over time, a process known as *secular evolution.* Because the timescale for evolution is shorter than the age of the Galaxy, the present distribution of periods must reflect this evolution rather than the initial distribution of periods. Histograms of the number of systems at a given orbital period thus supply the most important constraints on theories of the secular evolution of close binaries. The CV period histogram (Fig. 1) has over 100 members (Ritter, 1987). The distribution is clearly far from uniform, and in particular has a highly significant gap between ~ 2 and ~ 3 hr. Explaining this gap, and some of the other features of the period histogram, is the major goal of theories of secular evolution.

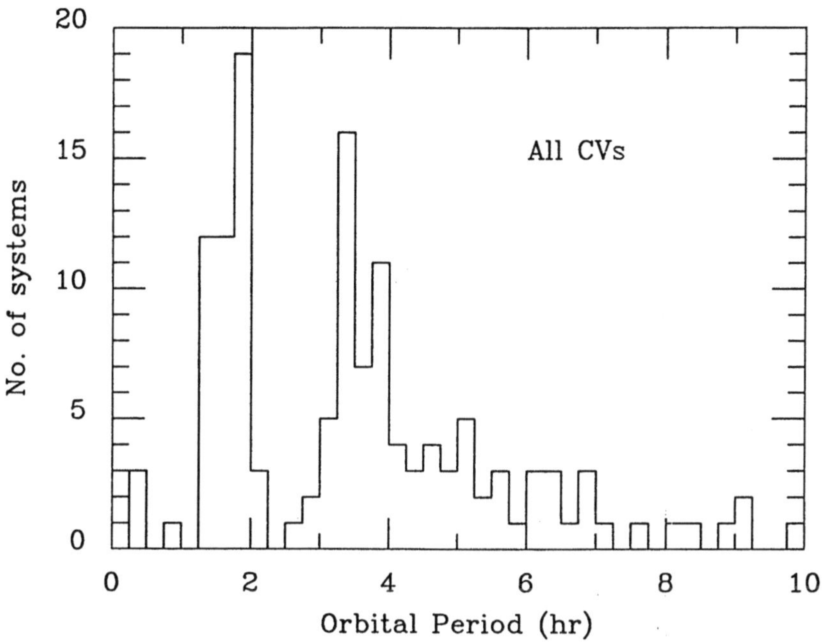

Fig. 1. the period histogram of cataclysmic variables

2 Secular Evolution

As we have seen, during orbital evolution angular momentum losses shrink the binary and cause mass transfer, reducing the binary period in the process. Current pictures of this evolution attempt to explain the main features of the period histogram in terms of a competition between the timescale of mass transfer and the secondary star's thermal time. Once these timescales become comparable, this star's radius then does not shrink rapidly in response to the mass loss. The secondary becomes progressively oversized for its mass compared to the main sequence, since it is out of thermal equilibrium. Because of the relation between Roche lobe size and the orbital period, this means that the binary period is somewhat longer than it would be if the star had its main-sequence radius. At a period $\sim$ 3 hr, angular momentum losses are apparently severely reduced; this may be connected with the fact that the secondary becomes fully convective at around this period. The secondary is now able to shrink faster than the Roche lobe, and mass transfer ceases (Rappaport, Verbunt and Joss, 1983; Spruit and Ritter, 1983). Mass transfer eventually resumes because the star cannot get smaller than its main-sequence radius, and the separation and Roche lobe radius are continually reduced by gravitational radiation and any other angular momentum losses. For a plausible degree of thermal disequilibrium at $P \sim 3$ hr, mass transfer restarts near $P \sim 2$ hr, as observed in the form of the period gap. The sides of the gap are not infinitely steep because this process occurs at slightly different values of the orbital periods, depending on binary parameters (especially the white dwarf mass, as we shall see). The gap itself is not entirely empty of systems, because CVs can be 'born' in it and not suffer the thermal disequilibrium which forms the gap.

A particularly interesting subset of CVs in this regard are the AM Herculis systems, which contain white dwarfs with magnetic fields $\gtrsim 10^7$ G. 17 such systems are known, and their period distribution shows an extraordinary feature: no less than 6 of them lie in the very narrow range 113.6 - 114.8 min. This highly significant accumulation, the 'period spike', lies close to the lower edge of the period gap for all CVs. Hameury *et al.* (1988) (hereafter HKLR) proposed to fit the AM Her period spike into the general scheme for secular evolution by identifying it as the point where mass transfer restarts for most AM Her systems after they cross the period gap. They showed that the relative discovery probability

$$p(P) \propto \dot{M}^{3/2}/|\dot{P}| \tag{1}$$

is considerably enhanced at this period. This happens because of a combination of two effects working in the same direction. Since the secondary is fully convective and has re-established thermal equilibrium in passing through the period gap, its initial response to the resumption of mass loss is to expand adiabatically. Once it is sufficiently far out of thermal equilibrium it will begin instead to shrink in response to the mass loss. Because of Roche geometry, this means that P will first *increase* slightly before decreasing. At the same time, the adiabatic expansion means that the mass transfer will occur at about twice the rate appropriate when the stellar radius shrinks instead. This follows from the evolution equation (e.g. King, 1988)

$$\dot{M}/M_2 = (-\dot{J}/J)[5/6 + \beta/2 - M_2/M_1]^{-1}, \tag{2}$$

where $M_2, J, \dot{J}$ are the secondary mass, orbital angular momentum and its loss rate respectively, and β is the index in the secondary's mass-radius relation $R_2 \propto M_2^\beta$. The mass ratio M_2/M_1 is small, so the denominator on the rhs of (2) is effectively given by $5/6 + \beta/2$. For adiabatic expansion, $\beta \simeq 1/3$, while for a star shrinking on mass–loss β has approximately the main-sequence value ~ 1, giving denominators $\sim 2/3$ and $\sim 4/3$ in the two cases. Thus the system spends rather longer near its initial period and is somewhat brighter than at shorter periods (Fig. 2), resulting in an enhanced discovery probability there (Fig. 3).

This argument shows that CVs are more likely to be discovered as they resume mass transfer than at smaller P. But to give the observed period spike, we must require that this resumption period is the same for a substantial fraction of AM Her systems (to within the width of the spike, i.e. $\Delta P \lesssim 1$ min.). This requires them all to have secondaries which become fully convective at the same period, and to show the same degree of thermal disequilibrium at this period, so that they all restart mass transfer near ~ 114 min. Since the latter is measured by comparing the star's thermal timescale with the mass-transfer timescale $t_{\dot{M}} = M_2/\dot{M}$, (2) shows that M_1 must be the same for all

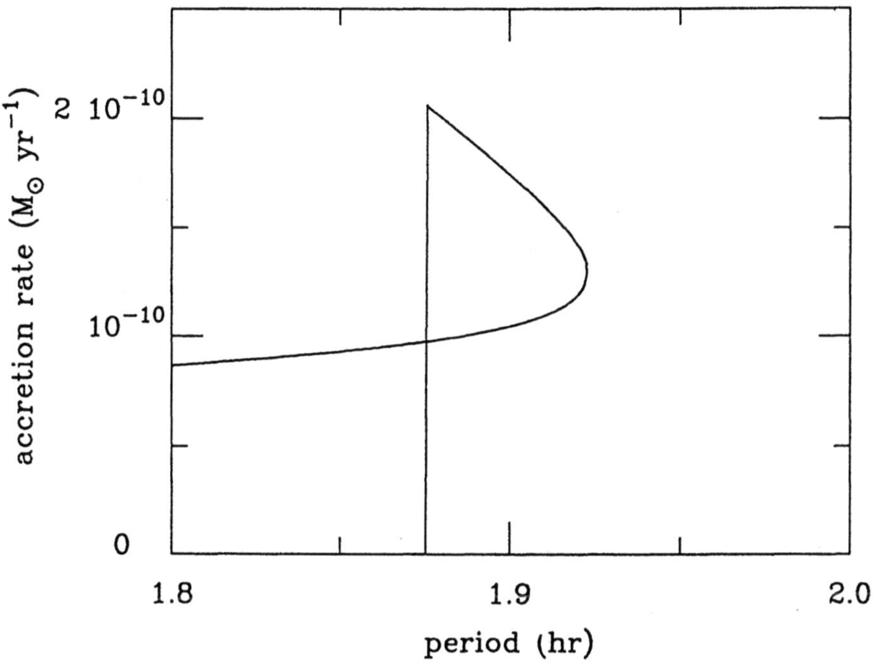

Fig. 2. the evolution of AM Her systems in the period spike

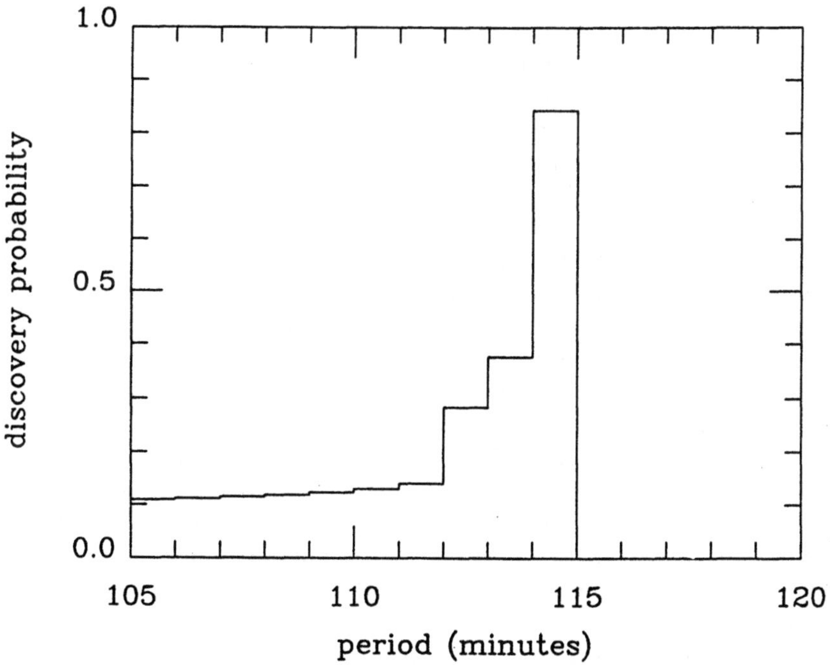

Fig. 3. theoretical discovery probability of AM Her systems in the period spike

the systems in the spike. (The only exceptions would be those rare systems which resumed mass transfer at longer periods and are now at 114 min purely by chance.) Since from (2) $J/\dot{J}$ must also be identical, we see that the angular momentum loss mechanism above the gap must be 'universal', in the sense that its long-term average value can only depend on M_1, M_2 and P.

After HKLR made their proposal, a new AM Her system (EXO 033319-2554.2, now called UZ For) was discovered by Giommi *et al.* (1987) and Beuermann and Thomas (1987), with the longer period of 127 min. Unless this was taken as one of the rare CVs 'born' in the period gap, it was clear that UZ For could only fit into HKLR's picture if it had passed through a rather narrower gap than the systems in the period spike. This requires its secondary to have had a smaller degree of thermal disequilibrium when it became fully convective, which in turn requires a longer mass transfer timescale $t_{\dot{M}} = M_2/\dot{M}$ above the period gap. For any reasonable form of angular momentum loss rate $\dot{J}$, (2) shows that the main sensitivity of $t_{\dot{M}}$ is to the factor $[5/6 + \beta/2 - M_2/M_1]$. Since M_2 and β are very similar for all systems just above the gap, we see that *a narrower period gap corresponds to a higher white dwarf mass.* The connection can be made quantitatively precise, and led Hameury, King and Lasota (1988) to predict a substantially larger M_1 for UZ For than in the spike, the most likely values being $\sim 0.6\ M_\odot$ for the spike and $\gtrsim 1.2\ M_\odot$ for UZ For. The latter prediction was confirmed by Beuermann, Thomas and Schwope (1988) and Ferrario *et al.* (1989). A recent survey (King, Hameury and Lasota, 1990) of the available mass estimates concludes that they are compatible with the theoretical predictions.

It is clear that the mass constraints described in the last paragraph place very tight restrictions on physical processes and evolution in AM Her systems. For example, nova explosions are frequently thought to carry off rather more mass from the white dwarf than is accreted between outbursts, thus gradually reducing the value of M_1. If this occurred in most AM Her systems however, the value of M_1 at any epoch would depend on how long the system had been transferring mass, and hence on the initial period, or equivalently the initial mass of the secondary. To get a large number of systems to pass through $P =$ 114 min with the same M_1 would require some kind of conspiracy between the initial white dwarf and secondary masses, which seems very unlikely. The only plausible way of arranging things is to assume that most AM Her systems have almost the same initial M_1, $\simeq 0.6 - 0.7\ M_\odot$, and that this remains unchanged (Hameury, King, Lasota and Livio, 1989). There must also be a smaller group with M_1 allowed to have larger values. The latter would contain UZ For and V1500 Cygni, the only known nova amongst AM Her systems. Nova theory provides tentative support for the idea that M_1 might stay fixed for the lower–mass group, since the magnetic field prevents shear mixing of the accreted matter with the white dwarf envelope and so inhibits mass loss. The lack of this constraint may explain why no period spike has yet been identified for non–magnetic CVs, and why the evolution of the AM Her systems is particularly easy to follow.

3 ROSAT and CVs

The AM Her systems are frequently discovered up by X-ray satellite experiments as they are the strongest soft X-ray sources amongst CVs. For example, the low-energy detectors on EXOSAT serendipitously discovered 3 new AM Her systems with only 5% sky coverage. The forthcoming ROSAT/Wide Field Camera combination will be at least as sensitive as EXOSAT in this energy range. A simple extrapolation then suggests that ROSAT will turn up at least 60 new AM Her systems; more sophisticated calculations give rather larger estimates.

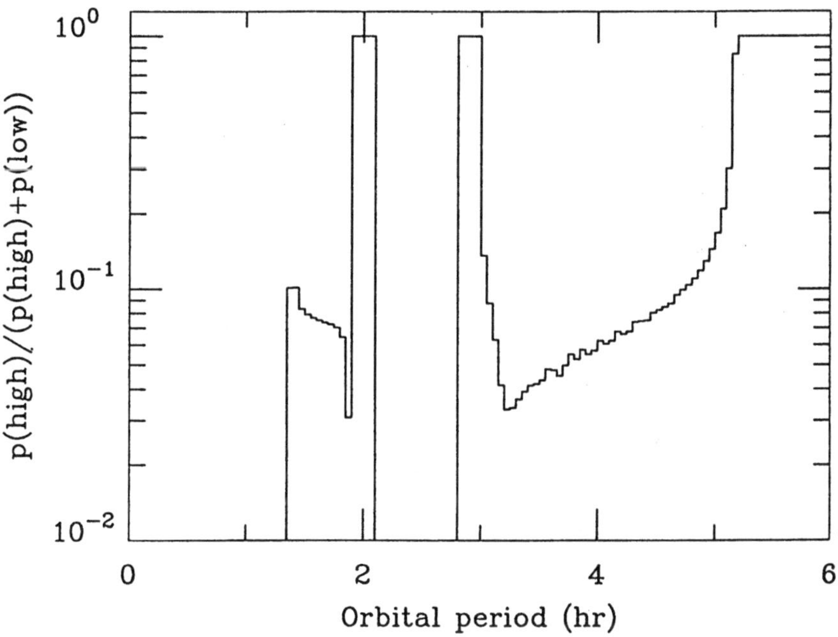

Fig. 4. expected relative frequency $p(\text{high})/[p(\text{high}) + p(\text{low})]$ of high-mass AM Her systems as a function of orbital period; the normalizations of $p(\text{high}), p(\text{low})$ are chosen to make high-mass systems contribute 10% of the observed total

It should be feasible to identify these sources in a reasonable time, as the Wide Field Camera is expected to detect only a few thousand sources in its very soft bandpass. The orbital period of an AM Her system is always found once the system is identified. This will make the period histogram of AM Her systems roughly as well-populated as that of all CVs is at present. Obviously this must transform our understanding of every aspect of these systems. This will certainly be true for studies of their evolution, and particularly of the suggested explanation of the period spike (which should have some 27 or more members!). The dependence of $t_{\dot{M}}$ on M_1 leads to a simple test, not only of this suggestion, but also of the basic picture that CVs cross the period gap in the manner suggested. For the larger sample expected from ROSAT, Hameury et al (1989) show that the M_1 – distributions on each side of the gap must be mirror images (Fig.4); those systems closest to the edges must have the largest values of M_1. This effect is potentially observable, and offers a clearcut test of the idea that CVs cross the period gap.

Acknowledgment

I thank Dr J. P. Lasota for valuable help in writing this article.

References

Beuermann, K. and Thomas, H.C., 1987. *IAU Circ.* No. 4486
Beuermann, K., Thomas, H. C. and Schwope, A. 1988, *Astron. Ap.* **195**, L15
Ferrario, L., Wickramasinghe, D. T., Bailey, J., Tuohy, I. R. and Hough, J. H. 1989, *Ap. J.* **337**, 832
Giommi, P., Angelini, L., Osborne, J., Stella, L. and Tagliaferri, G., 1987. *IAU Circ* No 4486
Hameury, J. M., King, A. R. and Lasota, J. P., 1988, *Astron. Ap.* **195**, L12
Hameury, J. P., King, A. R. and Lasota, J. P., 1990, *M. N. R. A. S.* **242**, 141
Hameury, J. P., King, A. R., Lasota, J. P. and Livio, M., 1989, *M. N. R. A. S.* **237**, 835
Hameury, J. M., King, A. R., Lasota, J. P. and Ritter, H. 1988, *M. N. R. A. S.* **231**, 535
King, A. R., 1988 *Q. Jl. R. astr. Soc.*, **29**, 1
King, A. R., Hameury, J. M. and Lasota, J. P. 1990, in *Proc. 11th North American Workshop on CVs and LMXRBs*, Cambridge University Press, ed. C. W. Mauche, in press
Rappaport, S. A., Verbunt, F. and Joss, P. C., 1983, *Ap. J* **275**, 713
Ritter, H., 1987, *Astron. Ap. Supp. Ser.* **70**, 335
Spruit, H. C. and Ritter, H., 1983, *Astron. Ap.* **124**, 267

Radiation Hydrodynamics of the Boundary Layer of Accretion Disks in Cataclysmic Variables

W. Kley

Institut für Astronomie und Astrophysik der Universität München, Scheinerstr. 1, D-8000 München 80, Fed. Rep. of Germany

1 Abstract

At the innermost part of the accretion disk where the disk grazes the surface of the central star frictional interaction decelerates the disk material to stellar rotational velocity and a transition zone (boundary layer, BL) is formed. In the case of a slowly rotating star, nearly half of the total accretion energy is liberated here. Therefore the inclusion of this boundary layer is very important and crucial for a complete and selfconsistent description of the accretion disk structure. The correct theoretical description of the BL structure is necessarily a two-dimensional problem and has to be solved selfconsistently. Therefore the author developed an algorithm which solves the coupled 2-D radiation-hydrodynamic equations for axisymmetric flows. It is a mixed explicit-implicit numerical method including viscosity, angular momentum and radiation transport treated in the flux-limited diffusion approximation. It was possible to study the detailed structure of the disk-star interaction for different parameters (stellar mass, viscosity, etc.).

2 The problem

In a close binary system like a Cataclysmic Variable (CV) the matter transferred from the secondary star is accreted by the White Dwarf (WD) via an accretion disk. The release of gravitational energy during the accretion process yields to the disk luminosity

$$L_{\text{disk}} = \frac{1}{2} L_{\text{acc}}, \tag{1}$$

where $L_{acc} = \frac{GM_{WD}\dot{M}}{R_{WD}}$ (G=gravitational constant, M_{WD}, R_{WD} = mass and radius of the WD, $\dot{M}$ = mass accretion rate) is the total accretion luminosity (if the distance to the secondary is large (infinite)). Obviously only half of the total available energy is liberated in the disk. The rest is still contained in the kinetic energy of the matter (Keplerian rotation at the stellar radius).

In the case of corotation of the WD with the orbital motion of the secondary (what is usually assumed) it rotates much slower that its critical or break-up velocity. Then frictional interaction decelerates the disk material to stellar rotational velocity and a transition zone or boundary layer (BL) is formed. Conservation of energy and angular momentum yield for the luminosity of this BL

$$L_{BL} = L_{disk} \left(1 - \frac{\Omega_{WD}}{\Omega_K(R_{WD})}\right)^2 , \tag{2}$$

where Ω_{WD} is the angular rotation of the WD and $\Omega_K(R_{WD})$ is the Keplerian angular velocity at the stellar surface (R_{WD}). If the star rotates slowly approximately *half* of the total accretion energy is liberated in the BL. It is for this reason, why the inclusion of the BL in theoretical considerations of Accretion Disks is so important.

Since the emitting area of the BL is much smaller than the size of the complete disk it will be a source of high energy radiation. A very few selected examples of observations confirm this view.

- About 70 % of all CVs show emission of hard (kT $\geq$ 2 keV) X-rays in quiescence (Cordova and Mason, 1983). It is assumed that this radiation comes from a non-thermalized optically thin BL or an extended corona around the WD (see e.g., Pringle and Savonije, 1979; King and Shaviv, 1984)

- Some systems (like SS Cyg, U Gem and others) emit soft (some few keV) X-rays during their outburst state (Cordova and Mason, 1984; van der Woerd, 1987). This soft radiation is assumed to be emitted by a thermalized, optically thick BL (e.g., Pringle, 1977)

- The WDs in CVs have on average higher temperatures than single WDs (Sion, 1985). This heating of the stellar surface can be a consequence of the accretion process (Pringle, 1988)

- During outburst a lot of systems show P Cygni line profiles in their UV spectra characteristic for stellar winds. Eclipse properties of the lines indicate that the wind originates close to the WD/BL

The above list of observations is not claimed to be complete, but should rather give some examples of points which are intimately connected with the structure of the BL region. In the following, results of numerical investigations of the properties of the optically *thick* (high $\dot{M}$) BL in CVs are presented.

3 Method of Solution

There have been several attempts to model theoretically the BL structure (for a review see Shaviv, 1987). Many of these are one-dimensional, i.e. a functional dependence only on the distance from the star (e.g. Pringle, 1977; Regev, 1983; Papaloizou and Stanley, 1986). But it is not known in advance whether it is permitted to neglect also (like in the disk) the vertical structure of the BL. Matter could flow towards the stellar poles to higher latitudes, and more complicated mixing processes might occur (e.g. Kippenhahn and Thomas, 1982). Consequently two-dimensional studies were started (Robertson and Frank, 1986; Kley and Hensler, 1987). These consisted of numerical hydrodynamical calculations and the results indicated a flow of the matter up to the poles. The accreted material surrounded the WD completely forming a shell like structure. But both attempts are also incomplete in that *no consistent* energy transport is taken into account.

Here the view is taken, that the problem of the BL structure is an inherently two-dimensional problem which has to be solved selfconsistently. The method of solution taken here is a numerical simulation solving the coupled 2-D radiation hydrodynamical equations for axisymmetric flows. The mixed explicit-implicit numerical method includes gravitation, the full tensor viscosity, the radiation transport in the flux-limited diffusion approximation and a continuum radiation force. The approximations used are constant kinematic viscosity, ideal gas law, Kramers opacity and no inclusion of any heat conduction. The method and the test calculations are described in more detail in Kley (1989*a*). The parameters which are fixed in all models during the calculations are

$$T_{\rm eff} = 1.5 \times 10^4 \ {\rm K}, \dot{M} = 3 \times 10^{-8} \ {\rm M_\odot yr^{-1}}.$$

The other parameters used are given in the following table

Name	Grid Size	$M_{\rm WD}$ ($\rm M_\odot$)	ν (10^{15} cm^2 s^{-1})	$\Omega_{\rm WD}/\Omega_{\rm K}(R_{\rm WD})$
LDTR3	61 × 61	1.0	1.0	0.0
LDTR9	61 × 61	1.0	1.0	0.3
crl3	85 × 85	1.0	4.0	0.0
crl5	85 × 85	0.6	0.0	0.0
crl7	85 × 85	1.0	1.0	0.0

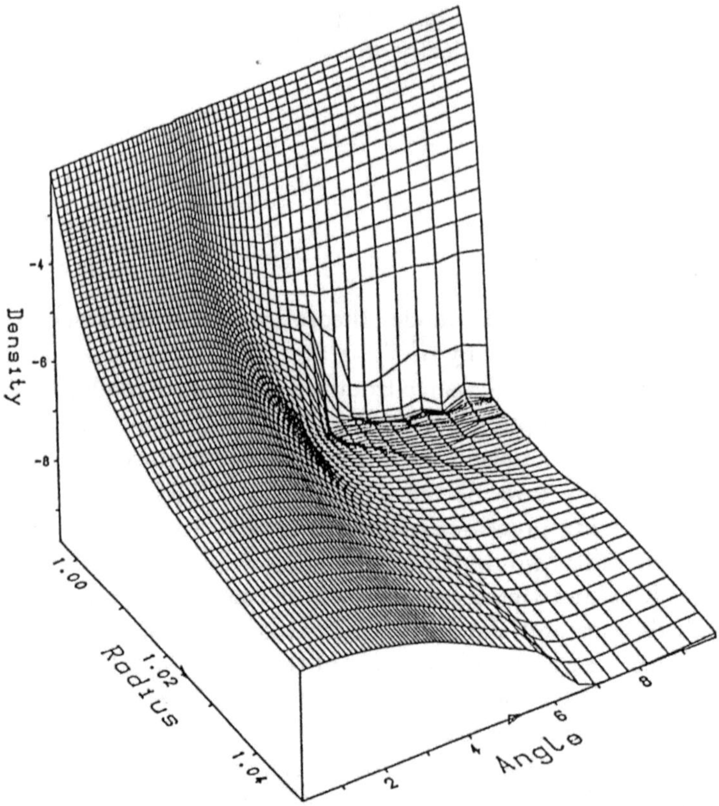

Fig.1: 3-D plot of the density in a logarithmic scaling. The radial coordinate is given in units of the stellar radius. The angular distance from the equator in degrees. Displayed is only a part of the computational domain. The plot refers to model crl5 after 8.38 Keplerian periods

Here some of the parameters are varied to study their influence on the BL structure. In the second model for example the star rotates with $\approx \frac{1}{3}$ of its break up velocity, while in the third model the viscosity has been increased four times. The calculations are done in spherical polar coordinates and the computational domain has the extent $R_{\rm WD} \leq r \leq 1.5 R_{\rm WD}$ and $0 \leq \alpha \leq \pi/2$, where r denotes the radial coordinate and α is the angular distance from the equator, i.e. $\alpha = \pi/2 - \theta$. Only in model crl7 was $r_{\rm max} = 2.0$. The boundary conditions are described in Kley (1989*a*).

As initial conditions in the radial direction a stellar atmosphere in hydrostatic equilibrium was taken around the WD. Near the equatorial plane a vertical equilibrium disk model was imposed. For consistency the initial models in both directions were obtained with a one-dimensional version of the program. The initial configuration is then evolved in time until a quasistationary state is reached (usually after 4-5 Keplerian periods $\Omega_{\rm K}(R_{\rm WD})$).

4 Results

4.1 General Structure

The results of the calculations are already described in more detail in Kley (1989*b*). Here we only mention the important fact that the incoming matter from the disk is accreted in an *equatorial belt* around the star. This is in contrast to the previous purely hydrodynamic calculations (e.g. Kley and Hensler, 1987) and is a result of the inclusion of radiative processes, specifically the cooling. Fig. 1 visualizes this feature in detail. In the 3-D plot (model crl5, $M_{\rm WD} = 0.6 {\rm M}_\odot$) it becomes obvious that the matter is not leaving the equator more than $\approx 7°$. The density increases strongly towards the star (logarithmic scaling).

We note here near the innermost part of the computational domain the initial distribution of the stellar layers is still undisturbed. This means that the results cannot be affected significantly by the exact boundary condition there (see also Kley, 1989*b*). Having in mind this small vertical extension of the disk, it might be possible, with the correct boundary conditions, to study some overall features (radial extension, rotation law) of the BL with a simplified one-dimensional model.

4.2 Radial Structure

Using the "classical" definition the BL thickness is given by that radius where the gradient of the rotational velocity vanishes, i.e. where $\Omega(r)$ has its maximum. For the studied case of a constant kinematic viscosity ν we find that the thickness $\delta_{\rm BL}$ (= the distance from $R_{\rm WD}$ to the maximum of Ω) does only depend on the absolute value of the viscosity and the mass (and radius) of the central star and *not* for example on the stellar rotation or the mass inflow rate $\dot{M}$. Indeed, from dimensional arguments it follows that the BL thickness scales like

$$\delta^{\rm v}_{\rm BL} \sim \sqrt{\frac{\nu}{\Omega_{\rm K}(R_{\rm WD})}}, \tag{3}$$

where the label v stands for *viscous*. Another important parameter is the absolute magnitude of the angular velocity at its maximum. Let us define the factor β as

$$\Omega(r_{\rm m}) =: \beta\ \Omega_{\rm K}(r_{\rm m}), \tag{4}$$

where $r_{\rm m} = R_{\rm WD} + \delta^{\rm v}_{\rm BL}$ is the location of the maximum, i.e. β gives the fraction of the Keplerian value at that radius. For the models we find:

Name	$\delta^{\rm v}_{\rm BL}/R_{\rm WD}$	β
LDTR3	0.121	0.958
LDTR9	0.111	0.952
crl3	0.228	0.870
crl5	0.115	0.953
crl7	0.115	0.955

For the low viscosity models the average values are

$$r_{\rm m} = 1.11 R_{\rm WD}, \ \beta = 0.955.$$

These results are in agreement with eq. (2) since a four times higher viscosity leads to a doubling of $\delta^{\rm v}_{\rm BL}$. The different masses of the stars do not lead to a different thickness because $\delta^{\rm v}_{\rm BL}/R_{\rm WD} \sim \sqrt{\nu/h_{\rm WD}}$, where $h_{\rm WD}$ is the specific angular momentum at the surface of the star $\sqrt{GM_{\rm WD}R_{\rm WD}}$. $h_{\rm WD}$ is approximately equal for the different stars as a consequence of the mass-radius relation for White Dwarfs.

In Fig. 2 the angular velocity is plotted for model crl7 and the values of $r_{\rm m}$ and β can be read off. The dashed-dotted curve refers to a first order correction to the angular velocity. It is obtained by solving approximately the angular momentum equation and the radial momentum equation (6), where only the major contributions have been retained (refer to Fig. 5, below). The details of the calculations are presented elsewhere. In Fig. 3 the radial infall velocity is plotted and the analytical approximation is superimposed. Near $r_{\rm m}$ the matter is accelerated strongly inwards, because centrifugal support is lost. Inside of $r \approx 1.03 R_{\rm WD}$ the material settles slowly onto the stellar surface. In both cases the approximations for $v(r)$ and $\Omega(r)$ fit reasonable well, but one has to keep in mind that they are *only valid for* $r \geq r_{\rm m}$. Around $r_{\rm m}$ the matter falls highly supersonic towards the star. At the maximum of $v(r)$ at $r \approx 1.03 R_{\rm WD}$ the velocity corresponds to Mach 7.2.

Inside $r_{\rm m}$ the rotational velocity of the disk decreases rapidly to the stellar rotation leading to a tremendous dissipation of energy with the maximum in the middle of the viscous BL. In Fig. 5 the specific dissipation rate (erg/g/s) vertically averaged over the disk is plotted for the different models. The higher viscosity leads to a spreading out, while a lower stellar mass shifts the whole curve by a constant factor and rotation leads only inside $r_{\rm m}$ to a reduction of the dissipation. Integration over the total computational domain gives for the luminosity of the models

$$\frac{L^{\rm rot}_{\rm BL}}{L_{\rm BL}} \approx 0.48 \ldots 0.51,$$

where $L^{\rm rot}_{\rm BL}$ refers to the rotating star model (LDTR3). This result is in very good agreement with the theoretical value of $(1-0.3)^2 = 0.49$ (eq. (2)). In this region the matter falls *supersonically* towards the star, because centrifugal support is lost.

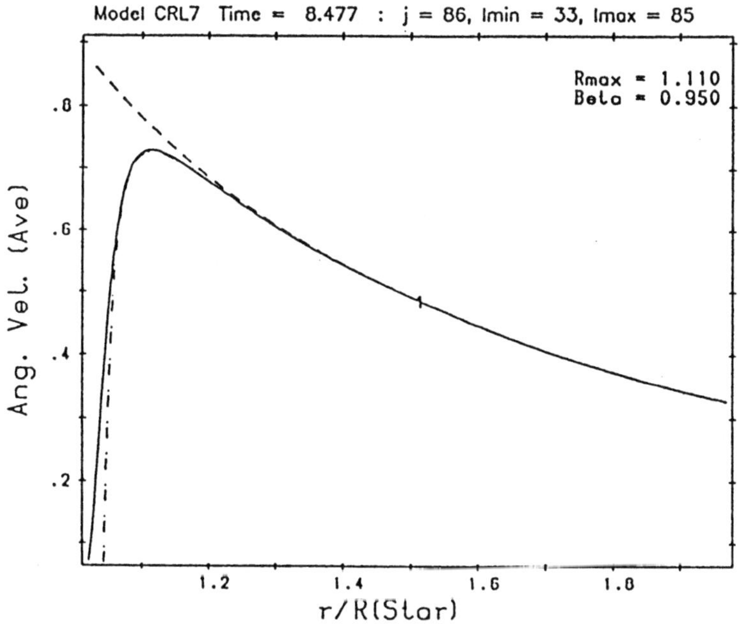

Fig.2: The average angular velocity (s^{-1}) in the BL/Disk as a function of the radius (for model crl7). The continuous line refers to the numerical model. The dashed line to Keplerian rotation and the dashed-dotted line to an analytical approximation (see text)

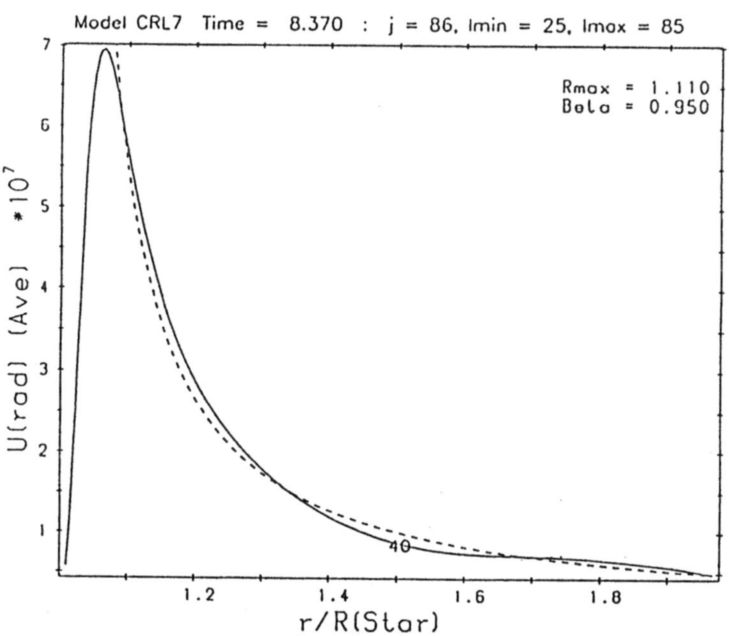

Fig.3: The average radial velocity (absolute value, in cm s^{-1}) in the BL/Disk as a function of radius. The continuous line refers to the numerical model. The broken line to an analytical approximation (see text)

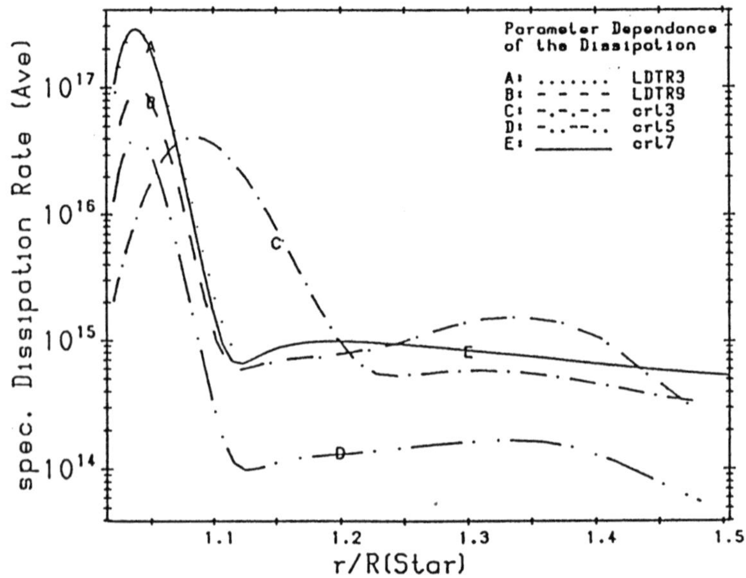

Fig.4: The average specific dissipation rate (erg/g/s) in the BL/Disk as a function of the radius. Five models are compared. See text

4.3 Theoretical considerations

The importance of β can be seen from the following considerations. Starting from the equation for the radiation flux from the surface of the disk

$$F = -\frac{1}{4\pi}\frac{\partial}{r\partial r}\left[-\mid \dot{M} \mid \left(\frac{1}{2}r^2\Omega^2 + \Psi\right) - 2\pi\nu\Sigma r^3\Omega\frac{\partial\Omega}{\partial r}\right], \tag{5}$$

where the terms in the square brackets are just the flow of kinetic and potential energy and energy flow caused by the viscosity. In eq. (5) Σ denotes the surface density and Ψ the gravitational potential (of the WD). In the derivation of eq. (5) it has additionally been assumed that terms involving the radial infall velocity are small. Using the angular momentum equation with the (boundary) condition (4) and assuming that the rotation of the disk is Keplerian throughout (in spite of the condition (4)), then one finds for the luminosity of the disk

$$L_{\text{disk}} = L_{\text{acc}}\left(\frac{3}{2} - \beta\right). \tag{6}$$

In deriving (6) we additionally assumed that the point of maximum rotation r_{m} coincides with the stellar radius, i.e. $r_{\text{m}} = R_{\text{WD}}$. This expression for the disk luminosity is identical to the one derived from Regev (1983, his equation 38), and his integration constant C just equals β. The limits on β are obviously $\frac{1}{2} \leq \beta \leq 1$. A value of β smaller than one yields to a higher disk luminosity than given by eq. (1) which is a result of an increased outward viscous energy flow, necessary for the lowering of the Keplerian rotation. In

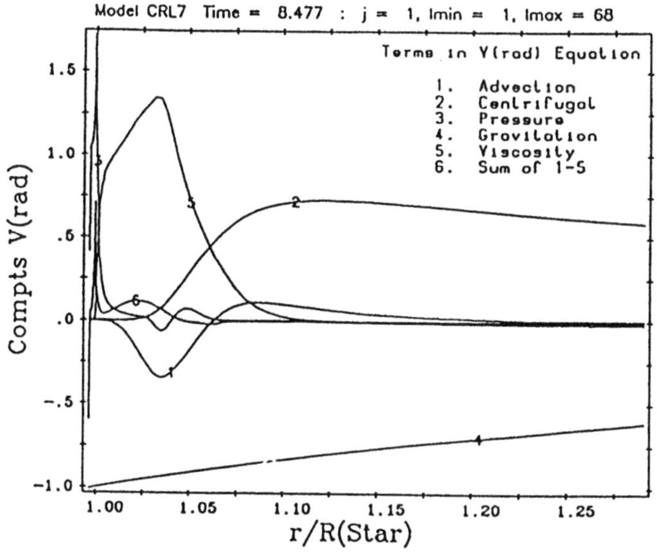

Fig.5: The contributions of the individual terms in the radial momentum equation as a function of the radius (see text)

contrast to Regev's one-dimensional result we find much higher values of β (in fact close to one). In other words: the disk indeed rotates nearly Keplerian.

Why did we name the BL considered so far *viscous* BL? Let us look at Fig. 5. There we have plotted the contributions of the individual terms to the radial momentum equation (for model crl2)

$$\frac{\partial v}{\partial t} = -v\frac{\partial v}{\partial r} - \frac{GM_{\mathrm{WD}}}{r^2} - \frac{1}{\rho}\frac{\partial p}{\partial r} + r\Omega^2 + \frac{1}{\rho}(\nabla \cdot \mathsf{T})_r , \qquad (7)$$

where v is the radial velocity, p the gas pressure, and T the viscous stress tensor. The terms on the right hand side are advection, gravitation , pressure force, centrifugal force, and viscous transport, respectively. Referring to Fig. 5, outwards of r_{m} the centrifugal force compensates against gravitation, i.e. the disk rotates Keplerian. Around and inside r_{m}, where the centrifugal balance is lost the viscous term (Note: this is *only* the radial part, *not* the contribution of the angular momentum equation) balances gravitation. Only a small contribution comes from the advection part which goes even in the same direction as the gravitation. So we see, that in the area called *viscous* BL, gravitation is balanced by viscosity (sic!) and *not* pressure.

Only very close to the stellar surface the pressure gradient begins to dominate. Since here the radial velocities are small this is approximately hydrostatic. So we call this part the *hydrostatic* BL, with a thickness

$$\delta^{\mathrm{h}}_{\mathrm{BL}} \ll \delta^{\mathrm{v}}_{\mathrm{BL}}.$$

$\delta^{\mathrm{h}}_{\mathrm{BL}}$ is about an order of magnitude larger (because of higher temperatures) than the pressure scale height of the WD. Pringle (1977) refers only to this

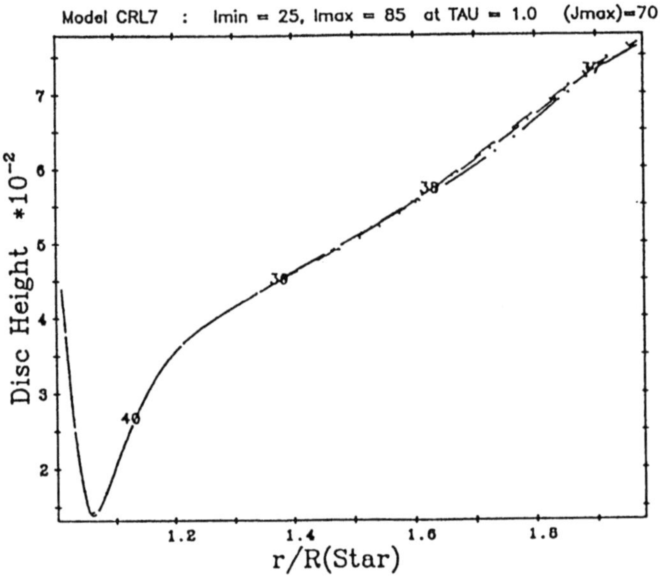

Fig.6: The height (in units of R_{WD}) of the BL/Disk versus the radius. The height is defined to be that point, where the optical depth is equal to one. Superimposed are 4 different time levels from ($t = 7.32$ to 8.37)

part as the BL, because in his derivation of the BL thickness δ_{BL} and temperature T_{BL} he does not take into account the viscous term, but equates just pressure gradient and gravitation which is the hydrostatic approximation. As an outcome of this calculation we have to distinguish between these two rather different length scales. If the viscosity is smaller (possibly resultant of a locally variable viscosity coefficient), the difference becomes less pronounced.

4.4 Surface

Near the star the disk height (see Fig. 6) decrease more rapidly than in the disk because of the strongly increased radial infall velocity. In Fig. 6 different time levels are superimposed to illustrate that the model has indeed reached a quasistationary state where the physical variables (Note: the disk height is a derived secondary quantity) do not vary any longer. The decline of the disk height leads to a decrease of the surface density Σ. The disk height is smallest in the middle of the viscous BL at $r = R_{WD} + \frac{1}{2}\delta^{v}_{BL}$ and increases again near the stellar surface (Fig. 6).

At the point of lowest disk height the surface temperature is highest, reaching from $T_{max} \approx 2 \cdot 10^5$ K to $T_{max} \approx 3.2 \cdot 10^5$ K depending on the mass and rotation of the WD. Inwards and outwards of $r = R_{WD} + \frac{1}{2}\delta^{v}_{BL}$ the temperature decreases. The temperature distribution clearly reflects the dissipation (Fig. 4).

Caused by the accretion the stellar surface is also heated up slightly. But a significant heating of the outer stellar layers is also restricted to an area close to the equator ($\leq 10°$in latitude). Above the BL and around the star a hot corona ($T \approx 10^8$ K) is formed, caused by the constant viscosity and a very small cooling rate (low densities). A comparison of a theoretical BL/Disk spectrum (Black Body) based on these model calculations with the observed EINSTEIN spectrum of SS Cyg is presented in Kley (1989*b*).

References

Cordova, F. A., Mason, K. O.: 1983, in *Accretion-Driven Stellar X-ray Sources*, eds. W. Lewin, E. van den Heuvel, CUP 1983, p. 147

Cordova, F. A., Mason, K. O.: 1984, *Monthly Notices Roy. Astron. Soc.*, **206**, 874

King, A. R., Shaviv, G.: 1984, *Nature*, **308**, 519

Kippenhahn, R., Thomas, H.-C.: 1982, *Astron. Astrophys.*, **114**, 77

Kley, W., Hensler G.: 1987, *Astron. Astrophys.*, **172**, 124

Kley, W.: 1989*a*, *Astron. Astrophys.*, **208**, 98

Kley, W.: 1989*b*, *Astron. Astrophys.*, **222**, 141

Papaloizou, J. C. B., Stanley, G. Q. R.: 1986, *Monthly Notices Roy. Astron. Soc.*, **220**, 593

Pringle, J. E.: 1977, *Monthly Notices Roy. Astron. Soc.*, **178**, 195

Pringle, J. E., Savonije, G. J.: 1979, *Monthly Notices Roy. Astron. Soc.*, **197**, 777

Pringle, J. E.: 1988, *Monthly Notices Roy. Astron. Soc.*, **230**, 587

Regev, O.: 1983, *Astron. Astrophys.*, **126**, 146

Robertson, J. A., Frank, J.: 1986, *Monthly Notices Roy. Astron. Soc.*, **221**, 279

Shaviv, G.: 1987, *Ap. Sp. Sci.*, **130**, 303

Sion, E. M.: 1985, *Ap. Sp. Sci.*, **130**, 47

van der Woerd, H.: 1987, *Astrophys. Space Sci.*, **130**, 225

Curious Observations of Cataclysmic Variables

F.V. Hessman

Max-Planck-Institut für Astronomie, Königstuhl 17,
D-6900 Heidelberg, Fed. Rep. of Germany

Summary: Observations of the two seemingly normal, non-magnetic cataclysmic variables , UU Aqr and OY Car, are presented. The behavior of the emission lines in both systems is not easily explained using the canonical picture of these close binary systems. Although there are several models which can be used to partially explain what is seen, all of them have serious problems. Various arguments suggest that what we see is due either to weak magnetic fields in systems with temporarily low accretion rates (e.g. OY Car), or to stronger fields roughly aligned with the orbital axis (UU Aqr). Nevertheless, there are still many aspects of the observations which should be difficult to explain using any simple model.

1. Introduction

The "canonical" model of disk-accreting cataclysmic variables developed during the 1960's and 70's has been very successful at explaining a multitude of observations (for a classical summary, see Robinson, 1976). In this picture, a late-type star (the "secondary") orbits around a white dwarf (the "primary") and, because it fills its Roche lobe, loses mass via its L_1 point. The specific angular momentum of the accretion stream prevents it from falling directly onto the white dwarf, resulting in the formation of a gaseous disk. Viscous processes of thus-far unknown origin enable a large fraction of the gas to be accreted, releasing the gravitation potential energy. The angular momentum lost by the accreted material is transfered through the viscous disk and is finally carried away by small amounts of gas or is transfered to the binary system *via* tidal torques. Slight variations on this basic model (e.g. relatively weak magnetic fields on the white dwarf, gross variations in the mass-transfer and -accretion rates, thermal instabilities in the secondary or the disk) can be used to explain practically all of the various types of cataclysmic variables . The major exceptions to this rule are the highly magnetic AM Her variables (where the magnetic field is so strong that no disk is allowed to form) and – perhaps – some of the magnetic DQ Her variables.

Given the low-mass secondary stars and accretion rates typically between 10^{-11} and 10^{-8} $M_\odot\ yr^{-1}$, the optical light of most cataclysmic variables is dominated by the accretion disk. The effective temperatures of the disks range from several thousand degrees at the outer edges to many tens of thousand degrees near the white dwarf. Although the white dwarfs have measured temperatures between about 15,000 and 30,000K, their small radii rarely allow them to compete against the optical light of the

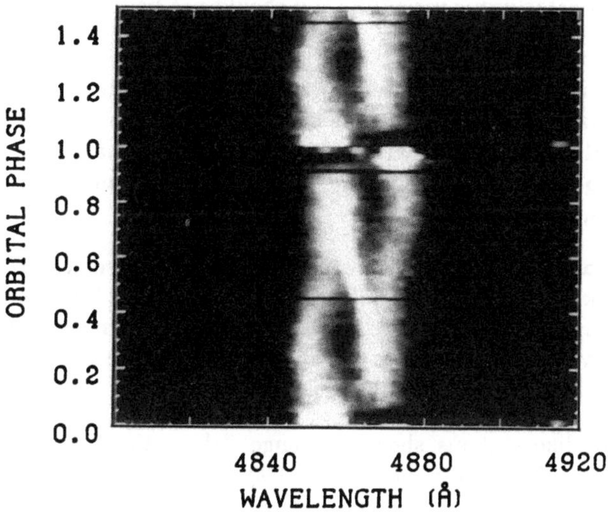

Fig. 1: Phase-binned Hβ observations of IP Peg. The lines in each phase bin have been normalized to the same total flux, producing exaggerated effects during the eclipse. The "S-wave" component is due to chromospheric emission from the secondary star rather than from the "hot spot."

accretion disks. The other luminous component which is often seen is the "hot spot", a shock caused by the impact of the accretion stream on the outer disk. The geometry of this shock causes the line and continuum emission from the "hot spot" to be emitted highly anisotropically, producing very characteristic "humps" in the light curves.

The most prominent features in the optical spectra of most cataclysmic variables are the broad Balmer emission lines. The broadening mechanism in some eclipsing cataclysmic variables can be directly deduced from the behavior of the lines during the eclipse. For example, consider the phase-binned Hβ profiles shown in Fig. 1 of the dwarf nova IP Peg (Hessman, 1989). During most of the orbit and at the beginning and end of the eclipse, the lines are clearly double-peaked and appear to show an orbital radial velocity variation (in addition, a strong chromospheric component is also visible). At the beginning of the eclipse, first the blue peak dissappears, followed by the blue and red wings. Towards the end of the eclipse, the process reverses itself, mirrored about the line center. This is exactly what one expects from the eclipse of a Keplerian disk: the outer regions rotate more slowly than the inner. The "double-peakedness" is simply the result of the outer, more slowly rotating disk having more area; it turns out that the half-separation of the peaks corresponds to the outer rotation speed in the disk (Smak 1969). Since IP Peg has beautifully double-peaked emission lines and also shows a photometric "hot spot", this system would appear to fit the canonical model quite well.

However, there are a multitude of signs which suggest that our simple picture of how cataclysmic variables "work" is failing to keep pace with the flood of new and better observations. The situation is worst for the inhomogenous subclass of cataclysmic variables about which we know the least: the so-called "novalike" variables. These systems often have strong continuua and a large ratio of high-excitation (He II, N III, C III) to Balmer line fluxes. The following list contains a few of the more pernicious problems *just* for this subclass of cataclysmic variables (and should not be interpreted as being complete!):

- A particularly annoying problem is the presence of different *photometric* and *spectroscopic* "orbital" periods. The best documented case is that of TT Ari (Thorstensen, Smak and Hessman, 1985), where the instantaneous phase difference between the two depends upon the photometric state of this "anti-dwarf nova" (Hutchings, Thomas, and Link, 1986).

- Although some dwarf novae show the double-peaked emission line profiles expected from a rotating disk, most novalike variables show very pointed lines, for which there is no obvious production mechanism. Although it has been suggested by Lin, Williams, and Stover (1988), pressure broadening does not seem to be the culprit (Dhillon, Marsh, and Jones, 1990).

- Many high-inclination novalike systems show a strange "absorption" feature around orbital phase 0.4 to 0.6 (Szkody and Piche, 1990). Although the phasing suggests some connection with the "hot spot", there is no obvious reason why significant absorption should be produced in the hot inner-to-middle disk gas against the relatively "warm" radiation of the "spot".

- The radial velocities measured from different lines do not necessarily agree (not even within the Balmer series!) and can change with time for the same line, so that it is often doubtful if the measured motions have anything to do with the orbital motion of the white dwarf (e.g. Shafter, Hessman, and Zhang, 1988).

- The line eclipses of novalike variables do not behave as one would expect for simple Keplerian disks (e.g. Honeycutt, Schlegel, and Kaitchuck, 1986). Although the fluxes decrease during the eclipse, the changes in the line profiles are relatively small and cannot be explained as being due to the eclipse of a rotating disk (Williams, 1989).

In the following sections, I present observations of two cataclysmic variables which illustrate some of these enigmas and add a few more items to the growing list of empirical problems waiting to be explained.

2. UU Aquarii

UU Aqr was discovered to be an eclipsing cataclysmic variable with an orbital period of 3.9 hours by Volkov, Shugatov, and Seregina (1986). A. Fiedler (Munich) and I observed this system during the two nights of 1988 November 11 & 12 using the Boller & Chivens spectrograph on the Calar Alto 3.5m and the Munich MMCP 15-channel photometer (Barwig, Schoembs, and Buckenmayer, 1987) on the 1.23m telescope. Simultaneous observations permitted a fraction of the narrow-slit spectra to be absolutely calibrated. Detailed analyses of all the data will be presented elsewhere

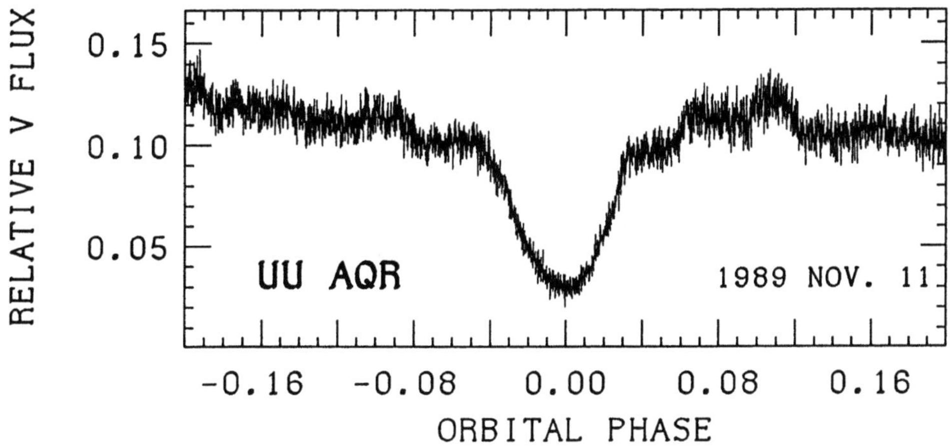

Fig. 2: V-filter light curve of UU Aqr.

(Wilfling, 1990; Wilfling, Hessman, and Fiedler, 1990). The eclipse light curve in the V filter on the first night is shown in Fig. 2. The mean light curve shows a deep, rounded, and smooth eclipse of a blue object, indicating a high inclination and an extended hot continuum source around the white dwarf.

The mean optical spectrum is shown in Fig. 3. Note the rounded appearance of the Balmer lines, the strong He I lines, and the weak high-excitation blend of N III λ46xx, He II λ4686, and He I λ4713 - typical of any number of dwarf novae and novalike variables. The orbital behavior of the Hβ profile can be seen best in the phased spectra of Fig. 4, where a strong peak component with an amplitude of over 400 $km\ s^{-1}$ can be seen. This feature is very reminiscent of the strong, high-velocity Hα component seen in the novalike DW UMa by Shafter, Hessman, and Zhang (1988).

In order to catch as much detail as possible in the emission lines during the eclipses, we trailed the star along the spectrograph slit, thereby avoiding the slow read-out time of the CCD between exposures. Using the photometry, we then flux-calibrated the spectra. The flux-calibrated spectra during two eclipses and the normalized Hβ profiles for one of the eclipses are shown in Figs. 5, 6, and 7. The most startling result can be seen in the shape of the normalized profile (roughly the same for both eclipses): absolutely no sign of a disk eclipse can be seen! Slow changes in the line profiles are visible – for example, the line became narrower during the eclipse on Nov. 11 – but there is no particular sign of the eclipse in the *shapes* of the Hβ lines.

The spectrum of UU Aqr during the eclipse would appear to be a simple continuation of the peaked profile seen outside of the eclipse. One explanation for this behavior would be to assign all of the peaked emission to the "S-wave" of the "hot spot" and to invoke an orbital inclination small enough so that the inner disk isn't eclipsed at all. However, the smooth light curve rules out this possibility: the object being eclipsed must be spatially extended, roughly symmetric, and centrally peaked – a behavior not

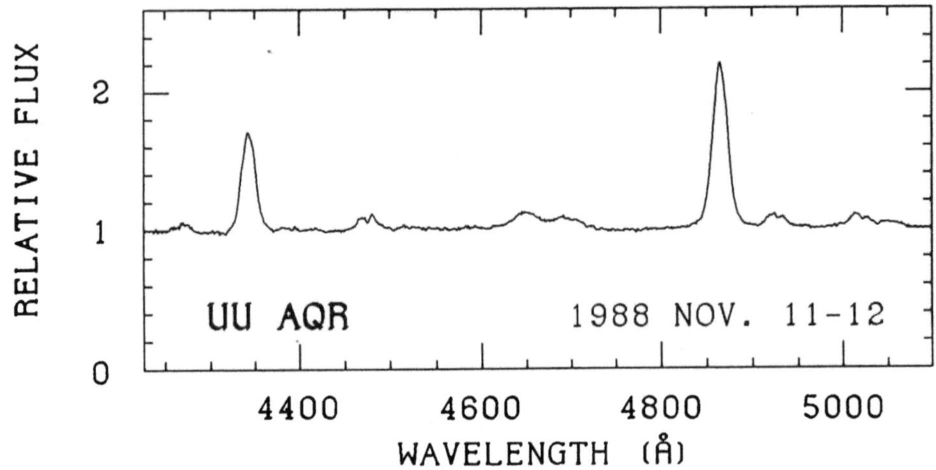

Fig. 3: The mean optical spectrum of UU Aqr. Note the broad, singly-peaked Hγ and Hδ lines, the weaker but doubly-peaked He I lines, and the blend of C III, N III, and He II λ4686.

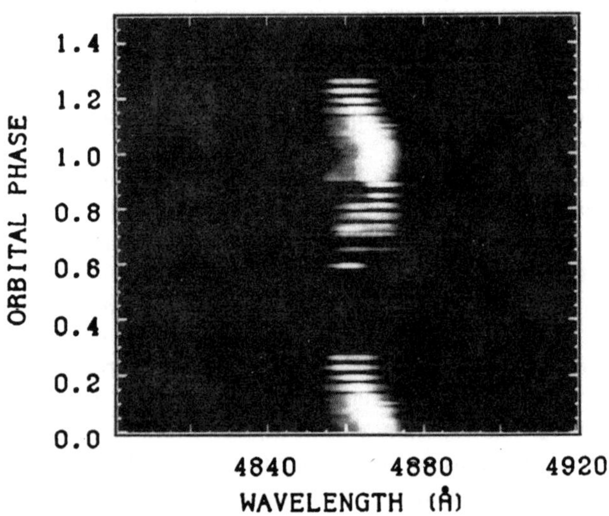

Fig. 4: The phased Hβ spectra of UU Aqr. The most obvious feature is a single, strong, sinusoidal component with an amplitude of over 400 km/s.

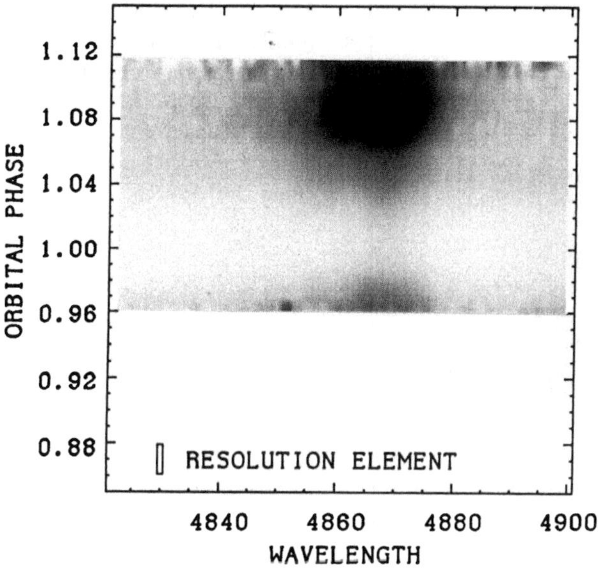

Fig. 5: Flux-calibrated spectrum of UU Aqr around Hβ during the eclipses on 1988 November 11. Note the general decrease of the continuum level during the central eclipse.

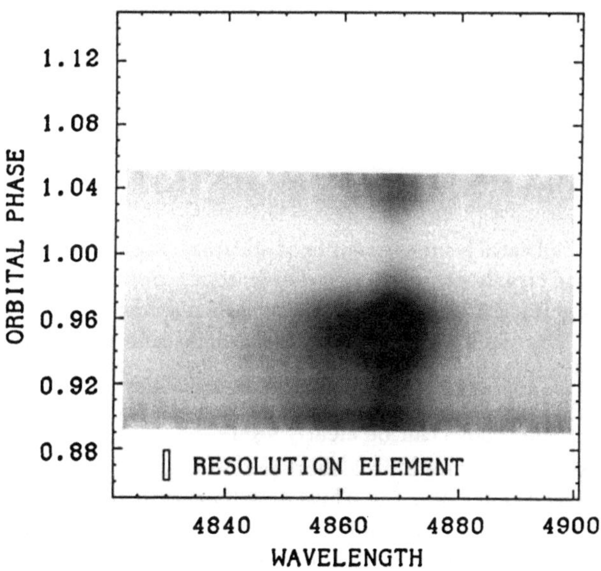

Fig. 6: Flux-calibrated spectrum of UU Aqr during the eclipse on 1988 November 12. The eclipse of the total line flux during both eclipses follows that of the continuum.

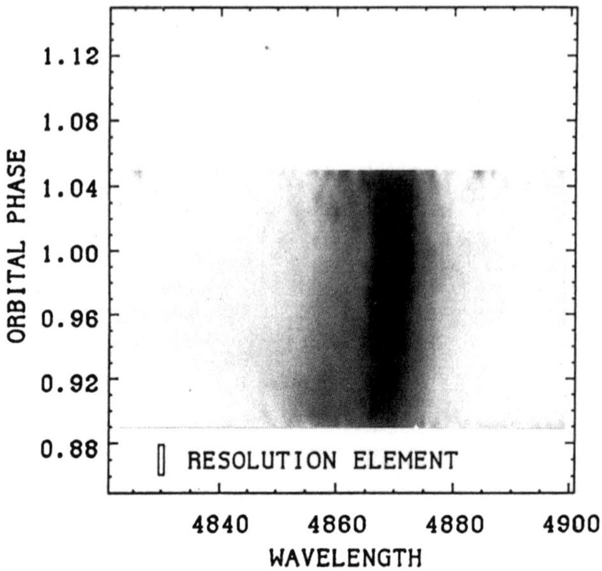

Fig. 7: Normalized Hβ line profiles of UU Aqr during during the eclipse on 1988 November 12. No obvious sign of the eclipse of a Keplerian disk (or, for that matter, anything else) is seen.

expected even for a complicated "hot stripe" (Royzczka and Schwarzenberg-Czerny 1987). Thus, what we are seeing in UU Aqr must be the eclipse of *something* non-disk-like at the center of the accretion disk. What makes UU Aqr all the stranger is that this occurs in a system which has an emission line spectrum like a normal dwarf nova, i.e. shows none of the spectral signatures typical for novalike variables other than a strong, blue continuum.

3. OY Carinae

OY Carinae is one of the best-studied cataclysmic variables of all time – mainly because the white dwarf and "hot spot" are clearly eclipsed during quiescence, permitting one to estimate all of the orbital parameters. Also, it is a SU UMa dwarf nova which shows "super-outbursts" during which "superhumps" are seen in the light curves. Thus, there have been many photometric studies by several groups (see Wood *et al.* 1989 for a nearly complete list). The mean light curve during the eclipse is shown in Fig. 8: the eclipse of the white dwarf and "hot spot" can be clearly seen.

There have been fewer spectroscopic studies of this short period (P_{orb} = 1.5 hours) system. Hessman *et al.* (1989) showed that, after the white dwarf's spectrum has been removed, what remains looks very typical of practically any dwarf nova in quiescence. R. Mundt and I obtained a little more than one orbital period of Hα spectra on 1987 February 24 using the MPG/ESO 2.2m on La Silla. R. Schoembs and H. Barwig observed OY Car simultaneously with the Munich MMCP photometer on the ESO 1.0m. The mean light curve during the spectroscopic observations is shown in Fig. 8, where one sees that the flux level is higher than normal: the observations were

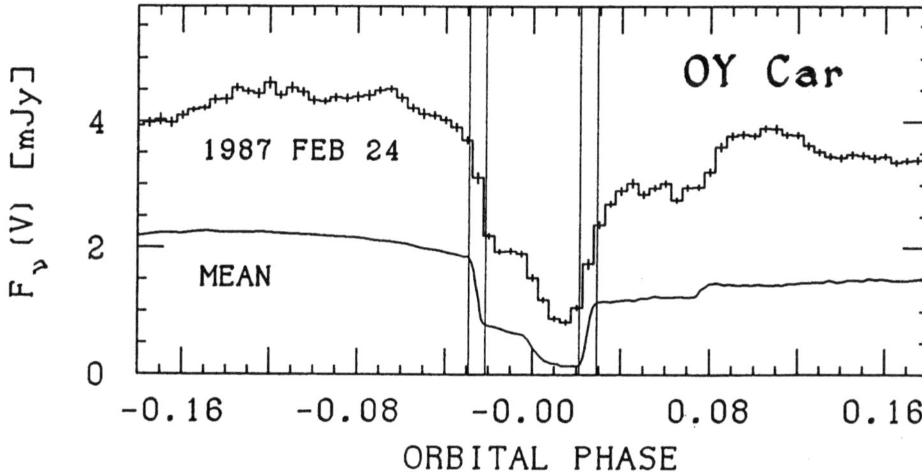

Fig. 8: Mean V-filter light curves of OY Car during quiescence (bottom) and immediately after a "super-outburst" in 1988 (top). The photometric hump due to the "hot spot" is clearly seen in both curves. The eclipses of the white dwarf (vertical lines) are sharp in the quiescent curve, but are somewhat broader and deeper in the 1988 observations, showing that a brighter, more extended object was being eclipsed.

taken at the end of a "super-outburst". Note that the eclipse of the white dwarf takes longer than usual in this state (most obvious during the exit), an effect seen before by Schoembs (1986).

The mean Hα profile is shown in Fig. 9. At first glance, it would appear to be rather normal: double-peaks with a half-separation of about 650 km/s and wings extending to over 2000 km/s. A picture of the phased spectra (Fig. 10) shows two oddities, however. The first oddity is that line profile during the eclipse is not symmetric and shows no substantial eclipse of the blue peak. Part of the reason is due to the second oddity: the red peak is *always* substatially stronger than the blue peak. A radial velocity analysis of the emission line wings and peaks shows that all components have the same orbital variations seen in earlier studies; specifically, all features show the same orbital amplitude and phasing (within the errors).

What could be causing the red-blue peak asymmetry? It's clear that we are not seeing the "hot spot": no matter what the form of the shock region is, the resulting emission should be Doppler shifted at the orbital period in the frame of the binary, i.e. we should see the extra peak wandering back and forth across the entire profile. Hessman *et al.* (1989) saw weak signs of such a component in their data during 1/2 of the orbit, but none is seen here.

Another obvious mechanism for producing a red-asymmetry is to invoke a strong, optically thick wind which could show P Cygni profiles. After all, such winds have been used to explain why the lines in many systems are not double-peaked. While this

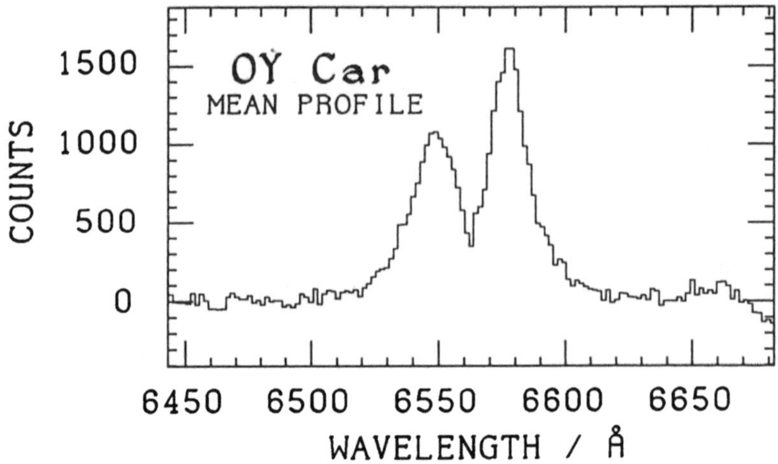

Fig. 9: The mean Hα line profile produced from data taken from just over an entire orbit right after a "super-outburst" in 1988. Note the strong emission feature – visible during the entire orbit – sitting on the underlying profile. The latter shows the double-peaked structure expected for a rotating disk.

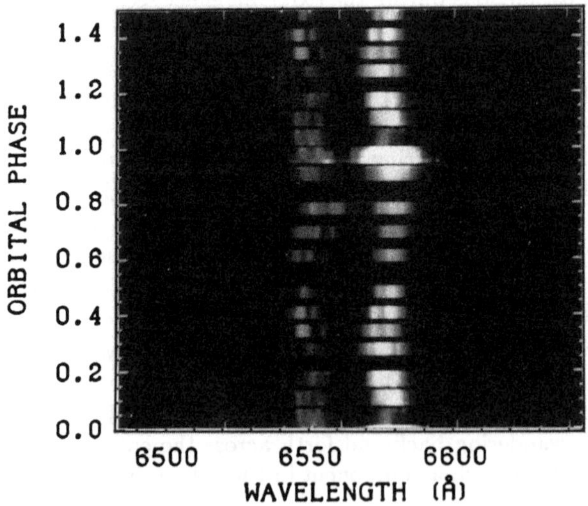

Fig. 10: The phase-binned Hα spectra of OY Car taken at the start of a quiescent period immediately after the occurance of a "super-outburst" in 1988.

model could reproduce the observed profiles in principle, there are several reasons why it seems unlikely that this is happening in OY Car:

- Any wind from the optically thin reminant of the outburst disk or white dwarf should be hot and optically thin in the Hα line, producing a narrow, roughly symmetric profile.
- The high inclination of OY Car should produce disk winds which are largely flowing perpendicular to the line-of-sight, destroying any classical (i.e. spherical) P Cygni profile.
- Hessman *et al.* (1989) saw much the same asymmetry in their quiescent spectra of the other Balmer lines – **but in the opposite sense** (i.e. the blue peak was *always* stronger than the red peak)!
- Although red and blue peaks are seen in Hα profiles from the winds in symbiotic stars (e.g. Robinson *et al.* 1989), the widths of the lines are much less than those seen in cataclysmic variables , despite the fact that the mass-accretion rates in symbiotic stars are several orders of magnitude higher than those in dwarf novae even at the height of their eruptions

The last obvious means of producing the Hα asymmetry (in either peak) is to have some fairly high-velocity flow – presumably in the disk or near the white dwarf – whose velocity is roughly constant during any observing campaign *in the frame of the observer*. This requires that the flow rotate much more slowly than the binary frame. The most obvious mechanism for producing such a flow is to have a magnetic field on a nearly non-rotating white dwarf. Given a complicated enough geometry near the white dwarf (where one might use the primary to hide other parts of the flow), one could conceive of being able to produce the observed profiles. The magnetic field would presumeable be so weak, that it can make itself apparent only during the time that the inner accretion disk has been nearly cleared of its material. Given field strengths of about 10^3 Gauss, any cyclotron emission would appear in the far infrared or even in the millimeter range. Radio emission has been seen in other dwarf novae immediately after eruptions, but it has not yet been proven that the emission is due to cyclotron processes near the white dwarf (Benz and Güdel, 1989).

Even if it is possible to have a magnetic field of the right strength and shape, is it reasonable to think the white dwarf rotates more slowly than the binary system? Naively, one would expect the material accreted from the disk over the lifetime of the system to speed up the white dwarf and, indeed, there are many systems with rapidly rotating primary stars. However, cataclysmic variables also have a dramatic means of removing the accreted material and, presumeably, the accompanying angular momentum: classical novae!. If this is the case, many white dwarfs may not be able to keep up with the evolution of the binary systems towards shorter orbital periods.

 $\Leftarrow \dot{M}$	$\Omega_\star \ll \Omega_{orbit}$	$\Omega_\star \approx \Omega_{orbit}$	$\Omega_\star \gg \Omega_{orbit}$
$\frac{B^2}{8\pi} \ll \frac{\dot{M}}{Rh}$	Most Dwarf Novae & Some Novalikes		
$\frac{B^2}{8\pi} \approx \frac{\dot{M}}{Rh}$	UU Aqr (high $\dot{M}$) ? OY Car (iff very low $\dot{M}$) ?		DQ Her's (iff $\vec{B}$ tilted!)
$\frac{B^2}{8\pi} \gg \frac{\dot{M}}{Rh}$	Unlikely ?	AM Her's	Unlikely ?

4. The "Big Picture"

The observations presented in the previous sections demonstrate that the classical picture cannot adequately explain the wide variety of cataclysmic variables and their behavior. Is it possible to construct a taxonometric framework within which the diverse characteristics can be qualitatively understood - if not in every detail for individual systems, at least for all cataclysmic variables as a whole? The classical cataclysmic variable genera are largely based either upon the appearance of the optical light curves (eg. the various types of dwarf novae - U Gem, SU UMa, or Z Cam - and the VY Scl stars, the "anti-dwarf novae") or that of the optical spectrum (the UX UMa systems and the novalike variables with strong emission lines). Since the novalikes sometimes resemble dwarf novae in outburst, this classification scheme fails to give us much physically interesting information if one changes the long-term (by human standards) mass-accretion rates. The AM Her systems can be easily identified by their strong magnetic fields (eg. X-ray emission), whereas the DQ Her systems are mostly classified based upon the presence of a stably and rapidly rotating white dwarf - a criterion suffering from obvious selection effects.

What we need is a taxonometric system which naturally provides niches for all of the observed systems and which depends not upon the vagaries of our observational possibilities but only upon the fundamental properties of the systems: masses, sizes, instantaneous mass-accretion rates, stellar rotation rates, and magnetic field strengths, and the inclinations of the magnetic axes to the rotation axes. A first attempt at freeing ourselves from the shackles of the old classification scheme is shown in Fig. 11, where only the parameters (1) the ratio of the magnetic field energy density to the instantaneous ram pressure of the mass-accretion at some appropriate radius (symbolically represented as $\dot{M}$ / radius / disk height) and (2) the ratio of the white dwarf rotation rate to the orbital frequency. *Any given system (i.e. with a given magnetic field and rotation rate) – is free to move up and down in this table according to the whims of the mechanism(s) which determine the instantaneous mass-accretion rate.* Interestingly, there is a part of the phase space without obvious examples amoung the known cataclysmic variables : the systems with moderate magnetic field strengths and accretion rates, aligned rotation axes but only slow rotation rates? Could UU Aqr be such an object?

Only further observations of these and similar systems will enable us to produce a new and more general "canonical model" for cataclysmic variables .

I would like to thank A. Fiedler, K. Mantel, R. Schoembs, and especially E. Wilfling for their collaboration, without which none of this work would have been possible. R. Mundt kindly provided some of his time on the MPG/ESO 2.2m and I enjoyed many fruitful conversations on this topic with V. Dhillon, A. King, H. Ritter, R.C. Smith, and R.E. Williams.

References

Barwig, H., Schoembs, R., and Buckenmayer, C.: 1987, *Astron. Astrophys.* **175**, 327
Benz, A.O., and Gödel, M.: 1989, *Astron. Astrophys.* **218**, 137
Beuermann, K., and Thomas, H.-C.: 1990, *Astron. Astrophys.* , in press
Dhillon, V.S., Marsh, T.R., and Jones, D.H.P. : 1990, *Proc. 11th North American Workshop on CV's & LMXRB's*, ed. C. Mauche, (Cambridge Univ. Press: Cambridge), in press
Hessman, F.V.: 1986, *Astrophys. J.* **300**, 794
Hessman, F.V.: 1989, *Astron. J.* **98**, 675
Hessman, F.V., Koester, D., Schoembs, R., and Barwig, H.: 1989, *Astron. Astrophys.* **213**, 167
Honeycutt, R.K., Schlegel, E.M., and Kaitchuck, R.H.: 1986, *Astrophys. J.* **302**, 388
Hutchings, J.B., Thomas, B., and Link, R.: 1986, *Publ. Astron. Soc. Pacific* **98**, 507
Lin, D.N.C., Williams, R.E., and Stover, R.J.: 1988, *Astrophys. J.* **327**, 234
Robinson, E.L.: 1976, *Ann. Rev. Astron. Ap.* **14**, 119
Robinson, K., Bode, M.F., Meaburn, J., and Whitehead, M.J.: 1989, in *Dynamics of Astrophysical Discs*, ed. J.A. Sellwood, (Cambridge Univ. Press: Cambridge), p. 85
Róyzczka, M., and Schwarzenberg-Czerny, A.: 1987, *Acta Astron.* **37**, 141
Schoembs, R.: 1986, *Astron. Astrophys.* **158**, 233
Shafter, A.W., Hessman, F.V., and Zhang, E.-H.: 1988, *Astrophys. J.* **327**, 248
Szkody, P., and Piche, F.: 1990, *Astrophys. J.* , in press
Thorstensen, J.R., Smak, J., and Hessman, F.V.: 1985, *Publ. Astron. Soc. Pacific* **97**, 437
Volkov, I.I., Shugarov, S.Yu., and Seregina, T.M.: 1985, *Astron. Tzirk.* **1418**, 3
Wilfling, E.: 1990, Diplom Thesis, Universität München.
Wilfling, E., Hessman, F.V., and Fiedler, A.: 1990, in preparation
Williams, R.E.: 1989, *Astron. J.* **97**, 1752
Wood, J.H., Horne, K., Berriman, G., and Wade, R.A.: 1989, *Astrophys. J.* **341**, 974

Accretion in AM Herculis Stars

A.D. Schwope

Institut für Astronomie und Astrophysik, TU Berlin,
Sekr. PN 8-1, Hardenbergstr. 36, D-1000 Berlin, Fed. Rep. of Germany

Summary: The optical continua of magnetic cataclysmic binaries (AM Herculis stars) are formed by cyclotron radiation in the hot accretion plasmas. Phase-resolved low-resolution spectrophotometry is one of the most important diagnostic tools to investigate the properties of cyclotron radiation. A review is given of those AM Her stars for which cyclotron spectroscopy is available and implications for the accretion geometry, the plasma parameters and heating/cooling processes in the accretion regions are outlined.

1 Introduction

The AM Herculis stars (polars) form the subgroup of cataclysmic variables where the mass-accreting star is a strongly magnetic white dwarf. The strong magnetic field fundamentally affects the dynamics and observational properties of these systems in a threefold manner: (a) Both stars are kept in synchronous rotation, $\omega_{\rm rot} = \omega_{\rm wd}$. (b) The formation of an accretion disk is suppressed. The matter from the secondary star is free-falling into the gravitational potential of the white dwarf until the magnetic pressure dominates the ram pressure of the acccretion stream (at some 10 $R_{\rm wd}$). Then the matter is threaded to magnetic field lines and guided into the polar regions where its kinetic energy is thermalized. (c) The heated accretion regions on the white-dwarf surface are the sources of intense cyclotron radiation which peaks, depending on field strength and optical depth, in the near-IR, optical or near-UV regime. The optical continua are, therefore, strongly polarized, linearly and circularly up to 30 %.

Observationally, the AM Her stars are further characterized by soft and hard X-ray emission from the accretion region and strong optical/UV emission lines of H, HeI, HeII and e.g. CIV from different places in the system. The multicomponent emission lines display a phase-dependent complex shape.

For recent reviews on observational as well as theoretical aspects of AM Herculis stars see Cropper (1990) and the monograph 'Polarized Radiation of Circumstellar Origin' edited by Coyne *et al.* (especially the contributions by Bailey, Beuermann, Lamb, Piirola, Schmidt, Stockman and Wickramasinghe).

2 Cyclotron spectroscopy

In this paper, I describe the results of phase-resolved, low-dispersion spectrophotometric observations of AM Her stars. This kind of observations may be refered to as 'cyclotron spectroscopy'. This discipline was founded 1979 by Visvanathan and Wickramasinghe who detected broad intensity humps in spectra of VV Pup separated by hundreds of Angstroms, later identified as cyclotron harmonics in emission (Wickramasinghe and Meggitt, 1982). VV Pup was only the third confirmed AM Her star. Although their number grew to 17 confirmed members during the last decade, the next detection of cyclotron lines was reported not before 1988 by Beuermann *et al.* and Ferrario *et al.* in spectra of UZ For (EXO033319-2554.2). Although it is clear that the optical continuum of every AM Her star is essentially a cyclotron spectrum (plus stellar contributions and an optically thin Paschen continuum from the accretion stream), the detailed understanding of the accretion processes requires the knowledge of the magnetic field strength. It may be determined in an optimal and very accurate way by resolving individual cyclotron lines. An up-to-date list of AM Her stars displaying cyclotron lines and their derived field strengths is compiled in Table 1. For the remaining 9 systems which are not included, either observations of sufficiently low resolution are missing (EF Eri, Grus V1, ST LMi, V1500 Cyg, AM Her, QQ Vul) or such observations did not reveal any sign of cyclotron line emission (BL Hyi: K. Beuermann (private communication), WW Hor: Beuermann *et al.* (1987), EXO032957-2606.9: Cropper *et al.* (1990)). For BL Hyi (Wickramasinghe *et al.*, 1984; Schwope and Beuermann, 1985), ST LMi (Schmidt *et al.*, 1983) and AM Her (Schmidt *et al.*, 1981; Wickramasinghe and Martin, 1985) magnetic field determinations resulted from observations of Zeeman-shifted photospheric absorption lines when accretion had ceased or when the accretion region was occulted by the white dwarf itself (ST LMi).

The value of the magnetic field strength is only one, but of course important, piece of information which may be derived from cyclotron spectroscopy. The full potential of this method is available only if the observations cover the whole binary cycle. This is the case only for VV Pup, V834 Cen and UZ For (Table 1). The observations and results obtained on these objects are reviewed in the following chapters and the implications for the magnetic field geometry, the accretion geometry, the plasma parameters and heating and cooling processes are discussed. Before doing so, I provide as a diagnostic tool some basic results of cyclotron theory.

Table 1. AM Herculis stars showing cyclotron lines

Name		$B_1^{a)}$ [MG]	$B_2^{a)}$ [MG]	Ref.
DP Leo (E1114+182)	$S^{b)}$	44		1
VV Pup	P	30-32	55-56	2,3,4
V834 Cen (E1405-451)	P	23		5
MR Ser (PG 1550+191)	S	25		1
EK UMa (1E1048.5+5241)	S	47		6
AN UMa	S	36		1
UZ For (EXO033319-2554.2)	P	53	75	7
BY Cam (H0538+608)	S	41		1

a) B_1 and B_2 are the field strengths in the main and secondary accretion region
b) S: single spectrum, P: phase-resolved spectroscopy

References: 1) Cropper et al. (1989); 2) Wickramasinghe and Meggitt (1982); 3) Wickramasinghe et al. (1989); 4) this work; 5) Schwope and Beuermann (1990); 6) Cropper et al. (1990); 7) Schwope et al. (1990)

2.1 Some elements of cyclotron theory

The accretion plasmas in AM Her stars are heated to mildly relativistic temperatures, $kT \simeq 1 - 30$ keV. In this regime the cyclotron absorption coefficient $\alpha_{\rm cyc}$ is a steep function of the harmonic number $m_{\rm H} = \omega/\omega_{\rm c}$ ($\omega_{\rm c} = eB/mc$: cyclotron frequency). Therefore, the cyclotron spectrum changes from optically thick to thin at optical or near-optical wavelengths making detailed radiative transfer calculations necessary. In the most simple case of a homogeneous isothermal plasma with large Faraday rotation, the radiative transfer equations for the two magnetoionic modes ($\pm$) decouple and their solutions become simply $I_\pm = I_{\rm RJ}\ [1 - exp(-\tau_\pm)]$, where $I_{\rm RJ}$ is the RJ-intensity per polarization mode and $\tau_\pm = \ell\alpha_\pm = \Lambda\phi_\pm$. The optical depth parameter Λ is given by $\Lambda = \ell\omega_{\rm p}^2/(c\omega_{\rm c}) = 2\times 10^5(\ell/10^5{\rm cm})(N_{\rm e}/10^{16}{\rm cm}^{-3})(B/30{\rm MG})^{-1}$ and the dimensionless absorption coefficient $\phi_\pm$ becomes a function of temperature T, angle to the magnetic field Θ and harmonic number $m_{\rm H}$ only (ℓ: geometric path length, $\omega_{\rm p}$: plasma frequency, c: speed of light). The cyclotron absorption coefficients $\alpha_\pm$ may be derived from the emissivity of a single gyrating electron averaged over the electron distribution function (see e.g. Chanmugam and Dulk, 1981). The coefficient ϕ_- for the dominant extraordinary mode normalized by division of a smooth fit to the cyclotron maxima is shown in Fig. 1 for $kT = 5$ and 20 keV, respectively. With increasing temperature individual harmonics become broader, they appear more redshifted and their angular dependence grows. From Fig. 1 it becomes clear that the detection of cyclotron lines constrains the spread in temperature, azimuth and lat-

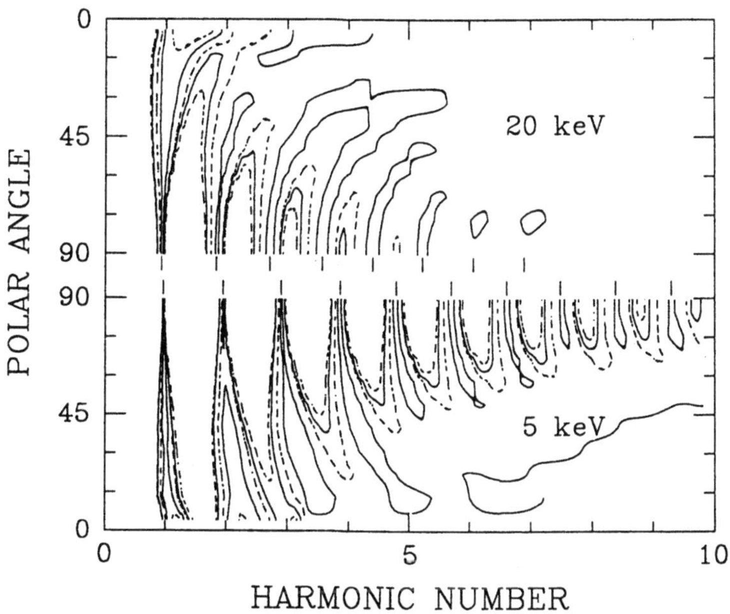

Fig. 1. Normalized cyclotron absorption coefficient dependent on temperature and polar angle calculated for a homogeneous plasma. The positions of line centers at $\Theta = 90°$ are marked by vertical tics. For high harmonic numbers and low polar angles the cyclotron harmonics broaden and the line structure is washed out.

itude (in magnetic coordinates) and field strength in the emission region in order not to smear out the line structure. On the other hand, the non-detection of cyclotron lines in the optically thin regime is suggestive of a structured, extended emission region with inhomogeneous temperature and density distribution.

In order to apply the cyclotron models to the observed motion of the peaks of individual lines, we fitted low-order polynomials to their calculated positions. The results for the 4th to 10th harmonics of a 10 keV plasma are shown together with the predictions of an approximate analytic formula presented by Barrett and Chanmugam (1985) in Fig. 2. The analytic formula has been used by Cropper *et al.* (1989) and Cropper (1990) to derive field strengths for some polars from a single spectrum only (see Table 1). The differences between the predictions made by the two approaches are considerable. In our calculation, the 7th harmonic appears at the same position as the 8th given by the approximate expression ($\Theta = 90°$). The deviations are caused by two reasons. First, our results are based on detailed numerical calculations and second, we have determined the positions of cyclotron maxima after normalization to a local continuum which resulted in a blueshift of the line position if compared to that determined from the unnormalized line

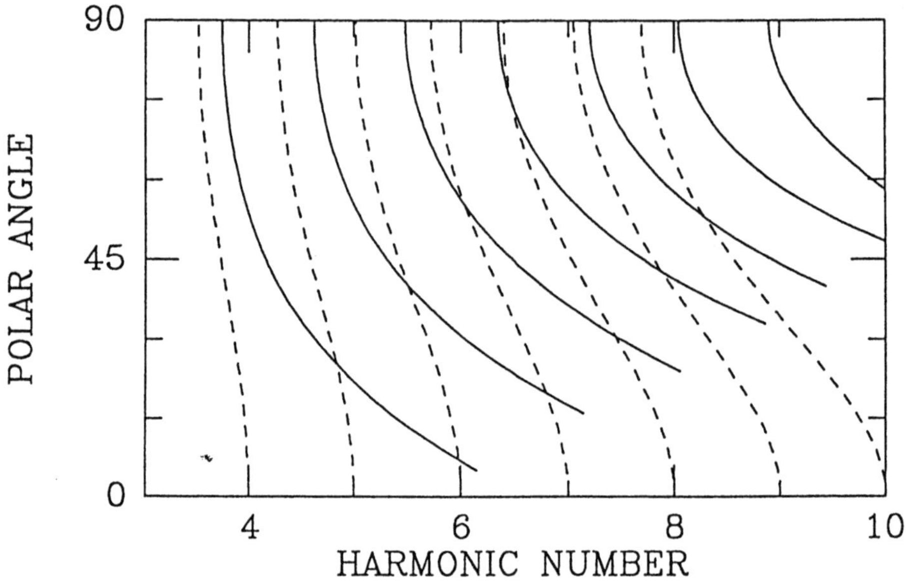

Fig. 2. Motion of the 4th to 10th cyclotron harmonics calculated for a homogeneous plasma with $kT = 10$ keV. Dashed lines are based on an analytic expression given by Barrett and Chanmugam (1985). Solid lines are based on detailed numerical calculations.

profile. This procedure has been used in order to treat the theoretical models as closely as possible to the observational data. The wavelength/frequencies of cyclotron lines obtained from observations may then be fitted dependent on field strength B, temperature T, orbital inclination i, inclination of the accreting field line in the hot accretion spot η, and orbital phase φ. The latter three quantities are related to the polar angle Θ by

$$\cos\Theta = \sin i \sin\eta \cos 2\pi(\varphi - \varphi_o) + \cos i \cos\eta .$$

3 Results and discussions on individual systems

3.1 UZ For (EXO033319-2554.2)

UZ For was discovered as a serendipitious EXOSAT X-ray source (Osborne *et al.*, 1988) and optically identified as an AM Her star by Beuermann *et al.* (1988). It became of special interest because it was found to be an eclipsing binary, the measured orbital period, $P = 126.5$ min, redefined the lower edge of the cataclysmic variable period gap, and it harbours an unusually massive white dwarf, $M_{\rm wd} \simeq 1.3 M_{\odot}$. A sketch of the accretion geometry, the observed visual light curve, and spectra for two selected phases are shown in Figs. 3–5. Here, as in the following graphs, the numbers (1) and (2) mark the phases when our line-of-sight is best aligned with and farthest away from the accreting field lines, respectively. The primary pole in UZ For is self-eclipsing, i.e. it is hidden behind the white dwarf for ~ 50 % of an orbital cycle. The spectrum during the faint phase is dominated at long wavelengths by the secondary star. The single M-dwarf G3-33 (Gl83.1) may serve as template for the secondary in UZ For and its spectrum is shown also in Fig. 5.

During the bright phase the spectrum is deeply modulated by cyclotron line emission. The maxima at 6570Å, 5050Å and 4130Å were identified as the 3rd to 5th harmonics in a field of $B \simeq 53$ MG (Beuermann *et al.*, 1988). The observed motion of the harmonics throughout the bright phase are shown in Fig. 6 along with results from model calculations. The quantity plotted is the wavelength of the blue wing of the harmonics at half maximum which is less affected by optical-depth effects than the sharp maximum. Suitable fits were found within the temperature range $kT = 5 - 10$ keV and yielded $\eta \simeq 155°$ for i fixed at $i = 85°$. The phase at which the harmonics appear maximally blueshifted agrees with the centers of the optical and X-ray bright phases. Consequently, the accreting field lines cannot be significantly inclined in azimuthal direction. The faint-phase intensity maximum centered on 4740Å is a persistent feature throughout the orbital cycle and was seen also by Ferrario *et al.* (1988) and Allen *et al.* (1990) during episodes with higher accretion rates. It always displays maximum intensity around eclipse phase $\sim 0.1 - 0.2$. At least in the data presented by Schwope *et al.* (1990), a slight orbital modulation of its position may be recognized with maximum blueshift again around $\varphi \sim 0.1 - 0.2$. This feature, therefore, most probably represents emission in the 3rd harmonic of a low-temperature cyclotron source at the opposite pole with $kT \simeq 1$ keV and $B \simeq 75$ MG. If this interpretation is correct, then (1) the field cannot be a centered dipole and (2) the emission regions are very likely not located at the footpoints of the same closed field line.

Schwope *et al.* (1990) have demonstrated that not only the motion of the cyclotron lines but also the flux distribution may be well described with a constant-Λ model, $\log \Lambda = 4 - 5$ and $kT = 5 - 10$ keV. The temperature derived is far below the temperature reached if the accreted material goes through a strong adiabatic shock, $kT_s \simeq 56$ keV for a white dwarf of $1.3 M_\odot$. The almost complete lack of atomic emission lines suggests that hard and soft X-ray emission from the pole caps was negligible at that time of observation. This justifies the assumption that the accretion luminosity $L_{acc} = GM\dot{M}/R$ is balanced by the cyclotron luminosity alone. Both, the total as well as the specific mass accretion rates deduced on that assumption were found to be very low, $\dot{M} \simeq 1.5 \times 10^{12}$ g s^{-1} and $\dot{m} \simeq 0.03$ g cm^{-2}s^{-1}. Therefore it appears likely that the emission regions in UZ For are heated via

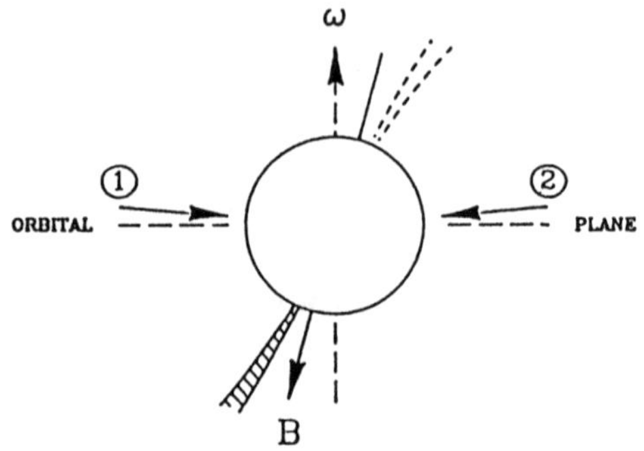

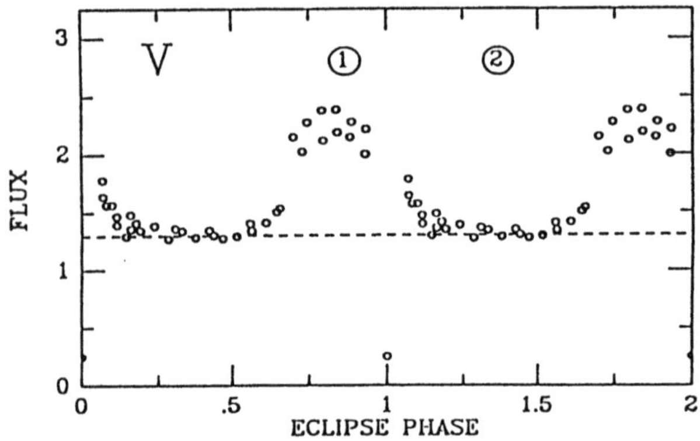

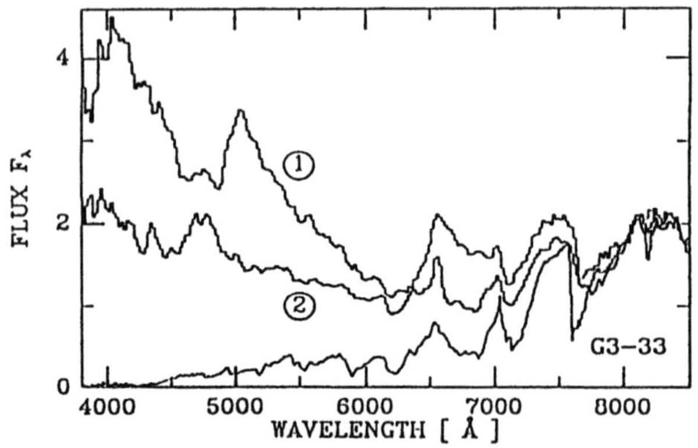

Fig. 3. (top): Sketch of the geometry of UZ For. Indicated are the orbital plane, the rotation axis, the dipole axis, field lines transferring matter (hatched) and the line-of-sight for two selected phases.

Fig. 4. (middle): Approximate V-band light curve of UZ For derived from spectrophotometry. The dashed line indicates the level of photospheric radiation from the white dwarf. Flux units are 10^{-16} ergs cm^{-2} s^{-1}Å^{-1}.

Fig. 5. (bottom): Representative bright-phase and faint-phase spectra of UZ For. Also shown is a suitably scaled spectrum of the dM4.5e star G3-33 (Gl83.1) which served as template for the secondary.

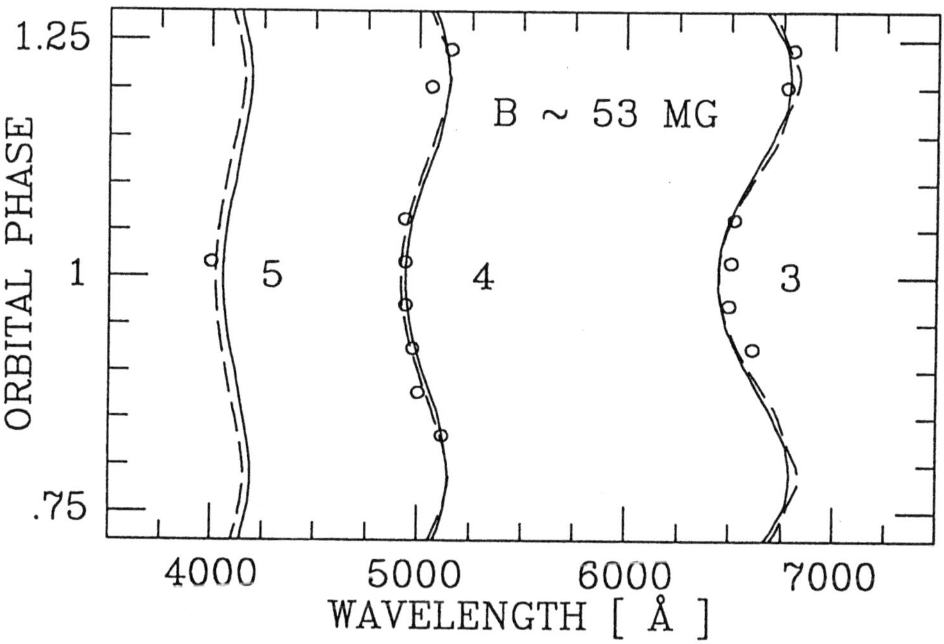

Fig. 6. Observed and predicted motion of the blue wings of the 3rd, 4th and 5th harmonic at half intensity observed in UZ For. The phase convention used here refers to the center of the bright phase. Model curves for $kT = 5$ keV (dashed line) and 10 keV (solid line) are calculated using $i = 85°$, $\eta = 155°$ and $B = 53$ MG.

particle bombardement where the incoming protons release their energy by collisions with photospheric matter, rather than via a hydrodynamic shock. This scenario, originally suggested by Kuijpers and Pringle (1982) and examined in some detail by Thompson *et al.* (1986) for the case of cooling via bremsstrahlung/line radiation still awaits a detailed numerical treatment.

3.2 V834 Cen (E1405-451)

As the majority of AM Her binaries, V834 Cen was originally discovered as a soft X-ray source (Jensen *et al.*, 1982). The photometric light curve with a 101.5 minute periodicity, strong and asymmetric line emission, and the high X-ray to optical luminosity led Mason *et al.* (1983) to classify V834 Cen as an AM Her system. The accretion region in V834 Cen is neither eclipsed by the secondary nor self-eclipsed by the white dwarf and the polar angle Θ varies between $\sim 0°$ and $\sim 90°$ during an orbital revolution (cf. Fig. 7). The visual light curve, shown in Fig. 8, is modulated mainly due to the anisotropy of cyclotron radiation, the so-called beaming. The cyclotron flux is at its minimum when we are looking down the accretion column at $\varphi = 0.0$. The asymmetry of the light curve and the secondary minimum at

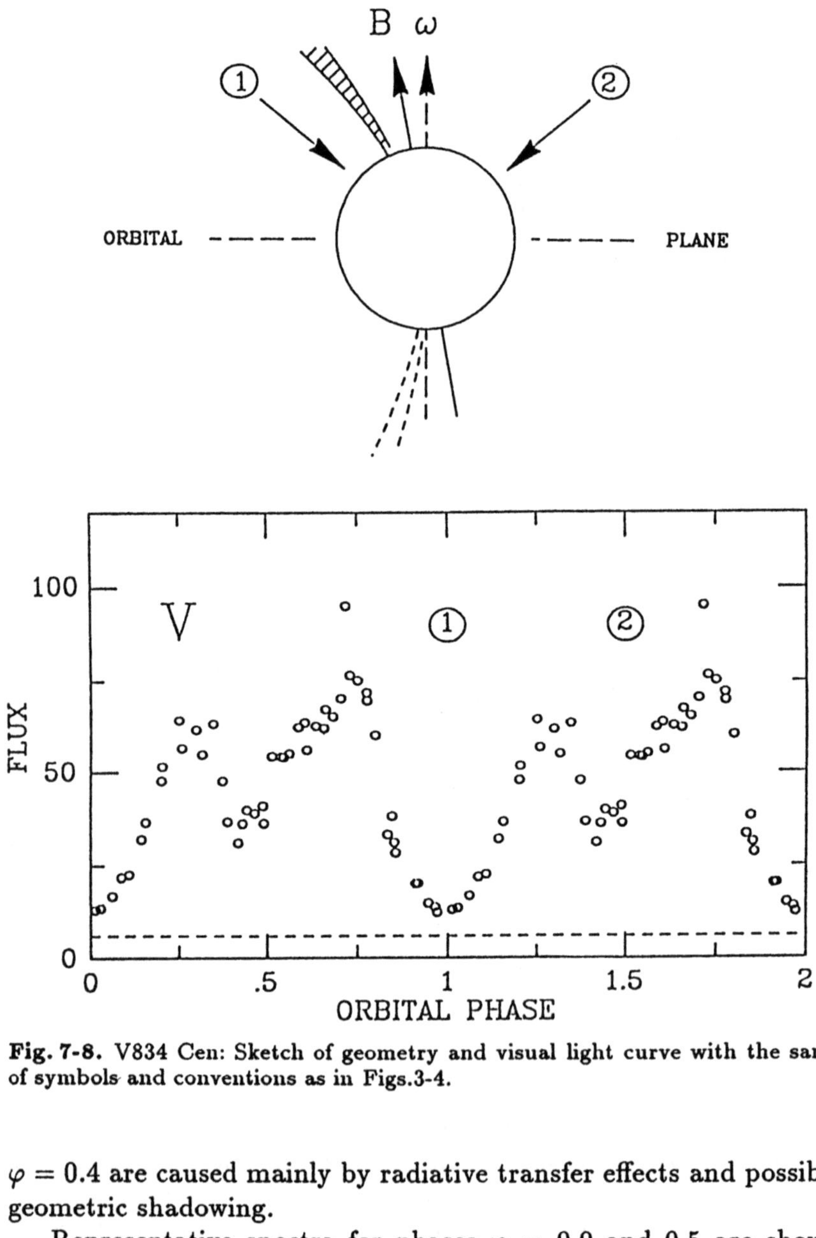

Fig. 7-8. V834 Cen: Sketch of geometry and visual light curve with the same meaning of symbols and conventions as in Figs.3-4.

$\varphi = 0.4$ are caused mainly by radiative transfer effects and possibly also by geometric shadowing.

Representative spectra for phases $\varphi = 0.0$ and 0.5 are shown in Fig. 9a together with a low-state spectrum which indicates the photospheric contributions of the stars. The spectrum at $\varphi = 0.0$ is flat in F_λ whereas low-frequent intensity modulations are obvious at $\varphi = 0.5$. Intensity maxima and minima could be identified after subtraction of the emission lines and normalization to a continuum (Fig. 9b) as high-harmonic cyclotron emission and Zeeman- and Doppler-shifted $H\alpha$-absorption (Schwope and Beuermann, 1990). The Zeeman absorption troughs appear most redshifted when our line-of-sight is best aligned with the accreting field lines at $\varphi = 0.0$

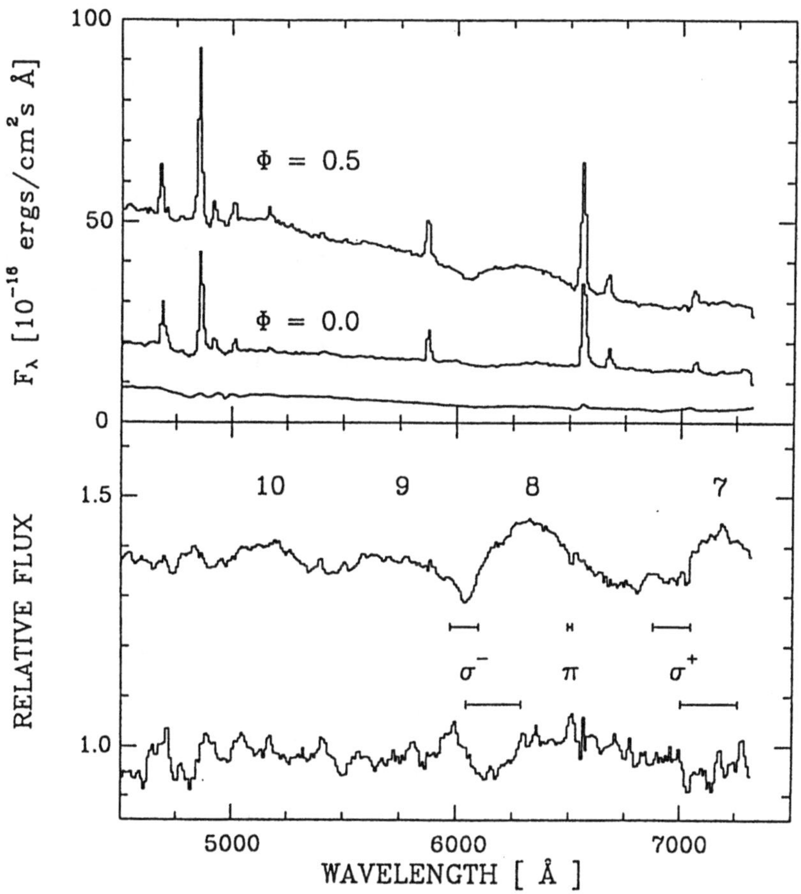

Fig. 9. (a) Upper panel: High-state spectra of V834 Cen at phases 0.0 and 0.5, obtained on 1986, March 1 and 2, and average low-state spectrum obtained on 1989, January 29. The low-state spectrum was used for removing the stellar contributions in the high-state spectra. (b) Lower panel: Same high-state spectra as above after removal of the emission lines and division by a smooth fit to the continuum. The normalized spectrum at $\varphi = 0.5$ was shifted vertically by 0.4 units. Cyclotron harmonics and Zeeman absorption components of Hα are indicated.

indicating that they arise in free-falling matter. The field strengths derived from both magnetic features, $B \simeq 23$ MG, agree within the limits and show the close coexistence of hot emitting plasma and cool absorbing matter. Values for the angles i and η were obtained from fits to the cyclotron lines and the relevant solid curves in Fig. 10 were drawn using a consistent set of data, $i = 50°, \eta = 40°, B = 23$ MG, $v_{\rm rad} = 4000$ km s^{-1} for the velocity of the infalling cool material. The maximum polar field strength consistent with the derived geometry is $B_{\rm p} = 26$ MG assuming that the field is a centered and aligned dipole.

The low-state observations (Beuermann *et al.*, 1989) revealed photospheric absorption lines at nearly the same positions as in the high state,

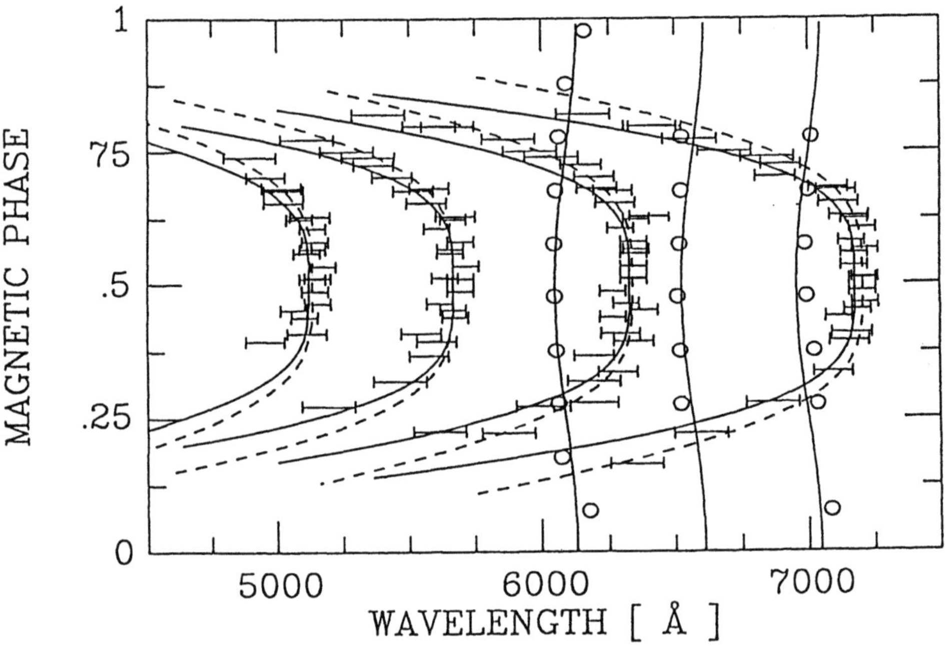

Fig. 10. Observed wavelengths of cyclotron maxima (bars) and Zeeman dips (circles). The phase convention used here (magnetic phase) refers to the maximum redshift of the cyclotron lines. The magnetic phase and the photometric phase used in Fig.8 agree within 0.02 phase units. Solid curves are predicted positions. Cyclotron models are shown for $kT = 5$ keV (dashed line) and $kT = 10$ keV.

implying a mean field strength of $B \simeq 22$ MG (Fig. 11). Model calculations assuming a centered dipole field and uniform flux distribution, $T_{\rm eff} \sim 15000$ K, yielded a polar field strength $B_{\rm p} \simeq 35$ MG, inconsistent with the results of cyclotron spectroscopy. It seems, therefore, necessary to assume a "warm" spot with $T_{\rm bb} \sim 30000$ K at the location of the former hot accretion region which gives the wanted bias towards lower field strength in spectra averaged over the observable hemisphere (Fig. 11). This idea is supported by the photometric variations in the low state where maximum intensity occured at the former photometric minimum around $\varphi = 0.0$. The proof of an inhomogeneous temperature distribution on the white dwarf in V834 Cen may have consequences for the interpretation of low-state Zeeman spectra of other AM Her stars. The low-state spectrum of, e.g., BL Hyi (Fig. 11) looks almost identical to that of V834 Cen but polar field strengths of $B_{\rm p} = 30-40$ MG are reported for this object (Wickramasinghe *et al.*, 1984, Schwope and Beuermann, 1985)!?

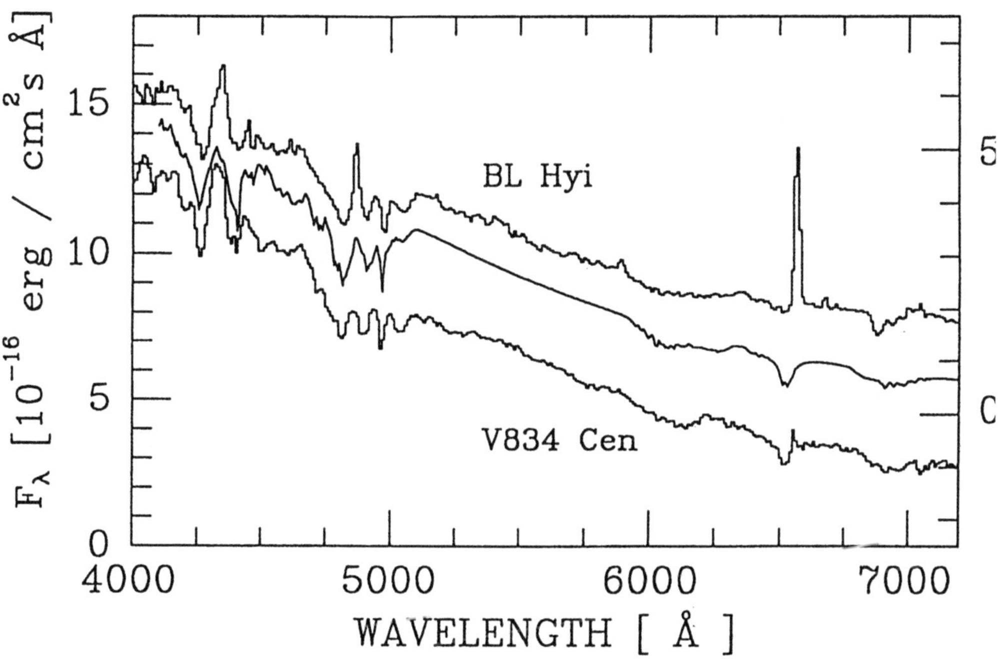

Fig. 11. Low-state spectra of V834 Cen (left scale) and BL Hyi (right scale). The synthetic spectrum (middle curve) was calculated using a two-temperature model with spot temperature $T_{spot} = 30000$ K and photospheric temperature $T_{phot} = 17000$ K (S. Jordan, private communication).

So far we have discussed only the optically thin line forming part of the cyclotron spectrum. Contrary to UZ For with nearly 100 % intensity modulation, the continuum in V834 Cen is depressed between the cyclotron maxima by maximally 5%. Whether the underlying continuum is optically thick radiation or more or less optically thin radiation with all line structure washed out due to the inhomogeneities described above is difficult to decide. However, homogeneous and especially constant-Λ models were not found to be able to produce spectra as flat as observed during the whole orbital cycle. In context with the asymmetric and variable shape of the optical light curve (Cropper, 1989), this is suggestive of a structured emission region, making detailed radiative transfer calculations necessary. Such calculations are, of course, difficult to perform and additional parameters have to be introduced which cannot be determined directly from observations. We note that published model spectra for structured emission regions (Wickramasinghe and Ferrario, 1988; Wu and Chanmugam, 1988, 1990) indeed tend to flatten the spectra.

The total cyclotron luminosity is $L_{cyc} \simeq 2 \times 10^{31} d_{100}^2$ ergs s^{-1}, with d_{100} being the distance to V834 Cen in units of 100 pc. The accretion luminosity is given by $L_{acc} = L_{SX} + L_{HX} + L_{cyc} \simeq 3.5 \times 10^{32}$ ergs s^{-1}

where the suspected blackbody component in the soft-X/XUV-regime is by far the most important term, $L_{SX} \simeq 10(L_{HX} + L_{cyc})$ (cf. Bonnet-Bidaud *et al.*, 1985) and $L_{HX} \simeq L_{cyc}$. Frank *et al.* (1988) have shown that the soft X-ray excess in AM Her stars is best explained by inhomogeneous and time-dependent accretion, where dense blobs of matter with a time-averaged specific accretion rate $\dot{m} \simeq 50 - 500$ g cm^{-2} s^{-1} manage to reach large optical depths before their energy is released. Such a scenario may work also in V834 Cen since the specific mass accretion rate is likely to be much larger than $\dot{m} \sim 3$ g cm^{-2} s^{-1}, obtained by relating the total mass accretion rate to the cyclotron emitting area. The latter quantity is known to be a factor of 10-100 greater than the bremsstrahlung emitting areas and, hence, the total area accreting dense blobs (Beuermann, 1988).

3.3 VV Pup

VV Pup was recognized as the third AM Herculis type binary system by the discovery of strong periodic circular and linear polarization (Tapia, 1977). Based on its photometric and polarization characteristics, an accretion geometry similar to that of UZ For was established (Fig. 3) with a slightly lower orbital inclination, $i \sim 75°$, just avoiding an eclipse by the secondary (Liebert *et al.*, 1978; Allen and Cherepashchuk, 1982; Brainerd and Lamb, 1985). Visvanathan and Wickramasinghe (1981) detected pronounced modulation of the continuum intensity during the bright phase which was, at first glance, considered as due to cyclotron absorption in a field of ~ 30 MG. Later, when detailed cyclotron models became available (e.g. Chanmugam and Dulk, 1981; Meggitt and Wickramasinghe, 1982), the intensity maxima were recognized as cyclotron emission lines and the field strength could be confined to $31.5 - 32$ MG (Wickramasinghe and Meggitt, 1982; Barrett and Chanmugam, 1985). Photometric variations and non-zero polarization during the faint phase when the primary pole is out of view were interpreted due to accretion on a second pole (Liebert and Stockman, 1979). The firm spectroscopic proof for accretion on both poles was given by Wickramasinghe *et al.* (1989) via the detection of two distinct systems of cyclotron lines. The somewhat surprising result was that the field strength at the second pole, $B_2 \simeq 56$ MG, exceeds that at the main pole by nearly a factor of 2. The newly determined field strength for the main accretion region, $B_1 \sim 30.5$ MG, was found to be slightly lower if compared with the initial measurements. These findings have been interpreted in terms of a dipole which is offset by $\sim 0.1 R_{wd}$ from the center of the white dwarf in the direction of the dipole axis with the emission regions located near the footpoints of the same closed field line(s). The lower value found for B_1 indicates a migration of the main emission region by $\sim 10°$ towards the magnetic equator.

In Fig. 12, we present the results of additional spectrophotometry obtained at two different epochs (Schwope and Beuermann, in preparation)

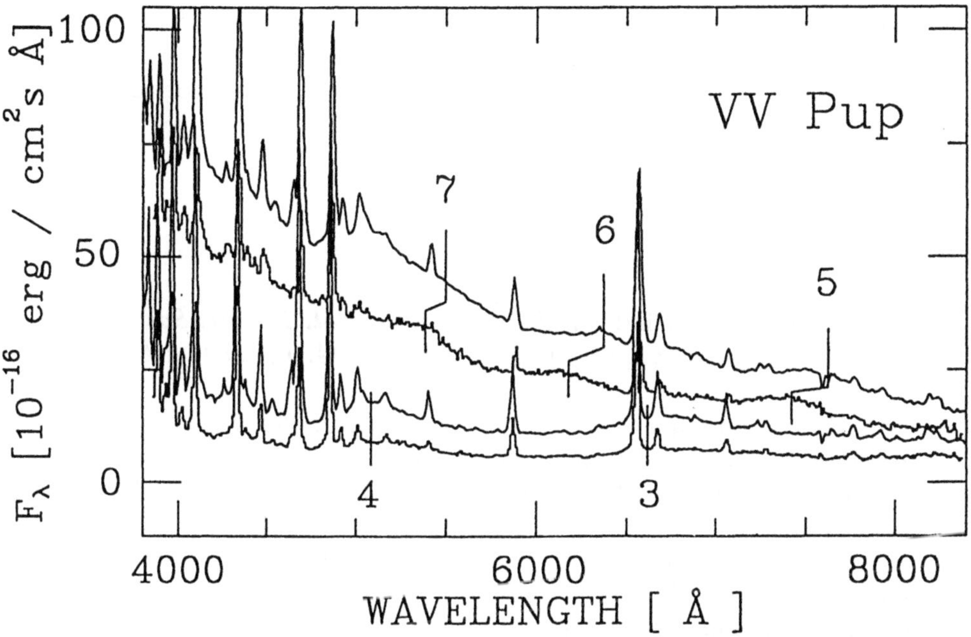

Fig. 12. Representative bright-phase (upper curves) and faint-phase spectra of VV Pup obtained at two different epochs. Positions and shifts of individual harmonics used for determination of the field strengths in both emission regions are indicated.

which confirms the detection of cyclotron lines from both poles. Although our results are a bit preliminary, some results may be regarded as certain facts founding the base for future work: (1) The mean intensity level during the faint phases differs by a factor of 2 between the different observations but the faint-phase spectra look perfectly identical with respect to the cyclotron-line pattern. The field strength derived by fitting the 3rd-4th harmonics agrees with that given by Wickramasinghe *et al.* (1989),$B_2 \simeq 54$ MG. (2) The second accretion spot is visible at all phases and the corresponding cyclotron lines are most prominent even during the bright phase when the system gains higher optical brightness and (?) mass accretion rate (upper curve in Fig.12). (3) The cyclotron lines from the primary pole are shifted between the two epochs by an equivalent of $\Delta B \simeq 1$ MG. Since the cyclotron pattern from the second pole remains unchanged, the emission regions cannot be located at the foot points of the same closed field line(s). Whether the field is dipolar or not remains open. (4) Maximum blueshift of both cyclotron-line systems caused by the approach of the line-of-sight to field lines occurs at a phase shift of $\sim 160°$ with the primary pole leading. It is an open question how the material manages to reach the second pole which is nearly in antiphase to the mass-donating secondary star.

Acknowledgements. I would like to thank K. Beuermann for introducing me into the field of research on AM Herculis stars, providing observational data and especially for many fruitful discussions. I thank S. Jordan for calculating the synthetic spectrum shown in Fig. 11.

References

Allen, D.A., Cherepashchuk, A.M.: 1982, *Mon. Not. R. astr. Soc.* **201**, 521

Allen, R.G., Berriman, G., Smith, P.S., Schmidt, G.D.: 1989, *Astrophys. J.* **347**, 426

Barrett, P.E., Chanmugam, G.: 1985, *Astrophys. J.* **298**, 743

Beuermann, K.: 1988, *Adv. Space Res.* **8**(2), 283

Beuermann, K., Thomas, H.-C., Schwope, A.D.: 1988, *Astron. Astrophys.* **195**, L15

Beuermann, K., Schwope, A.D., Thomas, H.-C., Jordan, S.: 1989, in *Proceedings of the 11th North American Workshop on CVs and LMXBs*, Santa Fe, October 9-12, in press

Beuermann, K., Thomas, H.-C., Giommi, P., Tagliaferri,G.: 1987, *Astron. Astrophys.* **175**, L9

Bonnet-Bidaud, J.-M., Beuermann, K., Charles, P., Maraschi, L., Motch, C., Mouchet, M., Osborne, J., Tanzi, E., Treves,A.: 1985, ESA SP-236, 155

Brainerd, J.J., Lamb, D.Q.: 1985, in *Cataclysmic Variables and Low Mass X-ray Binaries*, eds. D.Q. Lamb and J. Patterson, Reidel, Dordrecht, p. 247

Chanmugam, G., Dulk, G.S.: 1981, *Astrophys. J.* **244**, 569

Coyne, G.V., Magalhães, A.M., Moffat, A.F.J., Schulte-Ladbeck, R.E., Tapia, S., Wickramasinghe, D.T.: 1988, *Polarized Radiation of Circumstellar Origin*, University of Arizona Press, Tucson

Cropper, M.: 1990, *Space Sci. Rev.*, in press

Cropper, M., Mason, K.O., Mukai, K.: 1990, *Mon. Not. R. astr. Soc.* **243**, 565

Cropper, M., Mason, K.O., Allington-Smith, J.R., Branduardi-Raymont, G., Charles, P.A., Mittaz, J.P.D., Mukai, K., Murdin, P.G., Smale, A.P.: 1989, *Mon. Not. R. astr. Soc.* **236**, 29P

Ferrario, L., Wickramasinghe, D.T., Tuohy, I.R., Bailey, J.: 1988, IAU Circ. No. 4523

Ferrario, L., Wickramasinghe, D.T., Bailey, J., Tuohy, I.R., Hough, J.H.: 1989, *Astrophys. J.* **337**, 832

Frank, J., King, A.R., Lasota, J.-P.: 1988, *Astron. Astrophys.* **193**, 113

Jensen, K., Nousek, J.A., Nugent, J.J.: 1982, *Astrophys. J.* **261**, 625

Kuijpers, J., Pringle, J.E.: 1982, *Astron. Astrophys.* **114**, L4

Liebert, J., Stockman, H.S.: 1979, *Astrophys. J.* **229**, 652

Liebert, J., Stockman, H.S., Angel, J.R.P., Woolf, N.J., Hege, K., Morgan, B.: 1978, *Astrophys. J.* **225**, 201

Mason, K.O., Middleditch, J., Cordova, F.A., Jensen, K.A., Reichert, G., Murdin, P.G., Clark, D., Bowyer, S.: 1983, *Astrophys. J.* **264**, 575

Meggitt, S.M.A., Wickramasinghe, D.T.: 1982, *Mon. Not. R. astr. Soc.* **198**, 71

Osborne, J., Giommi, P., Angelini, L., Tagliaferri, G., Stella, L.: 1988, *Astrophys. J. Lett.* **328**, L45

Schmidt, G.D., Stockman, H.S., Grandi, S.A.: 1983, *Astrophys. J.* **271**, 735

Schmidt, G.D., Stockman, H.S., Morgan, B.: 1981, *Astrophys. J. Lett.* **243**, L157

Schwope, A.D., Beuermann, K.: 1985, ESA SP-236, 172

Schwope, A.D., Beuermann, K.: 1990, *Astron. Astrophys.*, in press

Schwope, A.D., Beuermann, K., Thomas, H.-C.: 1990, *Astron. Astrophys.* **230**, 120

Tapia, S.: 1977, IAU Circ. No. 3045

Thompson, A.M., Brown, J.C., Kuijpers, J.: 1986, *Astron. Astrophys.* **159**, 202

Visvanathan, N., Wickramasinghe, D.T.: 1979, *Nature* **281**, 47

Wickramasinghe, D.T., Ferrario, L.: 1988, *Astrophys. J.* **334**, 412

Wickramasinghe, D.T., Martin, B.: 1985, *Mon. Not. R. astr. Soc.* **212**, 353

Wickramasinghe, D.T., Meggitt, S.M.A.: 1982, *Mon. Not. R. astr. Soc.* **198**, 975

Wickramasinghe, D.T., Ferrario, L., Bailey, J.: 1989, *Astrophys. J. Lett.* **342**, L37

Wickramasinghe, D.T., Tuohy, I.R., Visvanathan, N.: 1987, *Astrophys. J.* **318**, 326

Wickramasinghe, D.T., Visvanathan, N., Tuohy, I.R.: 1984, *Astrophys. J.* **286**, 328

Wu, K., Chanmugam, G.: 1988, *Astrophys. J.* **331**, 861

Wu, K., Chanmugam, G.: 1990, *Astrophys. J.*, in press

X-ray Diagnostics of Accretion Disks

G. Hasinger

Max-Planck-Institut für Extraterrestrische Physik,
D-8046 Garching, Fed. Rep. of Germany

1 Introduction

Within this article the very wide field of accretion phenomena observed in X-rays is restricted to bright low-mass X-ray binaries (LMXB), where recent observational evidence leads to a global picture that for the first time sheds light on the innermost regions of an accretion disk. LMXB are actually very similar to cataclysmic variables, one just has to replace the central white dwarf by a neutron star. Given a small enough neutron star magnetosphere the disk then reaches to much smaller radii and therefore much higher temperatures. Consequently, radiation pressure can become important in the inner part of accretion disks around neutron stars.

Table 1 from the recent review by van der Klis (1989b) shows the known LMXB sorted according to apparent X-ray brightness and marked as "Z-sources" and "Atoll-Sources" according to the classification scheme by Hasinger and van der Klis (1989). The class of LMXB we are concerned here – the Z-sources – belong to the brightest X-ray objects in the sky. They are characterized by Z-shaped X-ray colour-colour diagrams and comparatively narrow peaks of quasi-periodic oscillations (QPO) at frequencies between 5 and 50 Hz in their power spectra.

In the recent years it turned out that in the handful Z-sources nature has provided us with a magnificent laboratory to study the innermost regions of accretion disks – not at least at X-ray wavelengths.

Table 1. The brightest low-mass X-ray binaries (adapted from van der Klis, 1989b)

Source Name	I_x [μJy]	P_{orb} [hr]	Type A/Z	Phenomenology
Sco X-1 (1617-155)	12400	19.2	Z	QPO
GX 5-1 (1758-259)	1200	—	Z	QPO
GX 349+2 (1702-363)	780	—	Z	QPO
GX 17+2 (1813-140)	680	19.8?	Z	QPO, (bu)
GX 9+1 (1758-205)	650	—	A	—
GX 340+0 (1642-455)	490	—	Z	QPO
GX 3+1 (1744-265)	430	—	A	(QPO),(Bu)
Cyg X-2 (2142+380)	430	235.	Z	QPO,(bu)
GX 13+1 (1811-171)	340	—	A	—
GX 9+9 (1728-169)	290	4.2	A	Mo
4U 1820-30 (NGC 6624)	260	0.2	A	(QPO),(Bu),Mo
4U 1705-44	260	—	A	Bu
4U 1636-53	220	3.8	A	Bu
Ser X-1 (1837+049)	200	—	-	Bu
GCX-1 (1742-294)	170	—	A	Bu?
4U 1728-33	170	—	A	Bu
GX 339-4 (1659-487)	160	14.8?	-	QPO,BH?
4U 1735-44	160	4.6	A	Bu

2 X-ray Spectrophotometry

The technique we utilize here is rather simple and similar to the UBV spectroscopy in optical astronomy. The raw X-ray spectrum is divided into three rather broad energy bands - typically 1-3 keV, 3-6 keV and 6-15 keV. From the ratio between the middle and the soft X-ray band a so-called *soft colour* is derived, the ratio between the hard and the middle X-ray band gives the *hard colour*.

Figure 1 shows X-ray data from Cyg X-2 measured over four days with the Japanese satellite *Ginga* during a multifrequency campaign in 1988 (from Hasinger et al., 1990, hereafter HA90). The lowest panel shows the X-ray intensity as a function of time, the middle one gives the hard colour. The uppermost panel shows the times of simultaneous observations in other wavebands (see below). Both X-ray intensity and X-ray colour vary quite erratically on a variety of

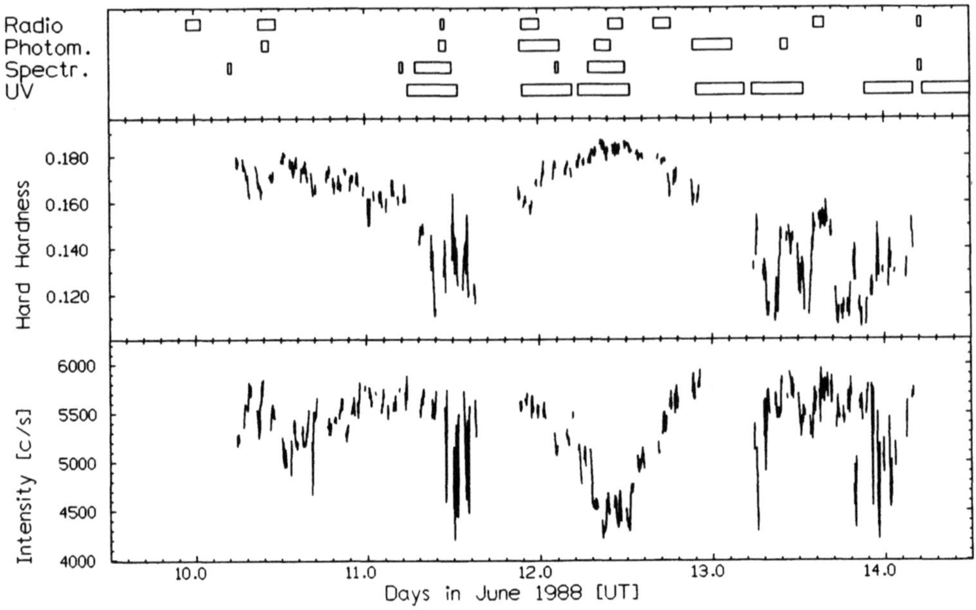

Fig. 1. X-ray lightcurve (lower panel) and hard colour (middle panel) of Cyg X-2 as a function of time for the 4-day Ginga observation in June 1988. The upper panel shows the time interval of observations in other wavebands. Taken from Hasinger et al., 1990

different timescales. However, the source has been observed in different modes of activity. Phases of slowly variing intensity and hardness are interrupted by periods of rapidly variing colour or intensity.

3 The "Z-Diagram"

The chaotic nature of the temporal variability changes into very well structured and characteristic patterns when one plots the same data into a colour-intensity diagram (similar to the Hertzsprung-Russel diagram of optical astronomy) or into a colour-colour diagram. These two diagrams are given in Figs. 1 and 2. Now three well separated "spectral branches" are visible which form an elongated "Z-shape" in the colour-colour diagram which gave these sources their name.

The three branches are called "horizontal branch" (HB; upper left) where the hard colour stays almost constant while the intensity varies, "normal branch" (NB; middle) where both soft and hard colour change in a correlated fashion while the X-ray intensity stays almost constant, and "flaring branch" (FB; bottom) which again is characterized by a correlated variation of soft and hard colour, however along a different track. (The dramatic intensity decrease along the flaring branch is peculiar to Cyg X-2, in most other Z-sources a similarly dramatic intensity increase has been observed, thus the term "flaring".)

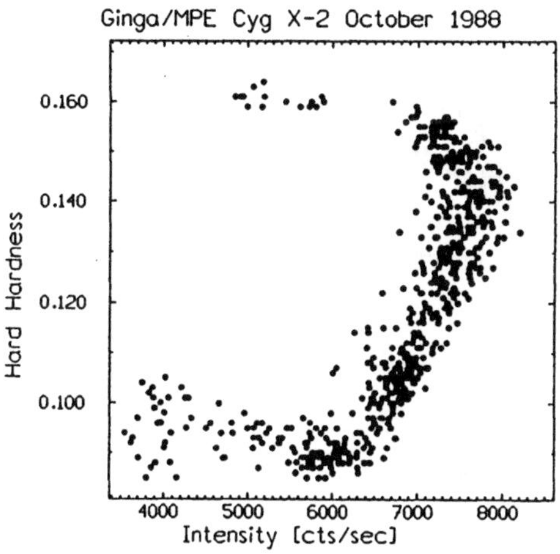

Fig. 2. Colour-intensity diagram from the data in Fig. 1 (HA90)

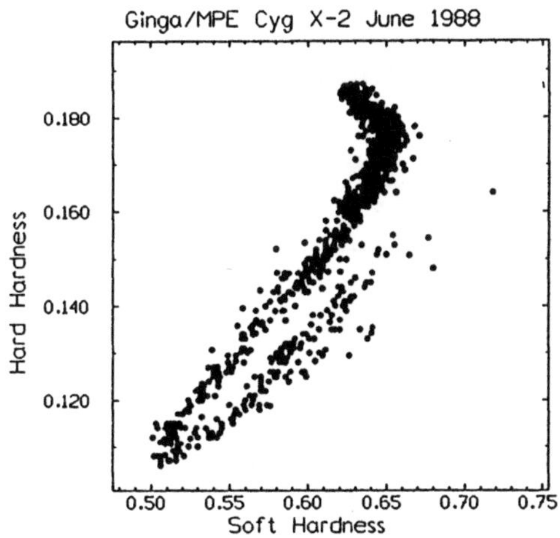

Fig. 3. Colour-colour diagram for the same data (HA90).

The same technique has been applied to a large body of data from the European X-ray satellite EXOSAT (Hasinger and van der Klis, 1989, hereafter HK89) and a total of 6 bright LMXB have been identified as Z-sources. Figure 4 shows their colour-colour diagrams which display a similar complete Z-pattern or at least fragments of it.

The existence of these characteristic, well confined tracks of spectral and intensity evolution calls for a simple explanation. One interpretation assumes that the observed changes are governed by the variation of a single parameter, which most likely is the mass-accretion rate $\dot{M}$ fed through the disk onto the central engine (Priedhorsky et al., 1986). If this is the case, then there are only two choices how $\dot{M}$ can vary along the "Z": either in the "positive" sense (from the upper left to the lower right) or in the opposite direction. In order to solve this question simultaneous observations in other wavebands had to be performed.

4 Observations in Other Wavebands

Simultaneously with the X-ray measurements two multifrequency campaigns with observations of Cyg X-2 in the UV, optical, and radio range have been performed (see figure 1). The IUE observations (Vrtilek et al., 1990), the optical spectroscopy from La Palma (van Paradijs et al., 1990) and the VLA radio measurements (Hjellming et al., 1990) will be published together with the Ginga X-ray data (HA90) soon. Luckily enough the source collaborated: it displayed all three spectral states during the simultaneous observations and

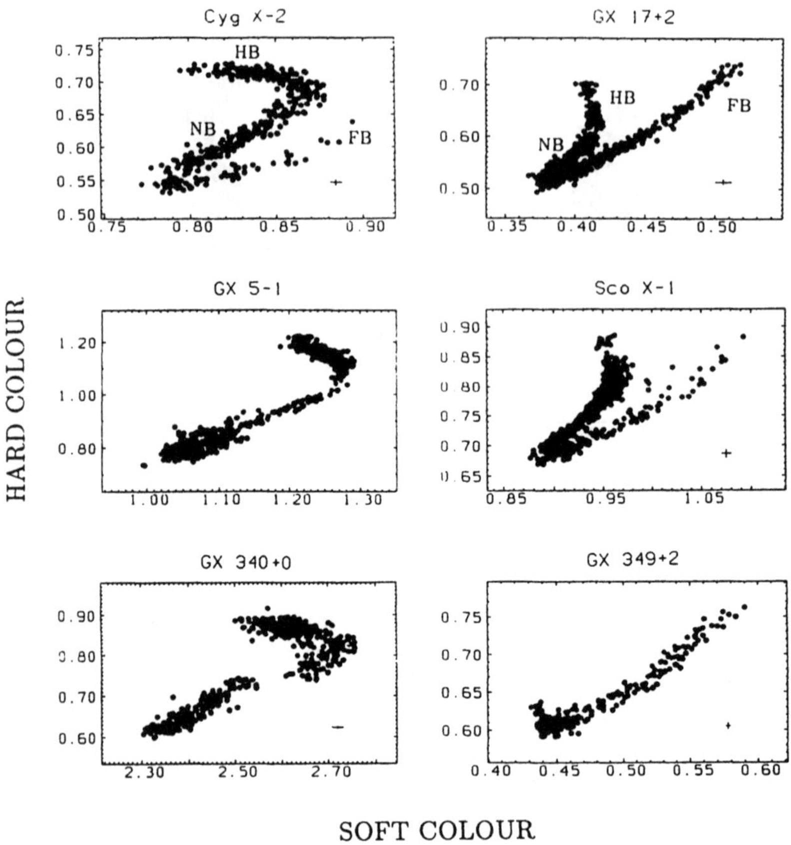

Fig. 4. Colour-colour diagram of EXOSAT data from 6 Z-sources (HK89)

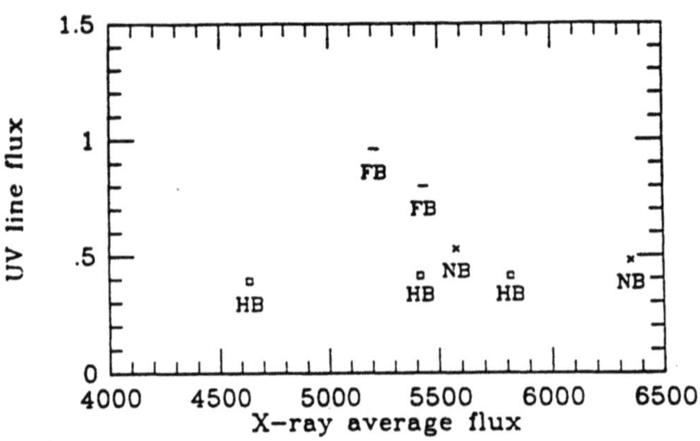

Fig. 5. UV continuum flux as (IUE) versus X-ray flux (Ginga). The X-ray spectral state is indicated for each measurement. The UV flux increases continuously from horizontal to normal branch. Taken from Vrtilek et al., 1990

in all wavebands clear correlations with the source position in the Z-diagram were found.

The UV continuum and line flux increase steadily by a factor of about 2 as the source moves from the horizontal branch to the flaring branch (Vrtilek et al., 1990, see fig. 5). The same is true for the equivalent width of the He II λ4686 and λ4640 Bowen emission lines as well as the Hβ and Hγ absorption lines (van Paradijs et al., 1990). A similar behaviour was found already in the seventies for Sco X-1, where the optical flux increases by about one magnitude from the top to the bottom of the normal branch normal and further up the flaring branch (Canizares et al., 1975). Since the optical and UV flux of the X-ray irradiated disk is dominated by reprocessed radiation, one can conclude that the X-ray (or EUV) illumination of the disk increases from HB to FB – and thus the mass accretion rate has to increase in the same way (Vrtilek et al., 1990, HA90).

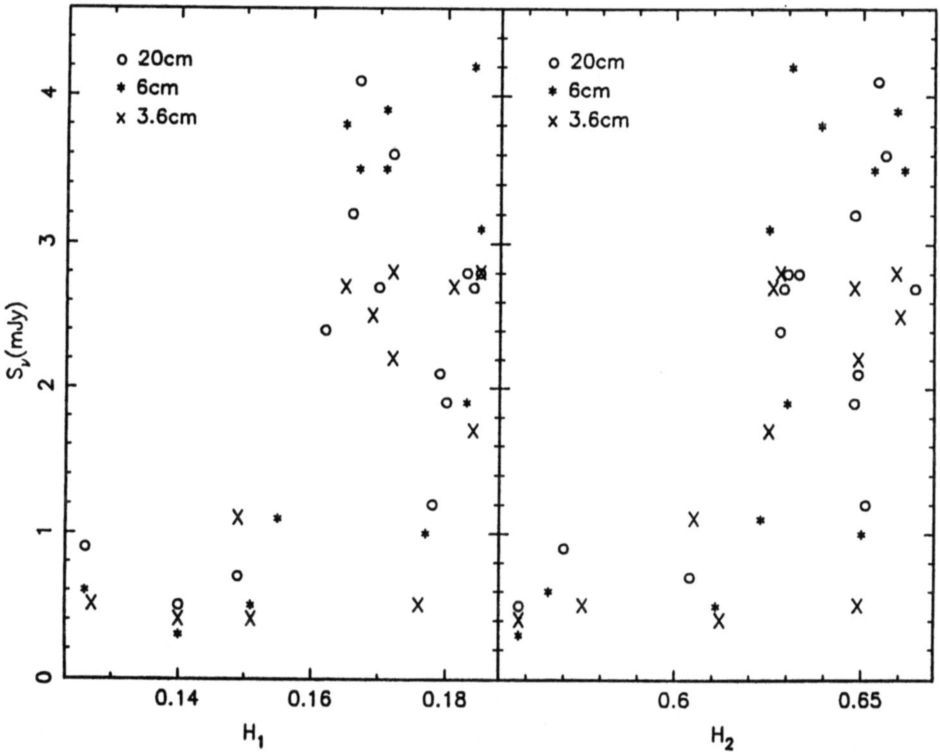

Fig. 6. Radio flux density (VLA) versus X-ray hardness (Ginga). Hardness is the same 'hard hardness' as in fig. 1-3, thus the source is on the horizontal branch for a hardness larger than 0.16. The radio flux density is high and strongly variable on the horizontal branch and low and quiescent on the normal and flaring branch.

The radio flux behaves in a completely different manner: as Figure 6 shows, the radio flux is loud and highly variable on the horizontal branch, and quiet and quiescent on the normal and flaring branch (Hjellming et al., 1990). This is similar as observed before in the radio/X-ray correlation in GX 17+2 (Penninx et al., 1988).

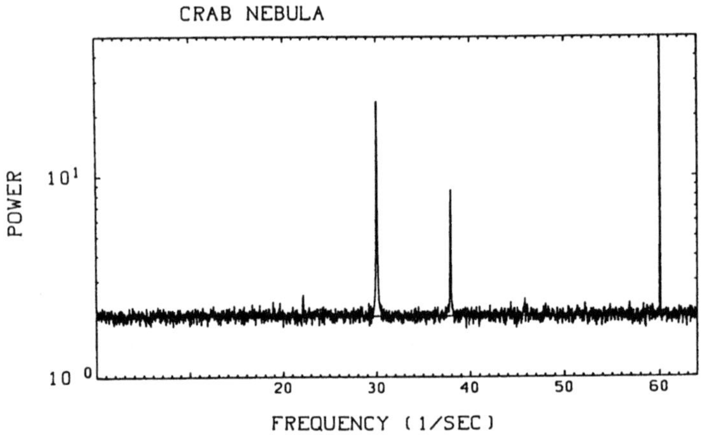

Fig. 6a. Power spectrum of the Crab Nebula and its pulsar as observed with EXOSAT. The fundamental pulsar frequency (30 Hz) and its first harmonic (60 Hz) are visible as sharp spikes. A third peak is seen at a frequency of 38Hz, it is the aliased signal of the second harmonic (90 Hz).

5 Quasi-Periodic Oscillations and Noise

The last pieces to solve the puzzle of the Z-sources comes from a systematic analysis of their X-ray flickering in the sub-second range. As diagnostic tool the segmented Fourier analysis technique (see van der Klis, 1989a) is utilized: the observation is divided into small continuous stretches of data (typically 8-32 seconds), the power spectra of which are averaged to enhance the statistical significance of the signal. If this method is applied to the 33ms- pulsar in the Crab Nebula a series of sharp spikes at the fundamental pulsar frequency and its harmonics is observed (see figure 6a)on top of a constant (white noise) signal which is due to the counting statistics of the data.

In a search for possible millisecond pulsars in low-mass X-ray binaries van der Klis et al. (1985) analysed data of the bright galactic bulge source GX 5-1. However, instead of several sharp spikes they found a single broad peak in the power spectra, signalling that not a strictly periodic, but a *quasi periodic oscillation* (QPO) is present in the data. Additionally a component called 'red noise' or 'low-frequency noise' (LFN), i.e. excess power which rises continuously towards the lowest frequencies, is visible in the power spectra (see figure 7).

Moreover, as the X-ray source intensity varies, the centroid frequency of the QPO peak is not stable but varies as a function of time, roughly in the frequency range 20-45 Hz – the two quantities go hand in hand.

It turns out, that the presence of QPO peaks in the power spectra is a class property of all Z-sources (see HK89). However, the rapid, intensity-dependent QPO are mainly visible in the horizontal branch. As soon as the source 'turns the corner' into the normal branch the power spectra change dramatically. Figure 8 shows a comparison of two power spectra of Cyg X-2, taken in the horizontal and normal branch, respectively. Compared to figures 6a and 7 the

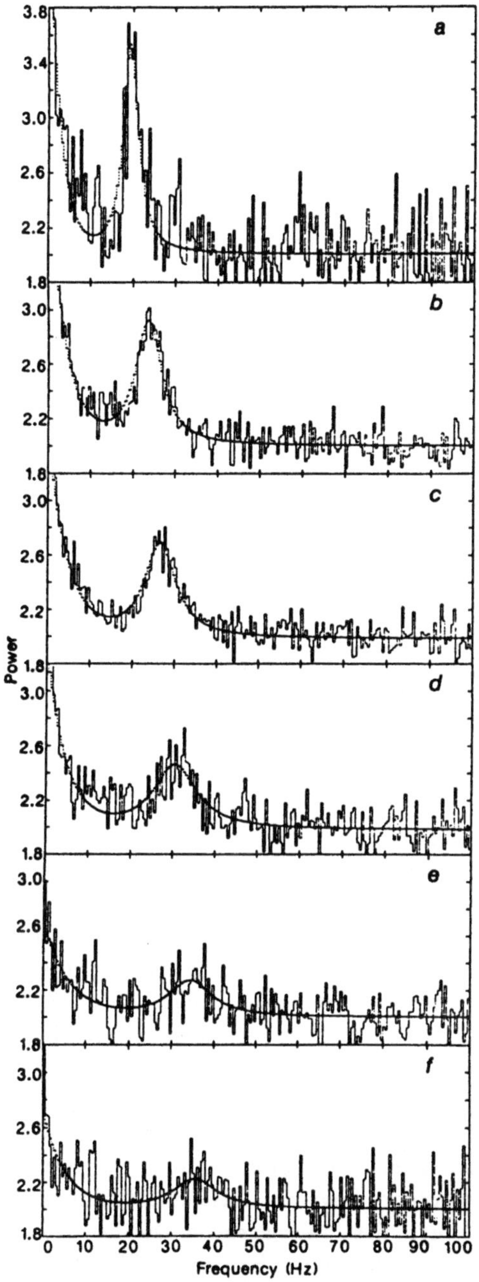

Fig. 7. Six different power spectra of the bright galactic bulge source GX 5-1 in the same representation as figure 6. The X- ray source intensity rises from approximately 2200 cts/s to 3800 cts/sec in the EXOSAT detector from top to bottom (from van der Klis et al., 1985).

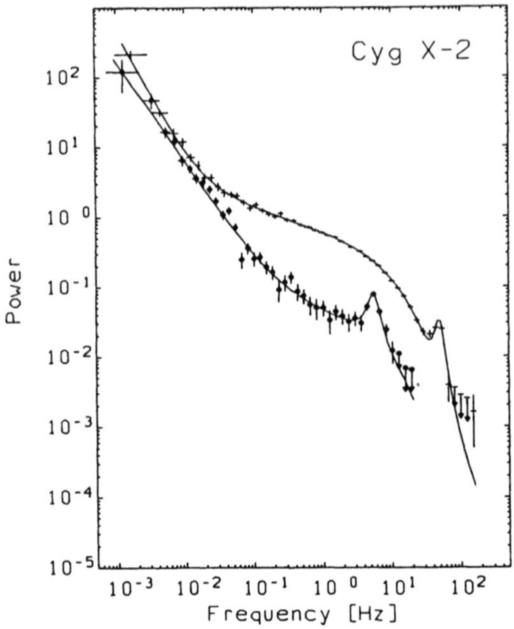

Fig. 8. Two power spectra of the Z-source Cyg X-2 taken in the horizontal (upper) and normal (lower) branch. The power spectra are displayed on a logarithmic scale, with the white noise level due to counting statistics removed. Note that the power spectra are not shifted with respect to each other (from Hasinger, 1987a).

power spectra here are displayed on a logarithmic scale and the white noise due to counting statistics has been subtracted. One sees, that the strong and characteristically shaped low-frequency noise component, dominating in the horizontal branch at frequencies 0.05 - 50 Hz, is almost completely gone in the normal branch. The same is true for the fast (50 Hz) QPO peak. Roughly in the middle of the normal branch another, slower QPO peak occurs, its centroid remains rather stable at frequencies 5-7 Hz throughout the lower part of the normal branch.

These dramatic changes of the different power spectral components are further highlighted in figure 9 where all Cyg X-2 power spectra from the EXOSAT and Ginga X-ray observations have been systematically analysed to trace the source behaviour throughout the whole Z-diagram. Particularly interesting seems the variation of the LFN, which first increases on the horizontal branch and the abruptly dies out close to the HB/NB-corner. This transition has become one of the most important ingredients in the interpretation discussed below. The changes in the HBO (horizontal-branch oscillation) and NBO (normal-branch oscillation) power are anticorrelated, while the former continuously decreases from HB to NB, the latter increases abruptly along the normal branch and is most prominent in the middle of the NB. There are actually NB power spectra which show both HBO and NBO peaks simultaneously (HA90).

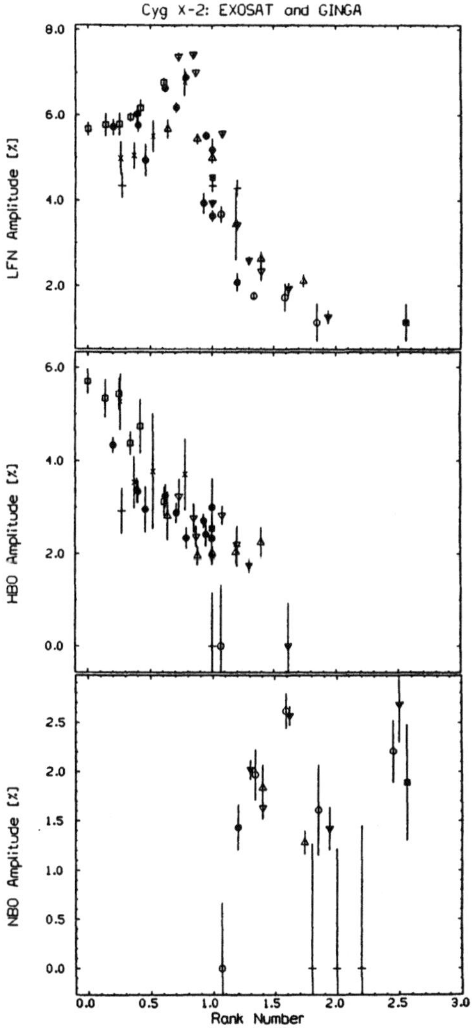

Fig. 9. Variation of the main power spectral components of Cyg X-2. The figures display the rms. variability in the red noise (upper), in the horizontal-branch QPO (middle) and in the normal- branch QPO (lower panel) as a function of position along the Z- diagram. The abscissa (called 'rank number') gives an arbitrary coordinate that increases monotonically from the left of the HB (0.0) over the HB/NB-transition (1.0) and the NB/FB-transition (2.0) to the upper right of the flaring branch (3.0), following the change in mass-accretion rate. Data from the EXOSAT and Ginga X-ray satellites have been merged (Hasinger, et al. 1990, in prep.

The same analysis applied to other Z-sources reveals almost carbon-copy results. Figure 10 shows a selection of power spectra from the same set of sources the colour-colour diagrams of which are given in figure 4 (HK89). Where available, power spectra have been accumulated separately for each branch. The sense and magnitude of variation in the different power spectral components is very similar to Cyg X-2.

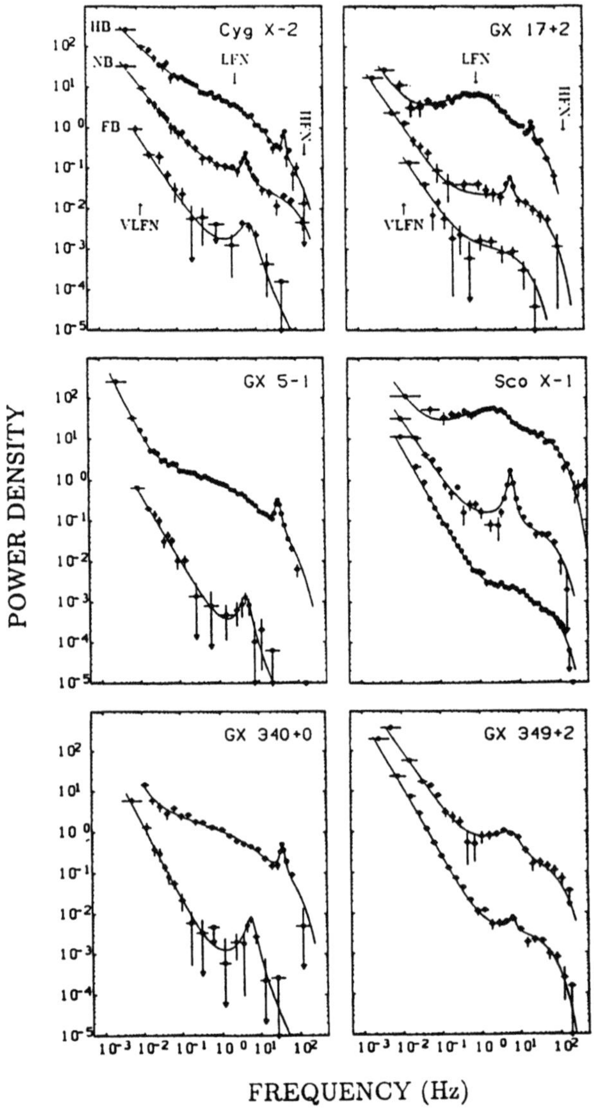

Fig. 10. Power spectra of six Z-sources in the different spectral branches, corresponding to figure 4. Power spectra are given on logarithmic scales with white noise background subtracted. For clarity power spectra from different branches (from top to bottom: HB, NB, FB) are shifted by two decades with respect to each other (from HK89).

6 The Big Picture: a Jumping Accretion Disk ?

Figure 11 summarises the complex variation of some of the observable quantities as a function of mass accretion rate along the Z-diagram: $\dot{M}$ increases monotonically from HB to FB by about a factor of 2 (panel 1). The UV and blue optical light and the emission lines follow hand in hand (panel 2). This is also true for the X-ray flux on the horizontal branch, however, at the HB/NB transition it starts to deviate from this one-to-one correspondence (panel 3). The latter three quantities all show marked transitions close to the corner between horizontal and flaring branch: horizontal-branch QPO and normal branch QPO change importance in the upper normal branch, where they can coexist simultaneously (panel 4). The LFN, which is strong on the HB, dies out at the same transition (panel 5; see also Figure 9), and finally, strong nonthermal radio flares occur only on the horizontal branch.

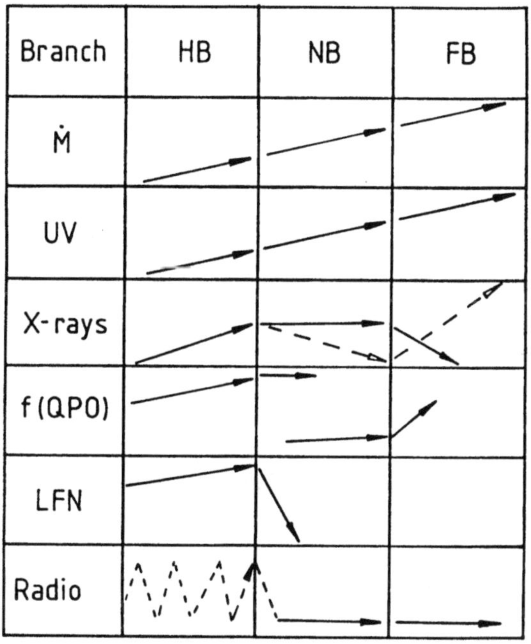

Fig. 11. Schematic display of the variation of important observables and the mass-accretion rate as a function of source position along the Z-profile (see text) (from HA90).

We have now collected the ingredients neccessary to to interpret the complex spectral, power spectral and multifrequency behaviour of Z-sources. Most important is the knowledge that the observed variations are most likely driven by a simple change of mass accretion rate. The second most important feature - at least in my judgement - is the dramatic phase transition when the source moves from the horizontal to the normal branch. X-ray spectrum, and intensity, noise and QPO as well as the radio flux all change dramatically and almost simultaneously there. These phenomena suggest a scenario (Hasinger 1988, HA90), which is sketched in Figure 12, and which is similar to the 'unified QPO model' by Lamb (1989, 1990) in some important features.

At low accretion rates (HB) the source starts out in a configuration with a geometrically thin inner disk which is cutting into a small magnetosphere as shown in figure 12a. This is the geometry required for the 'accretion modulated beat frequency model' for the horizontal-branch QPO (Alpar and Shaham, 1985; Lamb et al., 1985). Turbulence at the magnetospheric boundary

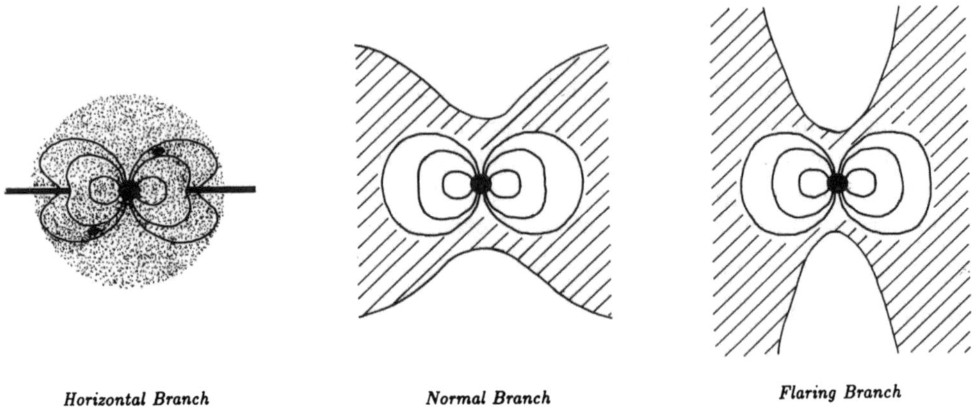

Fig. 12. Schematic display of the accretion geometry in the adopted scenario for the (a) horizontal, (b) normal, and (c) flaring branch (from HA90).

produces clumps, causing an instationary accretion which is responsible for the LFN. Interaction between the clumps in Keplerian motion and between the rotating neutron star magnetic field cause quasi-periodic intensity enhancements leading to HBO. The turbulent magnetospheric boundary may also be the site of electron acceleration which we observe through the HB radio flares.

When a critical value of mass accretion rate is exceeded, but still well below the neutron star Eddington limit, the inner disk is suddenly flaring up to form a thick torus (see Fig. 12b), engulfing the whole magnetosphere (Hasinger, 1988; HK 1989). All the dramatic changes observed at this transition can be explained by this postulated instability. The additional cool disk material in the line of sight causes the softening of the observed spectrum by Compton degradation (Hasinger, 1987b; Lamb, 1990). The interface between the disk torus and the magnetosphere is probably much less turbulent than in the horizontal branch and matter can accrete smoothly onto the neutron star. Therefore LFN and HB QPO are reduced in strength (although the existence of weak HB QPO in the upper normal branch indicates that some inhomogeneities are still present). The flux of energetic electrons producing the HB radio flares may be quenched in two ways: first, the reduced turbulence at the magnetospheric boundary may be less efficient in accelerating particles, and second, the thick inner disk torus will trap the electrons.

Down the normal branch the mass accretion rate is increasing even further and approaches the Eddington limit. At this moment the effective gravity is reduced by the strong radiation pressure, enough to slow down the dynamic timescales so that sound waves (Hasinger 1987a, Alpar et al., 1990) or other resonances in the flow (Lamb 1989, 1990) will become important and possibly form the 6 Hz normal-branch QPO.

Finally, on the flaring branch the mass accretion rate has increased above the Eddington limit and matter will be ejected, probably through the narrow funnel formed by the thick inner disk torus (see Fig. 12c). The funnel might also effectively beam the X-ray radiation, so that the line-of sight inclination becomes important for a distant observer. This may explain the different intensity behaviour on the flaring branch between e.g. Sco X-1 and Cyg X-2 (Hasinger et al., 1989)

References

Alpar, M.A., Shaham, J.: 1985, *Nature* **316**, 239
Alpar, M.A., Hasinger, G., Shaham, J., Yancopoulos, S., 1990 (in prep.)
Canizares C.R. et al.: 1975 *Ap. J.* **197**, 457
Hasinger, G.: 1987 a, *Astron. Astrophys.* **186**, 153.
Hasinger, G.: 1987 b, *IAU Symp.* **125**, 333.
Hasinger, G.: 1988 Proc. Symp. *"Physics of Neutron Stars and Black Holes"*, Tokyo: Universal Academy Press, p. 97
Hasinger, G., Priedhorsky, W.C., Middleditch, J.: 1989, *Ap. J.* **337**, 843.
Hasinger, G., van der Klis, M.: 1989, *Astron. Astrophys.* **225**, 79
Hasinger, G., van der Klis, M., Ebisawa, K., Dotani, T., Mitsuda, K.: 1990 *Astron. Astrophys.* (in press)
Hjellming, R.M., Han, X.H., Cordova, F.A., Hasinger, G.: 1990 *Astron. Astrophys.* (in press)
Lamb, F.K., Shibazaki, N., Alpar, M.A., Shaham, J.: 1985 *Nature* **317**, 681
Lamb, F.K.: 1989, Proc. 23rd ESLAB Symposium, ESA SP-296, Part 1., p. 215
Lamb, F.K.: 1990, *Ap. J.* (in press)
Penninx, W., Lewin, W.H.G., Zijlstra, A.A., Mitsuda, K., van Paradijs, J., van der Klis, M.: 1988 *Nature* **336**, 146
Priedhorsky, W., Hasinger, G., Lewin, W.H.G., Middleditch, J., Parmar, A., Stella, L., White, N.: 1986, *Ap. J. Lett.* **306**, L91.
Shakura, N.I., Sunyaev, R.A.: 1973 *Astron. Astrophys.* **24**, 337
Van der Klis, M., Jansen, F., van Paradijs, J., Lewin, W.H.G., van den Heuvel, E.P.J., Trümper, J.E., and Sztajno, M.: 1985 *Nature* **316**, 225.
Van der Klis, M.: 1989a, Proc. NATO ASI *"Timing Neutron Stars"*, Çeşme, Turkey. Dordrecht: Reidel p.27
Van der Klis, M.: 1989b, *Ann. Rev. Astron. Astrophys.* **27**, 517
Van Paradijs, J., Allington-Smith, J., Callanan, P., Charles, P.A., Hassal, B.J.M., Machin, G., Mason, K.O., Naylor, T., Smale, A.P.: 1990 *Astron. Astrophys.* (in press)
Vrtilek, S.D., Raymond, J.C., Garcia, M.R., Verbunt, F., Hasinger, G., Kürster, M.: 1990 *Astron. Astrophys.* (in press)

Accretion Phenomena at Neutron Stars

A. Rebetzky, H. Herold, U. Kraus, H.-P. Nollert, and H. Ruder

Lehr- und Forschungsbereich Theoretische Astrophysik,
Universität Tübingen, Auf der Morgenstelle 10,
D-7400 Tübingen, Fed. Rep. of Germany

1. Introduction

The basic picture of an X-ray pulsar is that of a strongly magnetized neutron star which accretes matter from its companion. The reason for mass transfer can be a strong wind of the companion or a Roche-lobe overflow from the companion through the inner Lagrangian point. In the latter case the matter forms an accretion disk around the neutron star, in which it spirals inwards and heats up due to viscous interaction. If the neutron star is strongly magnetized, the disk does not reach to the surface of the neutron star, because at the Alfvén radius the matter is dominated by the magnetic forces. The physics of this region, where the matter couples to the magnetic field lines, is not yet fully understood. The ionized matter which is on the field lines is accelerated by the strong gravitational field and falls along the magnetic field lines towards the magnetic poles of the neutron star, where it reaches velocities of about $c/2$ ($c =$ is the velocity of light). The fully ionized matter is stopped in the atmosphere of the neutron star and its kinetic energy is converted into radiation energy. This region, where the radiation is produced, is called the "hot spot".

Generally the axis of the magnetic field is not parallel to the rotational axis. Therefore, the observer detects a signal pulsating with the rotation frequency of the neutron star. The pulse shape depends strongly on the energy, and the spectrum depends strongly on the pulse phase, at which it is observed. This indicates that the formation of the pulse-phase dependent X-ray spectra is a very complicated problem. The reasons are the extreme parameters of those systems. The neutron star has a radius of only about 10 km, approximately twice its Schwarzschild radius. The free-fall velocity of the hot plasma at the magnetic pole is about half the speed of light, and the magnetic field strength at the pole is about $10^{8\cdots 9}$ T. The accretion rate is in the range of $10^{-9}\ M_\odot/\mathrm{yr}$. The area where the matter crashes onto the surface is only about 1 km^2.

On the theoretical side, there are many problems which are still unsolved. First there is the problem of how the matter at the inner edge of the accretion disk gets onto the magnetic field lines. A second unsolved problem is the production of X-rays in the hot spot. It is assumed that the matter is stopped by Coulomb interaction, and in the case of high accretion rates, additionally by radiation pressure. The atmosphere is heated up to some 10^7 K. The bremsstrahlung process produces X-ray photons which are then comptonized and finally have to leave the hot spot. A third complicated problem is the radiative transfer through the infalling plasma. There are a lot of research projects in this area, but the observed spectra and pulse shapes are not yet understood in detail.

For luminosities beyond the Eddington luminosity the dynamics of the infalling matter is strongly influenced by the radiation pressure. In this case one has to consider both the radiative transfer and the dynamics of the infalling plasma.

As a consequence of the strong gravitational field (the ratio between the radius of the star and the Schwarzschild radius is about 2) the relativistic effects on the propagation of light cannot be neglected. Photons follow null geodesics in the given metric. In curved space-time these lines differ strongly from straight lines in the Minkowski space-time. A photon starting near the surface can be deflected by more than 45°. It is obvious that the gravitational effects have a strong influence on all radiative quantities, i.e. on the light curves and pulse phase spectra.

In this article we try to point out the most important physical effects which influence the formation of spectra and light curves of X-ray pulsars. First we will present some observational data which show the complexity of the problem. In a further chapter we will discuss the dynamics of the hot spot plasma. Finally the radiative transfer in the accretion column is discussed, considering the effects of the fast moving plasma and the strong gravitational field. In the last chapter we give a short outlook on future work.

2. Some observational data

Observations of X-ray pulsars were made by a lot of satellites. The best information consists of energy-dependent pulse shapes and pulse-phase dependent spectra. Some typical light curves of the most prominent X-ray pulsar Her X-1 are shown in Fig. 1 ([1],[2],[3],[4],[5]).

The light curves at low energies (1–10 keV) are more complicated than those at higher energies. In Fig. 2 some phase averaged spectra are shown. At about 38 keV a cyclotron absorption feature is visible. The spectrum in the low energy region is relatively flat, at about 20 keV we can notice a cut-off and a strong decrease of the radiative flux. The total flux of Her X-1 is about 4×10^{30} W.

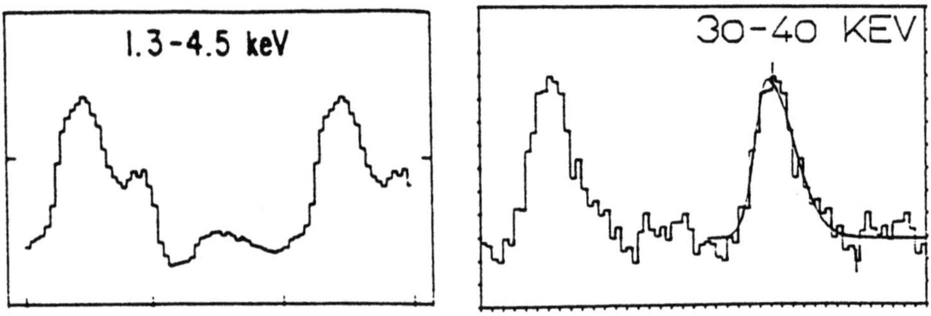

Fig. 1. Light curves of Her X-1 at two different energies ([2],[3])

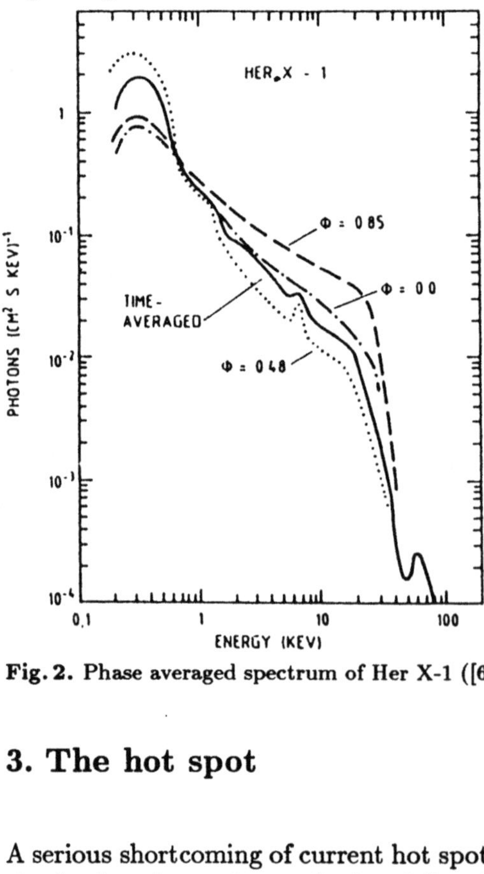

Fig. 2. Phase averaged spectrum of Her X-1 ([6])

3. The hot spot

A serious shortcoming of current hot spot models is the inconsistent coupling of the deceleration region to the free-fall regime of the accretion column. The high-energy protons are stopped in the atmosphere of the neutron star within only a few cm. The main physical process for the deceleration is the collision between the infalling protons and the electrons of the atmosphere. The occurrence of a radiative shock seems to be unavoidable. Some authors have presented models with a shock front building a mound over the hot spot ([7],[8],[9],[10],[11]). In the lower region the plasma is falling with a smaller velocity, it sinks down slowly to the neutron star's surface [12] and is decelerated to thermal velocities. In the shock region photons gain energy by the effect of Fermi acceleration. Thermal bremsstrahlung becomes a very important source of photons in the sinking region, because the efficiency of bremsstrahlung is proportional to ρ^2, where ρ is the density of the matter. Concerning the opacity, bremsstrahlung also dominates the scattering processes, which are only proportional to ρ. The still unsolved problem in modelling the hot spot of neutron stars is the calculation of the correct frequency-dependent radiation hydrodynamics, which must include the effects of the shock region.

As an example, in Fig. 3 [9] the energy of the infalling protons is plotted versus the column density of the atmosphere. The plasma is stopped after passing about 20 g/cm^2. A spectrum resulting from thermal bremsstrahlung is presented in Fig. 4. The strong peak is the resonant cyclotron emission. At low energies there is the well-known power law spectrum for thermal bremsstrahlung.

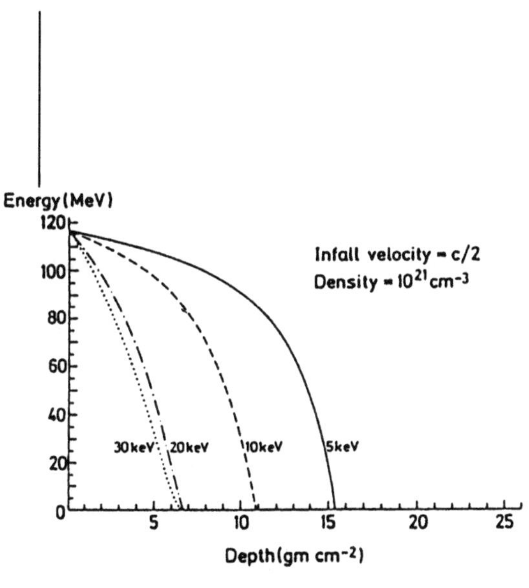

Fig. 3. The stopping length of a plasma with different temperatures for the atmosphere. The infalling plasma has a velocity of $c/2$ ([11])

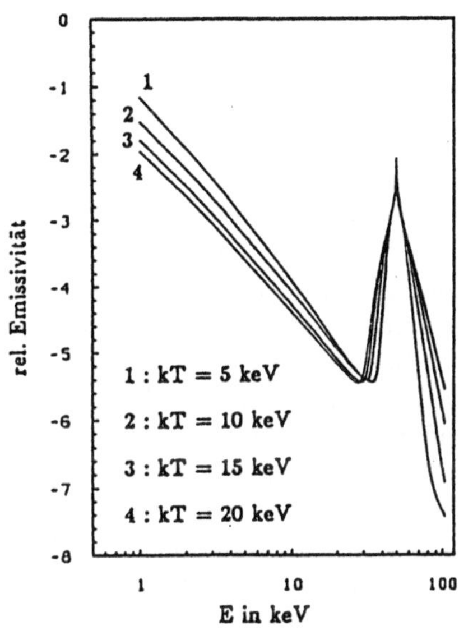

Fig. 4. Thermal bremsstrahlung spectrum in a strong magnetic field

4. The free-fall region

The photons which are produced in the hot spot have to be transferred through the infalling plasma. As already mentioned, the plasma has a velocity of about $0.5\,c$. One method to treat the radiative transfer in the free-fall region is the diffusion approximation ([13],[14],[15],[16],[17],[18],[19],[20],[21],[22],[23]), which has been applied by several authors. The basic requirement for the validity of the diffusion approximation is the matter being optically thick in order

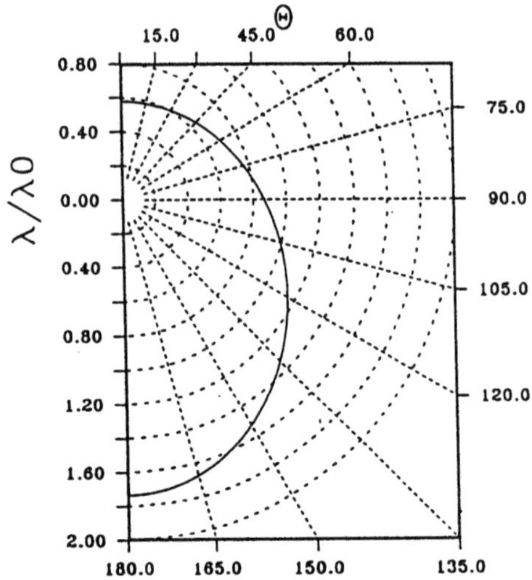

Fig. 5. Ratio between the mean free path in the observer's frame and the mean free path in the local rest frame of matter ([19])

to ensure enough scattering events for each photon such that the photon intensity is almost isotropic. The advantage of the diffusion approximation is that the application is straight forward and numerical computations are fast. A great disadvantage is the fact that the diffusion approximation is not valid in the boundary layers because there photons can escape outwards. Just in these regions the results of the approximation are wrong. One way to avoid these difficulties is the use of flux limiters in the diffusion approximation.

In principle the diffusion approximation is valid for arbitrary velocities, because the equation of radiative transfer is Lorentz invariant. In the case of an accretion column, however, the validity of the approximation is restricted. A photon scatters only a few times in the free-fall region before it is scattered back into the hot spot, because the scattering coefficients are very anisotropic due to the high velocity. The consequence is that the intensity does not become quasi-isotropic, violating an important requirement for its validity. In the following we will discuss in detail the fact that there are only few scattering events in the free-fall region.

First we have to consider the important effects of the high velocity on the radiative quantities [24]. The mean free path l_0 of a photon in the local frame of the moving plasma must be transformed into the observer's frame (quantity l). If the velocity of the plasma in units of the velocity of light is β, the transformation reads

$$l = \frac{1}{\gamma(1+\beta\cos\vartheta)} l_0 \tag{1}$$

where ϑ is the angle between the velocity of the plasma and the direction of the photon propagation in the observer's frame. In Fig. 5 we present the ratio between the mean free path in the observer's frame and the mean free path in the rest frame of the matter.

But the scattering cross section also changes drastically in the observer's frame, as presented in Fig. 6. In this case, the cross section in the direction of the plasma flow is six times larger than the cross section against the flow. Therefore the probability of a photon to be scattered into the direction of the plasma flow is much larger than the probability that it is scattered against the plasma flow. As a consequence the photons emitted by the hot spot cannot be scattered very often in this free-fall region before they have returned into the hot spot. This effect is generally known as the advection effect.

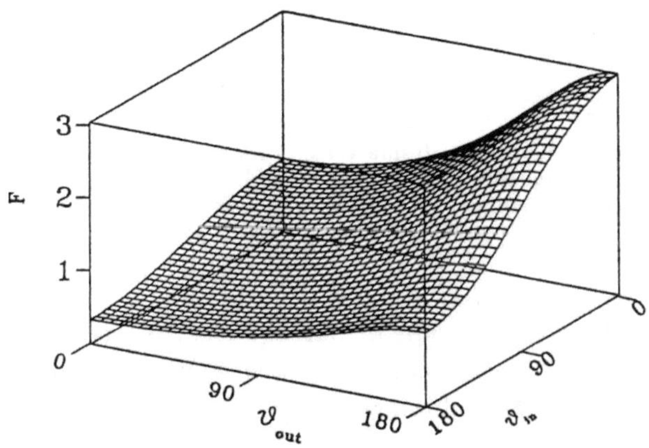

Fig. 6. Ratio between the cross section for photon-electron scattering in the observer's frame and the cross section in the local rest frame of matter ([19])

In the following we briefly describe another method to solve the equation of radiative transfer, the so-called iterative scattering approximation (ISA). The equation of radiative transfer for the distribution function f can be written in integral form as

$$\begin{aligned} f(x_1,p_1) =& f(x_0,p_0)e^{-\tau(0,\lambda_1)} \\ &+ \int_0^{\lambda_1} d\lambda E_0(x,p)e^{-\tau(\lambda,\lambda_1)} \int_{4\pi} d^2k_0'\sigma_0(x,p_0' \to p_0)f(x,p') \end{aligned} \tag{2}$$

where the specific intensity can be calculated by

$$I_\nu = \frac{h^4}{c^2}\nu^3 f \tag{3}$$

and the quantities p and E are given by

$$p = \frac{dx}{d\lambda}\ , \quad E = c^2\frac{dt}{d\lambda} = h\nu \tag{4}$$

The suffix "$_0$" means that a quantity is written in the rest frame of the matter, otherwise it has to be considered in the observer's frame. λ is the affine path parameter and τ is the optical depth between two points characterized by λ_1 and λ_2. In the free-fall regime scattering is the dominant process, therefore

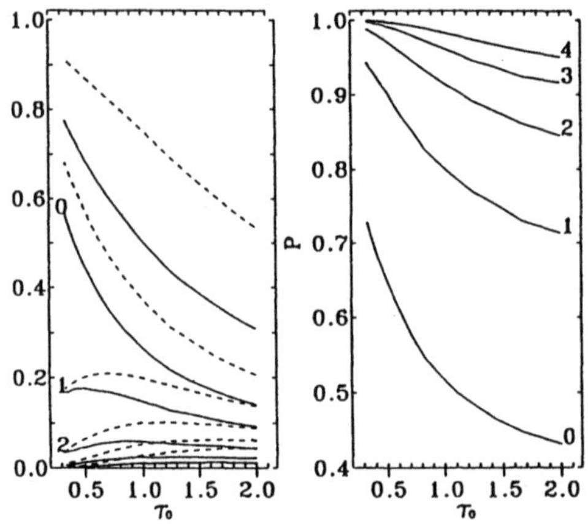

Fig. 7. Convergence of the ISA method. Due to the plasma velocity more photons are scattered towards the neutron star. As a consequence the escape probability decreases. Taking into account the high velocity field the convergence is rapid: After four scattering events up to 95 % have escaped ([19])

we do not consider any absorption or emission in our equation of radiative transfer. A solution of this integral equation can be written as an absolutely and uniformly convergent series, each term obtained by an iterative procedure:

$$f = f^{(0)} + f^{(1)} + f^{(2)} + \ldots \tag{5}$$

The zeroth term is given by the first term on the right hand side of (2). All higher orders are obtained by the following recursive formula :

$$f^{(n)}(x_1, p_1) = \tag{6}$$
$$\int_0^{\lambda_1} d\lambda E_0(x,p) e^{-\tau(\lambda,\lambda_1)} \int_{4\pi} d^2 k_0' \sigma_0(x, p_0' \to p_0) f^{(n-1)}(x, p')$$

In this equation $f^{(n)}$ is the distribution function of photons scattered exactly n times, which have been scattered into the direction given by p_1 in the last scattering process. An observer detects unscattered photons from the hot spot, plus those which are scattered once into the line of sight, twice, and so on.

A first simple model with a homogeneous magnetic field, cylinder geometry, and a constant velocity $c/2$ of the plasma shows several interesting features: In comparison to the case without a velocity field, the total escape probability including the velocity is smaller. The reason for that is the already mentioned fact that the photons are scattered preferably in the direction of the plasma flow. Therefore more photons are scattered back into the hot spot. The convergence of the ISA method is very rapid. At $\tau = 2$ only 4 iterations are necessary to get more than 95 % of the total flux out of the accretion column. This is also an effect of the high plasma velocity (see Fig. 7).

We have seen that the high velocity makes it easier to apply ISA which is a rather time-consuming method.

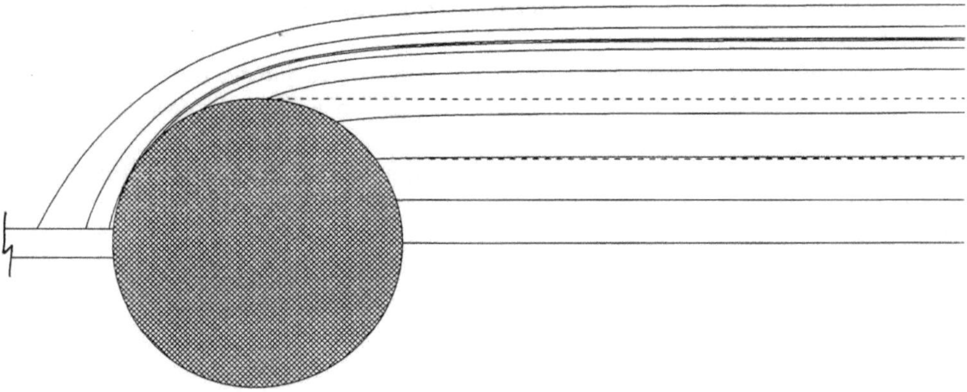

Fig. 8. Relativistic light deflection near neutron stars. The ratio between the star's radius and its Schwarzschild radius is 2

As already mentioned, a neutron star has a very strong gravitational field, so that we have to consider relativistic effects on the propagation of light ([24],[25],[26]). The ratio between radius and Schwarzschild radius is about 2, which leads to strong effects on the light deflection, as shown in Fig. 8.

One effect is that we can see part of the back side of the neutron star, a second one, that the neutron star seems to be larger than in the case of the light rays being straight lines (Fig. 9).

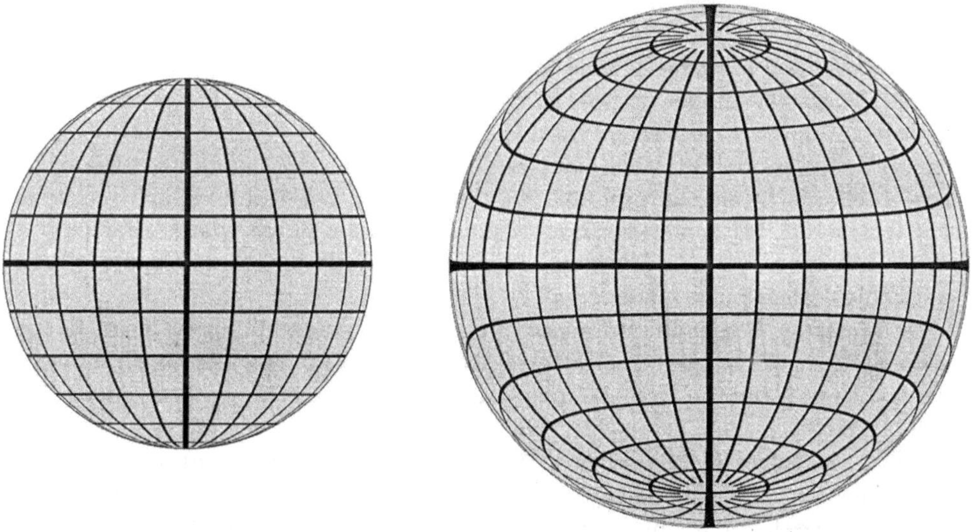

Fig. 9. A neutron star with spherical coordinate grid in Minkowski space-time and in Schwarzschild space-time

A photon travels along null geodesics in vacuum. To calculate the radiative transfer in curved space-time, we can simply apply the ISA method described above. The integration path then is no longer a straight line, but a curved one in the given space-time. We neglect rotational effects on the metric and use a Schwarzschild metric.

It is obvious that this is a computer intensive procedure, but it is the only possible way to calculate the radiative transfer in a curved space-time in an accurate way.

As a first result we present the total escape probability of unscattered photons (Fig. 10).

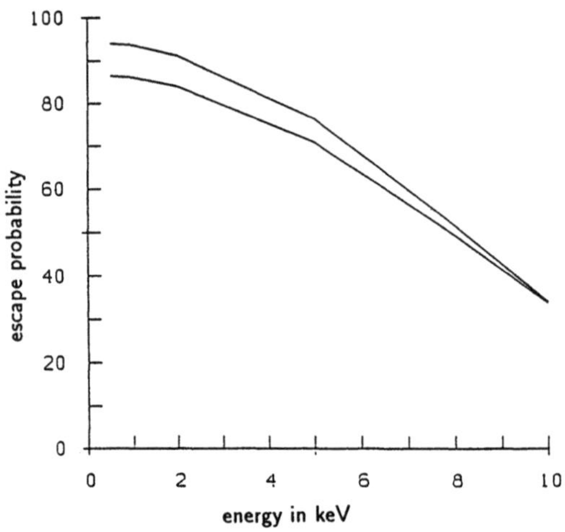

Fig. 10. Total escape probability of unscattered photons for both polarization modes of photons in the strongly magnetized plasma

We assumed a magnetic field strength corresponding to a cyclotron energy of 50 keV at the magnetic pole of the neutron star. In the low-energy range almost all photons can escape from the system unscattered, whereas at higher energies this probability decreases rapidly. For a photon energy of 10 keV in the observer's frame the probability is only 50 %. The resulting light-curve is plotted in Fig. 11. It shows a double pulse just at a phase where we expect a single hot spot pulse, because we are looking exactly on one hot spot. Where does this double pulse come from?

The answer lies in the influence of the strong magnetic dipole field on the cross sections of the photon-electron scattering. In Fig. 12 we present the mean free paths of a photon in a strongly magnetized plasma as angle-averaged diffusion lengths.

The mean free path far below the cyclotron energy is about 1 km. In the resonance the mean free path decreases to only a few cm ! For comparison the mean free path for an unmagnetized plasma is also plotted (Thomson scattering). The magnetic field strength in the vicinity of a strongly magnetized neutron star varies strongly with the radial distance, because there is a dipole-like field. In Fig. 13 we plot the dipole field in the Schwarzschild metric and lines of constant cyclotron energy. In a distance of only 10 km the cyclotron energy decreases to very low energies, assuming a polar field strength corresponding to 50 keV. A photon with an energy of about 1 keV in the observer's

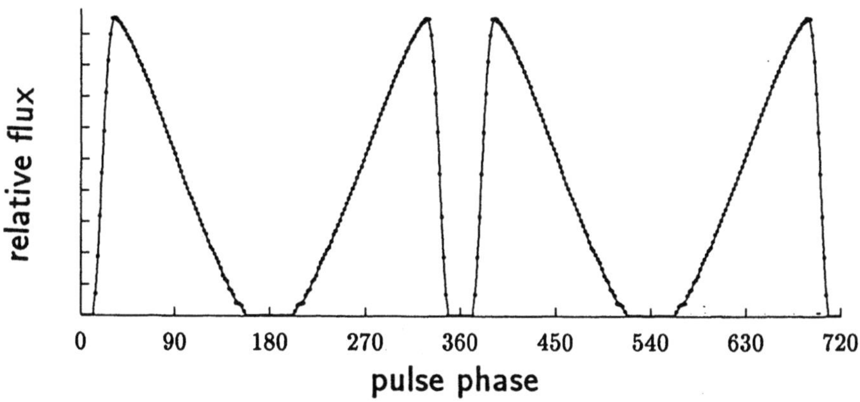

Fig. 11. Light curve of unscattered photons, including relativistic light bending and the velocity of the infalling plasma. The rotational axis is perpendicular to the axis of the magnetic field

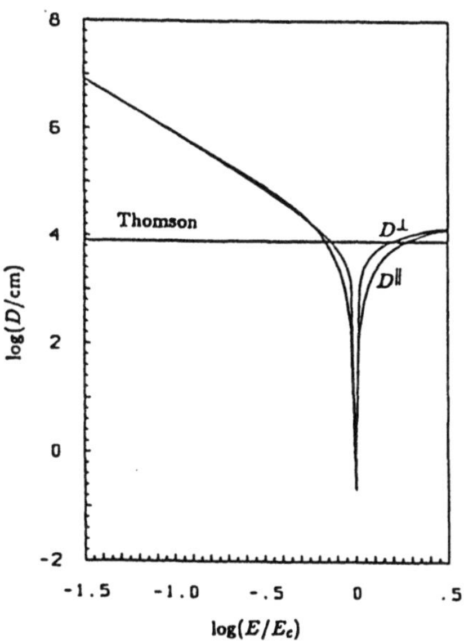

Fig. 12. Mean free path of photons (angle-averaged) as a function of the photon energy in the local rest frame

frame starts at the hot spot with a local enery of about 1.5 keV due to the gravitational redshift. At a height of two kilometers the cyclotron resonance energy is about 25 keV, and the optical thickness is large enough to prevent the photon from escaping. The combined effect of the decreasing cyclotron energy and the Doppler shift due to the high velocity increases the optical thickness in comparison to the case with a homogeneous magnetic field without velocity, which has been calculated by many authors to produce spectra and light curves of X-ray pulsars.

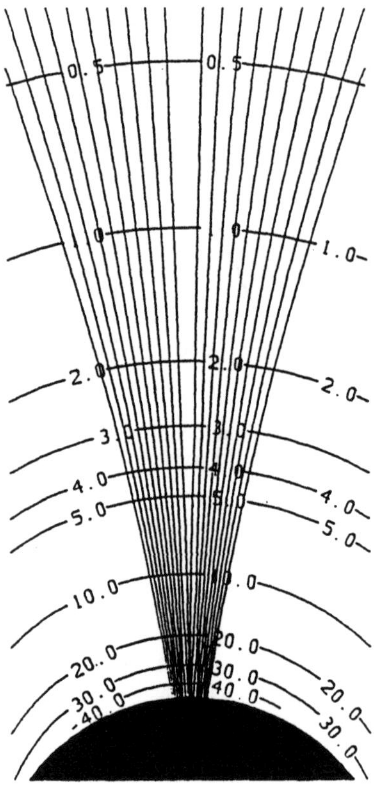

Fig. 13. Magnetic dipole field in Schwarzschild space-time. The cyclotron energy decreases faster than $1/r^3$

5. Summary and outlook

In this article we have reported on some main features of modelling the emission regions of X-ray pulsars. The plasma is falling with high velocities onto the polar caps of the strongly magnetized neutron star. In the atmosphere it is decelerated, and the radiation is produced. The photons have to traverse through the down-falling plasma. The Doppler shift and effects of aberration have to be considered as well as the fact that photons travel not on straight lines, because of the strong gravitational field.

One major focus of our future work lies in the correct modelling of the hot spot region, where we try to perform frequency-dependent radiation hydrodynamics neglecting effects of the strong gravitational field. A second major area of research is the correct radiative transfer in the free-fall region of the accretion column, including effects of the high velocity and the strong gravitation.

Coupling these two regions we hope to understand the observed features of X-ray pulsars better than it is possible until now.

References

1 Trümper,J., *et al.*: *Ap.J. Lett.*, **219**, L105 (1978)
2 Voges, W.: MPE Report 191, (1985)
3 White,T.R., Lightman,A.P., Zdziarski,A.A.: *Ap.J.*, **331**, 939 (1988)
4 Lightman,A.P., Lamb,D.Q., Rybicki,G.B.: *Ap.J.*, **248**, 738 (1981)
5 Rappaport,S., Joss,P.C.: in R.Giacconi (ed.) *X-ray Astronomy with the Einstein Satellite*, Reidel, Dordrecht, Holland (1981)
6 McCray, R., Shull, J., Boynton, P., Deeter, J., Holt, S, White, N.: *Ap.J.*, **262**, 301 (1982)
7 Davidson,K.: *Nature*, **246**, 1 (1973)
8 Basko,M.M., Sunyaev,R.A.: *M.N.R.A.S.*, **175**, 395 (1976)
9 Kirk, J.G., Galloway, K.: *Plasma Physics*, **24**, 359 (1982)
10 Kirk,J.G.: *A.&A.*, **158**, 305 (1986)
11 Kirk,J.G., Mészáros,P.: *Ap.J.*, **241**, 1153 (1980)
12 Wang,Y.-M., Frank,J.: *A.&A.*, **93**, 255 (1981)
13 Arons,J., Klein,R.I., Lea,S.M.: *Ap.J.*, **312**, 606 (1987)
14 Blandford,R.D., Payne,D.G.: *M.N.R.A.S.*, **194**, 1041 (1981)
15 Burnard,D.J., Klein, R.I., Arons,J.: *preprint*, UCLR-100802, Univ. of Cal. (1989)
16 Castor,J.I., Abbott,D.C., Klein,R.: *Ap.J.*, **195**, 157 (1975)
17 Nagel,W.: *Ap.J.*, **251**, 278 (1981)
18 Nagel,W.: *Ap.J.*, **251**, 288 (1981)
19 Maile,T.: *Dissertation, Universität Tübingen* (1989)
20 Maile,T., Herold, H., Rebetzky, A., Ruder, H., Ventura, J., Wolf, K.: *A.&A.*, **223**, 251 (1989)
21 Rebetzky,A., Herold,H., Maile,T., Ruder,H., Wolf,K.: *A.&A.*, **205**, 215 (1988)
22 Rebetzky,A., Bock, U., Herold,H., Kraus, U., Maile,T., Nollert, H.-P., Ruder,H., Wolf,K.: *A.&A.*, **225**, 137 (1988)
23 Mészáros,P., Nagel,W.: *Ap.J.*, **298**, 147 (1985)
24 Riffert,H.: *Ap.J.*, **327**, 760 (1988)
25 Kraus, U., Rebetzky, A., Herold, H., Nollert, H.-P., Maile, T., Ruder, H.: *ESA Proc. SP-296, 23nd ESLAB Symposium*, Bologna (1989)
26 Nollert, H.-P., Kraus, U., Herold, H., Maile, T., Rebetzky, A., Ruder, H.: *ESA Proc. SP-296, 23nd ESLAB Symposium*, Bologna (1989)

A Torus-Dynamo for Magnetic Fields in Galaxies and Accretion Disks

D. Schmitt

Universitäts-Sternwarte, Geismarlandstr. 11,
D-3400 Göttingen, Fed. Rep. of Germany and
High Altitude Observatory, National Center for Atmospheric Research [1],
P.O. Box 3000, Boulder, CO 80307, USA

Summary. Accretion disks around compact objects as well as the gaseous components in galaxies often have the form of a torus. To study the structure and behaviour of magnetic fields generated in such rings, a kinematic $\alpha\omega$–dynamo is investigated which is working inside a torus embedded into vacuum. Modes symmetric and antisymmetric with respect to the equatorial plane are obtained separately. By far the most easily excited dynamo mode is stationary and of quadrupolar symmetry. This basic mode appears for negative dynamo numbers and could be responsiable for the axisymmetric field observed in spiral galaxies like M 31 or IC 342. Quadrupolar modes for positive dynamo numbers as well as all excited dipolar modes are of an oscillatory nature. They need considerably higher excitation than the stationary solution and are not expected under galactic conditions, but they might play a role in accretions disks around compact objects. Here we consider only axisymmetric $\alpha\omega$–dynamos. The observation of bisymmetric fields in some spiral galaxies calls for the treatment of nonaxisymmetric dynamos. The strong poloidal fields observed in the central parts of our own Galaxy and of mildly active edge–on galaxies might be generated by α^2–dynamos operating in partially ionized gaseous rings that are found in those regions. Research on both these lines of inquiry is presently under way.

1. Introduction: Magnetic Fields in Spiral Galaxies

Knowledge of the magnetic field structure of spiral galaxies can be best obtained through observations of the total and linearly polarized radio continuum synchrotron emission [1]. These observations indicate that the large–scale magnetic field of spiral galaxies is located within the galactic plane and orientated mostly along the spiral arms in the disk [2]; but sometimes it is strongest outside the optical spiral arms [3]. The field is dominant in the regions of the interstellar gas which may occupy the whole galactic plane, but often has the form of a ring around the center. The field structure can be axisymmetric like

[1] The National Center for Atmospheric Research is sponsored by the National Science Foundation.

in M 31 [4] or IC 342 [5] or bisymmetric like in M 51 [6] or M 81 [7]. The field is mainly toroidal with pitch angles of the order of 10°. The field strength varies from 3 to 15 μG. Illustrations of the field structure can be found in [8]. Tables of magnetic field data of various galaxies are compiled in [9]. There is no clear observational indication about the symmetry of the field with respect to the equatorial plane: only IC 342 has been examined and it appears to be symmetric (M. Krause, private communication).

In contrast to the predominately toroidal field in spiral galaxies the fields in the center of our Galaxy [10] and in the nuclei of mildly active edge–on galaxies like M 82 [11] are perpendicular to the galactic plane, i.e. predominately poloidal. It has been suggested that these fields are connected with molecular rings rotating around the nucleus [12] and with jet–like outflow phenomena.

Magnetic fields in galaxies can be generated by plasma motions in the electrically conducting (i.e. the ionized components of the) interstellar gas [13]. The gaseous components of galaxies often have the form of a torus. This can be very clearly seen in HI pictures of the Andromeda galaxy M 31 [14]. The ionized gas components of the interstellar matter of this galaxy are in the same region and look very similar [15]. The magnetic field of M 31 is concentrated in that ring at about 10 kpc distance from the center [4]. Therefore, we are motivated to consider the structure and behaviour of a magnetic field generated by the action of a dynamo in a torus.

In this contribution we present a study of an axisymmetric $\alpha\omega$–dynamo in a torus with a circular meridional cross–section [16]. Since the toroidal field component of an $\alpha\omega$–dynamo is much larger than the poloidal field, this study is relevant for galaxies like M 31 or IC 342 where the large scale magnetic field is mainly elongated along the spiral arms. The strong poloidal fields observed in the central parts of our Galaxy and of mildly active edge–on galaxies like M 82 might be generated by an α^2–dynamo in partially ionized gaseous rings found in these regions. Bisymmetric fields like those of M 51 or M 81 call for the study of nonaxisymmetric $\alpha^2\omega$–dynamos in thick rings.

2. An Axisymmetric $\alpha\omega$–Dynamo Model in Torus–Geometry

Consider a torus of rotating, turbulent, electrically conducting matter embedded into vacuum. The radius of the circular meridional cross section may vary, and accordingly, we have studied quite slender and rather thick tori as illustrated in Fig. 2. Appropriate to the geometry is the introduction of toroidal coordinates (η, φ, θ) — see Fig. 1 and [17]. The coordinate η varies from zero to infinity and surfaces of constant η are the tori. Small values of η_0 correspond to thick, larger η_0 values to thinner tori. Asymptotically, $\eta \to 0$ encircles a whole half–sphere, and $\eta \to \infty$ degenerates to a circle of radius a, the equatorial axis of the torus. The coordinate θ varies from $-\pi$ to π and constant θ surfaces are spherical bowls through the equatorial axis. The ignorable third coordinate φ is the azimuth. The surfaces of the torus is most conventienty expressed by $\eta = \eta_0 = \text{const.}, 0 \leq \varphi \leq 2\pi, -\pi \leq \theta \leq \pi$. The radius of the meridional cross section is given by $\rho = a/\sinh\eta_0$.

Inside the torus the $\alpha\omega$–dynamo equations are solved for the axisymmetric field. Splitting $\boldsymbol{B}$ into its toroidal and poloidal components, described by the scalars B and A according to $\boldsymbol{B} = B\boldsymbol{e}_\varphi + \text{rot}\, A\boldsymbol{e}_\varphi$, these equations are

$$\frac{\partial B}{\partial t} + D\,(\text{rot rot}\, B\boldsymbol{e}_\varphi)\cdot\boldsymbol{e}_\varphi = \text{rot}\,(\omega r\boldsymbol{e}_\varphi \times \text{rot}\, A\boldsymbol{e}_\varphi)\cdot\boldsymbol{e}_\varphi \tag{1}$$

$$\frac{\partial A}{\partial t} + D\,(\text{rot rot}\, A\boldsymbol{e}_\varphi)\cdot\boldsymbol{e}_\varphi = \alpha B, \tag{2}$$

where D stands for the diffusivity, which is assumed constant. Both the α–effect, generating poloidal from toroidal field, and the differential rotation, generating toroidal from poloidal field, are prescribed, meaning we are dealing with a kinematic dynamo. We used $\omega(r) = \omega_0 + \omega_0' r$, where the radial derivative of ω is assumed constant here for convenience, and estimated by the Keplerian rotation on the horizontal axis of the torus. For the α–effect, an induction effect of rotating turbulent matter, and estimated to first order in the approximation of mean–field electrodynamics [13], we used $\alpha = \alpha_0 \sin\theta$, and alternatively $\alpha = \alpha_0 z/\rho$, simple forms satisfying antisymmetry with respect to the equatorial plane. With these choices the system of linear equations splits into two independent parts, describing fields antisymmetric and symmetric with respect to the equator. The first are called dipolar, the latter quadrupolar solutions, though they also contain higher multipoles. Introducing dimensionless variables the induction effects enter into the equations by the dimensionless dynamo number P, the product of two Reynolds numbers measuring the ratio of induction to diffusion:

$$P = R_\alpha R_\omega = \frac{\alpha_0\rho}{D}\frac{a\omega_0'\rho^2}{D}. \tag{3}$$

Outside the torus we assume vacuum. The toroidal part of the magnetic field vanishes at the boundary and the poloidal field must fit continuously and smoothly to the outside vacuum field, satisfying

$$\text{rot rot}\, A\boldsymbol{e}_\varphi = 0. \tag{4}$$

This equation can be readily solved to give (see [17])

$$A_n = (\cosh\eta - \cos\theta)^{1/2} P^1_{n-\frac{1}{2}}(\cosh\eta) \begin{cases} \cos n\theta & \text{for the dipole} \\ \sin n\theta & \text{for the quadrupole} \end{cases}. \tag{5}$$

For the solution of the dynamo equations inside the torus the magnetic field is expanded according to

$$A = e^{\Omega t} \sum_{n=0,l=1}^{\infty} a_{nl}(\cosh\eta - \cos\theta)^{1/2} e^{-\eta/2} J_n\left(r_{nl}e^{\eta_0-\eta}\right) \begin{Bmatrix} \cos n\theta \\ \sin n\theta \end{Bmatrix}, \tag{6}$$

$$B = e^{\Omega t} \sum_{n=0,l=1}^{\infty} b_{nl}(\cosh\eta - \cos\theta)^{1/2} e^{-\eta/2} J_n\left(s_{nl}e^{\eta_0-\eta}\right) \begin{Bmatrix} \sin n\theta \\ \cos n\theta \end{Bmatrix}. \tag{7}$$

These expansion functions form a complete and orthogonal set of functions which — and this is most important — already satisfy the boundary conditions when r_{nl} is the l–th zero of

$$r_{nl}J_{n-1}(r_{nl}) - (n - \frac{1}{2} + \frac{\mathrm{d}}{\mathrm{d}\eta} \ln P^1_{n-\frac{1}{2}}(\cosh\eta)|_{\eta_0})J_n(r_{nl}) = 0 \tag{8}$$

and s_{nl} is the l–th zero of

$$J_n(s_{nl}) = 0. \tag{9}$$

The motivation of these functions is that they represent the spatial part of the decay modes for a slender torus $\eta_0 \gtrsim 3$ [17]. In the case of a thick torus these functions are not longer the decay modes, but nonetheless an appropriate complete set satisfying the boundary conditions. The zeros in meridional direction are counted by n, whereas $l-1$ is the number of zeros in radial direction. Examples of the different decay modes can be found in [17].

Putting these expansions into the dynamo equations and specifying α and ω as above there follows a system of homogeneous linear equations for the two–dimensional set of expansion coefficients. The vanishing secular determinant determines the eigenvalues Ω where the factor $\exp \Omega t$ describes the time dependence: If $\mathrm{Re}\,\Omega > 0$ the magnetic field is maintained against its Ohmic losses, i.e. the dynamo is working; if $\mathrm{Im}\,\Omega \neq 0$ the magnetic field is oscillating in time. Typically values of n and l up to 10 to 15 are used in (6) and (7), in order to achieve eigenvalues correct to within 5 percent.

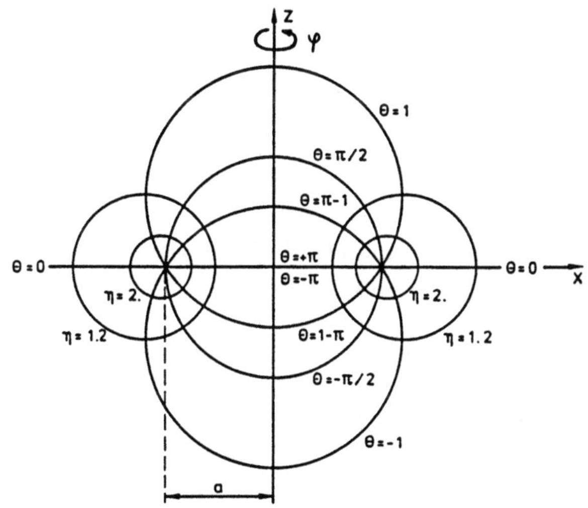

Fig. 1. Toroidal coordinates (η, φ, θ): $x = a \sinh\eta \cos\varphi/c$, $y = a\sinh\eta \sin\varphi/c$, $z = a\sin\theta/c$, $c = \cosh\eta - \cos\theta$.

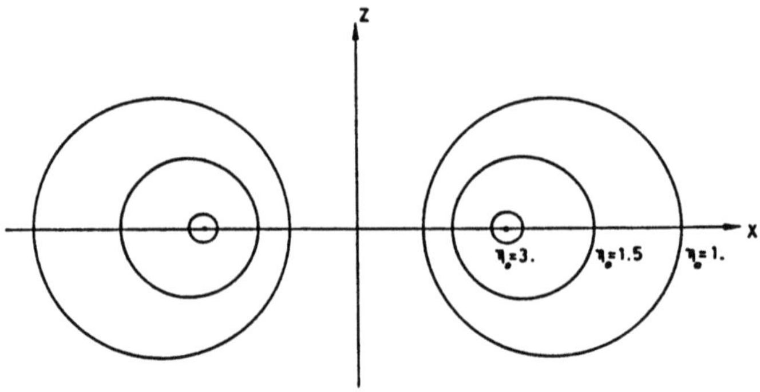

Fig. 2. Meridional cross-sections of three tori with different diameter for $\eta_0 = 1, 1.5, 3$.

3. Results: Excitation Conditions and Field Structures

A dynamo is excited, if for the value of P corresponding to the actual physical situation there exists at least one eigenvalue Ω with $\operatorname{Re}\Omega > 0$. Since realistic values of P are difficult to obtain, it is varied and the resulting paths of Ω are plotted in the complex plane – see Figs. 3, 4, describing the time behaviour of the system. In general four cases must be distinguished: the dipolar– and quadrupolar–type dynamos, each with $P > 0$ and $P < 0$, according to the different signs of ω_0' and α.

Each Ω–path starts for $P = 0$ at Ω belonging to one of the different decay–modes, i.e. somewhere along the negative Ω_R–axis. If $|P|$ increases there are two possibilities: Either Ω remains single and real or two Ω's are moving towards each other along the Ω_R–axis, eventually join each other and start a path through the complex plane as complex conjugate pairs. As long as $\operatorname{Re}\Omega < 0$ the (non–oscillating or oscillating) magnetic fields are damped in time. Dynamo action begins, if there exists a critical value P_c, for which $\operatorname{Re}\Omega \geq 0$. Among all the cases where critical dynamo numbers P_c exist at all, those with the lowest values are most interesting, because they belong to the most easily excited dynamos.

Although the details vary for tori of different thickness, the qualitative picture is independent and we discuss the case $\eta_0 = 1.5$ in more detail. Further the spatial distribution of α does not seem to be of much importance for the results. The discussion will therefore concentrate on the case $\alpha \sim \sin\theta$.

By far the most easily excited dynamo mode is of quadrupolar symmetry and has $P_c = -55$. It is of the non–oscillating type and remains excited until $P = -337$, whereupon its $\operatorname{Re}\Omega$ becomes again negative. After further increase of $|P|$ it joins another mode and the Ω–path moves off the real axis. It reaches (apart from a short interlude, where it is once again excited) $\operatorname{Re}\Omega > 0$ only for values of P as large as $P = -1131$ and remains an oscillating dynamo thereafter – see Fig. 3. Other modes of quadrupolar symmetry with $P < 0$ need even much larger dynamo numbers to become excited. The quadrupolar solutions for $P > 0$ are of the oscillating type only; excitation requires at least $P = 475$.

The dynamos of dipolar symmetry require values for $|P|$ of the same high order of magnitude as above to become excited. This is surprising, because the decay modes with the two lowest decay rates are moving towards the right along the Ω_R–axis, as P assumes positive values. But in both cases Ω is attracted to negative real values, and the dynamo is not excited. The reason seems that for increasing dynamo numbers the field is concentrated near the boundaries of the torus where it is obviously more and more unable to trigger the necessary induction effects. Paths starting from higher decay modes move off the real axis, eventually reaching the positive half–plane on straight lines for the large values of P already mentioned – see Fig. 4. Similar high values of P are required for excitation of dipolar dynamos for $P < 0$, which are then always of the oscillating type.

Similar results are found for dynamos in thin disks [18] or oblate spheroids [19].

The dynamos investigated here are $\alpha\omega$–dynamos. This means that differential rotation is assumed to be much more efficient in producing magnetic flux than the α–effect. Hence the toroidal field is of much higher strength than the poloidal field and the main part of the flux is contained within the torus. This should be kept in mind in looking at the field configurations in Figs. 5, 6.

In Fig. 5 we show an example of a non–oscillating dynamo, the quadrupolar mode for $P < 0$, excited — as already pointed out — under the most favorable conditions ($P_c = -55$). Hence it is a field of this kind which should be expected in the cases of M 31 and the like. As mentioned above there is no corresponding stationary dipolar mode.

The fields of a typical oscillating dynamo are displayed in a time sequence over half a period in Fig. 6. The field of this dipole for $P_c = 903$ evolves in the vertical direction. This reflects the property of dynamo waves to propagate preferentially along the surfaces $\omega =$ constant, which are cylinders in the case investigated. The same is true for the oscillating quadrupolar modes. For $P > 0$ this evolution proceeds away from and for $P < 0$ it proceeds towards the equatorial plane.

In Table 1 the critical dynamo numbers P_c, where $\Omega_R = 0$ and the dynamo evolves from a damped to an excited solution, are listed for the three different torus sizes shown in Fig. 2, for the first (second) oscillating dynamo mode and the stationary quadrupole. The first excited mode is the stationary quadrupole. All others are oscillatory and require considerably higher excitation by a factor of 5 to 20. The second oscillatory dynamo mode needs an approximately three times larger P_c than the first.

The critical dynamo numbers are roughly independent of η_0, describing the thickness of the torus. This is only because we have used the meridional cross section radius ρ as a typical length scale. The numbers would increase considerably with η_0 if we had used the equatorial radius a instead.

Table 1. Critical dynamo numbers P_c for three values of η_0, both kinds of symmetry and signs of P. Listed are the values of the first (second) excited oscillatory dynamo-modes. Numbers marked by an asterisk represent stationary modes.

	Dipole		Quadrupole	
η_0	$P > 0$	$P < 0$	$P > 0$	$P < 0$
1.0	960, (2437)	-793, (-2027)	526, (2002)	-65*, -1150
1.5	903, (2386)	-769, (-2067)	474, (1917)	-55*, -1131
3.0	862, (2413)	-763, (-2102)	442, (1841)	-47*, -1129

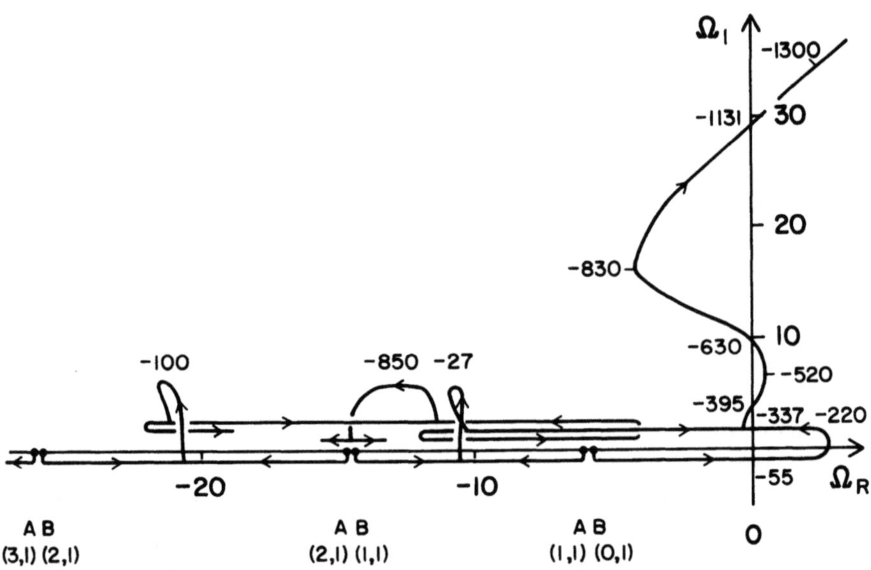

Fig. 3. Evolution of the eigenvalues Ω of the quadrupolar symmetry in the complex (Ω_R, Ω_I)-plane for varying dynamo number $P < 0$. A torus with $\eta_0 = 1.5$ was chosen. $|P|$ increases along the curves as indicated by the arrowheads. Due to the position of Ω in the diagram the dynamo mode is growing ($\Omega_R > 0$) or decaying ($\Omega_R < 0$), non-oscillating ($\Omega_I = 0$)) or oscillating ($\Omega_I > 0$). The thick dots on the real axis give the position of the decay modes ($P = 0$). They are characterized by their poloidal (A) or toroidal (B) structure, respectively, and by the numbers (n, l).

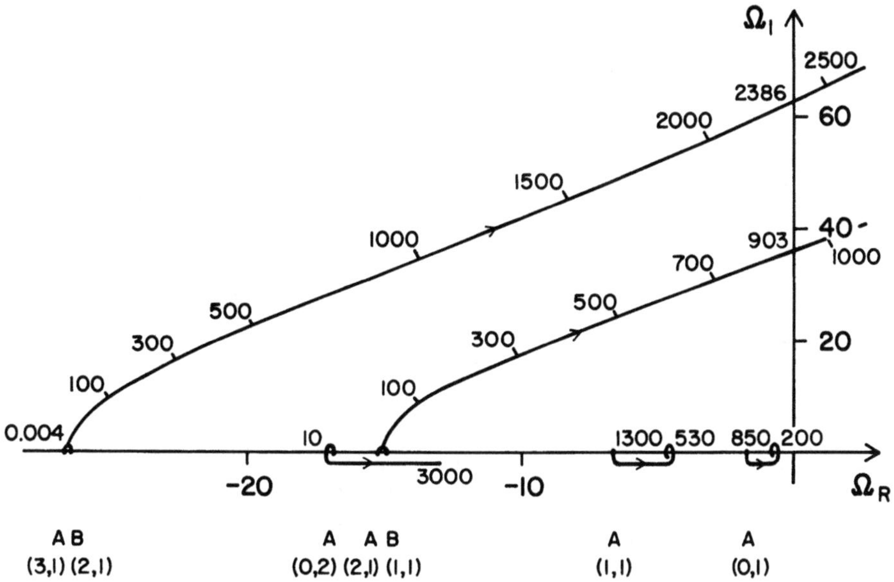

Fig. 4. Same as Fig. 3 for dipolar symmetry with $P > 0$.

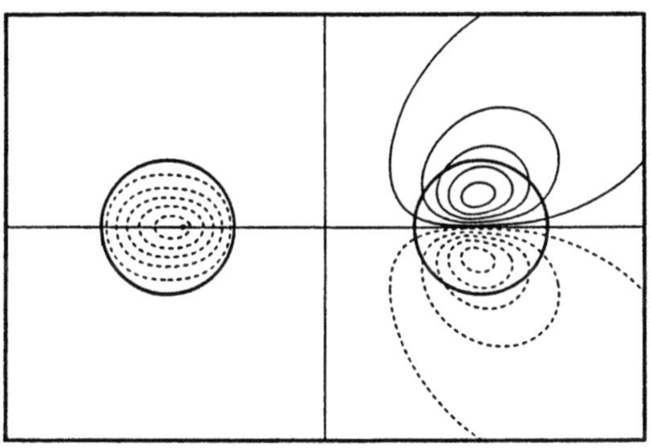

Fig. 5. Field configuration of the first excited mode, a stationary quadrupole at $P = -55$. Shown are meridional cross-sections of the torus for $\eta_0 = 1.5$, lines of constant toroidal field strength to the left, and poloidal field lines to the right. Solid and dashed lines indicate opposite directions.

4. Discussion: Application to Galaxies

We now estimate the conditions in galaxies in order to see which mode could be expected there. Typical values for the turbulent velocity and correlation length in the interstellar gas of galaxies are $v \approx 10\text{km/s}$ and $l \approx 100\text{pc}$ [20]. In M 31 the value $a\omega_0'$, which is of the order of ω itself, is $a\omega_0' \approx -\omega \approx -10^{-15}\text{s}^{-1}$ [20], corresponding to a revolution in 200 million years. The equatorial radius a is $a \approx 10\text{kpc}$ and the half-thickness of the disk, corresponding to the meridional cross-section ρ, is approximately $\rho \approx 700\text{pc}$ for M 31 [20].

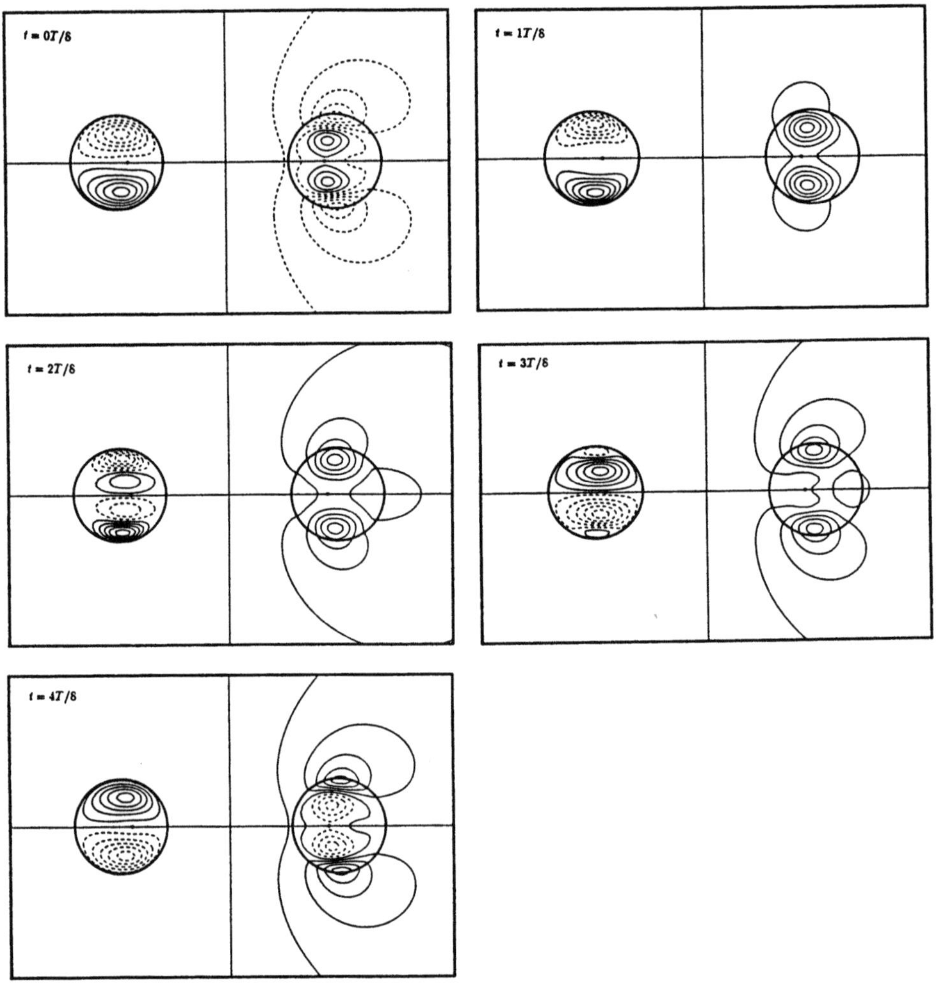

Fig. 6. Time-sequence over a half-period of the first excited oscillating dipolar mode for $P = 903$. The dimensionless frequency is 36. Otherwise as Fig. 5.

We use the formula $D \approx \frac{1}{3}lv$ for the turbulent diffusivity and $\alpha \approx \omega l^2/\rho$ for the α–effect [20]. This results in a Reynolds number of $R_\alpha \approx 1$ for the α-effect and of $R_\omega \approx -40$ for the differential rotation, giving a dynamo number of $P \approx -40$. This is just of the order of what is needed to excite the lowest mode, the stationary quadrupole, but far less for all other dynamo modes.

The ratio of poloidal to toroidal field strength is given by $B_r/B_\varphi \approx |R_\alpha/R_\omega|^{1/2} = l/\rho \approx 0.15$, giving a pitch angle of $i = \arctan B_r/B_\varphi \approx 8°$. The field is mainly toroidal, the flux mainly contained within the torus. The pitch angle corresponds fairly well with the observed values [3].

We therefore propose that the fields observed in M 31 and IC 342 are due to the excitation of a stationary quadrupole under the action of an axisymmetric

$\alpha\omega$–dynamo. There is an observational hint that the field of IC 342 is symmetric with respect to the equator, i.e. of quadrupolar symmetry.

All other modes discussed earlier are not expected under galactic conditions. They might play a role in accretion disks [21] around compact objects to which the model is also applicable.

5. Conclusion

We conclude with some remarks on possible generalizations.

Clearly the bisymmetric field structure observed in some spiral galaxies like M 51 or M 81 calls for nonaxisymmetric dynamo investigations [22]. Under most conditions explored so far the axisymmetric dynamo modes are much easier excited than the nonaxisymmetric modes [20]. The dominance of the first nonaxisymmetric mode therefore poses a challenge for dynamo theory. It is not at all clear whether the bisymmetric mode is due to an underlying nonaxisymmetry like spiral arms [23] nor whether it can be excited prior to the axisymmetric mode even in the case of a specific axisymmetric background [24], which might possibly be anisotropic [25].

Assuming axisymmetric induction effects and diffusion each degree of nonaxisymmetry can be treated separately and the above formalism can be extended. The key is to find expansion functions of the generalized toroidal and poloidal field components which already satisfy the boundary conditions.

Also the strong fields perpendicular to the galatic plane found in the center of our Galaxy and in the nuclei of a couple of mildly active edge–on galaxies like M 82 are not covered by the above $\alpha\omega$–dynamo. These magnetic fields seem to be connected with partially ionized gaseous rings of HI, CO and OH surrounding the nucleus at a distance where the rotation changes from rigid to differential. The field might be due to an α^2–dynamo acting in this turbulent accretion ring. Jet–like outflow phenomena perpendicular to the equatorial plane could be due to an alignment by a strong poloidal magnetic field.

Disk cyclides, flat ring cyclides [26], or oblate spheroids [19] are further appropriate geometries for the different galaxies, and clearly different rotation laws should be applied [27].

Further studies should embed the dynamo into a conducting outside medium rather than into vacuum [28]. An ideal conducting outside can be easily treated by an appropriate change of the boundary conditions.

Finally nonlinear effects like magnetic buoyancy or α–quenching [29] that limit the growth of the magnetic field strength should be included into the models. Magnetic buoyancy might play an important role also for the explanation of the halo fields [30].

Acknowledgement: This work has been done in collaboration with W. Deinzer and H. Grosser of the Univ.–Sternwarte Göttingen. It has been supported by the Deutsche Forschungsgemeinschaft under grant De 177/9–1. The author thanks T. Bogdan of the High Altitude Observatory for improving the manuscript.

References

1. R. Beck: IEEE Trans. Plasma Sci., **14**, 740 (1986);
   Y. Sofue, M. Fujimoto, R. Wielebinski: Ann. Rev. Astron. Astrophys **24**, 459 (1986);
   A. A. Ruzmaikin, A. M. Shukurov, D. D. Sokoloff: *Magnetic Fields of Galaxies* (Kluver, Dordrecht, 1988)
2. R. Beck, M. Krause: Sterne und Weltraum **28**, 440 (1989)
3. R. Beck, N. Loiseau, E. Hummel, E. M. Berkhuijsen, E. M. Gräve, R. Wielebinski: Astron. Astrophys., in press (1990)
4. R. Beck: Astron. Astrophys. **106**, 121 (1982);
   R. Beck et al., see 3.
5. R. Gräve, R. Beck: Astron. Astrophys. **192**, 66 (1988);
   M. Krause, E. Hummel, R. Beck: Astron. Astrophys. **217**, 4 (1989)
6. M Tosa, M. Fujimoto: Publ. Astron. Soc. Japan **30**, 315 (1978);
   Y. Sofue, T. Takano, M. Fujimoto: Astron. Astrophys. **91**, 335 (1980)
7. M. Krause, R. Beck, E. Hummel: Astron. Astrophys. **217**, 17 (1989)
8. R. Beck, R. Gräve (Eds.): *Interstellar Magnetic Fields* (Springer, Berlin, 1987)
9. R. Beck, see 1.;
   Y. Sofue et al., see 1.;
   A. A. Ruzmaikin et al., see 1.;
   R. Beck: Geophys. Astrophys. Fluid Dyn., in press (1990)
10. R. Wielebinski: Geophys. Astrophys. Fluid Dyn., in press (1990)
11. U. Klein, R. Wielebinski, H. W. Morsi: Astron. Astrophys. **190**, 41 (1988)
12. H. Lesch, A. Crusius, R. Schlickeiser, R. Wielebinski: Astron. Astrophys. **217**, 99 (1989);
    R. Wielebinski: Geophys. Astrophys. Fluid Dyn., in press (1990)
13. H. K. Moffatt: *Magnetic Field Generation in Electrically Conducting Fluids* (Cambridge University Press, 1978);
    E. N. Parker: *Cosmical Magnetic Fields* (Oxford Univerity Press, 1979);
    F. Krause, K. H. Rädler: *Mean–Field Magnetohydrodynamics and Dynamo Theory* (Pergamon, Oxford, 1980);
    A. A. Ruzmaikin et al.: see 1.
14. S. C. Unwin: Mon. Not. R. A. S. **192**, 243 (1980);
    R. Beck, R. Wielebinski: Sky and Telescope **61**, 495 (1981)
15. E. M. Berkhuijsen: Astron. Astrophys. **57**, 9 (1977)
16. W. Deinzer, H. Grosser, D. Schmitt: Astron. Ges. Abstr. Ser. **3**, 108 (1989);
    W. Deinzer, H. Grosser, D. Schmitt: *Torus–Dynamo*, to appear in: Proc. IAU-Symp. **140** *Galactic and Intergalactic Magnetic Fields*, held in Heidelberg, June 1989;
    W. Deinzer, H. Grosser, D. Schmitt: to be submitted to Astron. Astrophys. (1990)
17. H. Grosser: *Magnetische Decay–Modes in Torusgeometrie* (Diplomarbeit, Univ. Göttingen, 1984);
    H. Grosser: Astron. Astrophys. **199**, 235 (1988)
18. E. N. Parker: Astrophys. J. **163**, 255 (1971);
    S. I. Vainshtein, A. A. Ruzmaikin: Sov. Astron. **15**, 714 (1972);
    A. A. Ruzmaikin, D. D. Sokoloff, V. I. Turchaninoff, Ya. B. Zel'dovich: Astrophys. Space Sci. **66**, 369 (1979);
    A. A. Ruzmaikin, D. D. Sokoloff, V. I. Turchaninoff: Sov. Astron. **24**, 182 (1980);
    A. A. Ruzmaikin et al., see 1.
19. M. Stix: Astron. Astrophys. **42**, 85 (1975); Astron. Astrophys. **68**, 459 (1978);
    M. P. White: Astron. Nachr. **299**, 209 (1978);
    A. M. Soward: Astron. Nachr. **299**, 25 (1978)

20 A. A. Ruzmaikin et al., see 1.

21. R. E. Pudritz: Mon. Not. R. A. S. **195**, 881, 897 (1981);
T. F. Stepinski, E. H. Levi: Astrophys. J. **331**, 416 (1988);
G. Belvedere (Ed.): *Accretion Disks and Magnetic Fields in Astrophysics* (Kluwer, Dordrecht, 1989)

22. Y. Baryshnikova, A. A. Ruzmaikin, D. D. Sokoloff, A. M. Shukurov: Astron. Astrophys. **177**, 27 (1987);
Y. Krasheninnikova, A. A. Ruzmaikin, D. D. Sokoloff, A. M. Shukurov: Astron. Astrophys. **213**, 19 (1989);
S. V. Starchenko, A. M. Shukurov: Astron. Astrophys. **214**, 47 (1989);
R. Meinel, D. Elstner, G. Rüdiger, F. Krause: to appear in: Proc. IAU–Symp. **140** *Galactic and Intergalactic Magnetic Fields*, held in Heidelberg, June 1989

23. M. Chiba, M. Tosa: to appear in: Proc. IAU–Symp. **140** *Galactic and Intergalactic Magnetic Fields*, held in Heidelberg, June 1989

24. M. Stix: Astron. Astrophys. **13**, 203 (1971)

25. G. Rüdiger: Geophys. Astrophys. Fluid Dyn. **50**, 53 (1990)

26. P. Moon, D. E. Spencer: *Field Theory Handbook* (Springer, Berlin, 1971)

27. A. A. Ruzmaikin, D. D. Sokoloff, A. M. Shukurov: Astron. Astrophys. **148**, 335 (1975);
Y. Baryshnikova et al., see 22.;
Y. Krasheninnikova et al., see 22.

28. R. Meinel: Astron. Nachr. **310**, 1 (1989);
D. Elstner, R. Meinel, G. Rüdiger: Geophys. Astrophys. Fluid Dyn. **50**, 85 (1990)

29. D. Schmitt, M. Schuessler: Astron. Astrophys. **223**, 343 (1989)

30. F. Takahara: Progr. Theor. Phys. **62**, 629 (1979);
U. Klein: Geophys. Astrophys. Fluid Dyn., in press (1990)

Winds from Hot Stars

S.P. Owocki

Bartol Research Institute, University of Delaware,
Newark, DE 19716, USA

1 Introduction

Stellar winds from hot stars (i.e., spectral types O, B, and WR) have been known and studied for about 20 years now. An excellent, succinct summary of both theory and observations was given recently by Abbott [1]. (This was one of Abbott's last active contributions before giving up astrophysical research, a decision which, given his many contributions to our understanding of hot stars, was most unfortunate for science.) More extensive discussions of both theory and observations are contained in the monograph edited by Conti and Underhill [2], and there have been numerous additional reviews of theory [3 – 8] or observations [9 – 11]. In nearly all of these the emphasis has been on steady-state models and time-averaged spectra. The general overview of Kudritzki in this volume emphasizes the now quite impressive sophistication of such models and their quantitative agreement with observations. Given this already rich collection of reviews, I will not attempt here to give any kind of exhaustive list of the literature. The next section gives my own general overview of the subject, but the bulk of the remaining discussion will concentrate on examining the physical driving mechanism – line-scattering in the wind of the star's continuum radiation field – and what it implies for both the basic wind properties (e.g., mass loss rate, $\dot{M}$ and terminal flow speed, v_∞) (Sect. 3), as well as for the wind structure (e.g., variability, clumpiness, embedded shocks, etc.) (Sect. 4). The emphasis on structure and variability admittedly reflects a personal bias, but it is also intended to complement the above-mentioned emphasis on steady-state properties in previous reviews. In particular, one goal will be to place in some perspective the successes of the steady-state approach.

2 Overview

Let us first consider what makes hot-star winds such an interesting subject for astrophysical research. First, there is the fact that, just as hot stars themselves are extreme in such basic properties as mass ($M \approx 10 - 50 M_{\odot}$), temperature ($T \approx 10,000-50,000\,\mathrm{K}$), and luminosity ($L_{*} \approx 10^{5}-10^{6} L_{\odot}$), so are their winds also extreme, with mass loss rates ($\dot{M} \approx 10^{-7} - 10^{-5} M_{\odot}/\mathrm{yr}$) that exceed that of the sun by some nine orders of magnitude, and flow speeds ($v_{\infty} \approx 1000 - 3000\,\mathrm{km/s}$) of the order of 1% of the speed of light. Indeed, even though such stars only live $\sim 10^{7}$ years on the main sequence, the large mass loss rate means that a substantial fraction (~50%) of the star's initial mass can be lost during its main sequence lifetime, implying important effects on the star's evolution. Such winds contribute an important component to the mass and energy balance of the interstellar medium [12]. In high-mass X-ray binary systems they also provide the source material accreting onto the collapsed companion [13 – 15], a topic which, of course, provides another focus of this conference and of the accompaning reviews in this volume. Finally, as is emphasized by the general review of Kudritzki (also this volume), the detailed agreement between theory and observations makes it possible to use observations of lines formed in the wind to infer stellar properties like mass, luminosity, and temperature, and so this opens the way to using the luminous O and B stars as standard candles to infer distances in other galaxies.

In my own view, however, the most interesting aspect of hot-star winds is the opportunity to study in detail aspects of the relatively unexplored field of radiation hydrodynamics. Often this term is used in a relatively broad sense to refer to the quite common case when radiation plays an important role in the energy balance of a plasma (e.g., [16]); but here it applies in the stricter sense that the stellar radiation imparts *momentum* (as well as energy) to the plasma, and so drives a supersonic outflow. The former situation is more common because photons are, in a sense, better carriers of energy than momentum, since their momentum is given by dividing their energy by the largest speed possible, namely c. But the latter typically applies to the most luminous objects – those near their Eddington limit – which are also often the most energetic and hence most interesting objects in the universe, e.g., quasars, active galactic nuclei, and accretion disks. The relative proximity of hot stars makes it possible to obtain high spectral and/or temporal resolution observations, and these both guide and test a detailed wind theory that includes radiation-hydrodynamical processes likely to have relevance in the more-distant, and hence less well-observed, luminous objects.

In the case of hot-star winds the specific coupling mechanism between matter and radiation – namely *line* opacity – might at first seem surprising, since line transitions can only occur for a very narrow range of photon frequencies. However, what makes this mechanism in fact quite efficient here is that, once a flow is initiated, the line frequency becomes Doppler shifted, so that a fresh supply of relatively unattenuated flux from the star's continuum can then interact with the wind material in the spectral line. In this way a single line can "sweep up" the momentum from a fraction $v_\infty/c \sim 1\%$ of the star's spectrum, vs. only $\sim v_{th}/c \sim 10^{-4}$ for the static case in which the scattering ion has a thermal velocity v_{th}. Thus if there are ~100 strong lines spread across the spectrum, then nearly every photon will impart momentum to the wind, and so typically there is rough equivalence between the wind and radiative momentum loss, i.e., $\dot{M} v_\infty \sim L_*/c$.

Historically, the basic idea of line-driving dates back to early work by Milne [17] and the pioneering work by Sobolev [18, 19]; but the specific application to hot-star winds was first proposed by Lucy and Solomon [20] shortly after the original rocket-based observations by Morton [21] of asymmetric "P-Cygni" profiles [22] in UV lines had indicated the existence of continuous mass outflows from such hot, luminous stars. The original Lucy and Solomon model included only a few lines and so yielded mass loss rates significantly smaller (~1%) than had been inferred observationally. The landmark work of Castor, Abbott, and Klein (hereafter CAK) [23] developed a simple but powerful formalism, still in use in models today, for including the ensemble effect of many lines, and thereby obtained good general agreement with the observationally inferred $\dot{M}$. The original CAK model had assumed radial streaming of radiation, as from a point source, and correct inclusion of the angular size of the stellar disk was found independently by Friend and Abbott [24] and by Pauldrach, Puls, and Kudritzki (hereafter PPK) [25] to yield substantially improved agreement with the observationally inferred terminal flow speed v_∞. A more fundamental approximation of CAK was to use Sobolev theory to compute the line force [18, 19, 26, 27], but this too has since been relaxed with wind models that compute the radiative force from full comoving frame solutions for the line transfer [28, 25]. PPK showed that the CAK/Sobolev models, modified to include the finite disk effects, yield wind parameters nearly identical to the full comoving frame results. Recent extensions of the comoving frame models (see, e.g., review by Kudritzki and Hummer [8], and references therein) now treat unified wind/atmosphere models with self-consistent solution of level populations in thousands of transitions, and the theoretical line profiles generated thereby are in impressive quantitative agreement with observed line spectra.

Despite this great success of steady-state, line-driven-wind models in quantitatively matching many time-averaged spectral observations, there are strong reasons, both observational and theoretical, for believing that hot-star winds are not at all smooth, time steady flows, but rather are both spatially and temporally highly variable. The observational evidence for this will be discussed in greater detail below (Sect. 4.1), but listed briefly, it includes *Superionization* [29], *X-ray emission* [30, 31], *Nonthermal radio emission* [32], *Black profiles* in saturated UV lines [33], and *Discrete absorption components* in unsaturated UV lines [34 – 36]. The theoretical evidence stems from the expectation, which dates back to Lucy and Solomon [20] (and even to Milne [17]), that such a line-driven flow would be dynamically unstable. A small-scale increase in radial flow speed Doppler-shifts the local line-frequency out of the absorption shadow of intervening material, leading to an increased line-force which then tends to further increase the flow speed. Formal stability analyses [37 – 40]

(See review by Rybicki [41] and Sect. 4.2 below.) yield a growth rate nearly a hundred times the wind expansion rate. This implies that even very small base perturbations should quickly become nonlinear, perhaps giving rise to a wind embedded with multiple shocks and dense clumps [42, 43] (Sect. 4.3). Current efforts carry out numerical hydrodynamics simulations of the nonlinear wind structure that results from applied perturbations [44] (Sect. 4.4), or that arises intrinsically (Sect. 4.5). A important question is how much this structure affects the mean wind properties inferred by steady-state models (Sect. 4.6). The simulations carried out so far are intriguing, but are still based on several simplifying assumptions (e.g., pure absorption, isothermal flow, spherical symmetry) that will have to be relaxed in future work (Sect. 5).

3 Physics of Line Driving

3.1 The Radiative Force from a Single Line

The radiative acceleration $g_{\rm rad}$ (i.e., force-per-unit mass) resulting from the interaction of a radiative flux F_ν at a photon frequency ν with material of opacity κ_ν is given quite generally by

$$g_{\rm rad} = \int_0^\infty d\nu \; \kappa_\nu \, \frac{F_\nu}{c}. \tag{1}$$

A simple but useful limiting case is that of an *optically thin* line, for which one may neglect the attenuation of an externally imposed flux, so that (1) becomes

$$\begin{aligned} g_{\rm thin} &= \frac{F_\nu}{c} \Delta\nu_{\rm D} \kappa_{\rm L} \\ &= \frac{\nu_{\rm L} L_\nu}{L_*} \, v_{\rm th} \kappa_{\rm L} \, \frac{L_*}{4\pi r^2 c^2}, \end{aligned} \tag{2}$$

where L_ν is the source luminosity spectrum, $\nu_{\rm L}$ is the line frequency, and $\kappa_{\rm L}$ is the total line opacity; $\Delta\nu_{\rm D} \equiv \nu_{\rm L} v_{\rm th}/c$ is the line width, assumed here to be set by Doppler broadening from thermal motion at speed $v_{\rm th}$. The latter expression applies at a distance r from a spherically symmetric radiative source with total luminosity L_*, and emphasizes that, just like gravity, the optically thin force falls off with an inverse square law. The force from an optically thin *continuum* opacity $\kappa_{\rm c}$ (e.g., electron scattering) also follows this scaling and so is often accounted for by simply considering an effective gravity that is reduced by a factor $1 - \Gamma_{\rm c}$, where $\Gamma_{\rm c} \equiv \kappa_{\rm c} L_*/4\pi G M_* c$.

For an optically thick opacity source, the simplicity of equation (1) is actually somewhat misleading, since the flux F_ν depends sensitively on the opacity κ_ν, and in general must be determined from solution of the full nonlocal radiative transport problem. For the winds from O and B stars, the optically thick

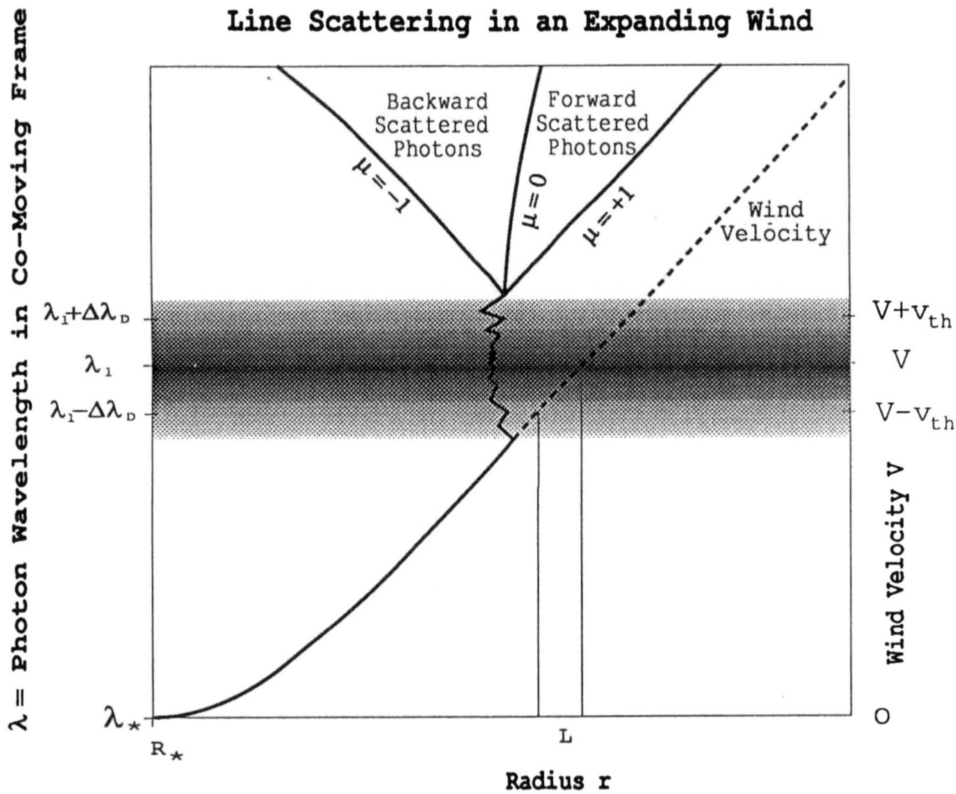

Fig. 1. Radial variation of photon wavelength as viewed in the local comoving frame of material in the wind. The shaded region represents a thermally broadened spectral line with which the outward propagating radiation interacts.

opacity is from numerous (~100) strong scattering lines. (For Wolf-Rayet stars the continuum can also be optically thick in much of the wind; see below.) Fortunately, for this case of line transfer in a supersonically expanding flow, methods first developed by Sobolev [18, 19] make it possible to obtain a good approximation to the force that depends only on *local* quantities.

The basic ideas of this Sobolev approximation are illustrated graphically in Fig. 1, which plots the radial variation of a photon's wavelength as viewed in a frame *comoving* with the mean flow in the expanding wind. The shaded band represents a line that is thermally broadened by a Doppler width $\Delta\lambda_D$ centered about the rest wavelength λ_l. A photon emitted in the stellar continuum at a wavelength $\lambda < \lambda_l$ propagates freely until red-shifted into the blue edge of the line, whereupon it is scattered. In addition to changing direction, the photon will in general be redistributed in frequency, most likely into the line core, but for purposes of this illustration, let us assume the scattering is coherent. The opacity on the blue edge is relatively weak, and so the scattered photon then travels some mean free path on the order of the Sobolev length $L \equiv v_{th}/(dv/dr)$, which in this context may be viewed as a local "opacity scale height".

A crucial point is that, because the wind is an *expanding* medium, distant material viewed in any direction from a given location always appears to be *receding*. Hence, whether a photon is scattered into the forward or backward directions, the subsequent distance traveled always causes its comoving wavelength to be *red-shifted*. After many scatterings this systematic drift toward the red brings the photon to the line's red edge, from which it's last scattering can, with roughly equal probability, be either into a fore or aft direction. Since the resulting diffuse radiation field thus has rough fore-aft symmetry, the net force associated with this diffuse radiation is small, i.e., $g_{\mathrm{diffuse}} \approx 0$. Thus the overall force associated with such line scattering is very nearly equal to what it would be for pure line *absorption*.

For simplicity let us assume (as did CAK) a point source of radially streaming radiation. The radiative force associated with an isolated line can then be written as

$$g_{\mathrm{rad}}(r) \approx g_{\mathrm{abs}}(r) = g_{\mathrm{thin}} \int_{-\infty}^{\infty} dx\ \phi(x - v(r)/v_{\mathrm{th}})\ e^{-t(x,r)}, \tag{3}$$

where ϕ is the line-profile function and $x \equiv (\nu - \nu_{\mathrm{L}})/\Delta\nu_{\mathrm{D}}$ is the frequency displacement from line center, measured in Doppler-units in the star's rest frame. The optical depth $t(x, r)$ at radius r is given by

$$t(x,r) = \int_{R_*}^{r} \kappa_{\mathrm{L}} \rho(r') \phi(x - v(r')/v_{\mathrm{th}}) dr'. \tag{4}$$

A crucial element of the Sobolev approximation is that the variation in the integrand of Eq. (4) is dominated by the Doppler-shift associated with changes in the velocity $v(r')$; by switching the variable of integration to the comoving frame frequency $x' \equiv x - v(r')/v_{\mathrm{th}}$, we thus find

$$t(x,r) \approx \tau \int_{x}^{\infty} \phi(x') dx', \tag{5}$$

where the Sobolev optical depth

$$\tau \equiv \frac{\kappa_{\mathrm{L}} \rho v_{\mathrm{th}}}{dv/dr} \tag{6}$$

represents a collection of spatial variables that are assumed to be approximately constant over a Sobolev length $L \equiv v_{\mathrm{th}}/(dv/dr)$. For example, in the case of a smooth, steady flow this means that the density scale length

$$H \equiv \frac{\rho}{d\rho/dr} \approx \frac{v}{dv/dr} \gg \frac{v_{\mathrm{th}}}{dv/dr} \equiv L, \tag{7}$$

and hence that $v \gg v_{\mathrm{th}}$. Since the ion thermal speed is on the order of the sound speed a, we can expect that the approximation will be well satisfied in the supersonic portions of a smooth flow. Applying Eq. (5) to Eq. (3) we see that both required frequency integrations can now be done analytically, yielding

$$g_{\text{rad}}(r) = g_{\text{thin}} \frac{1 - e^{-\tau}}{\tau}. \tag{8}$$

Note that for a weak line $\tau \ll 1$ we recover the optically thin expression (2), while for a strong line $\tau \gg 1$ we obtain the optically thick form

$$g_{\text{rad}}(r) = \frac{g_{\text{thin}}}{\tau} = \frac{\nu_{\text{L}} L_{\nu}}{L_*} \frac{L_*}{4\pi \rho r^2 c^2} \frac{dv}{dr}. \tag{9}$$

The latter is independent of the opacity κ_{L}, depending instead on the velocity gradient dv/dr. This can be understood physically from Fig. 1. Once a line is optically thick, all photons shifted into resonance with it are scattered, and further increasing the opacity only pushes the first scattering further into the blue wing. The number of photons scattered per unit length is thus independent of the opacity, and depends instead on the slope of the photon red-shift, which depends on the velocity gradient. Normally, such a velocity gradient is part of an inertial term that represents the acceleration that *results* from a driving force, but here we see that it helps *determine* the driving force. In general terms, just as the optically thin line force helps reduce the effect of gravity on the fluid, so does the optically thick line force reduce the effect of *inertia*.

A simple consideration of just the inertial force balance enables us to estimate the mass loss rate that results from such line-driving. The active acceleration resulting from such driving is just

$$v \frac{dv}{dr} = g_{\text{rad}} = \frac{\nu_{\text{L}} L_{\nu}}{L_*} \frac{L_*}{4\pi \rho r^2 c^2} \frac{dv}{dr}, \tag{10}$$

which implies that the mass loss rate $\dot{M}_{\text{l}}$ associated with driving by a single, isolated line is

$$\dot{M}_{\text{l}} = 4\pi \rho v r^2 = \frac{\nu_{\text{L}} L_{\nu}}{L_*} \frac{L_*}{c^2}. \tag{11}$$

If the line has a frequency near the peak of the stellar luminosity, the factor $\nu_{\text{L}} \text{L}_{\nu}/L_*$ is of order unity, implying that the *wind* mass loss rate driven by a single line is only about the *photon* mass loss rate! It is essentially for this reason that the original Lucy and Solomon [20] model, which considered the effect of driving in only a few strong lines, yielded such low mass loss rates.

3.2 The CAK Model: Driving by an Ensemble of Nonoverlapping Lines

A major advance of the CAK [23] model was to develop a formalism for efficiently including the cumulative effect of a large ensemble of lines, which they effectively assumed to have a flux-weighted number distribution that was a power law in opacity, i.e., $N(\kappa) \sim \kappa^{\alpha-2}$, where $0 < \alpha < 1$. Assuming the lines do not significantly overlap, the cumulative force can then be computed by integrating the expression (3) over this number distribution [44, 75]. Applying, as before, the Sobolev approximation, one obtains the CAK line-force

$$g_{\rm CAK} = \frac{KL_*}{r^2}\left(\frac{1}{\rho}\frac{dv}{dr}\right)^{\alpha}, \tag{12}$$

where K is a normalization constant for the line-distribution (related to the CAK constant k by $K = k(\kappa_e^{1-\alpha}/4\pi c v_{\rm th}{}^{\alpha})$, where κ_e is the electron scattering opacity.) Since $0 < \alpha < 1$, this force can be viewed as kind of a "geometric-mean" mixture of the optically thin and thick expressions (2) and (9). But note that the presence of the density in this expression means that whether any given line in the distribution is optically thick or thin is not determined *a priori*, but rather results from a self-consistent solution of the steady-state equations of motion including this force.

CAK derived their wind solutions by applying this expression in both the subsonic and supersonic regions of a steady wind with a finite temperature. However, strictly speaking, this Sobolev limit can only be attained for real driving ions in the limit of vanishing temperature, implying also a vanishing sound speed a. Furthermore, since such gas pressure forces play a relatively minor role in driving the wind, we can derive the basic wind characteristics by considering the balance among inertial, gravitational, and radiative forces in this limit $a \to 0$ of neglible gas pressure terms. Applying Eq. (12) for the radiative force, the requirement of momentum balance reduces to the condition,

$$F \equiv w' + 1 - C w'^{\alpha} = 0, \tag{13}$$

where $w' \equiv r^2 v v'/GM_*$ and

$$C \equiv \frac{K}{GM_*}\left(\frac{4\pi GM_*}{\dot{M}c}\right)^{\alpha}. \tag{14}$$

Fig. 2 illustrates that, depending on the value of C, equation (13) has either 0, 1, or 2 solutions. The critical case with one solution has the property that $F = \frac{\partial F}{\partial w'} = 0$, which implies the critical values $w'_c = \alpha/(1-\alpha)$ and $C_c = \alpha^{-\alpha}(1-\alpha)^{1-\alpha}$. Since Eq. (13) has no explicit spatial dependence, this critical value w'_c must hold throughout the flow. Integrating w'_c from the stellar radius R_*, we thereby obtain the usual "CAK velocity law",

$$v_{\rm CAK}(r) = v_\infty \sqrt{1 - \frac{R_*}{r}}, \tag{15}$$

where the asymptotic wind speed is given by

$$v_\infty \equiv \sqrt{\frac{2\alpha GM_*}{(1-\alpha)r_o}} = \sqrt{\frac{\alpha}{1-\alpha}}\, v_{\rm esc}. \tag{16}$$

Likewise, if we solve the definition (14) for the mass loss rate and apply the critical value C_c, we obtain the CAK mass loss rate formula,

$$\dot{M}_{\rm CAK} = 4\pi\alpha \left(\frac{GM_*}{1-\alpha}\right)^{1-1/\alpha} (KL_*)^{1/\alpha}. \tag{17}$$

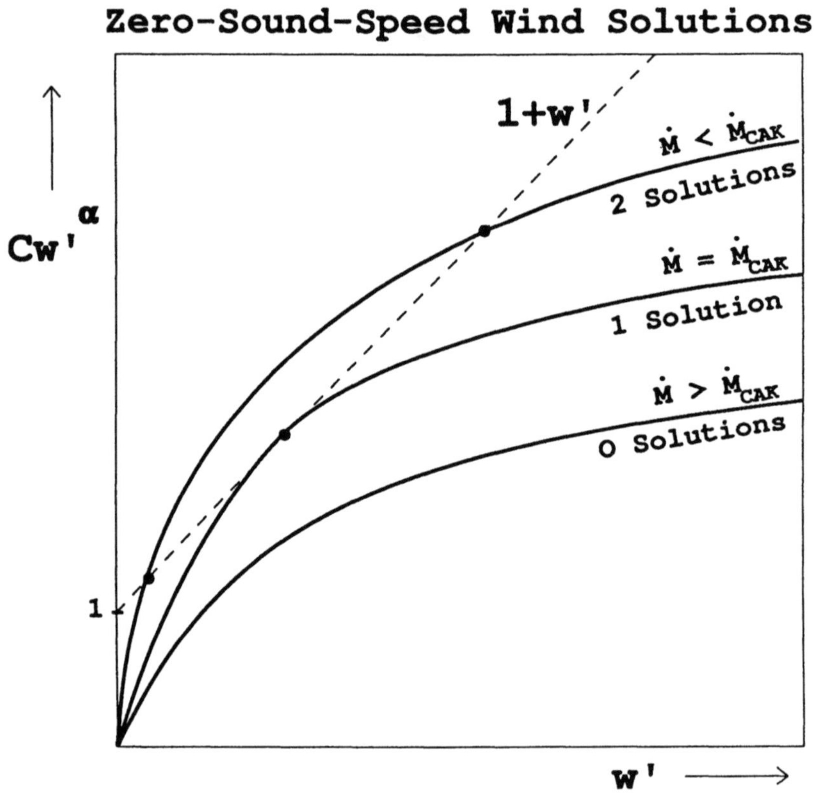

Fig. 2. Graphical solution of a CAK/Sobolev line-driven wind model in the limit of zero sound speed. The CAK mass loss rate represents the critical case with one solution; for $\dot{M} > \dot{M}_{CAK}$ there are no solutions, but for $\dot{M} < \dot{M}_{CAK}$ there are two solutions.

These are just the relations derived by CAK for the case of a point-star; taking account of the finite angular size of the stellar disk tends to make the mass loss rate somewhat lower and the velocity law somewhat shallower [24, 25].

It is important to realize that, unlike in the finite-sound-speed model of CAK, the *mass loss rate in this zero-sound-speed case is not unique*, since solutions with $\dot{M} < \dot{M}_{CAK}$ are now also allowed. Indeed, for these "subcritical" $\dot{M}$, there are *two* solutions, one steep, and one shallow. In the zero-sound-speed case, the CAK mass loss rate $\dot{M}_{CAK}$ thus now only represents a *maximum* mass loss rate for which solutions exist. As we discuss in greater detail in Sect. 4.5, a similar kind of solution "degeneracy" exists for finite-gas-pressure models that do not use the Sobolev approximation to compute the line-force.

3.3 Multi-Line Scattering in Overlapping Lines

This CAK mass loss rate is typically ~100 times the single line value, implying that there are effectively about 100 optically thick lines spread throughout the spectrum. On the other hand, note that the terminal flow speed v_∞ roughly scales with surface speed as $v_\infty \sim v_{esc}$ and is pretty much independent of the number of strong lines. Typically this is on the order of 1% of the speed of light, so that, quite coincidentally, in the CAK model

$$\dot{M} v_\infty \approx \frac{L_*}{c} \left(N_{\text{thick}} \frac{v_\infty}{c} \right) \approx \frac{L_*}{c}, \tag{18}$$

which is often referred to as the "single scattering limit". This means that, fortuitously, the CAK model is just barely self-consistent in ignoring line overlap, if one assumes the lines are smoothly spread throughout the stellar spectrum.

In reality, of course, there is always some line overlap, arising both from chance frequency coincidences of independent lines as well as from the tendency for lines to occur in multiplets. An early but quite elegant examination of the dynamical effects of such overlap was carried out by Friend and Castor [45], who derived an analytic extension of the CAK model to include overlap under the assumption that the line frequency spacing is Poisson distributed. Although this ignored the tendency of lines to occur in multiplets, it demonstrated quite clearly the major effect that overlapping lines can drive mass loss in excess of the single scattering limit, with $\dot{M} v_\infty \approx (2-5) L_*/c$. Similar results were obtained for more realistic line lists by Abbott and Lucy [46] from a Monte-Carlo calculation, and by Puls [47], who carried out the quite formidable calculation of solving self-consistently the coupled system of statistical equilibrium equations.

Fig. 3 illustrates graphically how multi-line scattering can result in the emergent stellar radiation imparting its radial momentum more than once to the wind. As in Fig. 1, this figure shows the radial evolution of a photon's comoving frame wavelength, but now including interaction with *two* scattering lines. The initially outward-propagating photons are first scattered in the bluer line, being reemitted with equal probability into the fore and aft directions and thus on the average imparting all their initial radial momentum to the wind. However, during their subsequent free flight through the spherical envelope, the component of the photon's momentum along the local radial direction systematically *increases*. Thus, if they don't strike the stellar core, the photons are eventually red-shifted into resonance with the redder line, but now again with an average net positive radial momentum, which is thus also imparted to the wind. The exact amount of extra momentum imparted depends, of course, on the line spacing and the wind velocity law.

For O and B stars such overlap effects are typically minor, but they probably play a much more fundamental role in the Wolf-Rayet stars, which are inferred observationally to have $\dot{M} v_\infty/(L_*/c) \approx 5-50$ [48]. Indeed the density in such winds is so high that the continuum remains optically thick well past the sonic point, and so models based on the usual "core-halo" treatment of the emergent continuum, which have proven so successful for O and B stars, completely fail for Wolf-Rayets. Instead, a suitable model must treat the diffuse transport of both line and continuum radiation within the expanding envelope, taking proper account of the substantial line overlap. This is sufficiently difficult that

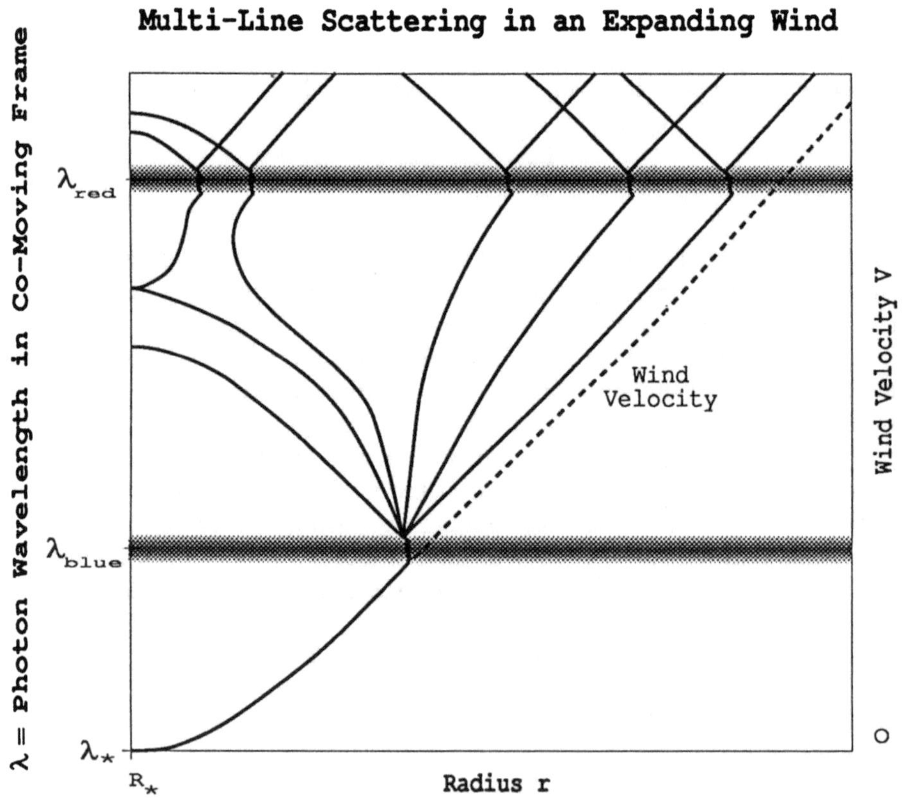

Fig. 3. Same as Fig. 1, except now with two spectral lines.

so far there is still no self-consistent, dynamical, Wolf-Rayet star wind model that can adequately explain the high mass loss rates inferred observationally.

3.4 Further Extensions of the Basic CAK Model

Finally, over the past decade there have been several other extensions of the original CAK wind model. These have largely been directed along three general lines: (1) Multi-dimensional effects, (2) Improved transfer theory, and (3) Instability and time-dependence.

Multi-dimensional effects have included rotation and/or magnetic fields, treated within the steady-state, CAK/Sobolev wind formalism. For typical O and B stars, MacGregor, Friend, and Gilliland [49] were able to place an upper limit $B_o < 100$ G on the open magnetic field near the equator from the apparent lack of rotational spin-down. Friend and MacGregor [50] found relatively minor effects of rotation on the basic wind parameters $\dot{M}$ and v_∞ unless the rotation is at a substantial fraction ($\gtrsim 70\%$) of breakup. Models for such rapidly rotating stars show a substantially different wind from pole and equator, and these have been applied to *Be* stars [51, 52] and (more speculatively) to Wolf-Rayets [53].

Improvements in radiative transfer include the above-mentioned (Sect. 2) PPK comoving frame models and Pauldrach's [54] treatment of the NLTE statistical equilibrium. Recent further efforts (e.g., [55]) have focussed on developing unified wind-atmosphere models that abandon the artificial distinction between "core" and "halo". A major overall goal of these efforts is to carry out "quantitative spectroscopy" [8], i.e., to compare quantitatively theoretical and observed spectra in order to infer more precisely and reliably the basic parameters of both wind and star. Further details on these efforts are given by the contributions of Kudritzki, Pauldrach, and Puls in this volume.

The next section will examine in some detail the last item, the question of the wind instability and variability.

4 Wind Instability and Variability

4.1 Observational Evidence for Wind Structure

Let us now review the observational evidence that hot-star winds are highly structured and variable. As noted briefly in Sect. 2, this includes:

1. Superionization, i.e., the presence in the wind of relatively high ionization stages like OVI, NV, and CIV not normally expected in material near the stellar effective temperature. Originally it was argued [29] that such ions arose from Auger ionization from X-rays, but recently Pauldrach [54] has argued that such stages can indeed be produced near the stellar effective temperature if one correctly computes the stellar UV photoionization from excited levels.

2. X-ray Emission. Nonetheless, the X-ray emission predicted by the Auger model was in fact subsequently observed by the *Einstein* Observatory. Spectra taken with the Solid State Spectrometer (SSS) typically show both a soft and hard component, corresponding respectively to temperatures of $\sim 3 \times 10^6$ K and $\sim 10^7$ K [31]. The original picture of a hot, geometrically thin corona at the wind base predicted the existence of K-edge absorption features that were not observed in the limited number of SSS spectra taken. This is generally viewed as implying that the X-rays must originate well above the wind base, e.g., from shocks embedded in the wind, although Waldron [56] argues that the coronal model is still viable if one self-consistently includes the effect of X-ray ionization in reducing the strength of the absorption edges. However, for O and B stars the observed X-ray luminosity tends to scale in proportion to the star's bolometric luminosity as $L_x \approx 10^{-7} L_*$, which is quite distinct from the scaling with rotation typical of late-type stellar coronae, and which implies an originating mechanism tied, as is the wind, to the star's radiative output. Chlebowski [57] notes that the deviation of L_x from this general relation correlates well with stellar environment, and he thereby infers that some X-rays may originate in the *very* distant wind, i.e., at the wind terminal shock.

3. Nonthermal Radio Emission. Of the 25 or so early type stars in a volume limited sample of radio detections with the VLA, about a quarter show a spectral variation that is distinctly nonthermal [58]. White [59] has proposed that this arises from gyrosynchrotron emission from nonthermal, high-energy particles that have undergone Fermi acceleration in embedded wind shocks arising from the wind instability. (An interesting, recent variation of this idea supposes that the high-energy tail of the observed X-ray spectrum results from inverse Compton scattering of the stellar UV radiation by the same high-energy particles. [60])

4. Black Profiles. Lucy [33] first noted that the observed blackness of the blue-shifted absorption troughs of saturated UV lines are a signature of a nonmonotonic velocity field. In a smooth outflow such blue-shifted absorption troughs should always contain a residual flux from radiation forward scattered toward the observer by the wind in the forward hemisphere. Lucy reasoned that this implied that the expanding medium had to predominantly *backscatter* radiation, and he showed how this was a natural consequence of the systematic reduction, roughly by a factor of two, of the forward scattered radiation at each one of multiple resonance layers that can line-scatter radiation in a nonmonotonic velocity outflow.

5. Discrete absorption components. These may be the most direct observational manifestations of wind structure and variability. They consist of relatively narrow absorption features superposed upon the absorption trough of the broad, P-Cygni profile of an unsaturated UV line. Spectra taken months or even years apart sometime show such a discrete component at the same velocity [34], but IUE monitoring observations [35, 36] taken about once an hour over several days show that the components typically form near a velocity $v \approx v_\infty/2$ and then narrow and shift to higher velocities over several days. Interestingly, the final velocity approached is typically smaller by 10-20% than the edge velocity defined in saturated lines, i.e., $v_\infty \approx 0.8 - 0.9\, v_{\text{edge}}$ [61, 62], and the recurrence and acceleration time-scales seem to correlate roughly with the stellar rotation period [63].

4.2 Linear Theory of Line-Driven Instability

As noted in Sect. 2, it was recognized quite early on [13,16] that the sensitive velocity dependence of the line-force might cause a line-driven flow to be dynamically unstable. However, the formal analysis of even the *linear* stability properties turned out to be quite subtle, and for a while the strength of the instability, and even whether it really existed, was not clear. The first formal perturbation analyses indicating instability [37, 38] were based on the assumption that perturbations were *optically thin*, meaning that one could ignore any change in the optical depth, and instead focus only on changes arising from the Doppler-shift associated with the perturbed velocity. Referring to the line-force expression (3), one can see that applying such a velocity perturbation and expanding to first order leads to a perturbed force that is proportional to the perturbed velocity, $\delta g_{\mathrm{abs}} \propto \delta v$. The net work done by such a force thus amplifies the perturbation, making it *unstable*.

However, in a study aimed at clarifying the nature of wave propagation through a CAK wind model, Abbott [64] took the quite different approach of simply assuming that the perturbed force would scale in the same way as the mean force, namely with the velocity *gradient*, i.e., $\delta g_{\mathrm{abs}} \propto \delta v'$. For the usual assumption of sinusoidal perturbations, this leads to a 90° phase difference between δg_{abs} and δv, with the consequence that *no net work* is done by the line-force, implying now that the perturbation is *stable*.

Owocki and Rybicki [39] reconciled these apparently contradictory results by determining the effect of a perturbation on the absorption-line-force (3) without making either of these assumptions. They thereby derived a simple "bridging" law,

$$\frac{\delta g_{\mathrm{abs}}}{\delta v} = \Omega_{\mathrm{b}} \, \frac{ik}{\chi_{\mathrm{b}} + ik}, \tag{19}$$

where the blue-edge line strength χ_{b} determines a bridging length, given roughly by half the Sobolev length, i.e., $\chi_{\mathrm{b}}^{-1} \approx L/2$. This length bridges the stable, long-wavelength ($k \ll \chi_{\mathrm{b}}$), Sobolev regime, for which $\delta g_{\mathrm{abs}} \propto ik\delta v \propto \delta v'$, and the unstable, short-wavelength ($k \gg \chi_{\mathrm{b}}$), optically-thin-perturbation regime, for which $\delta g_{\mathrm{abs}} \propto \delta v$. The instability growth rate is $\Omega_{\mathrm{b}} \approx 2 g_{\mathrm{abs}}/v_{\mathrm{th}}$, i.e., roughly twice the rate at which the mean flow is accelerated by a thermal speed. Since $g_{\mathrm{abs}} \approx v\, dv/dr$, we see also that $\Omega_{\mathrm{b}} \approx 2v/L$, i.e., the instability grows at twice the flow rate through a Sobolev length. Comparison with the wind expansion rate through a typical scale length $H \equiv \rho/(d\rho/dr) \approx v/(dv/dr)$ shows that $\Omega_{\mathrm{b}}/\Omega_{\mathrm{exp}} \approx 2H/L = 2v/v_{\mathrm{th}}$, which can approach 100 in these highly supersonic winds. An infinitesimal (i.e., linear) perturbation advected outward through the wind is thus predicted to have a cumulative growth on the order of $e^{v/v_{\mathrm{th}}} \approx e^{100}$! In practice this means, of course, that any small, but finite, amplitude perturbation would quickly become nonlinear in the wind. Efforts to determine the likely form of this nonlinear structure are discussed below (Sects. 4.3-4.5).

Although the wind driving is primarily by line-scattering, the above perturbation analysis assumes that, just as in the mean flow (See Sect. 3.), one may ignore the dynamical effect of the diffuse radiation field, and treat only the direct *absorption* component of the perturbed line force. However, Lucy [65] showed that, unlike the fore-aft symmetry seen in the comoving frame of the mean flow, the diffuse field seen by a small-scale velocity perturbation δv is stronger by a factor $\delta v/v_{\mathrm{th}}$ against the direction of the perturbed velocity. This diffuse field thus exerts a net *drag* force on such perturbations that is on the same order as the perturbed direct force that gives rise to the instability. In fact, in Lucy's coherent scattering analysis, this drag effect *exactly cancelled* the instability near the wind base, where the flow is nearly plane parallel. Under the more realistic assumption of complete redistribution, Owocki and Rybicki [40] found complete cancellation occurs only when the flow is plane parallel *and* a non-limb-darkened stellar core fills an entire hemisphere. Away from the wind base, the combined effects of spherical expansion and the shrinking of the solid angle subtended by the stellar core quickly reduce the relative importance of the drag term. At $r = 1.5R_*$, for example, the net growth rate is about half that obtained from a pure-absorption analysis, but this still implies a strong instability, with cumulative growth now about 50 (vs. 100) e-folds. This strong instability is also found not to be substantially reduced by any of several other potential stabilizing effects, e.g., line-overlap [40], radiative cooling, or thermal conduction [39].

In addition to reducing somewhat the growth rate of *radial* velocity perturbations, this line-drag effect tends to strongly dampen *horizontal* velocity perturbations (for which the countervailing direct term is much weaker) [66]. Hence any perturbations that are initially unpolarized at the wind base will become predominantly radially polarized not far into the wind. In this regard the wind and its variations can thus be reasonably treated within a simple 1-D, spherically symmetric model, although in general there will still be horizontal phase variations that will require a 2-D or 3-D treatment.

4.3 Heuristic Models of a Wind with Embedded Shocks

Given the supersonic nature of the wind outflow and the large instability growth rate, it seems inevitable that, whatever its detailed form, the nonlinear wind structure arising from this instability will include shocks. (A useful review of the properties of such wind shocks has recently been given by Castor [67]). Lucy [33] has proposed a heuristic model that assumes that this detailed structure consists, in fact, of a quasi-periodic train of *forward* shocks. As shown in Fig. 4, the assumed velocity structure has a ramped sawtooth character; each tip represents relatively fast material that sees unshadowed stellar flux and so is strongly driven, while each trough represents relatively slow material that is shadowed by the fast material and so is only weakly driven. As the fast material is pushed by the radiation force against the slower material, a forward shock forms that sweeps up and accelerates this slower material, thereby adding it to the fast material. A crucial point is that the fast material at each tip represents *post-shock* flow, with its associated *high density* and (at least initially) *high temperature*. To maintain the structure, this material must quickly cool and reform the driving ions that line-absorb the radiative momentum. Assuming that it does so, material at this tip then drifts backward, due to the influence of inward gravitational, inertial, and pressure forces that counter the outward radiative acceleration. At a velocity about a sound speed below the tip, the radiative term becomes sharply diminished by shadowing effects, dramatically increasing the inward drift and quickly bringing material toward the next trough, whereupon it can again be accelerated in the next shock.

The question of the ionization and energy balance behind the shock was analyzed by Krolik and Raymond [68] for a single, forward shock that propagates outward through the wind. They found that the column depth required to cool shocked material back to the ambient wind temperature scaled with the shock velocity Δv as

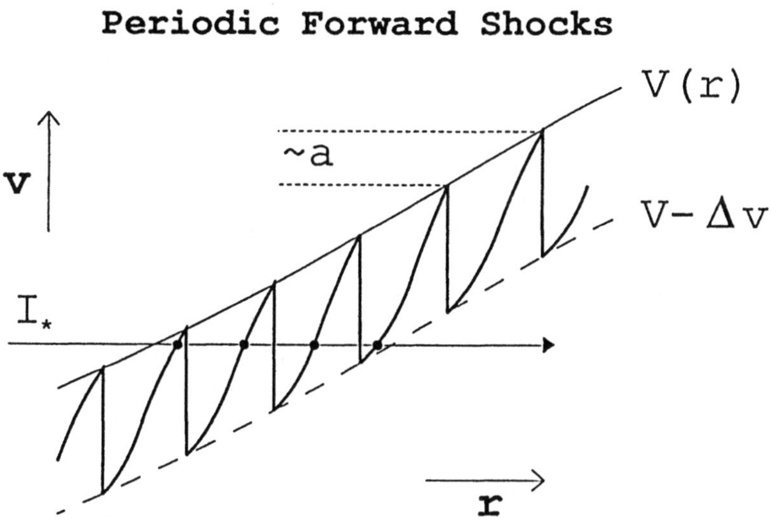

Fig. 4. Velocity versus height in Lucy's model of periodic forward shocks.

$$N_{\text{cool}} \approx 7 \times 10^{18} \left(\frac{\Delta v}{100\,\text{km/s}} \right)^4 \text{cm}^{-2}. \tag{20}$$

Comparing this to a characteristic value $N_{\text{wind}}(r) = nr$ for the total wind column depth at a radius r, we obtain an expression for the critical radius at which the assumption of cooling must break down,

$$r_c \approx 10^{14}\,\text{cm} \left(\frac{\dot{M}}{10^{-6} M_\odot/\text{yr}} \right) \left(\frac{v}{1000\,\text{km/s}} \right)^{-5} \left(\frac{\Delta v}{v/2} \right)^{-4}. \tag{21}$$

Krolik and Raymond used rather elaborate arguments to obtain the approximate expression (See their Eq. (11).),

$$\Delta v \approx v \left(\frac{N_{\text{thick}} L_*}{\dot{M} c^2} \right), \tag{22}$$

relating the shock strength to basic stellar parameters and the number of optically thick lines (f in their notation). But recall from Sect. 3 that the factor in parenthesis in Eq. (22) is necessarily of order unity, implying that $\Delta v \approx v$. In reality, of course, the line driving is divided between accelerating the mean flow and amplifying the instability, and so typically $\Delta v \approx v/2$. If one takes these points into account, then the somewhat different expression derived by Krolik and Raymond for r_c reduces to Eq. (21). Actually, since wind structure can be expected to form on a scale much smaller than the entire wind, breakdown of the cooling should occur at a radius much smaller than r_c, perhaps at $\sim r_c/10$.

Recently, MacFarlane and Cassinelli [69] studied the evolution of wind shocks with a phenomenological numerical-hydrodynamics model, meant to apply to the relatively weak wind from the main sequence B0 star τ Sco. The wind is assumed initially to be in a smooth and relatively slow state, but then a driving force is applied that accelerates the material to a much higher speed. The compression between the fast and slow material quickly forms a dense shell bounded on the front and back sides by a forward and reverse shock pair. Both the driving force and the initially slow flow-speed represent purely phenomenological free parameters, the former chosen to accelerate the fast wind to the observed terminal speed, and the latter chosen to give shocks of the strength necessary to match X-ray observations. Nonetheless, the temporal evolution of the temperature in the resulting shock-heated, dense shell is computed from a reasonably complete energy balance, including an optically thin, radiative cooling function. Ionization lag effects are ignored and so the radiative emission is assumed to be a known function of temperature, which is used to derive the X-ray emission spectrum from the shocks. For parameters chosen to give initial and final state terminal speeds of 500 km/s and 2500 km/s, the forward and reverse shocks each attain velocity amplitudes approaching 1000 km/s, and this turns out to be just what's needed to reproduce the observed X-ray properties.

Another heuristic wind-structure model worth mentioning is Mullan's application [70, 71] of the well-known solar wind phenomena of "co-rotating interaction regions" [72, 73] to the case of hot-star winds. The winds here again contain dense shells bounded by forward/reverse shock pairs, but now this is not thought of as arising from the line-driven instability, but rather from an assumed azimuthal variation in the wind flow speed and the interaction between fast and slow wind streams that results from the stellar rotation. Mullan argues that the resulting shocks and dense shells can give rise to the observed X-ray emission and discrete absorption components in UV lines. Although there is no independent evidence that O star winds have such an azimuthal stream structure, the effect of rotation in laterally extending any structure forming from the wind instability is likely to be quite similar.

4.4 Radiation-Hydrodynamical Simulation of Non-Linear Structure

Owocki, Castor, and Rybicki (hereafter OCR) [44] have recently developed a numerical, radiation-hydrodynamics code aimed at directly simulating the dynamical evolution of the line-driven wind instability, and thereby determining the likely nature of the resulting nonlinear wind structure. Because the instability occurs for perturbations with a length scale near and below the Sobolev length, OCR had to develop a method for computation of the line force which did not use the Sobolev approximation, but which still avoided the inordinate computational expense of solving the full line-transfer problem at each time step. The crucial simplification they adopted was to ignore the diffuse, scattered radiation, and to assume that the flow is driven by an ensemble of nonoverlapping, *pure-absorption* lines. By integrating the pure-absorption force expression (3) over a power-law number distribution $N(\kappa) \sim \kappa^{\alpha-2}$, we see that this combined force is of the form

$$g_{\rm tot}(r) \sim \frac{L_*}{r^2} \int_{-\infty}^{\infty} dx\ \phi(x - v(r)/v_{\rm th})\ \eta^{-\alpha}(x,r) \tag{23}$$

where $\eta(x,r)$ is a profile-weighted mass-column-depth given by

$$\eta(x,r) = \int_{R_*}^{r} \rho(r')\phi(x - v(r')/v_{\rm th})dr' + \eta(x, R_*). \tag{24}$$

In a smooth flow with all variations on a scale much larger than a Sobolev length L, this force appropriately reduces to the CAK form (12) (See Sect. 3); but in the presence of structure on a scale smaller than L, it also correctly includes the local Doppler-shift effects that give rise to the line-driven instability. Although much simpler than solving scattering-line transfer, this pure-absorption force is still much more complicated than the usual body forces assumed in hydrodynamical simulations, requiring at each time step a numerical evaluation of the double integral in x and r that dominates the code timing.

In order to focus on such dynamical terms, OCR avoided treatment of detailed energy balance, and simply assumed that radiative heating and losses would keep the wind nearly isothermal. This means they could not directly compute effects that depend heavily on the temperature, e.g., X-ray emission. In emphasizing dynamics but neglecting energy balance, the OCR calculation is in a sense the complement of that of MacFarlane and Cassinelli [69], who treated a detailed energy balance but assumed phenomenological force terms. (See Sect. 4.3.)

For a model with typical O-star parameters, Fig. 5 shows the wind spatial structure that results long (10^5 s) after introduction of a 1% amplitude, 4000 s period sound wave propagating outward from the wind base. The strong correlation of the driving force with flow speed quickly amplifies this initially small perturbation, giving rise to velocity variations $\Delta v \gtrsim 500\,$km/s by $r \approx 1.5R_*$. The base perturbation is initially *outward* propagating, and thus has velocity and density variations that are *in phase*, but the resulting wind structure has velocity and density with *opposite phase*. This turns out to be a very robust result, stemming directly from the linear instability property [39] that such opposite-phased waves are much more unstable. In fact, expected nonlinear dynamical effects like line-shadowing actually play little role, and so the waves grow at the rate predicted by linear theory up to the highly nonlinear amplitude $\Delta v \gtrsim 500\,$km/s, at which they become steepened by the simple kinematic effect of faster flow overrunning the slower. Although advected away from the star by the supersonic flow, these unstable waves actually propagate *inward* relative to the fluid, and so as they steepen they naturally evolve into *reverse*, not forward, shocks.

It is important to emphasize that, although the velocity structure in the hydrodynamical model of Fig. 5 appears quite similar to that assumed in the Lucy periodic shock model of Fig. 4, the density structure of the two models, and hence their entire nature, is actually fundamentally different. As described in Sect. 4.3, in Lucy's model *forward* shocks were assumed to abruptly *accelerate* ambient wind material as it is rammed by a dense, strongly driven flow. In

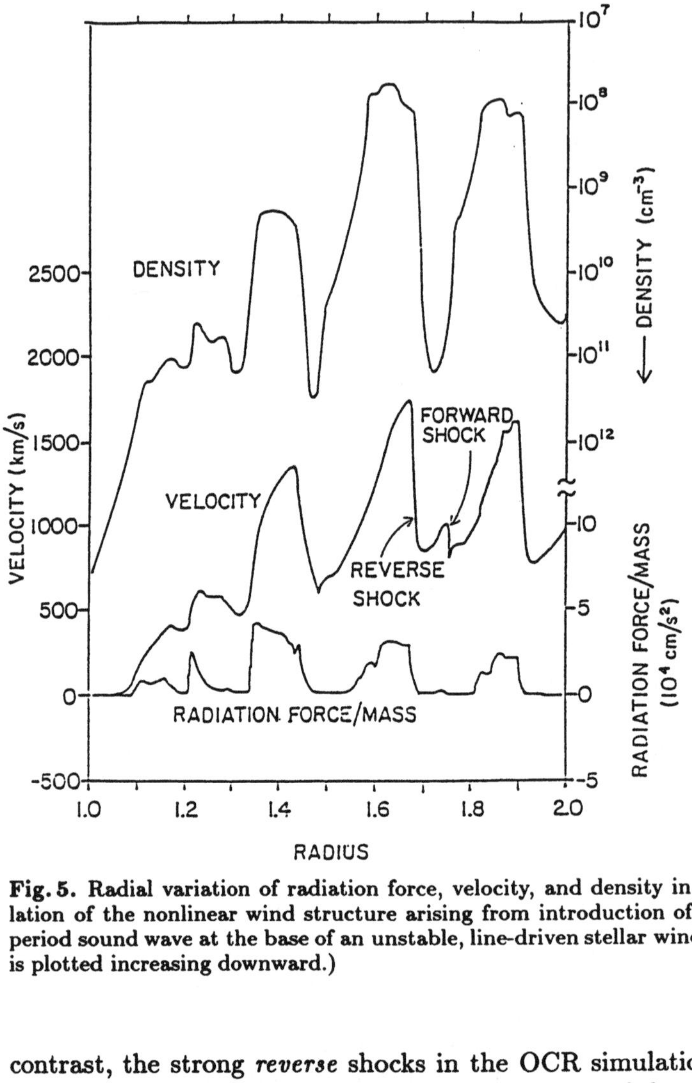

Fig. 5. Radial variation of radiation force, velocity, and density in OCR's numerical simulation of the nonlinear wind structure arising from introduction of a 1% amplitude, 4000 s period sound wave at the base of an unstable, line-driven stellar wind. (Note that the density is plotted increasing downward.)

contrast, the strong *reverse* shocks in the OCR simulation arise to *decelerate* high-speed, rarefied flow as it impacts slower material that has been compressed into dense shells. Since the high-speed flow is accelerated before being shocked, there is no need to assume rapid cooling to reform driving ions. In fact, the very low density of this high-speed flow implies that, unlike Lucy's model, only a very small fraction ($\sim 10^{-3}$) of the wind material ever undergoes a shock near the maximum amplitude $\Delta v \approx 1000\,\mathrm{km/s}$.

These overall characteristics of this dynamically computed wind structure seem in good qualitative correspondence with several observational properties. For example, the resulting shock strengths $\Delta v \lesssim 1000\,\mathrm{km/s}$ are roughly what's needed to obtain the 10^7 K plasma inferred from the high-energy tail observed in X-ray spectra, and the small fraction of strongly shocked material implies that a similarly small fraction $\lesssim 10^{-3}$ of the total flow kinetic energy $\dot{M} v_\infty^2/2 \approx 10^{-3}\, L_*$ goes into shock heating, which agrees roughly with the observed X-ray luminosity scaling $L_x \approx 10^{-7} L_*$. The dense shells should naturally give rise to narrow line absorption features similar to the discrete absorption components observed in unsaturated lines, and the high-speed, rarefied flow should give rise to non-sharp, variable edges in saturated lines, which is again what's often observed [61]. Although it remains to be shown in detail, it also seems likely that the nonmonotonic velocity field will give rise to the observed black absorption troughs in saturated lines.

4.5 Intrinsic Nature of Wind Variability

The structure shown in Fig. 5 arises from amplification of an explicit perturbation introduced at the wind base, but subsequent work [74, 75] has shown that such winds can also often exhibit an *intrinsic* variability that persists even in the absence of such explicit perturbations. The incidence of this intrinsic variability was found empirically to depend on the assumed value of v_{th}/a, the ratio of the ion thermal speed to sound speed. For the usual case of driving by CNO ions, an appropriate value of this ratio is $v_{th}/a \approx 0.3$. However, in order to study the effect of variability arising from explicit perturbations without the complication of a background model that was itself intrinsically variable, OCR found it necessary to assume an artificially high value $v_{th}/a \gtrsim 1/2$. The situation is graphically illustrated in Fig. 6, which shows the spatial and temporal variation of velocity in two unperturbed models that differ only in the assumed v_{th}/a. In both cases the flow is initially disrupted because the assumed CAK/Sobolev initial condition is not an appropriate steady state for either non-Sobolev model. The model with $v_{th}/a = 1/2$, however, quickly relaxes to a somewhat steeper [75] steady solution, whereas the model with slightly smaller $v_{th}/a = 3/8$ never settles down, but exhibits a nearly periodic variability.

Poe, Owocki, and Castor (POC) [75] showed that the different variability properties of these two cases reflect a difference in the nature of the corresponding steady solutions. Fig. 7 illustrates the steady-state solution topology near the critical (sonic) point. Unlike the usual saddle- or X-type solutions that apply, e.g., to the solar wind (Fig. 7a), the solution topology in this case is of the *nodal* type, with not one but *two* positive slope critical solutions (Fig. 7b). Note that along the steeper slope there is only one distinct solution passing though the critical point, whereas along the shallower slope there is a *degenerate family* of solutions that pass through this point. POC showed that, for reasonable boundary conditions, the distinct, steeper solution applies when $v_{th}/a \gtrsim 1/2$, whereas the degenerate, shallower solution applies when $v_{th}/a \lesssim 1/2$. Apparently, the existence of a well-defined steady solution in the former case is sufficient to enable a time-dependent model without explicit perturbations to relax to this steady state. On the other hand, the lack of a well-defined steady solution for the latter case leads in the time-dependent model to an intrinsic variability in which the flow, roughly speaking, continuously varies among this degenerate family of possible steady solutions. (Recall from Sect. 3 that a similar solution degeneracy exists for Sobolev models in the limit of vanishing sound speed $a \to 0$.)

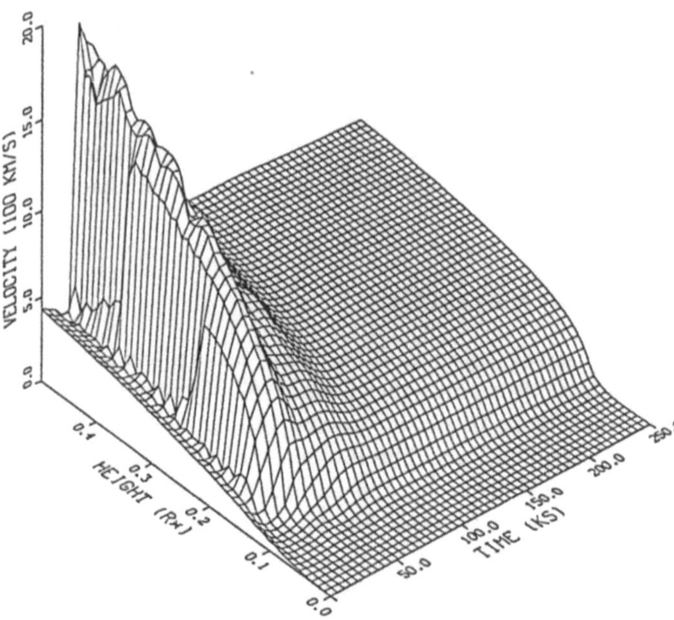

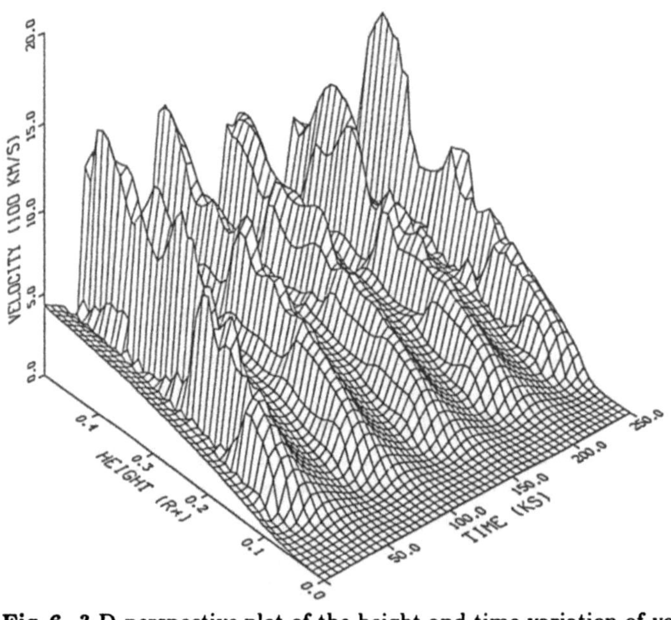

Fig. 6. 3-D perspective plot of the height and time variation of velocity in two unperturbed wind models which differ only in the assumed values of the ratio of thermal speed to sound speed.

Fig. 6 shows that there are brief intervals when the latter flow approaches a nearly smooth, CAK-like state, but this becomes disrupted by the strong amplification of small scale structure, which is then advected away, again allowing the flow to settle temporarily into a nearly smooth state. The time scale (~1/2 day) for this *relaxation oscillation* is much longer than the insta-

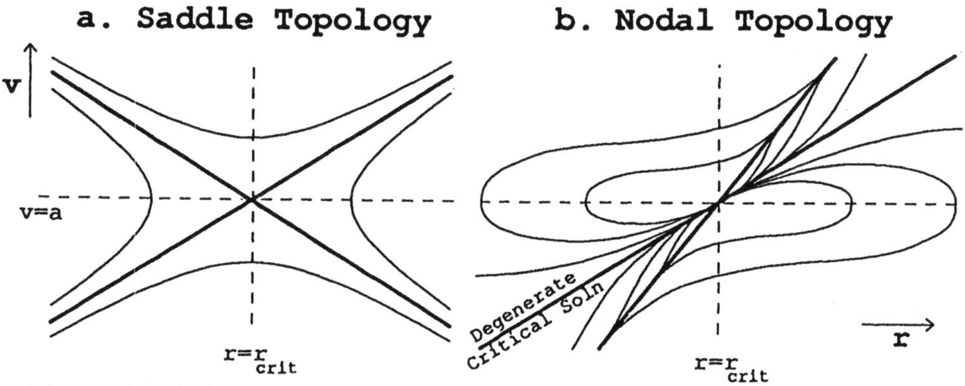

Fig. 7. The velocity near the sonic radius for solution topologies of (a) the usual saddle type that applies to the solar wind and (b) a nodal type that applies to absorption-line-driven winds. Note in the nodal case how a large number of solutions converge as they approach the sonic point along the shallower of the two positive critical solutions (solid lines). The implied solution degeneracy leads to the intrinsic variability seen in the right panel of Fig. 6.

bility growth time (~1 hour) of the individual small scale instabilities, and corresponds roughly with the observed repetition time of discrete absorption features (Sect. 4.1) [61, 62]. Indeed, the time-variations of synthesized absorption profiles show moving narrow absorption features quite reminiscent of such discrete absorption components [35, 36]. These arise from outward accelerating dense shells whose formation is repeatedly triggered by the quasi-periodic disruptions. It is interesting that, all by itself, such a unstable wind seems to give rise to structure with many of the qualitative features needed to explain such observational signatures of the wind structure.

4.6 Implications for Steady-State Models

If hot-star winds are indeed as temporally and spatially structured as implied by Figs. 5 and 6, how is it possible that steady-state models, which completely ignore this structure, can have been so successful in quantitatively matching most properties of time-averaged spectra?

In this regard, it is first worth noting that, despite the extensive wind structure, the gross wind properties like the *terminal* flow speed and *time-averaged* mass loss rate turn out to be in quite good agreement with those derived in steady models. Moreover, as noted above, only a small fraction of the mass in this wind is actually accelerated in a highly time-dependent way; indeed, the acceleration history of most of the mass is quite similar to what it would be in a smooth wind. From this point of view, the major effect of the instability is thus simply to spatially clump material without otherwise altering its properties. Given the intrinsic mass-weighting of spectral formation, and the extensive temporal and spatial averaging involved, it thus does not seem surprising that most spectral features would be quite similar in the smooth and structured flows.

A second point is that it is not at all clear whether the intrinsic variability characteristic of this pure-absorption model is really applicable to actual hot-star winds, for which the driving is actually by line-scattering. As noted in Sect. 3, the near fore-aft symmetry of the diffuse radiation field means that the associated net force is typically small in the supersonic regions of a smooth flow; but in the region near and below the sonic point, the radiation escape probability is rapidly increasing, leading to a diffuse field asymmetry and hence to a kind of a radiative viscous force. This can alter or even "break" the solution degeneracy found in pure-absorption models [76], and so might explain why comoving frame models that include scattering appear to have a well-defined solution [25]. Through the line-drag effect (Sect. 4.2), scattering might also stabilize the wind base enough to suppress variations that govern the intrinsic variability. The wind instability would then have an *advective* rather than *absolute* character [77, 78, 74], for which variability would persist only with some explicit driving from the underlying star, and would only become large amplitude in the supersonic wind.

5 Future Directions

Clarifying the dynamical role of scattering should thus be a primary area of focus in future work on hot-star winds. This refers to both its possible effect in regulating the wind instability, as well as its likely direct role in driving the optically thick winds from Wolf-Rayet stars, for which there is still no satisfactory dynamical theory of even the mean wind outflow. For OB stars, additional refinements in the steady-state wind models can be expected to further improve the already good quantitative agreement with time-averaged spectral observations. In my view one particularly important refinement will be to consider, if only in a phenomenological way, how some of the consequences of wind structure – e.g, X-ray heating, ionization – affect the overall properties of the time-averaged flow [79].

With regard to the direct study of the wind structure and variability itself, there is obviously a great deal of work still to be done to include the many potentially important effects neglected so far, e.g., scattering, detailed energy and ionization balance, shock X-ray emission, 2-D or 3-D structure, and stellar rotation. In addition to clarifying the role of scattering in regulating the instability, there is the question of what dynamical role it may play in the non-linear wind structure. Inclusion of a realistic energy balance in the dynamical wind models is needed to determine the temperature structure and thereby the shock X-ray emission. Likewise, inclusion of ionization balance is needed to determine whether ionization from shock heating and/or X-rays can effect the line-driving and hence the dynamics of the wind and instability. Consideration of rotation and other 2-D (and ultimately 3-D?) effects is needed to determine the likely lateral scales of the dense clumps. For example, can rotation string out such structures in longitude in a manner similar to corotating interaction regions in the solar wind [70, 71] (Sect. 3), and thus make it possible for a given dense clump to cover a large enough fraction of the stellar disk to produce the observed discrete absorption components? Finally, perhaps one of the most urgent needs is to develop methods to compare more closely predictions from the theoretical simulations with available observational diagnostics, and thereby guide and test the further theoretical development in this complex but fascinating problem in radiation hydrodynamics.

This work was supported in part by NSF grant AST 88-14580 and NASA grant NAGW-1487. Many of the computations were carried out using an allocation of supercomputer time from the San Diego Supercomputer Center. I thank G. Cooper and A. Fullerton for many helpful discussions, questions, and comments.

References

1. D.C. Abbott: In *Proceedings of the Sixth International Solar Wind Conference, Vol. 1*, ed. by V.J. Pizzo, T.E Holzer, and D.G. Sime (NCAR, Boulder 1988) p.149
2. P.S. Conti, A.B. Underhill: *O Stars and Wolf-Rayet Stars* (NASA, Washington 1988)
3. J.P. Cassinelli: Ann. Rev. Astron. Astrophys. **17**, 275 (1979)
4. A.G. Hearn: In *The Origin of Nonradiative Heating/Momentum in Hot Stars*, ed. by A.B. Underhill and A.G. Michalitsianos (NASA, Washington 1984) p.188
5. D.C. Abbott: In *Relations Between Chromospheric-Coronal Heating and Mass Loss in Stars*, ed. by R. Stalio and J.B. Zirker (Tabographics-TS, Trieste 1985) p.267
6. L.B. Lucy: In *Radiation Hydrodynamics in Stars and Compact Objects, IAU Colloq. No. 89*, ed. by D. Mihalas and K.-H.A. Winkler (Springer, Berlin 1986) p.75
7. R.P. Kudritzki, A. Pauldrach, J. Puls: In *New Insights in Astrophysics* (ESA, Paris 1986) p.247
8. R.P. Kudritzki, D.G. Hummer: In *Luminous Stars and Associations in Galaxies, IAU Symp. No. 95*, ed. by C.W.H. de Loore, A.J. Willis, D. Laskarides (Reidel, Dordrecht 1989) p.3
9. P.S. Conti: Ann. Rev. Astron. Astrophys. **16**, 371 (1978)
10. M.J. Barlow: In *Wolf-Rayet Stars: Observations, Physics, and Evolution, IAU Symp. 99*, ed. by C.W.H. de Loore and A.J. Willis (Reidel, Dordrecht 1982) p.149
11. J.P. Cassinelli, H.G.J.L.M. Lamers: In *The Scientific Accomplishments of the IUE*, ed. by Y. Kondo *et al.* (Reidel, Dordrecht 1987) p.139
12. D.C. Abbott: Astrophys. J. **263**, 723 (1982)
13. J.M. Blondin, T.R. Kallman, B.A. Fryxell, R.E. Taam: submitted to Astrophys. J. (1989)
14. P.P. Eggleton and J.E. Pringle: *Interacting Binaries*, (Reidel, Dordrecht 1985)
15. W.H.G. Lewin and E.P.J. van den Heuvel: *Accretion in Stellar X-ray Sources*, (Cambridge University Press, Cambridge 1983)
16. D. Mihalas, B.W. Mihalas: *Foundations of Radiation Hydrodynamics*, (Oxford University Press, Oxford 1984)
17. E.A. Milne: Monthly Notices Roy. Astron. Soc. **86**, 459 (1926)
18. V.V. Sobolev: Soviet Astr. J. **1**, 678 (1957)
19. V.V. Sobolev: *Moving Envelopes of Stars*, (Harvard University Press, Cambridge 1960)
20. L.B. Lucy, P.M. Solomon: Astrophys. J. **159**, 879 (1970)
21. D.C. Morton: Astrophys. J. **150**, 535 (1967)
22. D. Mihalas: *Stellar Atmospheres (2nd edition)*, (Freeman, San Francisco 1978)
23. J.I. Castor, D.C. Abbott, R.I. Klein: Astrophys. J. **195**, 157 (1975)
24. D.B. Friend, D.C. Abbott: Astrophys. J. **311**, 701 (1986)

25. A. Pauldrach, J. Puls, R.P. Kudritzki: Astron. Astrophys. **164**, 86 (1986)
26. L.B. Lucy: Astrophys. J. **163**, 95 (1971)
27. J.I. Castor: Monthly Notices Roy. Astron. Soc. **169**, 279 (1974)
28. S.V. Weber: Astrophys. J. **243**, 954 (1981)
29. J.P. Cassinelli, G.L. Olson: Astrophys. J. **229**, 304 (1979)
30. F.R. Harnden *et al*: Astrophys. J. **239**, L65 (1979)
31. J.P. Cassinelli, J.H. Swank: Astrophys. J. **271**, 681 (1983)
32. D.C. Abbott, J.H. Bieging, E. Churchwell: Astrophys. J. **280**, 671 (1984)
33. L.B. Lucy: Astrophys. J. **255**, 278 (1982)
34. H.J.G.L.M. Lamers, D.C. Morton: Astrophys. J. Suppl. **32**, 715 (1976)
35. R.K. Prinja, I.D. Howarth, H.F. Henrichs: Astrophys. J. **317**, 389 (1987)
36. R.K. Prinja, I.D. Howarth: Monthly Notices Roy. Astron. Soc. **233**, 123 (1988)
37. K.B. MacGregor, L. Hartmann, J.C. Raymond: Astrophys. J. **231**, 514 (1979)
38. R.G. Carlberg: Astrophys. J. **241**, 1131 (1980)
39. S.P. Owocki, G.B. Rybicki: Astrophys. J. **284**, 337 (1984)
40. S.P. Owocki, G.B. Rybicki: Astrophys. J. **299**, 265 (1985)
41. G.B. Rybicki: In *Instabilities in Luminous Early Type Stars*, ed. by H.J.G.L.M. Lamers and C.W.H. de Loore (Reidel, Dordrecht 1987) p.175
42. L.B. Lucy, R.L. White: Astrophys. J. **241**, 300 (1980)
43. L.B. Lucy: Astrophys. J. **255**, 286 (1982)
44. S.P. Owocki, J.I. Castor, G.B. Rybicki: Astrophys. J. **335**, 914 (1988)
45. D.B. Friend, J.I. Castor: Astrophys. J. **272**, 259 (1983)
46. D.C. Abbott, L.B. Lucy: Astrophys. J. **288**, 679 (1985)
47. J. Puls: Astron. Astrophys. **184**, 227 (1987)
48. D.C. Abbott, P.S. Conti: Ann. Rev. Astron. Astrophys. **25**, 113 (1987)
49. K.B. MacGregor, D.B. Friend, R.L. Gilliland: submitted to Astron. Astrophys. (1990)
50. K.B. MacGregor, D.B. Friend: Astrophys. J. **312**, 659 (1987)
51. C.H. Poe, D.B. Friend: Astrophys. J. **311**, 317 (1986)
52. C.H. Poe, *The Effects of Rotation on the Winds from Hot Stars*, (PhD. Thesis, Univ. Wisconsin), (1987)
53. C.H. Poe, D.B. Friend, J.P. Cassinelli: Astrophys. J. **337**, 888 (1989)
54. A. Pauldrach: Astron. Astrophys. **183**, 295 (1987)
55. R. Gabler, A. Gabler, R.P. Kudritzki, J. Puls, A. Pauldrach: Astron. Astrophys. **226**, 162 (1989)
56. W.L. Waldron: Astrophys. J. **282**, 256 (1984)
57. T. Chlebowski: Astrophys. J. **342**, 1091 (1989)
58. J.H. Bieging, D.C. Abbott, E.B. Churchwell: Astrophys. J. **340**, 518 (1989)
59. R.L. White: Astrophys. J. **289**, 698 (1985)
60. W. Chen, R.L. White: submitted to Astrophys. J. (1990)
61. R.K. Prinja: Monthly Notices Roy. Astron. Soc **231**, 21P (1988)
62. H.F. Henrichs, L. Kaper, G.A.A. Zwarthoed: In *A Decade of UV Astronomy with the IUE Satellite, Vol. 2*, ed. by E.J. Rolfe (ESA, Paris 1988) p.145
63. R.K. Prinja, M.J. Barlow, I.D. Howarth: Astrophys. J., in press (1990)
64. D.C. Abbott: Astrophys. J. **242**, 1183 (1980)
65. L.B. Lucy: Astrophys. J. **284**, 351 (1984)
66. G.B. Rybicki, S.P. Owocki, J.I. Castor: Astrophys. J. **349**, 274 (1990)
67. J.I. Castor: In *Instabilities in Luminous Early Type Stars*, ed. by H.J.G.L.M. Lamers and C.W.H. de Loore (Reidel, Dordrecht 1987) p.159
68. J.H. Krolik, J.C. Raymond: Astrophys. J. **298**, 660 (1985)
69. J.J. MacFarlane, J.P. Cassinelli: Astrophys. J. **347**, 1090 (1989)
70. D.J. Mullan: Astrophys. J. **283**, 303 (1984)
71. D.J. Mullan: Astron. Astrophys. **165**, 157 (1986)
72. A.J. Hundhausen: J. Geophys. Res. **78**, 1528 (1973)
73. V.J. Pizzo: In *Solar Wind Five*, ed. by M. Neugebauer (NASA, Washington 1983) p.675
74. S.P. Owocki, C.H. Poe, J.I. Castor: In *Properties of Hot Luminous Stars*, ed. by C.D. Garmany (Astronomical Society of the Pacific, San Francisco 1990) p. 283
75. C.H. Poe, S.P. Owocki, J.I. Castor: Astrophys. J., **355**, in press (1990)
(Springer: New York 1972)
76. S.P. Owocki, G. Zank: submitted to Astrophys. J.

77. A. Bers: In *Handbook of Plasma Physics, Vol. I: Basic Plasma Physics*, ed. by A.A. Galeev and R.N. Sudan (North Holland, Amsterdam 1983) p.451
78. S.P. Owocki, G.B. Rybicki: Astrophys. J., **309**, 127 (1986)
79. A. Pauldrach: In *Properties of Hot Luminous Stars*, ed. by C.D. Garmany (Astronomical Society of the Pacific, San Francisco 1990) in press

Radiation Driven Winds of Hot Luminous Stars. Applications of Stationary Wind Models

A.W.A. Pauldrach and J. Puls

Institut für Astronomie und Astrophysik, Scheinerstr. 1,
D-8000 München 80, Fed. Rep. of Germany

ABSTRACT

In this paper we describe the status of a continuing effort to calculate NLTE models of rapidly expanding hot star atmospheres including the hydrodynamics selfconsistently. In order to test our theoretical concept, we calculated a wind model for the O4I(n)f star ζ-Pup. An additional test was performed by calculating stationary wind models along evolutionary tracks. The fact that such models satisfactorily reproduce the observed luminosity dependence of the Si IV resonance line indicates that the theory can soon be used for quantitative analysis of stellar wind lines. We show further that the calculated dynamical parameters can already be used for a direct spectroscopic determination of stellar masses and radii. This is done not only for massive stars in our galaxy, but also for an O-star in the SMC.

I. INTRODUCTION

Massive hot stars do not preserve their mass. As a consequence of their intense radiation field, which is scattered by thousands of UV metal lines, they loose a substantial fraction of their initial mass via rapidly expanding stellar winds. Lucy and Solomon (1970) were among the first to show that the radiative line force exerted by the metal lines is sufficient to initialize and to maintain stellar winds and Castor, Abbott and Klein (1975), Abbott (1979) and Abbott (1982) formulated the theory of radiation driven winds in a self-consistent manner for the first time. Due to many simplifications their approach was, however, only qualitative, so that severe discrepancies with the observations remained: for OB-stars the mass loss rates ($\dot{M}$) were systematically too large by a factor 2-3 (Abbott, 1982) and, far more significantly, the terminal velocity (v_∞), a quantity which is measurable quite easily, was predicted to be a factor 2-4 too small. Moreover, the observed ionization ratios were not matched by the calculations.

The latter discrepancies were especially pronounced for highly ionized species like N V and O VI (Pauldrach, 1987) pointing to the problem of "superionization" (Cassinelli and Olson, 1979).
Although the work of Castor, Abbott and Klein was ground-breaking, these results made clear that a quantitative description of the radiation driven wind theory requires a careful solution of the complete equations of radiative transfer and atomic statistics including hydrodynamics at least in a one-dimensional, steady state model. There are many reasons to want to tackle this problem, but perhaps the most important one regards the evolution of massive stars, which is strongly affected by the mass loss due to stellar winds. Note that the rates of mass loss are still considered as a free parameter in evolutionary calculations. Of similar importance is the need to determine accurate and complete sets of stellar parameters in order to determine the evolutionary status of the stars studied and to check stellar evolutionary scenarios. Such parameters include effective temperature, surface gravity and chemical composition, for the determination of which the methods of photospheric quantitative spectroscopy (Kudritzki, 1988) is an excellent tool, and the stellar luminosity and mass, which can be obtained from the analysis of stellar wind lines (Kudritzki, 1990; Pauldrach and Puls, 1990; Pauldrach et al., 1990). Since the radius follows directly from this parameter set, the stellar distance can in principle also be obtained purely spectroscopically, i.e. without the need for any other assumptions. Massive hot stars can thus be used as ideal distance indicators.
We describe our effort to achieve this result in sect. IV, and partly in sect. III, where it is also shown that our calculations reproduce the observed changes of the Si IV spectral morphology. For this purpose it seems to be convenient to summarise the concept and the status of our computational method, which is done in sect. II.

II. CONCEPT OF THE COMPUTATIONAL METHOD

As in the basic framework for computing line driven winds (Castor, Abbott and Klein, 1975) we assume a radially symmetric, stationary, one-component flow, which should be appropriate to describe the time average mean of all important spectral features correctly. The primary aim is to develop accurate simulations of the basic physics in order to determine mass-loss rates, masses and luminosities for hot stars by only comparing observed and calculated UV P-Cygni profiles. For these calculations the only input parameters required are the effective temperature, the chemical composition and a guess of the surface gravity. In order to extract the desired information from this very small set of parameters we just need a 'black box' which solves the

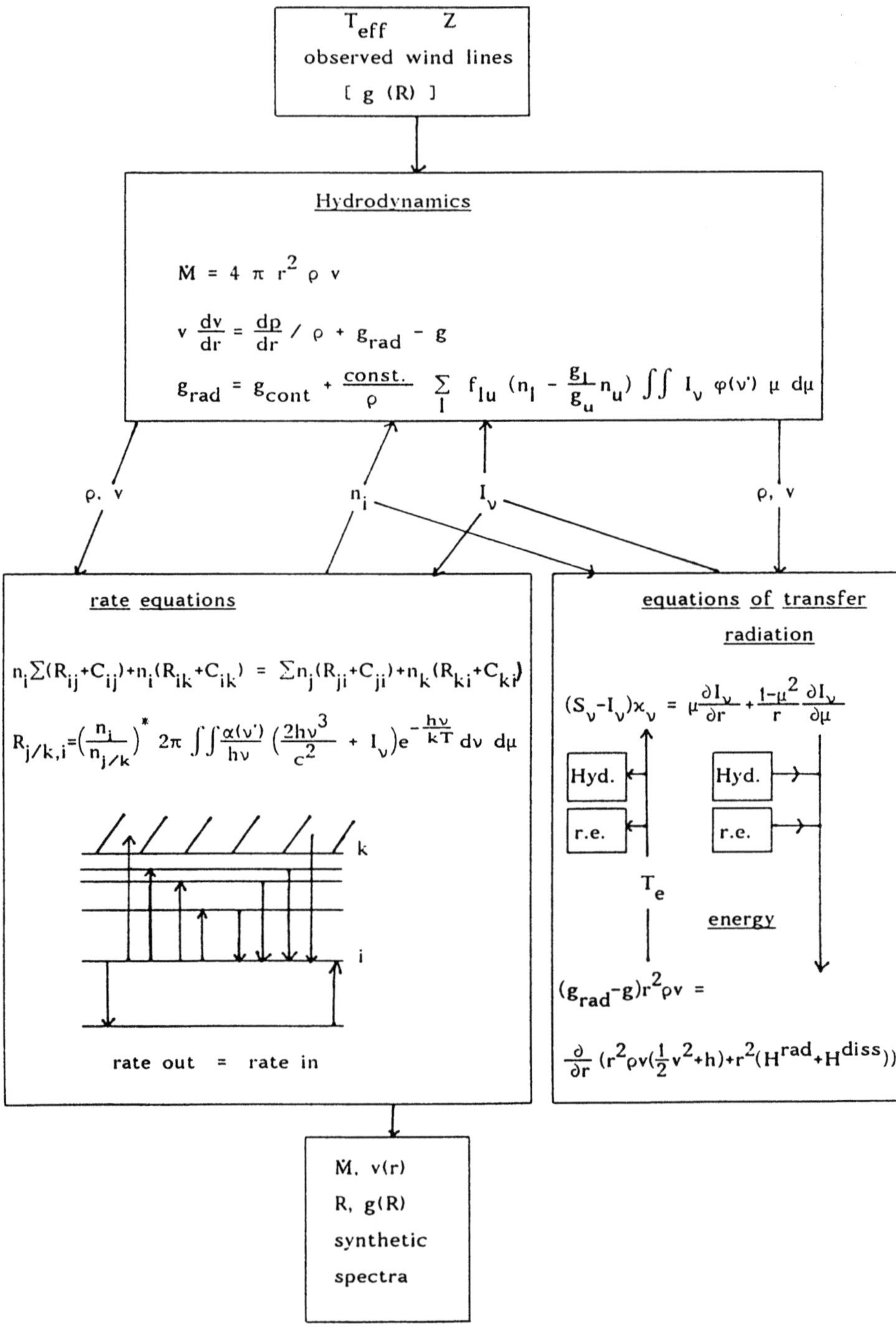

Fig. 1:
Schematic sketch of the extrem non-linear system of integro-differential equations which form the basis of radiation driven wind theory (symbols as usual - see e.g. Mihalas, 1978).

problem of time-independent radiation-hydrodynamics. As sketched in Fig. 1, this requires the simultaneous solution of several thousands of NLTE rate equations for the most abundant elements, of the line and continuum radiative transfer, the energy balance and the hydrodynamics. Since this is not an easy task, but a process requiring continuing effort, a final description is not yet available. However, as the theory has been drastically improved over the last years, the state of the art has reached a level which could be called "the first quantitative approach".

It is appropriate to summarise the present situation which led to this approach:

1. Detailed statistical equilibrium calculations are performed, (see Pauldrach, 1987) including:

   * 26 elements, 133 ionisation stages, 4000 levels, 10 000 radiative bound-bound transitions and electron collisions (Fig. 2a gives an

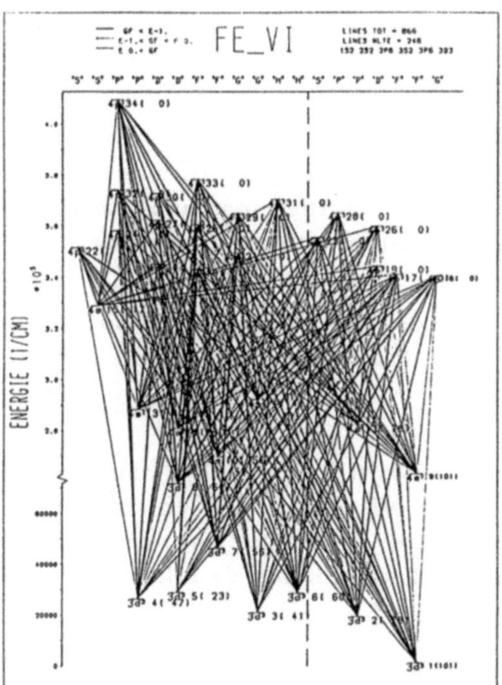

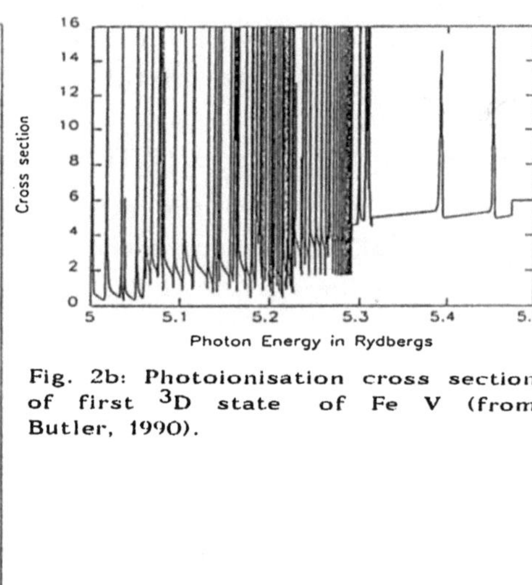

Fig. 2b: Photoionisation cross section of first ^3D state of Fe V (from Butler, 1990).

Fig. 2a: Grotian diagram of the used atomic model of Fe VI. On the right hand side of each level indicated: the excitation number and (in brackets) the number of lines for which only lower levels are considered, since the upper level lies above $5 \cdot 10^5$ cm^{-1} (data from K. Butler, private communication).

example of the atomic models used in our calculations. It should be noted that we have also begun to incorporate very detailed photoionisation cross sections (see Butler, 1990, and references therein) with all their resonances for all levels of the ions (see Fig. 2b)).

* the *correct* continuum radiative transfer with respect to bound-free opacities (up to 100 levels), free-free opacities of H, He and all metals and Thomson scattering. Since the NLTE-occupation numbers are a complicated function of the radiation field, an iteration cycle is required. For this iteration we used a modified form of the "accelerated lambda iteration" method, adapted to the case of O star winds (Pauldrach and Herrero, 1988).
* dielectronic recombination and autoionisation for C III and N III (data for other ions are on their way), using the method described by Mihalas and Hummer (1973). (Note that the wind conditions of O stars, which are not comparable to those of Planetary Nebulae, require the inclusion not only of the individual transition probabilities of each stabilizing downward transition considered, but also of the autoionisation process, which is not negligible)
* 100 000 lines in NLTE for the line force

2. In cases where the effect of multiple scattering turn out to be crucial, this is considered consistently for the most important 2000 or so lines (Puls, 1987; Puls and Pauldrach, these proceedings).
3. The hydrodynamical structure is computed by means of the improved theory of radiation driven winds which considers also stellar rotation in a simplified way (see Pauldrach, Puls and Kudritzki, 1986) and the whole procedure (1. to 3.) is iterated until convergence.

Very recently we took another step forward by extending our numerical concept of *wind* models to *atmospheric* models, which means that the wind is treated now together with the underlying photosphere (for details see Berger and Pauldrach, 1990; a short description is also given by Pauldrach et al., 1990b).

This approach to the theory of radiation driven winds is already self-consistent. However, one point should be borne in mind: our present treatment is still somewhat imperfect, because the energy equation is not solved consistently in the wind part, since in the winds of O-stars radiative equilibrium does not seem to be the proper assumption to make for the solution of the energy balance equations (Pauldrach et al., 1990b). The results of Pauldrach et al. indicate that energy is not only transported by radiation, but also by other dissipative mechanisms (H^{diss} in Fig. 1), which might be connected to instabilities in the wind (Owocki et al., 1989; Owocki, these proceedings), inferred from observations (e.g. Henrichs, 1988). Hence, up to now our procedure has been to choose a temperature structure based on empirical results (see Pauldrach et al., 1990, 1990b). The neglection of the presence of X-rays (see Cassinelli and Olson, 1979) and of NLTE line blocking and blanketing (the inclusion of these processes would require a NLTE temperature structure to be constructed also for the photospherical region) are additional shortcomings (see Pauldrach, 1990b).

Nevertheless, detailed comparisons with observations show that the basic ionization structure is not severely affected by these approximations and that, therefore, the dynamics of the average flow is treated correctly (see Pauldrach, 1987; Pauldrach et al., 1990a, 1990b). To demonstrate this, we report some results of our calculations for the O4f star ζ-Pup. For these calculations we adopted as input the parameters T_{eff} = 42 ± 2 kK, log g = 3.5 ± 0.15, R = 19 ± 8 $R_\odot$, determined by Kudritzki et al. (1983) and Bohannan et al. (1985) via detailed photospheric NLTE analyses. From our self-consistent wind calculations we obtained dynamical parameters (v_∞ = 2200 km/s, $\dot{M}$ = 3.8 · 10 $M_\odot$/yr) that agree quite well with the observed ones (v_∞ = 2200 ± 60 km/s, Groenewegen et al. (1989), $\dot{M}$ = 4.5 · 10^{-6} $M_\odot$/yr, Bieging et al. (1989)). The same agreement was found for our calculated wind profiles, which are a by-product of our self-consistent treatment, when compared to the observed ones (see Fig. 3). This example demonstrates clearly that the various improvements in calculating hot star winds led to models which are also useful in a quantitative sense.

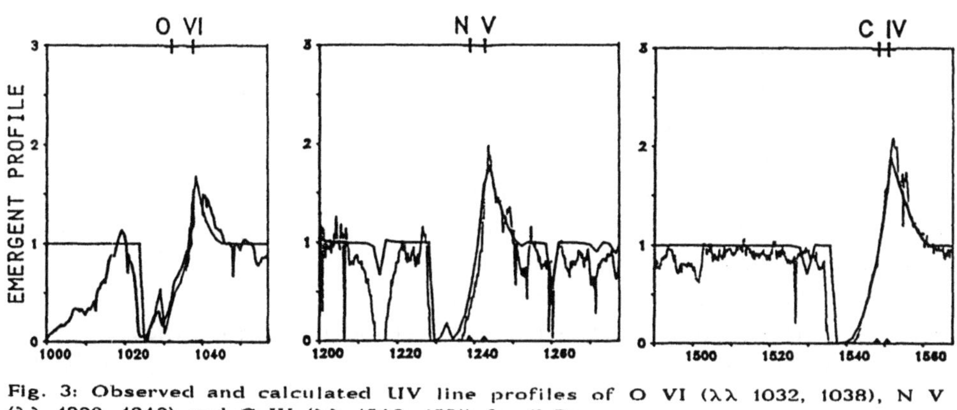

Fig. 3: Observed and calculated UV line profiles of O VI (λλ 1032, 1038), N V (λλ 1239, 1243) and C IV (λλ 1548, 1551) for ζ-Pup.

III. THE SPECTRAL MORPHOLOGY OF THE Si IV LINE - A MEANS FOR DIRECTLY DETERMINING THE LUMINOSITY

An important test of the reliability of the improved theory is to investigate whether it can properly describe the observed systematic changes of the spectral morphology as a function of stellar parameters (this work was recently fully presented by Pauldrach et al., 1990; a preliminary version was also given by Kudritzki, Pauldrach, Puls, 1987). One such systematic change of O star wind lines is implied by the observed *correlation* between *strength* of the Si IV ($\lambda\lambda$ 1394, 1403) *wind profile* and *luminosity class* (Walborn and Panek, 1984a,b, 1985; Walborn and Nichols-Bohlin, 1987; Walborn et al., 1985). The observations indicate the complete *absence of stellar wind features in Si IV* on the *main sequence*, and their gradual development from the intermediate luminosity class III into *saturated wind profiles* for *supergiants* (see Fig. 4 - note that all the features shown in this figure belong to objects of same spectral type (i.e. O 6.5), which means that the effective temperature decreases only slightly from class V to class Ia (see Kudritzki et al., 1989). It is, of course, crucial for a quantitative stellar wind theory to reproduce this effect. In order to investigate the behaviour of synthetic spectra of the Si IV line we performed self-consistent calculations of wind models along the evolutionary tracks of Maeder and Meynet (1987). The stellar parameters of the selected models are shown in Fig. 5. Fig. 6 demonstrates that the strong dependence of the Si IV profile on luminosity class is in principle reproduced by the theory. For early O dwarfs only weak photosperic lines are produced in agreement with the observations, where it is not clear whether lines are interstellar or photospheric and whether additional blends contribute (Fig. 6a). For the supergiants typical P-Cygni structures are obtained that match the observations in surprising detail (Fig. 6b) - for details see Pauldrach et al. (1990).

This encouraging result poses now the question, whether it is possible to determine luminosities directly from a comparison between observed and synthetic spectra of the Si IV resonance doublet. The sequence of calculated lines for models of equal effective temperature (T_{eff} 35.5 kK) and slightly increasing luminosity shown in Fig. 7 (from bottom to top) indicates that after a phase of calibration by detailed quantitative spectral analysis of individual standard stars this is surely possible. Note that even objects where $\Delta \log L/L_\odot$ differs by only *0.05* can be distinguished clearly with this method (remember that the error of the luminosity of cluster members is at least $\Delta \log L/L_\odot = 0.24$).

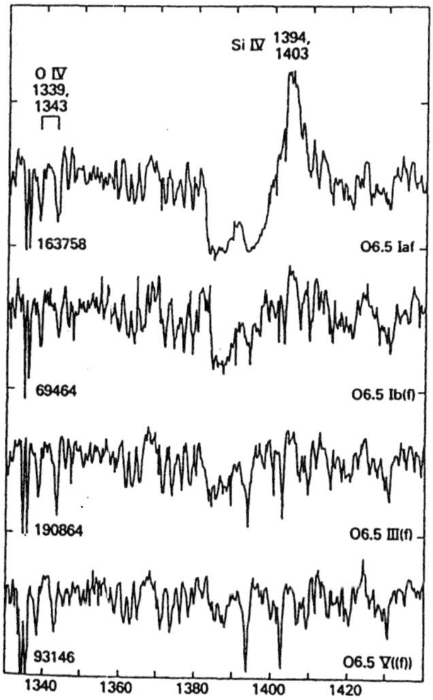

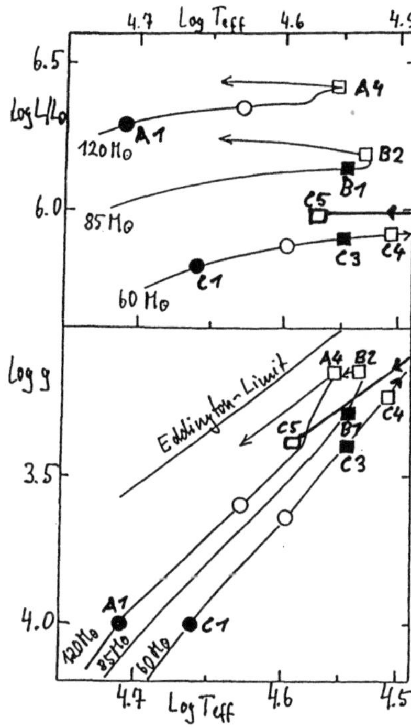

Fig. 4: Luminosity effect of Si IV (λλ 1397, 1403) at spectral type O 6.5 (from Conti, 1988).

Fig. 5: The HR-diagram and the (log g, log Teff) diagram for wind models along evolutionary tracks of initial masses 120 (A), 85 (B), 60 (C) and 40 $M_\odot$ (D). Filled circles correspond to early O dwarfs (V) and open squares to middle O supergiants (Ia/b).

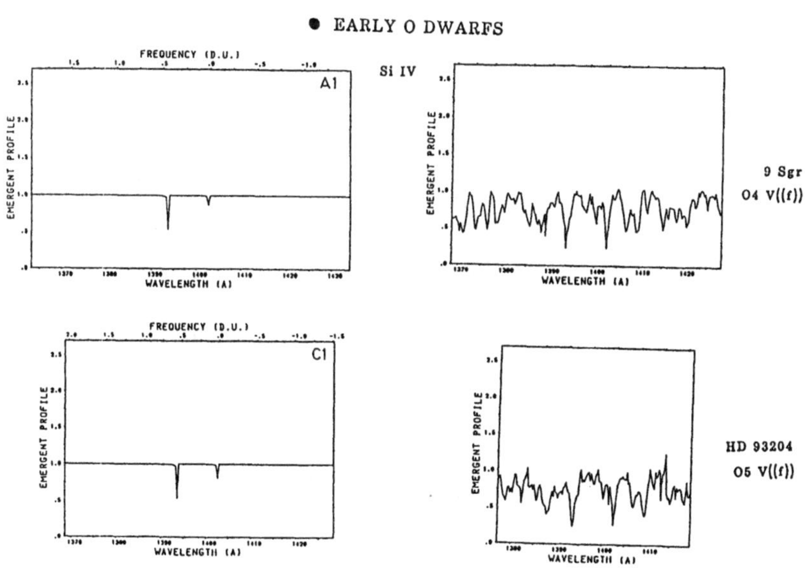

Fig. 6a

Fig. 6: Theoretical Si IV profiles versus observations displaying the strong observationally luminosity effect (Fig. 6a O dwarfs, Fig. 6b middle O supergiants).

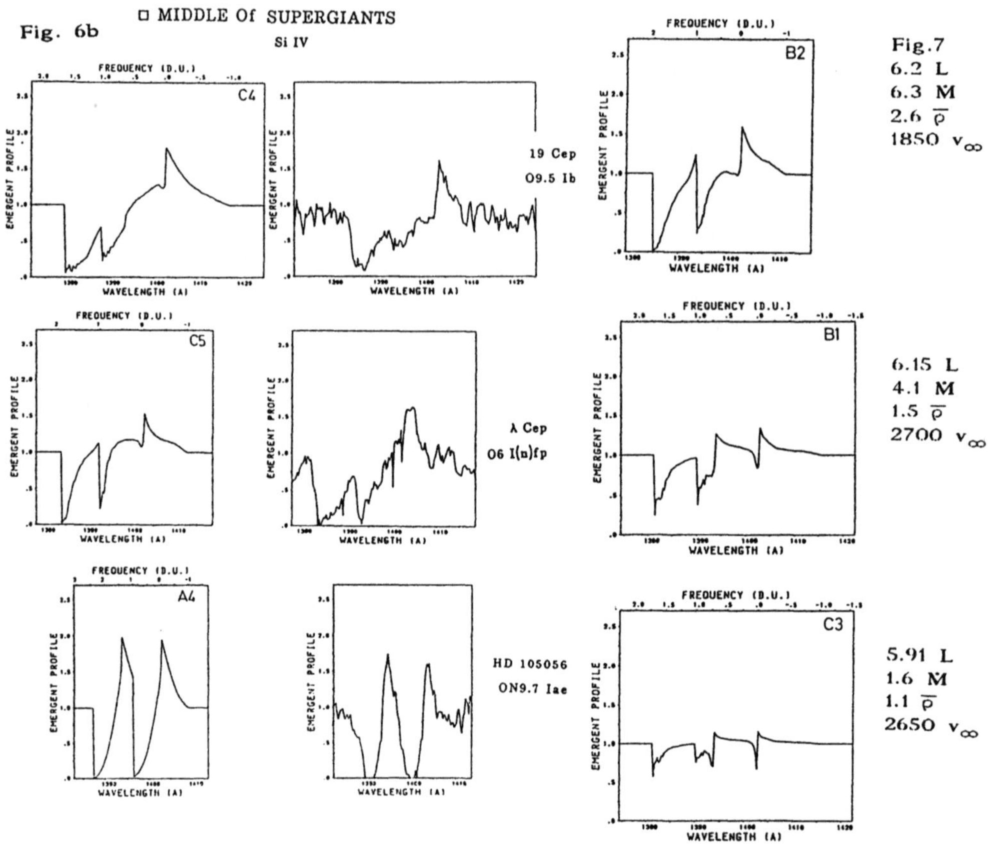

Fig. 7: Si IV profiles for a sequence of theoretical models at constant Teff (~35kK) and slightly increasing luminosity. The luminosity ($\log L/L_\odot$), the mass loss rate (10^{-6} $M_\odot$/yr), the wind mean density (10^{-12} $(M_\odot/yr)/(R_\odot^2$ km/s)) and the terminal velocity (km/s) are indicated in sequential order.

IV. MASSIVE HOT STARS AS DISTANCE INDICATORS

Apart from the successes already described, the theory of radiation driven winds can also be applied to one of the most important goals of stellar spectroscopy, the direct determination of stellar masses and radii and, hence of distances. This possibility is offered by the strong dependence of the theoretical terminal velocity (v_∞), which is also a precisely measurable quantity, on the surface gravity and radius

$$v_\infty \propto (g \cdot R)^{1/2} \propto (M/R)^{1/2} \tag{1}$$

To illustrate this behaviour, Fig. 8 shows this dependence for the case of ζ-Pup (see sect. II), within the error intervals of log g (± 0.15) and $R/R_{\odot}$ (± 8). The terminal velocity ranges from 290 km/s to 3070 km/s within the uncertainty of log g (where R was fixed at its mean value) and from 1140 km/s to 2243 km/s within the uncertainty of $R/R_{\odot}$ (here log g was fixed at its mean value). In principle this very large variation of v_{∞} can be used to reduce the uncertainties on log g and $R/R_{\odot}$ and thus to determine stellar masses and radii by a comparison with the observed value (v_{∞} = 2200 ± 60 km/s), however, we first have to show that this method yields reliable results. This test is possible for objects with known distance, because the mass can be determined directly by model calculations (see eq. (1)) which can then be compared to the mass obtained from log g, which is available from photospheric line analyses (see also Kudritzki, 1990). Fig. 9 gives an example for the first of these methods. For values of L comprised in the range of uncertainty of the luminosities, v_{∞} is shown as a function of mass. A comparison with the observed terminal velocities yields directly the stellar mass and its uncertainty. Table I, which gives for a sample of O-stars a comparison of masses

TABLE I Stellar masses from v_{∞} and log g for objects with known radii

star	log $M/M_{\odot}$ from v_{∞}	from log g	source
HD 93129A	2.08 ± 0.10	2.07 ± 0.17	(1,3,4)
93250	2.01 ± 0.11	2.07 ± 0.18	(1,3,4)
303308	1.61 ± 012	1.75 ± 0.21	(1,3,4)
ζ Puppis	1.68 ± 0.10	1.62 ± 0.20	(1,3,4)
λ Cep	1.60 ± 0.09	1.54 ± 0.20	(3)
HD 15629	1.40 ± 0.16	1.36 ± 0.26	(2)
15558	1.70 ± 0.15	1.70 ± 0.26	(2)
34656	1.42 ± 0.28	1.42 ± 0.40	(2)
193514	1.46 ± 0.16	1.44 ± 0.26	(2)
192639	1.37 ± 0.15	1.26 ± 0.26	(2)
Sk 80 = AV232	1.54 ± 0.09	1.50 ± 0.20	(this work,3,4)

(1) Kudritzki (1990)

(2) Herrero, Vilchez, Kudritzki (1990)

(3) Kudritzki, Pauldrach, Puls, Hummer (1990)

(4) Pauldrach, Kudritzki, Puls (1990)

Fig. 8

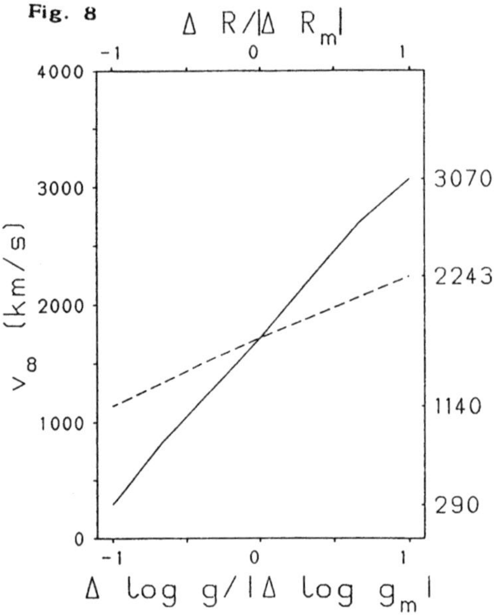

Fig. 9

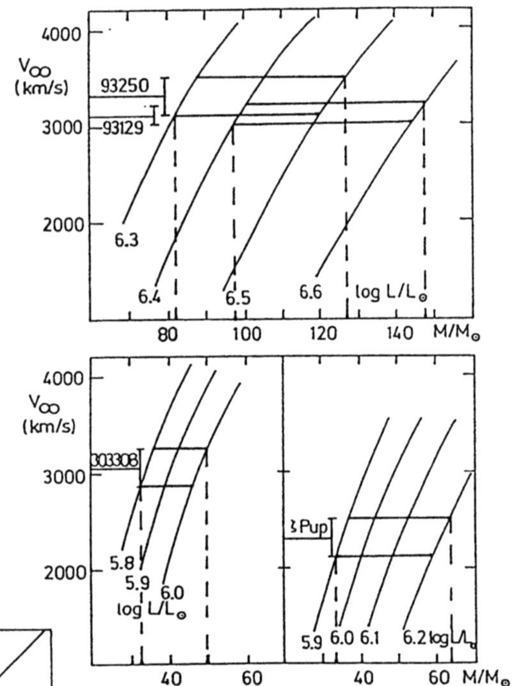

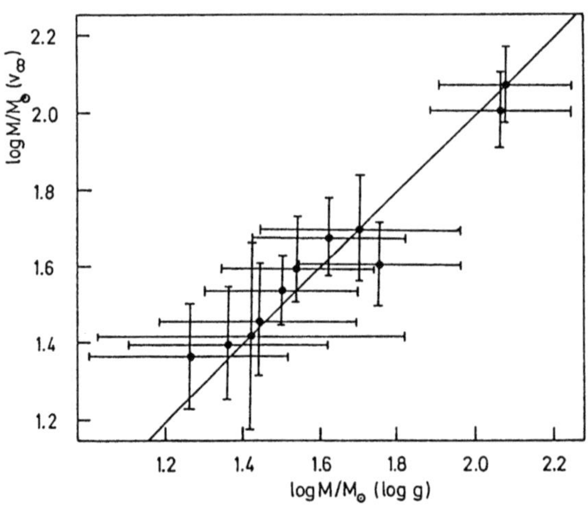

Fig. 10

Fig. 8: v_∞ vs. log g = 3.5 + Δlog g (fully drawn, Δ log g_m = ± 0.15) and $R/R_\odot$ = 19 + ΔR (dashed, ΔR_m = ± 8 $R_\odot$) for the case of ζ-Pup (from Pauldrach, 1985).

Fig. 9: Terminal velocity as function of stellar mass for four typical O-stars, where a comparison with the observed v_∞ yields the mass within the range of uncertainty of log $L/L_\odot$ (from Kudritzki, 1990).

Fig. 10: Stellar masses from v_∞ versus stellar masses from log g for objects with known distances and radii.

obtained in this way and masses derived from log g, shows strikingly that *the two methods agree almost perfectly* (see also Fig. 10). However, there are two reasons why the v_∞-method should be preferred: first, the v_∞-masses have smaller uncertainties, a factor of two in logarithmic scale, secondly, log g cannot be determined with enough precision for objects close to the Eddington limit (e.g. P-Cygni). It should also be borne in mind that *the two methods* used are *completely independent* - one (the log g-mass) has its origin in the photosphere and the other (the v_∞-mass) in the wind. In view of the discrepancies between log g-masses and evolutionary masses found by Groenewegen et al. (1989), Herrero et al. (1990) and Blomme (1990) this indicates most probably that the *evolutionary tracks even for main-sequence stars are in error*. Nevertheless, it is also conceivable that the agreement found in Fig. 10 is just by chance.

After this convincing result we turn to the suggestion that luminous hot stars can be used as ideal distance indicators. This possibility rises from the fact that not only v_∞ but also the mass loss rate ($\dot{M}$) is a strong function of T_{eff}, log g and $R/R_\odot$. Hence, masses, luminosities, radii and distances can be determined using the theory of radiation driven winds, where only the knowledge of the effective temperature is required. This is of particular interest for objects close to the Eddington limit, since for these stars T_{eff} is indeed the only piece of information that can be obtained from the photospheric lines.

The method is very simple (Kudritzki, 1990): once T_{eff} has been derived from photospheric spectroscopy and v_∞ and $\dot{M}$ are determined from wind lines (the latter value can also be obtained from H_α, He II (λ 4686) - Leitherer (1988), Gabler et al. (1990) - or radio data - e.g. Bieging et al. (1989)) log g and $R/R_\odot$ can be directly determined from a comparison with calculated values of v_∞ and $\dot{M}$. Hence M, L and the distance are obtained (note that in cases where log g can also be determined from photospheric lines the method can again be checked). As an example, this method is applied in Fig. 11 to the O4(f) star ζ-Pup, the BIa^+ star P-Cygni and the SMC star Sk 80 (AV 232). ζ-Pup was chosen as a standard object, and P-Cygni due to its large absolute visual magnitude (M_V = −8.5 ± 0.6, Lamers et al., 1983; see also Pauldrach and Puls, 1990) which makes it an object of special interest, because in the near future similar objects will be observable not only in the Local Group galaxies but also further beyond. (A comprehensive theoretical study of the expanding atmosphere of P-Cygni, which includes the determination of stellar parameters, was recently presented by Pauldrach and Puls, 1990). Such an extension of the method to other galaxies was performed with the study of Sk 80. For this object we adopted a composition of $Z = 0.1\ Z_\odot$ (Dufour, 1984). Table II shows the results obtained in the described way for all three objects.

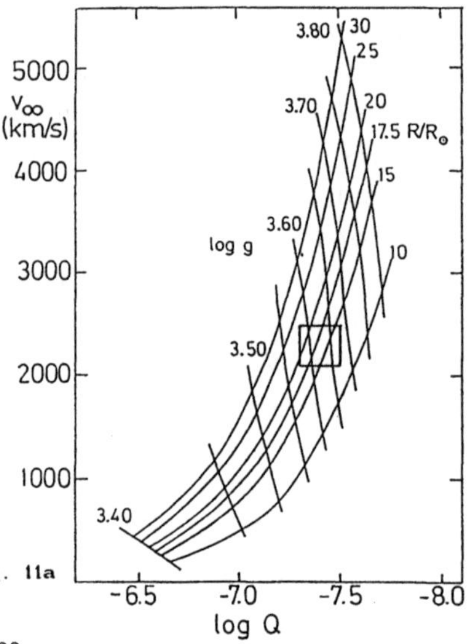

Fig. 11a

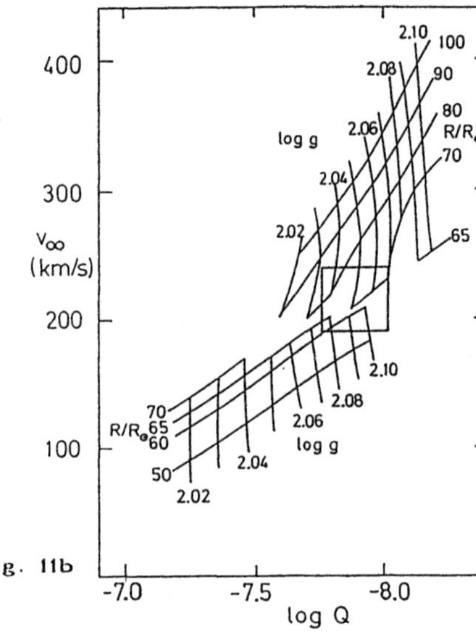

Fig. 11b

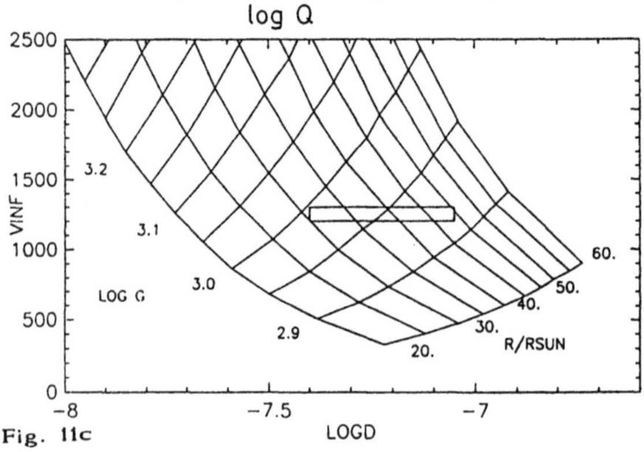

Fig. 11c

Fig. 11: v_∞ and $\dot{M}$ calculated for wind models with different log g and $R/R_\odot$; observed values are indicated by rectangular boxes (Fig. 11a: ζ-Pup; Fig. 11b: P-Cyg – log Q = log $\dot{M}$ ($M_\odot$/yr) – 3/2 log $R/R_\odot$ (from Kudritzki, 1990); Fig. 11c: the SMC-star Sk80 – log D = log $\dot{M}$ ($M_\odot$/yr) – log $R/R_\odot$).

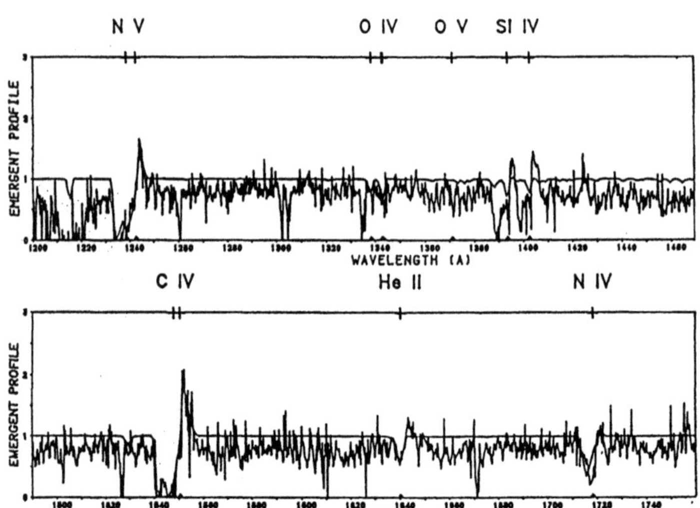

Fig. 12: Calculated and observed UV-spectrum for Sk80.

TABLE II: Log g, $R/R_\odot$ and error of distance modulus determined from v_∞ and $\dot{M}$. Log g from photospheric lines are also given.

star	log g	$R/R_\odot$	$\Delta(m_V-M_V)$	log g (phot.)
ζ-Pup	2.62 ± 0.05	17 ± 5	± 0.5	3.55 ± 0.1
P-Cyg	$2.05^{+0.05}_{-0.02}$	75^{+10}_{-15}	± 0.5	——
Sk 80	$2.95^{+0.05}_{-0.03}$	34^{+11}_{-9}	± 0.5	2.9 ± 0.1

Again, the agreement between photospheric-log g and wind-log g is striking, but the latter values have much smaller uncertainties. It is also encouraging that the error in the distance modulus, which is obtained directly from the error in radius, is not larger than ±0.5 mag, and thus comparable to values derived from other methods (e.g. δ-Cephei). This uncertainty in distance will possibly be further reduced if additional information obtained from a comparison of individual wind lines is considered. The UV-spectrum shown in Fig. 12 for the case of Sk 80 (for this model the input parameters T_{eff} = 34 kK, log g = 2.95, $R/R_\odot$ = 33, Z = 0.1 $Z_\odot$ yield $\dot{M} = 1.9 \cdot 10^{-6}$ $M_\odot$/yr and v_∞ = 1240 km/s) indicates that this will soon be possible.

V. CONCLUSIONS

Although our numerical concept of a quantitative description of stationary models of radiative driven winds is still in need of improvement, the results presented here have shown that the "evolutionary status" of this effort is already well advanced and ready for applications. These applications include the possibility *of using massive stars as distance indicators* by either comparing the observed and calculated Si IV resonance line, or comparing the observed and calculated dynamical parameters. As was shown by our encouraging first results, this method can be applied not only to galactic blue stars, but also to massive stars of large absolute visual magnitude in all Local Group Galaxies, and, with the telescopes of the near future - like HST and ESO-VLT - to stars in galaxies *"as far as the eye can reach"*.

Another important result of this paper concerns the discrepancy between spectroscopic and evolutionary masses. Since it was shown that two different and completely independent methods for determining spectroscopic masses yield values which agree nearly perfectly, it cannot be excluded that evolutionary tracks of massive stars are in error not only for evolved objects.

ACKNOWLEDGEMENTS: We wish to thank our "Tai-Pan" (Prof. Dr. R.-P. Kudritzki), who is not a coauthor of this paper only for reasons of modesty and possible misinterpretation. Special thanks go also to Dr. P. A. Mazzali for the careful reading of the manuscript. This work was supported by grants of the Deutsche Forschungsgemeinschaft (Ku 474/11-3).

REFERENCES

Abbott, D.C.: 1979, IAU Symp. 83, p. 237
Abbott, D.C., 1982, Astrophys. J. 259, 282
Berger, Ch., Pauldrach, A.W.A.: 1990, Astron. Astrophys, in preparation
Bieging, J.H., Abbott, D.C., Churchwell, E.B.: 1989, Astrophys. J. 340, 518
Blomme, R.: 1990, Astron. Astrophys., 229, 513
Bohannan, B., Abbott, D.C., Voels, S.A., Hummer, D.G.: 1986, Astrophys. J., 308, 728
Butler, K.: 1990, Astronomical Society of the Pacific Conference Series Vol. 7, "Properties of Hot Luminous Stars", ed. C.D. Garmany, p. 108
Cassinelli, J.P., Olson, G.L.: 1979, Astrophys. J. 229, 304
Castor, J., Abbott, D.C., Klein, R.: 1975, Astrophys. J. 195, 157
Conti, P.S.: 1988, NASA SP-497 "O Stars and Wolf-Rayet Stars", eds. P.S. Conti, A.B. Underhill, p.81
Dufour, R.J.: 1984, IAU Symp. 108, p. 353
Gabler, A., Gabler, R., Pauldrach, A.W.A., Puls, J., Kudritzki, R.P.: 1990, Astronomical Society of the Pacific Conference Series Vol. 7, "Properties of Hot Luminous Stars", ed. C.D. Garmany, p. 218
Groenewegen, M.A.T., Lamers, H.J.G.L.M., Pauldrach, A.W.A., 1989. Astron. Astrophys., 221, 78
Henrichs,H.: 1988, NASA SP-497, S.199
Herrero, A., Vildez, J.M., Kudritzki, R.P.: 1990, Astronomical Society of the Pacific Conference Series Vol. 7, "Properties of Hot Luminous Stars", ed. C.D. Garmany, p. 50
Kudritzki, R.P., Simon, K.P., Hamann, W.R.: 1983, Astron. Astrophys. 118, 245
Kudritzki, R.P., Pauldrach, A.W.A., Puls, J.: 1987, Proc. IAU Coll. 108 "Atmospheric Diagnostics of Stellar Evolution: Chemical Peculiarity, Mass Loss and Explosion"
Kudritzki, R.P., 1988, 18th Advanced Course Swiss Society of Astronomy and Astrophysics, Leysin, March 1988, invited lectures
Kudritzki, R.P., Gabler, A., Gabler, R., Groth, H.G., Pauldrach, A.W.A., Puls, J.: 1989, Proc. IAU Coll. 113 "Luminous Blue Variables", ed. K. Davidson, p.67
Kudritzki, R.P.: 1990, PASP "Frontiers of stellar evolution", Alpine

Texas ed. D. Lambert, in press
Kudritzki, R.P., Pauldrach, A.W.A., Puls, J., Hummer, D.G.: 1990, Astron. Astrophys., in preparation
Lamers, H.J.G.L.M., de Groot, M., Cassattella, A.: 1983, Astron. Astrophys., 128, 299
Leitherer, C.: 1988, Ap. J. 326, 356
Lucy, L.B., Solomon, P., 1970, Astrophys. J. 159, 879
Maeder, A., Meynet, G., 1987, Astron. Astrophys. 182, 243
Mihalas, D., Hummer, D.G.: 1973, Astrophys. J. 179, 872
Mihalas, D., "Stellar Atmospheres", W.H. Freeman and Company, San Francisco, 1978, 2nd edition
Owocki, S.P., Castor, J.I., Rybicki, G.B.: 1988, Astrophys. J. 335, 914
Pauldrach, A.W.A.: 1985, Diplomarbeit, LMU München
Pauldrach, A.W.A., Puls, J., Kudritzki, R.P.: 1986, Astron. Astrophys. 164, 86
Pauldrach, A.W.A.: 1987, Astron. Astrophys. 183, 295
Pauldrach, A.W.A., Herrero, A.: 1988, Astron. Astrophys. 199, 262
Pauldrach, A.W.A., Puls, J., Kudritzki, R.P., Butler, K.: 1990, Astron. Astrophys. , 228, 125
Pauldrach, A.W.A., Puls, J.: 1990, Astron. Astrophys., in press
Pauldrach, A.W.A., Puls, J., Gabler, R., Gabler, A.: 1990b, Astronomical Society of the Pacific Conference Series Vol. 7, "Properties of Hot Luminous Stars", ed. C.D. Garmany, p. 171
Pauldrach, A.W.A., Kudritzki, R.P., Puls, J.: 1990, Astron. Astrophys., in preparation
Puls, J.: 1987, Astron. Astrophys. 184, 227
Walborn, N.R., Panek, R.J., 1984a, Astrophys. J. 280, L27
Walborn, N.R., Panek, R.J., 1984b, Astrophys. J. 286, 718
Walborn, N.R., Panek, R.J., 1985, Astrophys. J. 291, 806
Walborn, N.R., Nichols-Bohlin, J., Panek, R.J., 1985, "IUE Atlas of O-Type Spectra from 1200 to 1900 Å", NASA Reference Publication 1155
Walborn, N.R., Nichols-Bohlin, J., 1987, PASP 99, 40

Theory of Radiatively Driven Winds of Hot Stars: II. Some Aspects of Radiative Transfer

J. Puls and A.W.A. Pauldrach

Institut für Astronomie und Astrophysik, Scheinerstr. 1,
D-8000 München 80, Fed. Rep. of Germany

ABSTRACT In this paper, we discuss the influence of detailed radiative transfer effects on radiatively driven stellar wind models. It is shown, that for "typical" O-star winds with moderate to large wind densities a realistic approach leads to only marginal changes (compared to an approximate treatment) in the overall structure, although the effect of line blocking is crucial in order to obtain the correct ionization equilibrium with respect to *trace ions* such as Si IV, CIII. For extremely thin winds (mainly luminosity class (LC) V stars), the line blocking can significantly lower the terminal velocity. In case of extremely dense winds (WR-stars), sophisticated transfer effects such as multi-line processes, the interaction line/continuum and the *diffuse* radiation force are *substantial* for a further understanding of the outflows of these objects.

I. INTRODUCTION

The typical ingredients necessary to construct reliable stellar wind models have to be found from the self-consistent solution of at least three different sets of equations:

i) the NLTE rate equations, which yield - for given atmospheric structure and radiation field - the occupation numbers. In case of radiatively driven models, these equations have to be solved for the total number of ions participating in building up the radiation pressure (typically of order 100).

ii) the equation(s) of radiative transfer, which yield - for given atmospheric structure and occupation numbers (→ opacities, emissivities) - the radiation field required to calculate both the radiative rates (lines and continuum) necessary for i) and one of the most important quantities of the theory presented here, namely the radiation pressure itself.

iii) the hydrodynamical equations which - for given radiation pressure - define the atmospheric model.

To obtain the correct temperature stratification, additionally the constraint of energy conservation (including both radiative and non-radiative processes) has to be accounted for.
From the self-consistent solution of these equations, we finally obtain the desired quantity to be compared to observations, the total emergent spectrum (ideally from the X-ray to the radio range) which only then gives a satisfying agreement if all variables such as mass-loss rate, velocity structure, occupation numbers and radiation field are consistent with or at least approximate the reality quite well.-

The present paper deals exclusively with item ii), the solution of the equations of radiative transfer and connected problems. Although one could argument that " *radiative transfer is just one of these things one should solve and that's it* ", there are actually a number of reasons which favour a more detailed discussion, especially in case of moving atmospheres (compared to the well-known problems in static ones):

1. The physics

Because of the complexity of the problem arising from the intrinsic coupling of the large number of lines (line overlap, cf. sect.II) of the same or different ions induced by the Doppler shift in moving atmospheres, it seems to be obvious not only to solve the equations but also to understand the results and the different importance of the various processes.

2. The computing time

Estimating the typical width of a line profile to be of the order of the thermal velocity (≈ 7.5 km/s for carbon ions and T_e = 40,000 K) and requiring a moderate resolution of 5 frequencies per line to calculate the profile-averaged mean intensity $\bar{J}$ and flux $\bar{H}$ for all lines, the total number of frequency points to be considered is given by (cf. Puls, 1987)

$$n = \log(\lambda_{max}/\lambda_{min}) \; / \; \log(1 + \Delta v/c) \approx 600{,}000 \ldots 900{,}000 \text{ frequencies}$$

for hot star winds (λ_{min}≈100Å, λ_{max}≈2,000...10,000Å, Δv=1.5 km/s). Due to the anisotropy of the problem, the inclusion of the angular information (≈80 different rays to be considered) yields a number of ca. $6*10^7$ ray-by-ray solutions to be performed in order to calculate the specific intensity $I(\nu,\mu,r)$ in sufficient resolution, from which all other quantities can be derived.
From this estimate, we see that a straightforward solution of the radiative transfer is beyond today's computer capacity, so that other, faster and more efficient algorithms have to be applied. Actually, the well-known Sobolev-approximation (SA) is an example which allows (at least for line processes) an enormous reduction in computing time, where, however, the additionally required spectral information makes the exact solution of the problem inevitable.

3. The historical reason

The first quantitative self-consistent approach to construct stellar wind models (Pauldrach, 1987) used a ("super"-) simplified method of radiative transfer to calculate the radiation field in the lines (actually neglecting almost totally the complexity of the problem), but nevertheless obtained results in nearly perfect agreement with observations. The following discussion will, among other points, clarify the reasons for this remarkable coincidence.

4. Present and future work

In the next few years, a number of new (and old) problems connected strongly with radiative transfer in moving atmospheres will have to be solved, e.g.

- the problem whether radition pressure is also sufficient to initiate and accelerate the extremely dense winds of WR-stars.
- the problems involved with line-formation in SNR's.
- the detailed simulation of radiatively driven instabilites in hot star winds (cf. the contributed paper by S. Owocki) including realistic driving forces.
- the problem of radiative transfer in moving media and arbitrary geometry (e.g. disks) including rotation and line overlap...

Only the knowledge of the various aspects of radiative transfer under somewhat simpler conditions (spherical symmetry, stationarity, monotonic flow) will finally allow to tackle these more complex problems successfully.

II. LINE OVERLAP IN EXPANDING ATMOSPHERES

1. A photon's path throughout the wind

In this section, we will briefly discuss the principle problem of line overlap where its effect on the line acceleration is given in sect. III (for a thorough discussion see Puls, 1987). We will restrict ourselves to stationary, spherically symmetric monotonic flows, which represent a good approximation for the time-averaged stellar winds of hot luminous stars.

The path of a photon begins when it leaves the stellar core or is emitted thermally somewhere in the wind. As it propagates, its oberver's frame frequency ν_{01} remains constant, while with respect to the comoving frame (the frame of matter flowing with velocity V(r)) it is shifted according to $\nu_{CMF}(r) = \nu_{01}(1 - \mu V(r)/c)$. The photon can be processed a first time (neglecting prior events of Thomson-scattering), when it reaches a so-called resonance zone. This is the zone where the CMF frequency lies in the range $\nu_{CMF}(r) = \underline{\nu}_1 \pm \Delta\nu_L$, with $\underline{\nu}_1$ a possible line transition frequency of an absorbing ion and $\Delta\nu_L$ the corresponding line width. Depending on the interaction probability IP $= (1 - \exp(-\tau_L))$ (with τ_L the line optical depth), the photon crosses this interaction zone undisturbed (IP $\approx$ 0), or, for IP $\approx$ 1, will be absorbed and thermally destroyed or reemitted. In the latter case, the photon can be absorbed again in the same line and resonance zone, provided the optical depth is large (multiple resonance scattering). For the *net* momentum transfer, however, only the first and last interaction is decisive, as the scatterings in between will lead to a neglegible contribution. A last thermal destruction then ends the photon's history, while a final reemission with following escape from the resonance zone leaves the photon with a rest frequency $\nu_{02} = \nu_{CMF}(r) / (1 - \mu' V(r)/c)$, where the prime refers to the last scattering event. The radial momentum transferred by this first process is given by

$$\Delta p_r = h/c\,(\mu\nu_{01} - \mu\nu_{02}) = h(\nu_{01} - \nu_{02}) / V(r) = \Delta E/V(r)$$

where ΔE is the energy loss/gain of the photon. Due to the assumption of a monotonic flow leading to a general redshift of the photon's frequency with respect to the CMF (*independent of the photon's direction*), the photon can be processed again in a second resonance zone of another line with $\underline{\nu}_2 > \underline{\nu}_1$. Again, momentum is transferred, and so on, until the photon finally leaves the envelope or hits the core.

2. ML and SL processes

This so-called multi-line (ML) process where a photon can transfer its momentum several times to the wind has to be compared to the so-called single-line (SL) approximation where the number of possible scatterings is restricted to unity. Two cases have to be investigated in order to understand the basic difference:

- a "realistic" treatment of the SL process (SL^R) allows for only *one* scattering per photon, but takes into account the *existence* of resonance zones which have been crossed undisturbed because of a too small IP. In terms of radiative transfer, the photons' distribution function at the beginning of the actual scattering zone is ~ $I_c \exp(-\Sigma\tau_L)$, where I_c is the core intensity and $\Sigma\tau_L$ the total optical depth of previous resonance zones. In other words, the SL^R approach takes into account the *line-blocking* effect.
- th SL^{PPK} approximation, used by a number of different authors (e.g. Abbott (1982), Pauldrach, Puls and Kudritzki (1986), Pauldrach (1987)) , in contrast to the SL^R approach, *neglects* line-blocking, so that in this case the photons' distribution function is *always* ~ I_c before entering the actual resonance zone.

3. Line density

It is obvious that ML and SL processes will yield the same acceleration if the lines are widely seperated, so that the photon has a priori only the chance to reach at most one interaction zone on its way throughout the wind. On the other hand, the higher the line density, the more important line overlap should become. The maximum number of lines a photon can interact with is given by the number of lines in the interval $\underline{\lambda}$ ... $\underline{\lambda}$ + $\Delta\lambda$, $\Delta\lambda = 2\, v_\infty/c$. Therefore, the higher the terminal velocity, the more lines can principally overlap. For O-stars, $\Delta\lambda$ is typically 8Å at 600Å. Fig. 1 gives the number of lines stronger than Thomson-scattering in dependence of log λ (λ = 228...5,000Å) and log r (r = 1...100 R_*) for the standard model of ζ Pup (O4If, T_{eff} = 42,000K, log g = 3.5, R_* = 19 R_o (Kudritzki et al., 1983) and $v_\infty \approx$ 2,200 km/s, $\dot{M} \approx 3.8\ 10^{-6}\ M_o$/yr, (Pauldrach, 1987)). Most strikingly, this function (for all radii) shows two extreme peaks for wavelength below 400Å with line densities beyond 150 per $\Delta\lambda$. Redwards from 400Å, the number of lines is 10...50, with another maximum at ~ 1,400Å, i.e. just in the IUE range. Only in the interval between 800-1,000Å and beyond 2,000Å the line density is small so that ML processes should be of minor importance. An additional remark should be made: From the irregularity of the line-density it is obvious that the line statistics

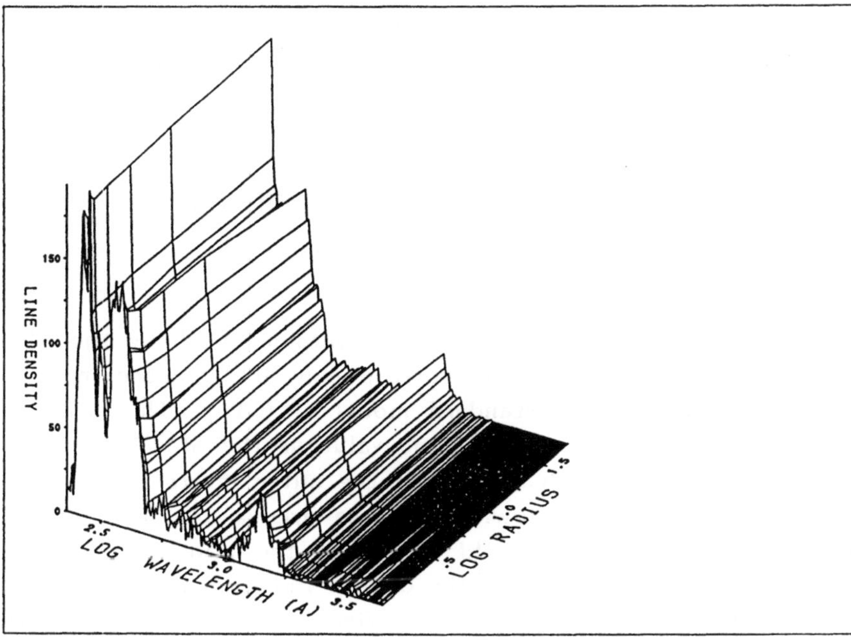

Fig.1 Line distribution function for ζ Pup

with respect to wavelength assumed by Castor, Abbott and Klein (1975, see also the contributed paper by S.Owocki) which reads $dN(\lambda) \sim d\lambda/\lambda$ is (absolutely) inconsistent with the actual circumstances, as this statistics would lead to a *constant* number of lines per overlap interval.

4. Interaction probability (IP)

Not only the line density, but also the IP is significant in order to investigate the efficiency of the ML process. Table 1 gives the number of lines with $\tau_L > 1$, i.e. with an IP > .63 for the standard model of ζ Pup at 3 significant radii:

Table 1: Number of lines with $\tau_L > 1$ for ζ Pup

r/R_*	1.022(~ sonic point)	1.5	6
$N_L(\lambda > 228Å)$	2,000	500	200

This is the number of lines which interacts *definitely in each transversed* resonance zone. On the other hand, one can show that for a terminal velocity of 2,200 km/s the critical number of lines where overlap effects become important, i.e. the SL approach unrealistic , is of the order *270* (cf. Puls, 1987). Therefore, we find from Table 1 that at least in the major acceleration regime ($r \lesssim 1.5\ R_*$) the inclusion of line overlap seems to be inevitable.

III. EFFECTS OF ML-PROCESSES ON THE HYDRODYNAMICS

1. "Typical" O-star winds

Fig 2. compares the total line acceleration obtained both by the SL^R and the ML approach in units of the corresponing SL^{PPK} approximation (defined in sect.II.2) for the standard model of ζ Pup. The *realistic* SL^R force (dashed-dotted) has the same value as the SL^{PPK} force in the photosphere, but is much smaller in the outer layers because of the line-blocking in zones below. The force arising by detailed ML calculations is shown by the bold curve. In the photospere, it is smaller than the SL^R value due to the diffuse radiation field scattered back from above, whereas in the outer wind it is much larger than the *real* single-scattering force due to the multiple momentum transfer – but still smaller than the SL^{PPK} approach. Hence, we find in this case – despite of the increase of the line force by ML scattering processes (a factor of *8* compared to the SL^R value) – a smaller line acceleration than the approximate value used e.g. by Pauldrach (1987) in *all* parts of the wind.

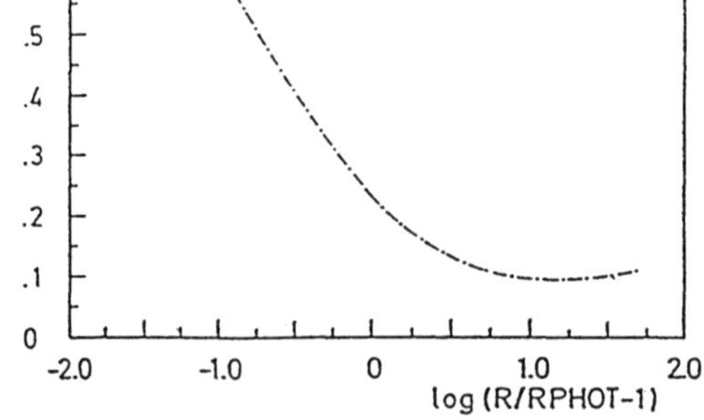

Fig. 2 Line acceleration in units of SL(PPK) approximation: *real* single-scattering (SL^R, –.–.) and multi-line scattering (ML, ——)

This *reduction* of the actual ML line force is typical for O-star winds with moderate to high mass loss rates (LC ~ I-III) and is generally of the order CFML = $g^{ML}/g^{PPK} \approx 0.75...0.85$ *throughout* the wind. As the line distribution function remains almost undisturbed by ML effects (in terms of the force-multiplier parameters : α, δ are *not* changed), the scaling relations of radiatively driven winds (Kudritzki et al, 1989) give a linear dependence of $\dot{M}$ on this correction factor while v_∞ is not affected. Therefore, we obtain for the *self-consistent ML model* of ζ Pup a slightly reduced mass-loss rate (dependent on the chosen parameters, especially $T_e(r)$) in the range $3.1...4.3 \cdot 10^{-6}$ M_O/yr with $v_\infty \approx 2{,}150$ km/s. Thus, the only *moderate and roughly constant* (with respect to r) change of the rather sophisticated ML line force compared to the simplified SL^{PPK} approach is the reason why the first models by Pauldrach (1987) where in such a good agreement with observations although the applied radiative transfer did not take into account the actual processes.

2. "Thin Winds"

A second class of hot star winds are those outflows where the mean wind density ($\sim \dot{M}/R_*^2 v_\infty$) is low, the so-called "thin winds" with its representative τ Sco (O9.5V, $\dot{M} \sim 5 \cdot 10^{-9}$ M_O/yr). For these (mainly LC V) objects the wind is accelerated to its rather large terminal velocities in the outer part almost only by optically *thin* lines, which has the following consequences:

- the true v_∞ is almost undetectable in the UV lines, as the ions begin to absorb only in regions with $v < v_\infty$. (The maximum edge velocity observed for τ Sco is of the order 2,300 km/s.)
- the IP of a photon to be scattered is small at larger radii ($r > 3R_*$), so that the CFML is dominated by the blocking in the inner part.

Although (for τ Sco) the CFML is 0.85 at the sonic point, it decreases monotonically to ~ 0.6 for $r \geq 3\ R_*$. In this case, we obtain (in addition to the small reduction in $\dot{M}$) a distinct reduced teminal velocity, where the difference is 3,600 km/s (ML) compared to 4,000 km/s (SL^{PPK}). The reader may note, however, that Pauldrach (1987) took into account an average CFML of 0.85 a priori and obtained a v_∞ of 3,880 km/s.

3. WR winds

The result for O/B stars that the correct ML line force is always smaller than the values obtained by the SL^{PPK} approach is not necessarily true under all circumstances and essentially connected with the question: what is the *maximum value of the line force* that can be theoretically achieved . In this context, the reader may note that the so-called *single-scattering limit* $\dot{M} v_\infty / Lc$ (≡ 1 means that the total momentum of the stellar radiation field is transferred *once* to the stellar wind) which is of order 0.1...0.5 for O/B stars can reach values of 3...50 for WR-stars (high $\dot{M}$ + high v_∞). Although it is (in principle) possible to obtain the large mass loss rates by either claiming that these objects are situated close to the Eddington limit (in contrast to today's evolutionary calculations) or by including the rather large continuum and diffuse line forces in the lower atmosphere (but far beyond the sonic point, see sect. V), the acceleration of this $\dot{M}$ to observed v_∞ 's is not feasible in the usual SL^{PPK} approach. On the other hand, a detailed investigation of the radiative acceleration including ML effects (Puls and Pauldrach, 1990) shows that this can be larger in the outer part of the wind if the lines themselves are not pure scattering lines but contain a significant thermal contribution. In this case, large line forces can be created by ML processes for $r \geq 2.5\ R_*$ where the requirement that the lines must *not* be *pure* scattering lines seems to be likely because of the extreme density and, in the outer (optically thin) parts, cooler temperatures (compared to O-stars) of the WR winds. (Cf. also the analysis of observed P-Cygni profiles, e.g. Castor, 1970.) The other possibility that most lines may behave as in typical O-stars (which cannot be excluded as self-consistent models have not been calculated yet) would lead to the conclusion that the large acceleration could not be understood in terms of the here discussed line processes alone. A (somewhat exaggerated) example of those possible large line forces is given by Puls (1987, Table 3a) where, assuming pure LTE-occupation, a ratio of g^{ML}/g^{PPK} of 50 at $r = 6\ R_*$ and of 17,000 at $r = 50\ R_*$ was found.

Thus, until further investigation, it seems to be probable that the final acceleration of WR star winds takes place far out in the wind. This hypothesis is strongly confirmed by newest observations (Auer and Koenigsberger (1990), who found actually an acceleration of the wind of HD 90657 (WN4 + O4-6) at distances larger than 50 R_*) and by the analysis of HeI/II lines in HD 50896 (WN5) which in comparison with UV data revealed that "the velocity must undergo a substantial increase (i.e., from 1,700 km/s to ≥ 2,400 km/s) beyond 50 R_{core}" (Hillier, 1987 , and references therein).

One critical remark with respect to the proposed mechanism should be stated here: If there is a significant thermal contribution to the line force in the outer parts, there will be also a large number of additional photons which can reach the inner part of the wind and will lead to a *significant reduction* of the acceleration there, especially in regions around the sonic point where the mass loss is created (CFML → 0.4). The question in how far the large $\dot{M}$ could be reached nevertheless will be discussed in sect. V.

IV. THE RADIATION FIELD

In this section, we will briefly discuss in how far additional effects induced by transfer processes in expanding atmospheres influence the overall model structure. Especially, we will investigate the effects of line blocking, back-scattering (albedo) and wind-blanketing.

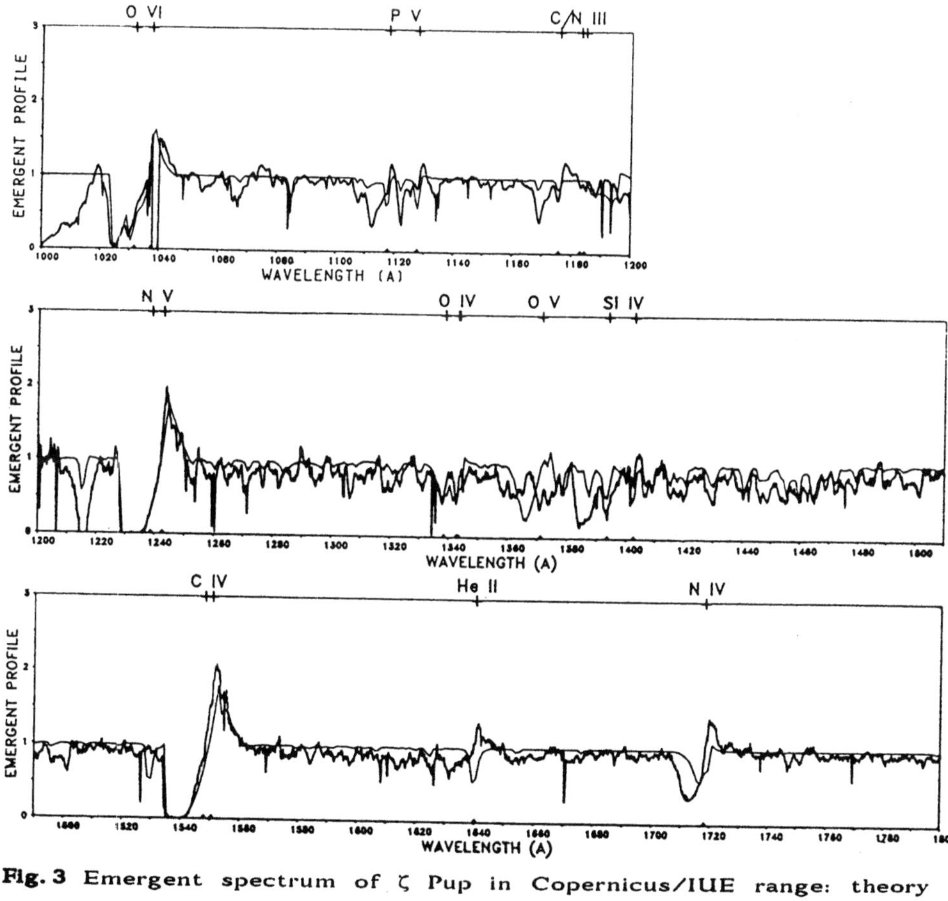

Fig. 3 Emergent spectrum of ζ Pup in Copernicus/IUE range: theory and observations

1. Line blocking

Fig. 3 shows the theoretical emergent spectrum of our standard model of ζ Pup ($T_e \leq 1.2\ T_{eff}$, see Pauldrach et al, 1990a) in comparison to observations (Copernicus+IUE). The only "fit" parameter in this calculation is an (up to now) adopted ("micro"-) turbulence of 150 km/s (Groenewegen et al., 1989). Strikingly, the resonance lines of the main ionization stages (NV, CIV) agree extremely well (cf. Puls, 1987), while for the trace ions differences exist. The resonance line of OVI is o.k. (i.e. the problem of "superionization" is solved in a *cool* wind model), while SiIV and CIII (not in this wavelength range) appear too weak. This result, however, is in *no* contradiction to the statement by Pauldrach et al. (1990), who claimed that the so-called SiIV luminosity effect is reproduced by our theoretical models, as in this paper model calculations were performed along the evolutionary tracks by Maeder and Meynet (1987), while a detailed comparison to *real* stars, as it is given in Fig. 3 , was not intended. The differential behaviour of OVI to SiIV/CIII is connected to the line blocking in the deep layers of the wind and will be explained in the following (cf. Pauldrach et al, 1990a). Fig. 4 gives the location of the radius corresponding to $\tau_\nu=1$ as function of $1/\lambda$ for our model of ζ Pup (from Pauldrach, 1987). Obviously, for λ > 228 Å the continuum is optically *thin* and the radiation field (and hence the ionization rates)

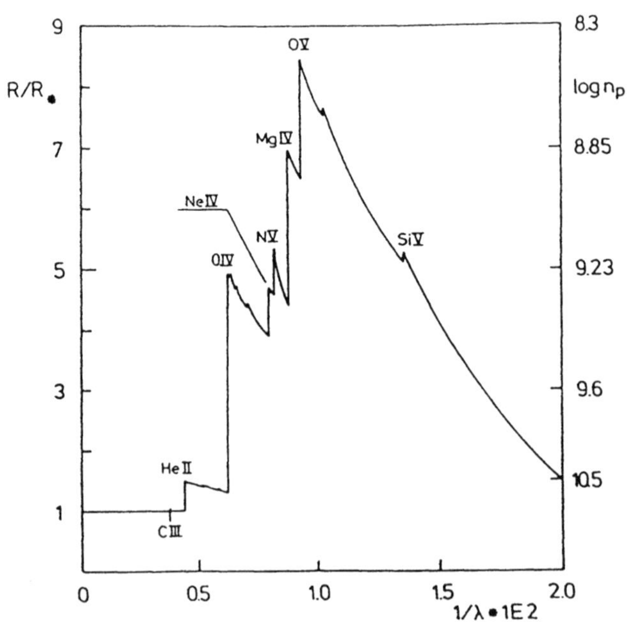

Fig. 4 Location and proton density of optical depth $\tau_\nu=1$

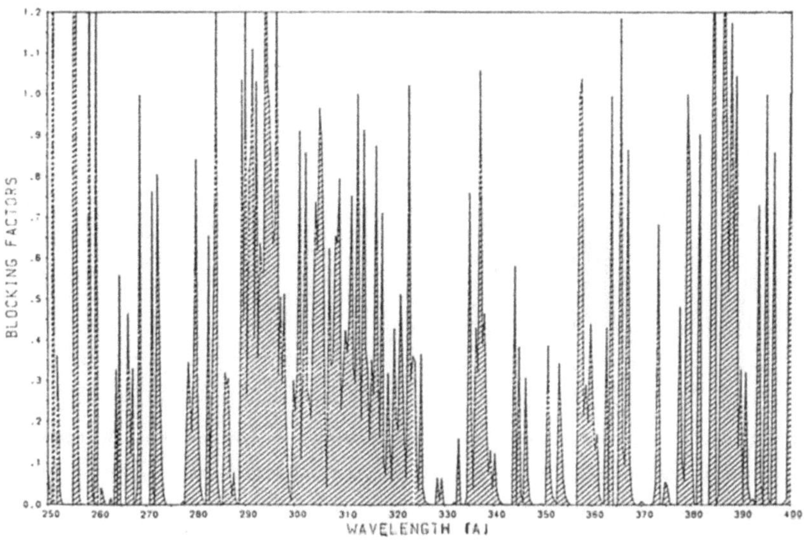

Fig. 5 Blocking factors at the sonic point

depend totally on line transfer effects. On the other hand, for λ < 228 Å the continuum becomes optically *thick* and reaches its "maximum" just at the OV edge. Consequently, the population of OVI depends solely on the adopted electron temperature and is reproduced by our model. For the trace ions with edges beyond 228 Å (SiIV edge at 278 Å, CIII edge at 258 Å), our current model takes into account the line blocking only roughly (photospheric blocking from plane-parallel line-blanketed LTE models (Kurucz, 1979), wind-blocking totally neglected) and the ionization ratios of SiIV etc. depend entirely on the photospheric radiation, which is then too large so that SiIV, CIII become underpopulated (main ionization stages are SiV and CIV/V). A realistic description of line blocking planed in forthcoming calculations (both photospheric and wind) should lead to a significant improvement as the line density in the interesting regions is large (cf. Fig.1) and the corresponding blocking factors (Fig.5) high.

2. Albedo and wind-blanketing

As is obvious from Fig.2 and Fig.5, there is a large number of photons which are scattered back from the wind *into* the photosphere, so that the standard assumption made in photospheric NLTE calculations : I^{-}(outer boundary) $\equiv 0$, is no longer valid if the mean wind density is not too small. In this way, as was stated the first time by Abbott and Hummer (1985), additionally to T_{eff} and log g the mass-loss rate

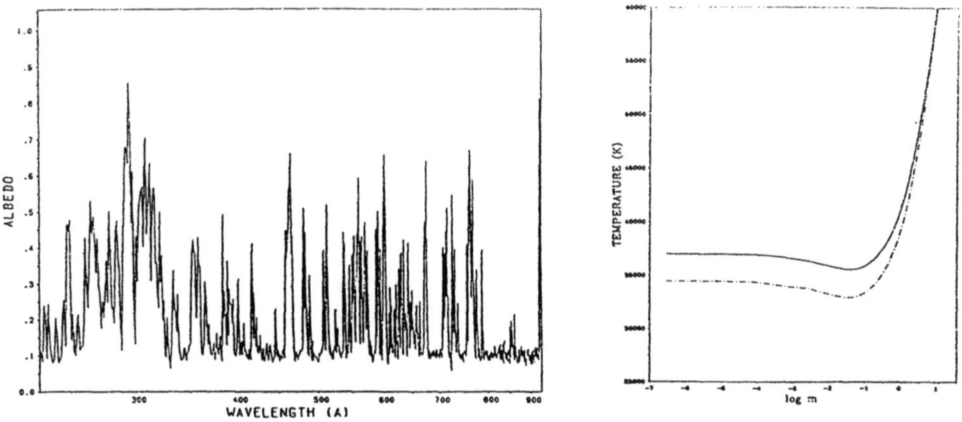

Fig. 6/7 Frequential albedo and corresponding temperature change for a plane-parallel NLTE model of an O4f supergiant (T_{eff} = 45,000 K, log g = 3.6, R_* = 21R_O)

Ṁ (more precisely: the mean wind density) defines the photospheric model structure as a *third* parameter. The most important influence of the back- scattered photons is an increase of the mean intensity which leads (via the constraint of radiative equilibrium, which should be valid at least below the sonic point) to an increase in the electron temperature in the outer regions where the strong lines of H/He (used as *the* diagnostic tool) are formed. Fig.6 and Fig.7 give for a typical O4f supergiant with large wind density the frequential albedo at the sonic point and the corresponding rise in T_e ($\Delta T_e \approx$2,500 K) . (For a more detailed discussion, see Abbott and Hummer, 1985 and Kudritzki et al.,1989a).

Finally, we want to point out the importance of a consistent treatment of radiation transfer in order to calculate correct numbers of ionizing photons needed in the analysis of planetary nebulae and the nebulae of WR's, not only with repect to H/He("unified models", cf. Gabler et al, 1989), but also with repect to other ionzation edges such as OII, NeII etc. which can be drastically modified due to wind blocking/ back-scattering/blanketing.

V. RADIATION PRESSURE IN OPTICALLY THICK CONTINUA

As was pointed out in sect. III.3, the question whether the extremely dense winds of WR stars can be initiated and accelerated by radiation pressure alone is still open. However, most of the assumptions concerning

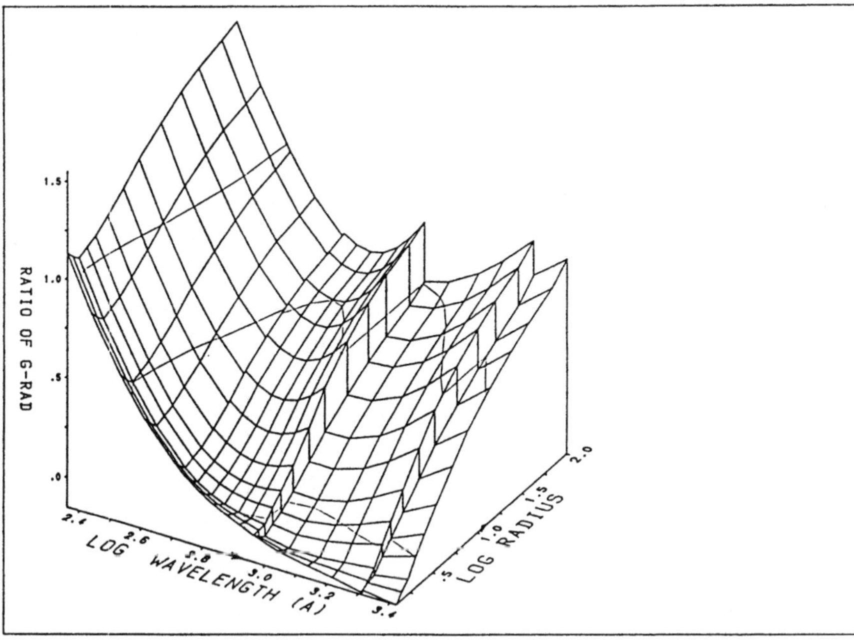

Fig. 8 *Logarithmic* ratio of total line acceleration to SL(PPK) approximation for the WN5 model by Hillier (1987); isocontours are shown for ratios 10, 3.16, 1

the radiative transfer which are appropriate for O-stars fail completely for these objects . This failure is directly caused by the fundamental difference in *both* the stellar *and* the wind parameters of WR's compared to O stars (at least for the types WNE/WC; for the WNL stars the situation looks somewhat better, cf. Langer, 1990) which is a consequence of the evolutionary scenario of the WR's as objects with core He-burning, where the chemical composition of the atmosphere is dominated by the products of complete H-burning (WNE's) or He-burning (WC's).
Due to the large mass loss rate (typically of order $3 \cdot 10^{-5}$ M_o/yr) and small radius (1-3 R_o) , the mean wind density is significantly increased and the velocity at the location of τ_e=1 (which is subthermal for O-stars) is here ≥300...600 km/s.
In other words, the main acceleration regime lies in regions where the continuum is *optically thick for all frequencies*, so that the interaction line/continuum (cf. Hummer and Rybick, 1985), which is neglegible for O-stars, becomes essential. Additionally, the anisotropic *diffuse* radiation field in the resonance zones (~μdS/dr, cf. Puls and Hummer, 1988) becomes important even in *supersonic* layers and leads to a further increase in the line force. As an example, Fig.8 gives the logarithmic ratio of the actual line force including these

two effects in comparison to the SL^{PPK} approximation taking into account an optically thin continuum. In this figure, we show a strong line (χ_L/σ_{TH} = 10^6) for a two-level atom with NLTE parameter $\varepsilon=10^{-6}$ for a WR model in accordance with the model by Hillier (1987) for the WN5 star HD50896 ($T_{eff}(3R_o)$ = 60,000 K, R_* = 2.5 R_o, $\dot{M}$ = $5 \cdot 10^{-5}$ M_o/yr, v_∞ ≈ 1600 km/s (before the final acceleration, cf. sect.III.3)). It is evident, that for a wide range in wavelength (maximum of flux at ~ 850 Å !) and radius the actual force is larger - up to a factor > 10 - than the simplified approach, so that these effects should lead to a significant improvement of our understanding of WR winds.

Finally, the radiation pressure by b-f, f-f processes in subsonic to sonic regions can become comparable to Thomson-scattering (cf. Table 1 in Langer, 1989), which reduces the effective gravity g_{eff} and leads to an additional enhancement of the mass loss rate ($\dot{M} \sim g_{eff}^{-(1-\alpha)/\alpha}$, α=0.6...0.7, see PPK, 1986).

Concluding, the mentioned processes may enlarge the total radiation force in the optically thick part of the wind drastically -more than compensating for the effect of force reduction due to ML-effects, see sect.III.3 - and give rise to mass-loss rates which cannot (or only under different conditions) be initiated in O-stars where the continuum is optically thin in the line-driving frequency range.

VII. CONCLUSIONS

Summarizing the results of the previous sections, we come to the following conclusions:

1) For "normal" O-stars with moderate to high wind densities, numerical models based on the theory of radiation driven winds agree - already in the present state of the art - well with the major observational features ($\dot{M}$, v_∞, main ionization stages). Detailed radiation transfer effects are of importance only indirectly and should lead in a future improvement of the model to satisfying results also with respect to trace ions. For the total line pressure and the occupation numbers (not discussed here, but cf. Puls, 1987), the changes from the simplified to the physical realistic approach appear - up to now - to be only marginal.
2) Under *extreme* conditions, i.e. thin or optically thick winds, sophisticated transfer effects (ML-processes, interaction line/continuum, diffuse radiation force) may become *extremely* important to explain even *1st-order* effects and may allow for a final understanding of still unsolved problems.

ACKNOWLEDGEMENTS

This work was supported by the Deutsche Forschungsgemeinschaft under grant Ku 474/11-3.

REFERENCES

Abbott, D.C.: 1982, Astrophys. J. 259, 282

Abbott, D.C., Hummer, D.G.: 1985, Astrophys. J. 294, 286

Auer, L.H., Koenigsberger, G.: 1990, Astronomical Society of the Pacific Conference Series 7, "Properties of Hot Luminous Stars", ed. C.D. Garmany, p.291

Castor.,J.: 1970, M.N.R.A.S. 149, 111

Castor, J., Abbott, D.C., Klein, R.: 1975, Astrophys. J. 195, 157

Gabler, R., Gabler, A., Kudritzki, R.P., Puls, J., Pauldrach A.: 1989, Astron. Astrophys., 226, 162

Groenewegen, M.A.T., Lamers, H.J.G.L.M., Pauldrach, A.W.A.: 1989, Astron. Astrophys. 221, 78

Hillier, D.J.: 1987, Ap.J. Suppl. 63, 965

Hummer, D.G., Rybicki, G.B.: 1985, Astrophys. J. 203, 258

Kudritzki, R.P., Simon, K.P., Hamann, W.R.: 1983, Astron. Astrophys. 118, 245

Kudritzki, R.P., Pauldrach, A., Puls, J., Abbott, D.C.: 1989, Astron. Astrophys.. 219, 205

Kudritzki, R.P., Cabanne, M.L., Husfeld, D., Niemela, V., Groth, H.G., Puls, J., Herrero, A. : 1989a, Astron. Astrophys. 226, 235

Kurucz, R.L.: 1979, Ap. J. Suppl. 40, 1

Langer, N.: 1989, Astron. Astrophys. 210, 93

Langer, N.: 1990, Trieste Series on Dynamical Phenomena in Stars and Circum- stellar Environments, "Angular Momentum and Mass Loss for Hot Stars", eds. L.A. Wilson, R. Stalio, in press

Maeder, A., Meynet, G.: 1987, Astron. Astrophys. 182, 243

Pauldrach, A.W.A., Puls, J., Kudritzki, R.P.: 1986, Astron. Astrophys. 164, 86

Pauldrach, A.W.A.: 1987, Astron. Astrophys. 183, 295

Pauldrach, A., Puls, J., Kudritzki, R.P., Butler, K.: 1990, Astron. Astrophys. 228, 125

Pauldrach, A.W.A., Puls, J., Gabler, R., Gabler, A.: 1990a, Astronomical Society of the Pacific Conference Series 7, "Properties of Hot Luminous Stars", ed. C.D. Garmany, p. 171

Puls, J.: 1987, Astron. Astrophys. 184, 227

Puls, J., Hummer, D.G.: 1988, Astron. Astrophys. 191, 197

Puls, J. , Pauldrach, A.W.A.: 1990, Astronomical Society of the Pacific Conference Series 7, "Properties of Hot Luminous Stars", ed. C.D. Garmany, p. 203

Winds of Late Type Stars

H.-P. Gail

Institut für Theoretische Astrophysik, Im Neuenheimer Feld 561,
D-6900 Heidelberg, Fed. Rep. of Germany

1. Introduction

Cool stars on the asymptotic giant branch (AGB) form a late stage of stellar evolution shortly before loosing their outer layers and moving into the state where a planetary nebula is excited. These stars show severe mass-loss rates between $10^{-8} M_{\odot}\mathrm{yr}^{-1}$ at the lower AGB, increasing to some $10^{-4} M_{\odot}\mathrm{yr}^{-1}$ as the star evolves to the top of the AGB while becoming slightly cooler and considerably more luminous. In the outflowing matter dust condenses from the gas phase and forms optically thin dust shells on the lower part of the AGB and massive dust shells at the top of the AGB where the stars are completely obscured by their dust shells.

The mass-loss rate steadily increases with ever increasing rate as the star evolves to the top of the AGB [1], indicating that this evolutionary phase is characterized by a runaway process where the mass-loss triggers further mass-loss. At the top of the AGB the outer layers of the star have been completely lost except for a thin envelope over an electron degenerated carbon-oxygen core. After completion of the AGB phase the star then turns to the left in the HR-diagram and rapidly evolves into a central star of a Planetary Nebulae.

The mechanisms which cause the mass-loss and drive the star into a desaster (or, perhaps, prevents the star from the catastrophe of becoming a supernova) are not yet definitely known. Essentially three mechanisms have been proposed: Dust driven winds, wave driven winds and pulsationally driven winds.

Dust condensation and radiation pressure on grains as the driving mechanism for stellar winds of cool stars has been proposed for theoretical reasons long before the observational detection of the existence of dust shells and winds around cool stars [35] and has extensively been discussed in the past two decades.

Pulsationally driven winds have been proposed [54, 56] since most if not all stars on the upper part of the AGB are observed to be Mira variables with pulsation periods between $\approx$ 150d and $\approx$ 600d or Long Period variables with pulsation periods between $\approx$ 500d and $\approx$ 2000d.

Wave driven winds have been proposed [31] since in subphotospheric convection zones waves are generated which propagate through the photosphere into the uppermost layers of the star where they steepen to shocks and deposit their momentum. The resulting pressure gradient possibly drives a wind.

In the following sections we disscuss these mechanisms. Related reviews on the subject are given by [33], [34] and [41]. A brief review on the obsered properties of circumstellar dust shells and winds of late type stars is found in [37].

2. Dust Driven Winds

The basic idea of this mechanism is that radiation pressure on grains generates an outwards directed force which is sufficiently strong to dominate over the gravitational attraction of the star. The momentum gain of the grains is efficiently transmitted to the gas and then the mixture of grains and gas is expelled (see Fig.1).

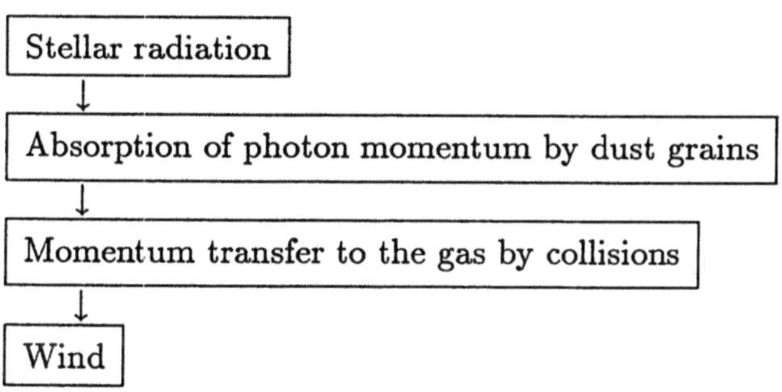

Fig. 1. The essential steps involved in the dust driven wind mechanism.

The idea that radiation pressure is the driving mechanism for winds of late type stars is old. Up to the early sixties, however, the possibility of formation of grains in the outer layers of a star was not recognized and the only opacity source known at that time, atoms and molecules, does not experience sufficient radiation pressure for driving an outflow of matter [52]. Then Hoyle and Wickramasinghe [35] proposed that carbon grains may condense from the gas in cool carbon stars and showed that they are sufficiently accelerated by radiation pressure to drift out of the star. Wickramasinghe et. al [53] then argued that grain-gas collisions transfer sufficient momentum to the gas to accelerate the mixture to velocities above the local escape velocity, thus driving a wind.

Gilman [29] showed on the basis of thermochemical calculations that in the outer layers of stars the conditions are favourable for formation of graphite and silicon carbide in C-stars and of silicates in M-stars. The rapid progress of IR-observations since the end of the sixties then made clear that the existence of dust in the environment of cool stars is a common and widespread phenomenon.

2.1 Kinematical Model of Dust Driven Winds

Gehrz and Woolf [28] suggested that dust condensation and radiation pressure on grains is responsible for the observed highly supersonic outflow velocities from late type stars. They considered the following simple model.

1) Dust condenses somewhere near the star. Details of this process are not considered.

2) The grains absorb momentum from the radiation field of the star and are accelerated. They start to drift through the gas and acquire some terminal velocity due to frictional coupling to the gas (for details see e.g. [2] and [40]. At mass loss rates above $\dot{M} \approx 5 \cdot 10^{-7} \mathrm{M}_{\odot} \mathrm{y}^{-1}$ the drift velocity turns out to be smaller than or at most equal to the sound speed and for most purposes can be neglected). Further absorption of momentum from the radiation field then is completely transmitted to the gas. Details of this process are not considered.

The resulting force density acting on the gas-dust mixture then is

$$F_{rad} = \frac{4\pi}{c} n_d \langle \sigma_d \rangle_H \, H \,, \tag{1}$$

where n_d is the number density of dust grains, $\langle \sigma_d \rangle_H$ the flux averaged extinction cross section of the grains and $4\pi H$ the total flux of the stellar radiation field. The resulting force generally exceeds the gravitational attraction of the star if the condensible material is completely condensed into grains.

3) Assuming a stationary and spherically symmetric outflow, the equation of motion for the gas-dust-mixture is

$$v \cdot \frac{dv}{dr} = \frac{4\pi}{c} \frac{\langle \kappa_d \rangle_H}{\rho} H = \frac{L_*}{4\pi c R_*^2} \frac{\langle \kappa_d \rangle_H}{\rho} \left(\frac{R_*}{r} \right)^2 . \tag{2}$$

Here the pressure is neglected (which is justified if the flow is supersonic) and it is assumed that radiation pressure strongly exceeds gravitational pull.

Assuming $\langle \kappa_d \rangle_H / \rho$ to be constant yields the solution

$$v^2 = \frac{2L_*}{4\pi c R_*} \frac{\langle \kappa_d \rangle_H}{\rho} \left(1 - \frac{r_c}{r} \right) , \tag{3}$$

where r_c is the the radius at which dust condenses.

This simple model yields terminal outflow velocities of the matter consistently to what has been observed for AGB-stars. It is sometimes used for simple model calculations.

2.2 Hydrodynamical Model of the Dust Driven Wind

The hydrodynamic aspects of the the problem have first been discussed by Salpeter [48], [49].

2.2.1 Stationary Winds. In a spherically symmetric and stationary outflow mass conservation requires

$$4\pi r^2 \rho v = \dot{M} = const \tag{4}$$

where $\dot{M}$ is the mass-loss rate of the star. The equation of state is

$$p = c_T^2 \rho \tag{5}$$

where c_T is the isothermal sound speed. The momentum conservation equation changes with (4) and (5) into

$$(v^2 - c_T^2)\frac{1}{v}\frac{dv}{dr} = \frac{2c_T^2}{r} - \frac{GM_*}{r^2}(1-\alpha) - \frac{dc_T^2}{dr} \tag{6}$$

where M_* is the mass of the star and

$$\alpha = \frac{L_*}{4\pi G M_* c}\frac{\langle\kappa_d\rangle_H}{\rho}\,. \tag{7}$$

is the ratio of radiative acceleration and gravitational deceleration.

If no dust is present, then $\alpha \ll 1$ since absorption by atoms and molecules transfers insufficient momentum to the gas for driving a wind. If, however, dust condenses from the gas then α exceeds unity for typical masses and luminosities of AGB-stars and for typical materials which may condense in the wind (graphite or dirty silicates).

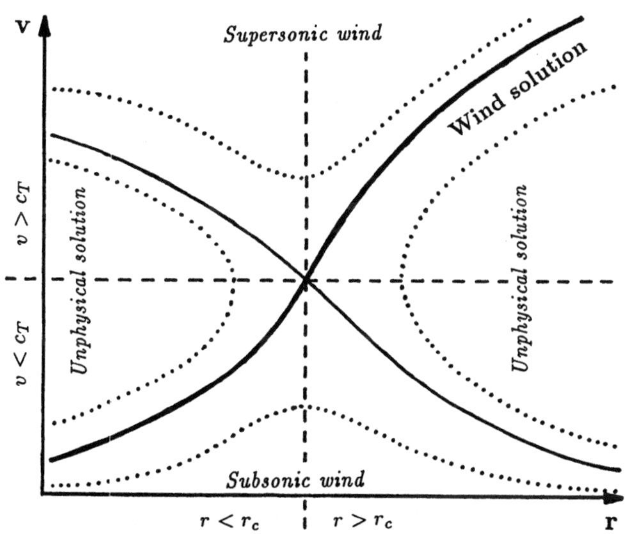

Fig. 2. Topology of the solutions of the wind equation. There exist four types of solutions: pure supersonic winds, pure subsonic winds (breeze solutions), two classes of double-valued solutions (only possible with stationary shocks) and two singular solutions passing through the singular point. The heavy drawn line is the wind solution.

A discussion of the topology of solutions of Eq.(6) (see Fig.2) shows that there exists one and only one solution which meets the requirement of being subsonic at the stellar photosphere, strongly supersonic at large distances from the star and everywhere continuous. This solution passes from subsonic to supersonic velocities through the singular point of Eq.(6) where the coefficient of the derivative and the r.h.s. of the equation vanish simultaneously

$$v^2 - c_T^2 = 0 \quad \text{and} \quad \frac{2c_T^2}{r} - \frac{GM_*}{r^2}(1-\alpha) - \frac{d\,c_T^2}{d\,r} = 0\,. \tag{8}$$

This defines the ***sonic point*** r_s of the flow. The solution in the supersonic region approximates the solution of the model of Gehrz and Woolf [28].

Neglecting the variation of temperature, the position of the sonic point is determined by the implicit equation

$$\frac{r_s}{R_*} = \frac{v_e^2}{c_T^2}\,\frac{1-\alpha(r_s)}{4}\,, \tag{9}$$

where

$$v_e = \sqrt{\frac{2GM_*}{R_*}} \tag{10}$$

is the escape velocity from the stellar surface. We always have $v_e \gg c_T$ since otherwise the star would be unstable against thermal evaporation.

If one always has $\alpha \ll 1$, then due to the decrease of the sound speed with distance from the star there exists a ***thermally driven*** wind, which owes its existence to the temperature gradient and the resulting pressure gradient. Consider the rhs. of Eq.(9) with α varying with r. If due to condensation of dust, α passes from $\alpha \ll 1$ (only a small fraction of the condensible material in grains) through the value $\alpha = 1$ to a final value $\alpha > 1$ (most of the condensible material in grains), then the r.h.s. of Eq.(9) passes from large positive values through zero close to the sonic point, where $\alpha \approx 1$, to large negative values for $\alpha > 1$. Thus, as dust formation occurs at some stellar radii, the position of the sonic point is determined by

$$\alpha(r_s) = 1 - 4\frac{r_s}{R_*}\frac{c_T^2}{v_e^2} \approx 1\,. \tag{11}$$

since $c_T^2/v_e^2 \ll 1$. This shows that the condensation of dust is the mechanism which drives the wind by forcing the matter to pass through the sonic point within the dust condensation zone.

2.2.2 Condition for the Existence of Dust Driven Winds. In addition, this shows that a necessary condition for the existence of a ***dust driven wind*** is that sufficient amounts of the condensible material goes into grains such that α rises during the condensation process to the critical value given by Eq.(11). Otherwise, one would only have a thermally driven wind which at best is supported by dust condensation.

This condition for the existence of dust driven winds can be interpreted in two ways. Assume that all condensible material is condensed into grains. Then the requirement that α is close to unity requires according to (7) that the stellar luminosity exceeds the ***Eddington limit*** luminosity

$$L_{edd} = \frac{4\pi GM_*\rho}{\langle\kappa\rangle_H}\,. \tag{12}$$

Only if the stellar luminosity L_* exceeds L_{edd} does radiation pressure on grains drive a stellar outflow. This limit depends on the grain material and the spectral energy distribution of the star. For cool stars on the AGB one obtains $L_{edd} \approx 5 \cdot 10^2 L_\odot$ for carbon stars and $L_{edd} \approx 2 \cdot 10^3 L_\odot$ for M-stars (see e.g. [23]). All AGB-stars have luminosities above this limit and,thus, their winds, in principle, can be driven by dust condensation.

If the condensation is incomplete, then for the actual value of α in the wind we have

$$\alpha = f \cdot \frac{L_*}{L_{edd}}, \tag{13}$$

where f denotes the fraction of the condensible material which actually has condensed into grains. The condition $\alpha > 1$ then is equivalent with

$$f > \frac{L_{edd}}{L_*}, \tag{14}$$

i.e., a small degree of condensation shifts the luminosity required for driving a wind by dust condensation to higher luminosities than are actually observed for AGB-stars.

2.3 Fixing the Mass-Loss Rate

Lucy [42] improved the model of the dust driven wind by considering radiative transfer. On the basis of a simple two-stream approximation, adapted for the spherical geometry of the problem, he showed that the gray temperature structure in radiative equilibrium is

$$T(r)^4 = \frac{T_*^4}{2}\left[1 - \sqrt{1 - \left(\frac{R_*}{r}\right)^2} + \frac{3}{2}\tau_L\right] \tag{15}$$

where τ_L is defined by

$$\frac{d\tau_L}{dr} = -\langle\kappa\rangle_H \left(\frac{R_*}{r}\right)^2 . \tag{16}$$

The optical depth τ_L comprises both opacity and geometric dilution effects in a spherically extended atmosphere. The boundary condition for τ_L at infinity is $\tau(\infty) = 0$. Contributions to τ_L come from the outer dust shell, where $\langle\kappa\rangle_H$ is due to the strong dust opacity and from the photosphere of the star, where $\langle\kappa\rangle_H$ is determined by the gas opacity. The region between the photosphere and the dust shell provides no substantial contribution to τ_L since the density in this region is very low compared to the photosphere.

At the stellar radius R_* the cumulated optical depth τ_L of the dust shell and the photosphere has to satisfy the condition

$$\tau_L(R_*) = \frac{2}{3} \tag{17}$$

to ensure that the temperature $T(R_*)$ is consistent with the definition of the effective temperature T_* and the radius R_* of the star

$$4\pi R_*^2 \sigma T_*^4 = L_* \,. \tag{18}$$

This additional boundary condition is a constraint for the mass-loss rate since $\langle\kappa\rangle_H$ is proportional to ρ and ρ is proportional to $\dot{M}$ (Eq. 4). Then, Eq.(17) can be satisfied only for one value of the mass-loss rate $\dot{M}$ which becomes a *characteristic value* of the problem.

This property also holds if the temperature is not determined by radiative equilibrium and has to be calculated from the general equation of energy conservation

$$c_V \cdot v \frac{dT}{dr} + p \frac{dv}{dr} = G - L \tag{19}$$

where G and L are the heating and cooling rates of the matter, respectively.

2.4 Dust Condensation

In the outstreaming gas the temperature drops from the photospheric temperature to very low temperatures in the outermost layers. If the partial pressure of those molecules in the gas which would be formed in the vaporization of a solid increases over their partial pressure in thermodynamic equilibrium, then a solid becomes stable against vaporization and in principle could be formed by condensation from the gas phase. From laboratory experiments it is known that condensation does not start until the temperature has dropped considerably below the equilibrium temperature, or, expressed in other words, if the *supersaturation ratio* (the ratio of actual pressure of the vapor molecules of the solid to their vapor pressure in in thermodynamic equilibrium) has considerably increased over unity (in a thermodynamic equilibrium state the supersaturation equals unity). This *supercooling* of the matter is required for the following reason. A large number of molecules from the gas phase cannot spontaneously associate to a single particle of macroscopic size but have to grow in a step by step process from molecules over intermediate states of molecular clusters with tens to hundreds of atoms to particles of macroscopic size. Usually the small clusters have much less binding energy per atom than a solid and the first members on the reaction path from molecules to solids are not formed until the temperature is considerably lower than the stability temperature of the bulk condensate.

In theories of the condensation process it is shown that the intermediate cluster of lowest equilibrium abundance on the reaction path forms the bottleneck of the whole process which determines the rate at which solid particles condense from the gas phase. Clusters with a size above the size of this *critical cluster* grow unlimited as long as the gas is not depleted from condensible material. The condensation process then can be considered to be a two-step process:

1) formation of critical clusters and

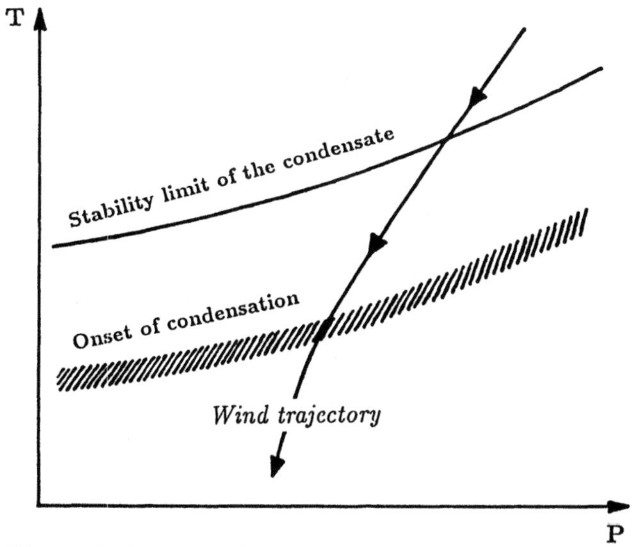

Fig. 3. Trajectory of the wind in the pressure temperature plane. As the wind trajectory crosses the stability limit of the solid from above, no condensation occurs until the initial molecules on the reaction chain from molecules to dust become stable. As the tracjectrory crosses this strip of dramatically increasing nucleation rate, a sudden onset of condensation is observed.

2) growth of these clusters to macroscopic sized particles.

The treatment of the second step is rather simple. The initial growth process up to cluster sizes of ≈ 1000 atoms is only a short fraction of the total growth history of a particle and can be neglected. Above ≈ 1000 atoms the grains can be treated as beeing solid particles and their growth is easily calculated from a macroscopic theory.

The most crucial point is the first step. The nucleation rate can be shown to be [25]

$$J_*^{-1} = \sum_i \frac{\tau_i}{\overset{o}{n}_i} \tag{20}$$

with J_* being the rate of formation of grains, $\overset{o}{n}_i$ the abundance of the intermediate clusters on the reaction chain in thermodynamic equilibrium and τ_i the mean time between reaction of cluster i on the chain to the next cluster $i+1$ on the chain.

As is easily seen the intermediate cluster of lowest abundance dominates the nucleation rate J_*:

$$J_* = \frac{1}{\tau_{N_*}} \cdot \overset{o}{n}_{N_*} \tag{21}$$

with N_* being the critical cluster on the reaction chain. Its partial pressure in equilibrium is according to the law of mass-action

$$\overset{o}{p}_{N_*} = \prod_i p_i^{\nu_i} \exp[-\frac{\Delta G_f}{RT}] \tag{22}$$

where p_i are the partial pressures of the molcules from which the critical cluster is formed and ΔG_f is its free enthalpy of formation from these molecules. ΔG_f usually is of the order of a few eV. Thus, the abundance of the critical cluster is a steeply increasing function of the temperature. For instance, if $\Delta G_f = 4$eV and T=1000K then a cooling of the gas by 25K results in an e-fold increase of the density of the critical cluster. The dependence on the pressure is only moderate; with increasing total pressure J_*, increases due to the increased partial pressures of the parent molecules of the critical cluster. Thus, there exists a narrow, slightly inclined strip in the pressure-temperature plane across which the nucleation rate increases dramatical from completely negligibly to extremely rapidly. Condensation occurs avalanche-like if the trajectory of an outmoving gas element in the temperature-pressure plane crosses this strip (cf. Fig.3).

A calculation of J_*, thus, requires a complete knowledge of the properties of all intermediate clusters along the reaction chain and their formation energies and reaction rates [12]. Presently, this problem is not yet solved due to the lack of data. For this reason one usually takes recourse to *classical homogeneous nucleation* theory which tries to extrapolate the binding energy of small clusters from the properties of the bulk condensate by using the concept of a surface tension for allowing for the lower formation energy of small clusters (the surface tension describes the defect of binding energy of a surface atom as compared to an interior atom in condensed materials). Then the equilibrium abundances $\overset{o}{n}_i$ are readily calculated from chemical thermodynamics and from this J_*. Dust formation based on classical nucleation theory has extensively been discussed by [13], [14], [15], [19], [24], [39], [50]. This approach may be considered as a zeroth order approximation as long as the data reqired for a more reliable calculation of J_* are not available (see, however, Donn and Nuth [11] who strongly critisize this point of view).

2.5 Model Calculation

Models for dust driven winds have been calculated by several authors in various degrees of approximation.

Some authors circumvented the difficulty of calculating the condensation of dust and its present large uncertainties by *assuming* a definite condensation temperature ([2], [30], [40], [45], [51]) or, what amounts to the same, a definite condensation radius. Since the subsonic-supersonic transition of the velocity is forced by dust condensation and growth and, thus, fixed to the inner edge of the dust shell, one can use information from model calculations for the IR-emission from circumstellar dust shells [46], [47] for fixing these unknown quantities.

A different approach has been used by [42], [43], which determines the condensation temperature by the requirement that the supersaturation ratio equals some prescribed value. Classical nucleation theory predicts that avalanche nucleation occurs if the supersaturation exceeds some critical value. Then the position of the sonic point can be selfconsistently determined in the model calculation.

Grain growth has usually been calculated in these model computations and in most calculations the drift of grains has been considered. In [42] the gas dynamical problem is simplified by assuming that dust condenses suddenly at the sonic point. Then one has $\alpha = 0$ for $r < r_s$ and α equals its final value for $r > r_s$. In principle, it is this approximation on which the model of Gehrz and Woolf [28] is based.

The first models using explicit calculations of the nucleation rate and grain growth have been calculated by Deguchi [8], Kozasa et al. [38] for M-stars and Gail et. al [19], [20] for C-stars. The model calculations of Deguchi and Kozasa et al. are based on a semianalytical treatment of dust formation and growth along the lines developed by Drain and Salpeter [15] and Yamamoto and Hasegawa [57] while Gail et al. [19] derived a method for calculating dust growth based on moments of the particle size distribution for the grains which allows a simple and efficient calculation of the properties of the dust formed in the circumstellar shell. A similar method has already been used by Fix [16], [17]. Improved versions of the method are derived by Gail and Sedlmayr [25], Dominik et al. [10] and Gauger et al. [27].

All these model calculations demonstrated clearly that the dust driven wind model can explain the essential properties observed for circumstellar shells and winds of AGB-stars. In a series of papers (Gail and Sedlmayr [22], Dominik et al. [9], [10]) the model calculations for C-stars have been considerably improved. It is now possible to calculate large grids of selfconsistent models considering the effects of hydrodynamics, radiative transfer and dust formation and growth of dust from the base of the stellar photosphere up to large distances from the star which depend only on mass, luminosity, effective temperature and element abundance of the central star (these parameters are determined by the internal structure and the evolutionary history of the star and not by the properties of the outer layers) and involve no freely adaptable unknown parameters. The results of such dust driven wind model calculations (see the contribution of C. Dominik, this volume) satisfy

1) the observed terminal wind velocities,
2) the observed mass-loss rates,
3) the observed optical depth' of the circumstellar shells and finally
4) the observed correlations between these quantities.

This strongly suggests that the mass-loss of cool AGB-stars is driven by dust condensation in the outflowing gas.

2.6 Limits of the Dust Driven Wind Model

There are some limitations of the dust driven wind model. A discussion of the growth process of grains and the resulting depletion of the gas phase from condensible material [22] shows that the condensation of dust drives a wind only if the mass-loss rate exceeds $2 \cdot 10^{-6} M_{\odot}\mathrm{yr}^{-1}$ for C stars and $5 \cdot 10^{-6} M_{\odot}\mathrm{yr}^{-1}$ for M stars. Below this *minimum mass-loss rate* the grains grow very slowly due to the low particle density in the wind and the final fraction f of condensible material in grains fails to meet condition (14) for the existence of a dust driven wind. For AGB-stars with an observed mass-loss rate below the minimum mass-loss rate, the dust condensation supports the wind and results in an additional acceleration. The wind itself, however, is driven by a different mechanism. Some C-stars and many M-stars, on or near the AGB, show mass-loss rates below the minimum value for dust driven winds. In these stars a different wind mechanism operates. Only the coolest AGB-stars with severe mass-loss and massive dust shells are candidates for dust driven winds.

A second shortcoming of the dust driven wind hypothesis is that it is not self starting. For a hydrostatic atmosphere in the outermost layers where the gas temperature is sufficiently low for condensation to become possible the density is too small to allow efficient condensation. Dust driven winds, thus, have to be kicked on by some different mechanism which levitates the atmosphere such that the density at low temperature is high enough to allow fast condensation. Then however, the gas is expelled by radiation pressure on grains and since no vacuum can be left behind, gas from the lower layers starts to flow outwards into the region where condensation becomes possible. In this gas flow, the energy expended by doing work against gravity is taken from its thermal energy content, which in turn is continuously replenished by absorption from the stellar radiation field. Thus, dust condensation sucks material from lower layers into the condensation zone from which it is blown to infinity by radiation pressure. This imposes a flow across the condensation zone with a velocity of the order of the sound speed (which would be the initial expansion velocity of a gas into a region of very low density). Hence, once kicked on, dust driven winds maintain themselves. The whole process resembles a pendulum-clock which only operates after someone has pushed the pendulum but then needs no further assistance.

A third problem is that all stars on the AGB seem to be variable and those showing the highest mass-loss rates and massive dust shells all are Mira variables or Long Period variables. The structure of the stellar photosphere and of the region between the photosphere and the inner edge of the dust shell then is determined by strong running shock waves and not by a continuous outflow of matter, as is assumed for dust driven winds. At the same time however, this offers a natural explanation for the mechanism which starts dust driven winds.

Generation of waves in the convection zone

↓

Steepening to shocks by the decreasing density

↓

Momentum deposition and heating of the gas

↓

Wind

Fig. 4. The essential steps of the mechanism of wave driven winds.

3. Wave Driven Winds

A different mechanism for driving a wind is dissipation of wave energy. In the convection zone acoustic and MHD waves (if magnetic fields are present) are generated and propagate outwards into high layers of the atmosphere. They dissipate and deposit their energy and momentum to the gas which is accelerated and heated and starts to move outwards due to pressure forces.

3.1 Time Dependent Hydrodynamics

Time dependent hydrodynamic calculations based on a numerical code tailored for handling a large number of shocks have been performed by Cuntz [7] for late K-stars (which, however, rarely show indications for dust formation). Acoustic shock waves with stochastically changing periods introduced at the base of the photosphere. The stronger shocks overtake weaker ones ahead of them due to their higher propagation velocity and merge with them which increases the strength of the strong shocks. Occasionally, very strong shocks are formed which accelerate the matter to velocities above the escape velocity. This process predicts episodic mass-loss events and may well be responsible for episodic mass-loss accompanied by dust condensation as it is observed [4] for stars on the lower part of the AGB where the stars are not Mira variables.

3.2 Time Averaged Hydrodynamics

A different approach is to study the average response of an atmosphere with many shocks and waves on the presence of these disturbances by taking the average with respect to time. The time averaged momentum conservation equation for sphericall geometry is [26]

$$\frac{d\langle p\rangle}{dr} + \frac{1}{r^2}\frac{d}{dr}r^2\langle \rho v^2\rangle = -g\langle\rho\rangle + \frac{1}{\mathcal{P}}\langle[\rho]U\rangle \tag{23}$$

where $\langle\ldots\rangle$ denotes time averages. The second term on the l.h.s. is the time average over waves and, if shocks are present, the time average at a fixed position over the variation of the hydrodynamical quantities *between* the passage of shocks. The second term on the r.h.s. is the average over the density jump $[\rho]$ across the shock front times the shock velocity U and the reciprocal average recurrence time $\mathcal{P}$ of shocks. This is the average impulsive force acting at a fixed position on the matter due to passage of shock waves.

Models of winds for AGB stars neglecting shocks and retaining waves only have been calculated by Hartmann and MacGregor [31] and Pijpers and Habing [44]. The pressure gradient then forces the gas to pass through the sonic point and to stream out of the star. For a model calculation Eq.(23) has to be supplemented by an equation for the dissipation of the wave energy $\langle\rho v^2\rangle$. Unfortunately at present it seems not to be possible to calculate the dissipation selfconsistently and the dissipation has to be modelld by introducing some dissipation length, which is a free parameter and has to be choosen by considerations based on plausibility and the outcome of the desired results. The model calculations of Pipers and Habing [44] demonstrated, that wave pressure is able to drive a wind in AGB stars with just the observed properties but it remains unclear whether real stars dissipate their wave energy just as is required in the model calculations. One point, however, is clearly demonstrated by these calculations: Wave pressure is an efficient process for increasing the scale height of an atmosphere which strongly increases the density in the outermost layers of the star and, thus, favours dust formation.

4. Pulsationally Driven Winds

According to observations most if not all AGB stars with dust shells are Mira variables or Long Period variables. Willson and Hill [54] proposed that shock waves generated in the atmosphere and outer layers of the star are responsible for the mass-loss. In principle, this mechanism works as follows. If a gas element is overrun by a shock, it experiences an outwards directed impulsive force. It starts to move outwards on a nearly ballistic trajectory in the gravitational field of the star and falls back. If the gas element is overrun by the next shock before it arrived at his former position, every shock effectively transports the gas element some distance outwards. The result is a wind (Fig. 5).

Pulsation of the star

↓

Generation of strong shocks in the photosphere

↓

Momentum deposition in the outer layers

↓

Wind

Fig. 5. The essential steps in the mechanism of pulsationally driven winds.

4.1 The Ballistic Limit

Semianalytical treatments of shock-driven winds are given by Willson and Hill [54] and Bertschinger and Chevalier [3].

Willson and Hill [54] calculated the time required for a gas element, starting at position r_0 to move upwards to a culmination point r_{max} and then to fall back to some position r. Neglecting pressure forces the particles follow a ballistic trajectory in the gravitational field of the star and the time required for falling back to its original position r_0 is twice the free-fall time from r_{max} to r_0. This recurrence time defines a natural *gravitational period* $\mathcal{P}_0$ for a gas element moving at r_0 with velocity v_0

$$\mathcal{P}_0 = \frac{2r_0}{\sqrt{v_e^2 - v_0^2}} \left[\arccos\sqrt{1 - \frac{v_0^2}{v_e^2}} + \frac{v_0}{v_e}\sqrt{1 - \frac{v_0^2}{v_e^2}} \right] . \tag{24}$$

The upper culmination radius of the trajectory is

$$r_{max} = r_0 \frac{v_e^2}{v_e^2 - v_0^2} \tag{25}$$

where v_e is the escape velocity at r_0 given by (10). If the real period $\mathcal{P}$ is shorter than the gravitational period $\mathcal{P}_0$ for given r_0 and v_0 then a mass element has not yet fallen back to its original position when it is overrun by the next shock. Then, each mass element is shifted upwards with each shock and a wind is generated. In the opposit case, each gas element returns to its original position and no mass transport is generated by shocks.

If $v_0^2 \ll v_e^2$ then (24) simplifies to

$$\begin{aligned} \mathcal{P}_0 &= \frac{4r_0}{v_e} \cdot \frac{v_0}{v_e} = \\ &12[\mathrm{d}] \cdot v_0[\mathrm{km} \cdot \mathrm{sec}^{-1}] \cdot \left(\frac{r_0}{R_*}\right)^2 \cdot \left(\frac{L_*}{L_\odot}\right) \left(\frac{3000\mathrm{K}}{T_*}\right) \left(\frac{M_\odot}{M_*}\right) . \end{aligned} \tag{26}$$

With increasing r_0, $\mathcal{P}_0$ increases and eventually exceeds the real period $\mathcal{P}$ of the star. Then, above this critical radius, the conditions for driving matter out of the star by repeated shock acceleration are satisfied and a wind is generated. For the shortest pulsation periods observed for AGB-stars of, say, 200d the critical radius is $r_0 \approx (4/v_0)R_*$ and for the slowest pulsating Long Period variables with periods of, say, 1500d the critical radius is $r_0 \approx (11/v_0)R_*$. Stronger shocks shift this critical point closer to the star but substantial modifications to the approximation (26) do not occur until v_0 is close to v_e.

Shocks of moderate strength, where the velocity jump $2v_0$ across the shock front is a few km$\cdot$sec^{-1} have their critical point close to the star for Mira-variables of $\mathcal{P} \approx 200$d, and only at a few stellar radii for Long Period variables. Shocks of intermediate strength will initiate a wind and mass-loss. Strong shocks shift the critical radius closer to the star and are even more efficient in driving a wind.

This wind, however, will lead to only a small mass-loss rate. In the limit $v_0 \ll v_e$ by solving the eqation of motion for constant gravitational acceleration (note that $r_{max} - r_0 < r_0$ in this limit) one obtains

$$\delta r = \frac{GM_*}{r_0^2} \mathcal{P}(\mathcal{P}_0 - \mathcal{P}) \tag{27}$$

for the increase in radial position which a gas parcel suffers due to the shocks if $\mathcal{P} < \mathcal{P}_0$. Assuming the propagation velocity of the shock roughly equal to v_0, then the distance between subsequent shocks is $\Delta \approx v_o \cdot \mathcal{P}$ and the fraction

$$\frac{\delta r}{\Delta} \approx \frac{\mathcal{P}_0 - \mathcal{P}}{\mathcal{P}_0} \tag{28}$$

of material contained between two shocks flows through the sperical shell of radius r_0 within one period.

The average outflow velocities of pulsationally driven winds is low. Well above the critical radius the average velocity is $\delta r/\mathcal{P}$, which after some algebra using (26) can be written as

$$\langle v \rangle = \frac{\delta r}{\mathcal{P}} = v_0 \left[1 - \left(\frac{r_c}{r} \right)^2 \right] \tag{29}$$

where r_c is the critical radius. Then, the terminal velocity at large distance from the critical radius is approximately one half of the velocity jump of the shock at the critical radius which is of the order of a few km·s^{-1} (Bowen [5], [6]). Thus, purely shock driven winds do not show the observed higly supersonic outflow velocities of $\approx$ 15km·s^{-1} of AGB stars.

4.2 The Role of Dust Condensation

Wood [56] and Jones et al. [36] have pointed out that the shock driven wind will be strongly supported by dust condensation. If the trajectory of an upwards moving gas element crosses in the temperature-pressure plane the boundary where efficient dust condensation is possible (see Fig.3) then by radiation pressure it is strongly accelerated to a velocity above the local escape velocity. It does not fall back but is removed from the star. This strongly increases the mass loss rate. Consider a position r_0 where the upper culmination point r_{max} of the trajectory enters into the region of dust condensation. At some instant, a shock passes this point and the gas element at r_0 is lifted into the dust formation zone and then leaves the star and this repeats at every subsequent shock at r_0. Thus, the whole material between two shocks is thrown within one pulsation period into the condensation zone and then is blown out of the star by radiation pressure on grains. The mass loss rate then is the mass contained in the shell between two subsequent shocks below the condensation zone, divided by the pulsation period and not merely the small fraction (28) of this rate as in a purely shock driven wind. If dust condensation is possible it strongly increases the mass-loss rate as compared to a pulsationally driven wind by imposing the *new boundary condition* to the problem that matter entering the condensation zone immediately leaves the star.

The dust condensation additionally increases the average outflow velocity since each gas element entering the condensation zone is accelerated to a high terminal velocity which is approximately given by (3).

4.3 Time Dependent Model Calculations

Numerical time dependent hydrodynamic calculations have been performed by Willson and Hill [54], Hill and Willson [32], Wood [56] and Bowen [5], [6]. The calculations of Willson and Hill and Hill and Willson where performed for purely shock driven winds in the limiting cases of isothermal and adiabatic shocks. The results showed that moderately strong shocks in the isothermal case initiate only insignificant amounts of mass loss, as is to be expected from the estimations based on the ballistic limit. The models with adiabatic shocks show unrealistic high mass loss rates due to very strong shocks. Adiabatic shocks, however, are not to be expected since the radiative cooling time in the atmosphere is short (Willson and Hill [54], Bowen [5]). Wood [56] obtained similar results. In this calculation, however, dust formation in the outer layers is considered by the simple approximation that dust condenses completely within a small temperature interval around some assumed condensation temperature. It was found that dust condensation dramatically increases the mass-loss rate.

Model calculations including calculation of the gas temperature and modelling dust formation similar as Wood [56] have been performed by Bowen [5]. In models without dust formation the shocks are closer to isothermal than to adiabatic shocks in and near the atmosphere due to efficient cooling. The resulting shock driven winds produce only low outflow velocities and mass-loss rates, just as the earlier calculations. The models which include dust formation (based on the same approximation as used by Wood [56]), however, show moderate ($10^{-7} M_{\odot}\mathrm{yr}^{-1}$) to large ($10^{-4} M_{\odot}\mathrm{yr}^{-1}$) mass-loss rates and outflow velocities well within the range of values observed for AGB stars. A comparison of the models with and without dust formation shows that the condensation of dust and subsequent radiation pressure on grains is the driving mechanism for the wind. The pulsation supports the dust formation by increasing the average scale hight of the atmosphere and, thus, the density in the condensation zone which enables massive dust condensation. The whole structure of the model is, however, completely changed, if dust condenses and forces a large mass outflow due to effectively removing each gas parcel from the star which enters the condensation zone. Thus, the combined action of shocks and dust condensation results in a mass-loss process which may be considered as a pulsationally supported, dust driven wind.

The models of Bowen [5] including dust formation for the more rapid pulsators with periods below $\approx$ 300d and mass-loss rates below $\approx 10^{-7} M_{\odot}\mathrm{yr}^{-1}$ should be considered with some caution. The model calculation assumes that dust condensation is complete once it becomes possible. This strongly overestimates the radiation pressure since at the corresponding low densities in the wind the grains grow slowly and the degree of condensation remains small (Gail and Sedlmayr [23]). For such stars one probably has to apply the models without dust formation since only a small fraction of the condensible material goes into grains and the radiation pressure is insufficient to cause significant dynamical effects.

5. Concluding Remarks

The mechanism driving a wind in the Mira-like variables at the upper part of the AGB seems now to be quite clear. Pulsationally supported, dust driven winds have all properties which are required to fit observations. Clearly, until more detailed models are available which predict reliable mass-loss rates, a lot of work still has to be done, especially to incorporate more realistic calculations of the dust formation process.

At the lower part of the AGB, the stars show only slight variations of a more irregular nature. There, a different mechanism has to be active. Neither can pulsations alone drive a wind, nor can the wind be driven by dust condensation since these stars show a low mass-loss rate below the minimum mass-loss rate for dust driven winds. Probably, this is the regime of wave driven winds where stochastic shocks grossly extend the atmosphere and shock merging occasionally ejects clumps of gas in which small amounts of dust condense.

References

1. Bedijn, P.J.: 1986, in *Light on Dark Matter*, F.P. Israel Ed., Reidel, Dordrecht, p.119
2. Berruyer, N., Frisch, H.: 1983, *Astron. Astrophys.* **126**, 269
3. Bertschinger, E., Chevalier, R.A.: 1985, *Astrophys. J.* **299**, 167
4. Bloemhof, E.E., Dauchi, W.C., Townes, C.H.: 1985, *Astrophys. J.* **299**, L37
5. Bowen, G.H.: 1988, *Astrophys. J.* **329**, 229
6. Bowen, G.H.: 1989 in *Evolution of Peculiar Red Giants*, H.B. Jonson and B. Zuckerman Eds., Cambridge University Press, p. 269
7. Cuntz, M.: 1987, *Astron. Astrophys.* **188**, L5
8. Deguchi, S.: 1980, *Astrophys. J.* **236**, 576
9. Dominik, C., Gail, H.-P., Sedlmayr, E.: 1989, *Astron. Astrophys.* **223**, 227
10. Dominik, C., Gail, H.-P., Sedlmayr, E., Winters, J.-M.: 1990, *Astron. Astrophys. in press*
11. Donn, B., Nuth, J.A.: 1985, *Astrophys. J.* **288**, 197
12. Donn, B., Hecht, J., Khana, R., Nuth, J., Stranz, D., Anderson, A.B.: 1981, *Surface Science* **106**, 576
13. Donn, B., Wickramasinghe, N.C., Hudson, J.R., Stecher, T.P.: 1968, *Astrophys. J.* **153**, 451
14. Draine, B.T.: 1979, *Astrophys. Space Sci.* **65**, 313
15. Draine, B.T., Salpeter, E.E.: 1977, *Journ. Chem. Phys* **67**, 2230
16. Fix, J.D.: 1969a, *Monthly Notices Roy. Astron. Soc.* **146**, 37
17. Fix, J.D.: 1969b, *Monthly Notices Roy. Astron. Soc.* **146**, 51
18. Fox, M.W., Wood, P.R.: 1985, *Astrophys. J.* **297**, 455
19. Gail, H.-P., Keller, R., Sedlmayr, E.:1984, *Astron. Astrophys.* **133**, 320
20. Gail, H.-P., Sedlmayr, E.: 1985, *Astron. Astrophys.* **148**, 183
21. Gail, H.-P., Sedlmayr, E.:1986a, *Astron. Astrophys.* **161**, 201
22. Gail, H.-P., Sedlmayr, E.:1987, *Astron. Astrophys.* **171**, 197
23. Gail, H.-P., Sedlmayr, E.:1987a, *Astron. Astrophys.* **177**, 186
24. Gail, H.-P., Sedlmayr, E.: 1987b, in *Physical Processes in Interstellar Clouds*, G.E. Morfill and M. Scholer Eds., Reidel, Dordrecht, p.275
25. Gail, H.-P., Sedlmayr, E.:1988, *Astron. Astrophys.* **206**, 153
26. Gail, H.-P., Cuntz, M., Ulmschneider, P.: 1990, *Astron. Astrophys. in press*
27. Gauger, A., Gail, H.-P., Sedlmayr, E.:1990 *Astron. Astrophys.* , *in press*
28. Gehrz, R.D., Woolf, N.J.: 1971, *Astrophys. J.* **165**, 285

29. Gilman, R.C.: 1969, *Astrophys. J.* **155**, L185
30. Goldreich, P., Scoville, N.: 1976, *Astrophys. J.* **205**, 144
33. Hartmann, L., MacGregor, K.B.: 1980, *Astrophys. J.* **242**, 260
34. Hill, S.J., Willson, L.A.: 1979, *Astrophys. J.* **229**, 1029
31. Holzer, T.E.: 1987, in *Circumstellar Matter*, I. Appenzeller and C. Jordan Eds., Reidel, Dordrecht, p.289
32. Holzer, T.E., MacGregor, K.B.: 1985, in *Mass Loss from Red Giants*, M. Morris and B. Zuckerman Eds., Reidel, Dordrecht, p.229
35. Hoyle, F., Wickramasinghe, M.C.: 1962, *Monthly Notices Roy. Astron. Soc.* **124**, 417
36. Jones, T.W., Ney, E.P., Stein, W.A.: 1981, *Astrophys. J.* **250**, 324
37. Knapp, G.R.: 1987, in *Late Stages of Stellar Evolution*, S. Kwock and S.R. Pottasch Eds. Reidel, Dordrecht, p.103
38. Kozasa, T., Hasegawa, H., Seki, J.: 1983, *Astrophys. Space Sci.* **98**, 61
39. Kozasa, T., Hasegawa, H.: 1987, *Progr. Theor. Phys.* **77**, 1402
40. Kwok, S.: 1975, *Astrophys. J.* **198**, 583
41. Linsky, J.L.: 1987, in *Circumstellar Matter*, I. Appenzeller and C. Jordan Eds., Reidel, Dordrecht, p.271
42. Lucy, L.B.: 1976, *Astrophys. J.* **205**, 482
43. Menietti, J.D., Fix, J.D.: 1978, *Astrophys. J.* **224**, 961
44. Pijpers, F.P., Habing, H.J.: 1989, *Astron. Astrophys.* **215**, 334
45. Phillips, J.P.: 1979, *Astron. Astrophys.* **71**, 115
46. Rowan-Robinson, M., Harris, S.: 1983a, *Monthly Notices Roy. Astron. Soc.* **202**, 767
47. Rowan-Robinson, M., Harris, S.: 1983b, *Monthly Notices Roy. Astron. Soc.* **202**, 797
48. Salpeter, E.E.: 1974a, *Astrophys. J.* **193**, 579
49. Salpeter, E.E.: 1974b, *Astrophys. J.* **193**, 585
50. Tabak, R.G., Hirth, J.P., Meyrick, G., Roark, T.P.: 1975, *Astrophys. J.* **196**, 457
51. Tielens, A.G.G.M.: 1983, *Astrophys. J.* **271**, 702
52. Weyman, R.: 1963, *Ann. Rev. Astron. Astrophys.* **1**, 97
53. Wickramasinghe, N.C., Donn, B.D., Stecher, T.P.: 1966, *Astrophys. J.* **146**, 590
54. Willson, L.A., Hill, S.J.: 1979, *Astrophys. J.* **228**, 854
55. Wodrow , J.E.J., Aumann, J.R.: 1982, *Astrophys. J.* **257**, 247
56. Wood, P.R.: 1979, *Astrophys. J.* **220**, 230
57. Yamamoto, T., Hasegawa, H.: 1977, *Progr. Theor. Phys.* **58**, 816

Analyses of Wolf-Rayet Stars

W.-R. Hamann[1], *U. Wessolowski*[1], *W. Schmutz*[2], *E. Schwarz*[1], *G. Dünnebeil*[1], *L. Koesterke*[1], *E. Baum*[1], *and U. Leuenhagen*[1]

[1]Institut für Theoretische Physik und Sternwarte der Universität, Olshausenstr. 40, D-2300 Kiel 1, Fed. Rep. of Germany
[2]Joint Institute for Laboratory Astrophysics, University of Colorado, Boulder, CO 80309, USA

1. Introduction

Wolf-Rayet stars represent an important stage in the evolution of massive stars, but are only poorly understood so far. For a better knowledge one might determine their effective temperatures, luminosities and atmospheric compositions. But the emission-line dominated Wolf-Rayet spectra were not accessible to a quantitative analysis for a long time, because "standard" model calculations for static, plane-parallel stellar atmospheres are not adequate to that type of stars.

In a systematic approach to that problem we developed elaborate computer codes for non-LTE radiation transfer in expanding atmospheres (Hamann and Schmutz 1987; Wessolowski *et al.* 1988). Basic features of these models are briefly summarized in the following section. With this tool we then analyzed the helium line spectra of WR stars in our Galaxy and in the LMC (Sect. 3). A special study concerns the WR+0 eclipsing binary V444 Cygni (Sect. 4). In the next step we started to determine the hydrogen abundances in WN atmospheres (Sect. 5). Recently we proceeded to the next generation of models which include further elements, namely nitrogen (for WN stars) and carbon (WC stars), accounting for very complex model atoms and for non-grey radiative equilibrium (Sect. 6).

2. The Model Atmospheres

Prerequisite for the intended analyses of WR spectra is the adequate calculation of the non-LTE spectrum formation. We assume that the WR atmospheres are expanding, spherically symmetric, homogeneous and stationary. (Of course, the validity of these simplifying assumptions can only be justified *a posteriori* from the achieved agreement with the observation.) The equation of continuity then reads

$$\dot{M} = 4 \pi r^2 \rho(r) v(r) \quad , \qquad (1)$$

where the mass-loss rate $\dot{M}$ is one of the essential free parameters of the models. The supersonic part of the velocity field is pre-specified by a law of the form

$$v(r) = v_\infty (1 - r_0/r) \quad . \tag{2}$$

The "terminal velocity" v_∞ is an adjustable parameter. At small velocities, a smooth transition towards static layers is ensured by suitable assumptions (for the definition of r_0 and other details see Hamann and Schmutz 1987). The "stellar radius" R_*, corresponding approximately to the location of the sonic point, is a further basic parameter which specifies a WR model.

The temperature structure is derived from the assumption of radiative equilibrium, until recently (cf. Sect. 6) evaluated in grey LTE approximation with the additional assumption that the temperature may never drop below 10 kK. A temperature parameter T_* is defined from the luminosity and the stellar radius R_* by

$$L = 4 \pi \sigma R_*^2 T_*^4 \tag{3}$$

and may be considered as an "effective temperature", related to the radius R_* of the hydrostatic stellar "core".

Hence a specific WR model is characterized by the four parameters R_*, T_*, $\dot{M}$, v_∞, and by the chemical composition. Note that no self-consistent hydrodynamics are included in our semi-empiric models. Actually, not even the physical nature of the accelerating forces is known at present.

The radiation transfer in the spherically expanding atmosphere is treated in our code with a co-moving frame formalism. The simultaneous solution of the equations of statistical equilibrium, which may represent very complex model atoms, is achieved by "iteration with approximate lambda operators" (cf. Hamann 1985, 1986, 1987). This new technique allows for the first time to treat such formidable non-LTE problems.

3. Analyses with Pure-Helium Models

In a first stage we worked with models which only account for pure helium. For a number of galactic WN stars we performed detailed fits of the helium line profiles and of the continuous energy distribution. A typical result is displayed in Fig. 1, concerning the WN5 star HD50896 alias WR6 (from Hamann *et al.* 1988). One sees that theoretical and observed helium lines agree in their basic properties. The fit of most He II lines is even excellent. Minor problems occur with the He I lines. A compromise is necessary in order to fit both lines, λ10830 and λ5876, with the same model. However, as the He I lines are very sensitive to the stellar temperature, the resulting uncertainty in T_* is unimportant. Moreover, the flat-topped profile of He I 5876 is poorly matched. Another systematic discrepancy concerns the absorption dips at maximum blueshift, which are produced by the outermost layers of the model, but are not observed.

Meanwhile we have performed eight "fine analyses" of the described kind, and we always arrived at similar conclusions. The models can reproduce well the general features of the observed helium lines. The theoretical continua are always in good agreement with observation, when accounting for the interstellar reddening. Thus we conclude that our models are basically adequate to describe real WR atmospheres. However, some discrepancies in details of the line profiles might indicate that the models are not yet perfect.

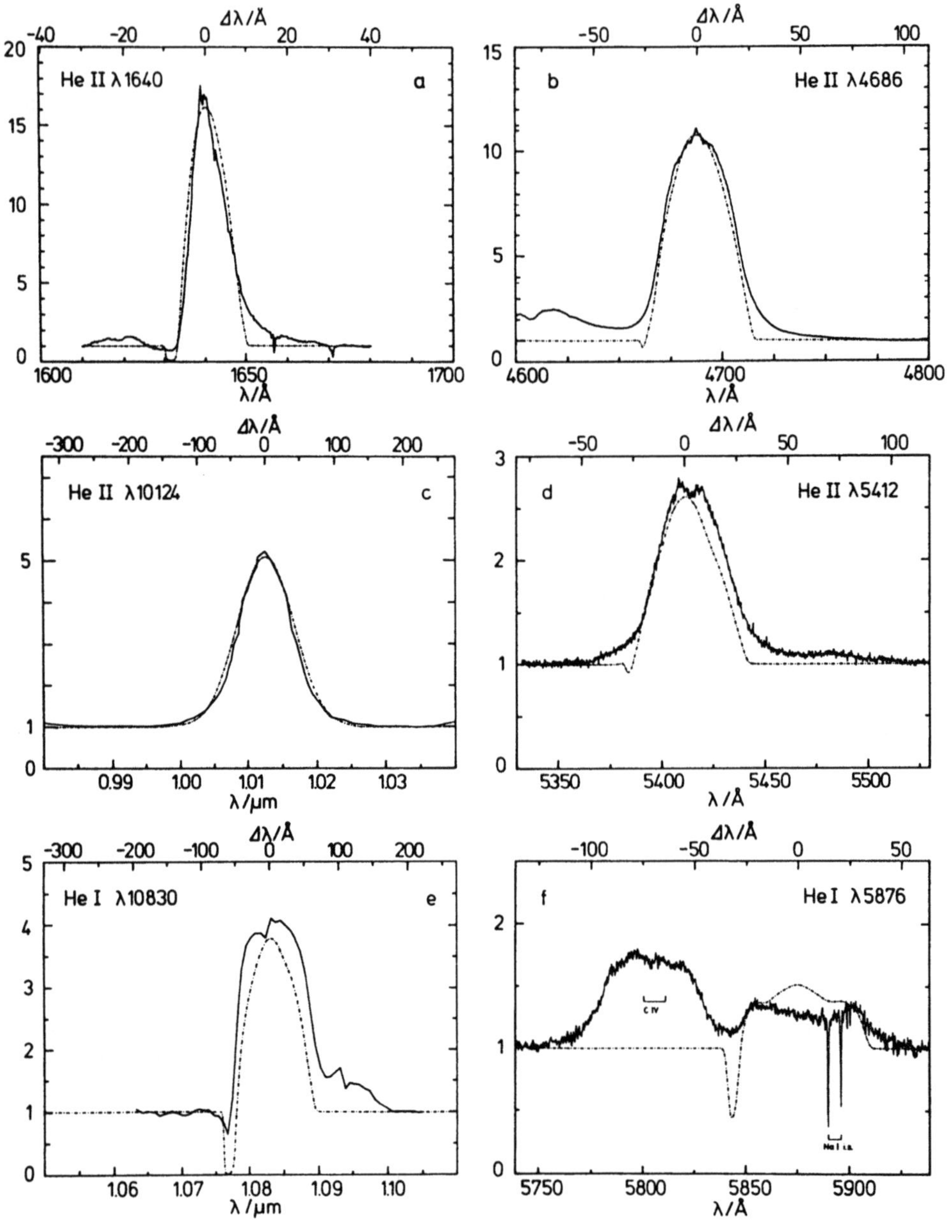

Fig. 1. Helium line profiles of WR6 (alias HD50896, subclass WN5), compared to theoretical profiles of a pure-helium model with $R_* = 1.5\ R_\odot$, $T_* = 63$ kK, $\dot{M} = 10^{-4.7}\ M_\odot$/yr, $v_\infty = 1700$ km/s. Observations are from IUE (λ1640: Howarth and Phillips 1986) and ESO, respectively.

Encouraged by these considerations, we established an extensive grid of models which now allows to determine the parameters of various WR stars with little effort. (For a "coarse analysis" it is sufficient to account for equivalent widths only, instead of a detailed line-profile fitting.) 30 galactic WR stars, covering the different subtypes, have been analyzed in this way (Schmutz *et al.* 1989). Fig. 2 (open symbols) gives the positions of the WN stars in the HR diagram. We find that the stellar temperatures and luminosities are smaller than most authors had expected previously. It turned out that there is no simple correlation between subtype (e.g. WN5, WN6 ...) and T_*, i.e. a one-dimensional classification is obviously insufficient. For instance, some of the WNE-A stars (i.e. early-type WN with weak lines) are actually as "cool" as WNL ("late") types, namely ≈ 35 kK.

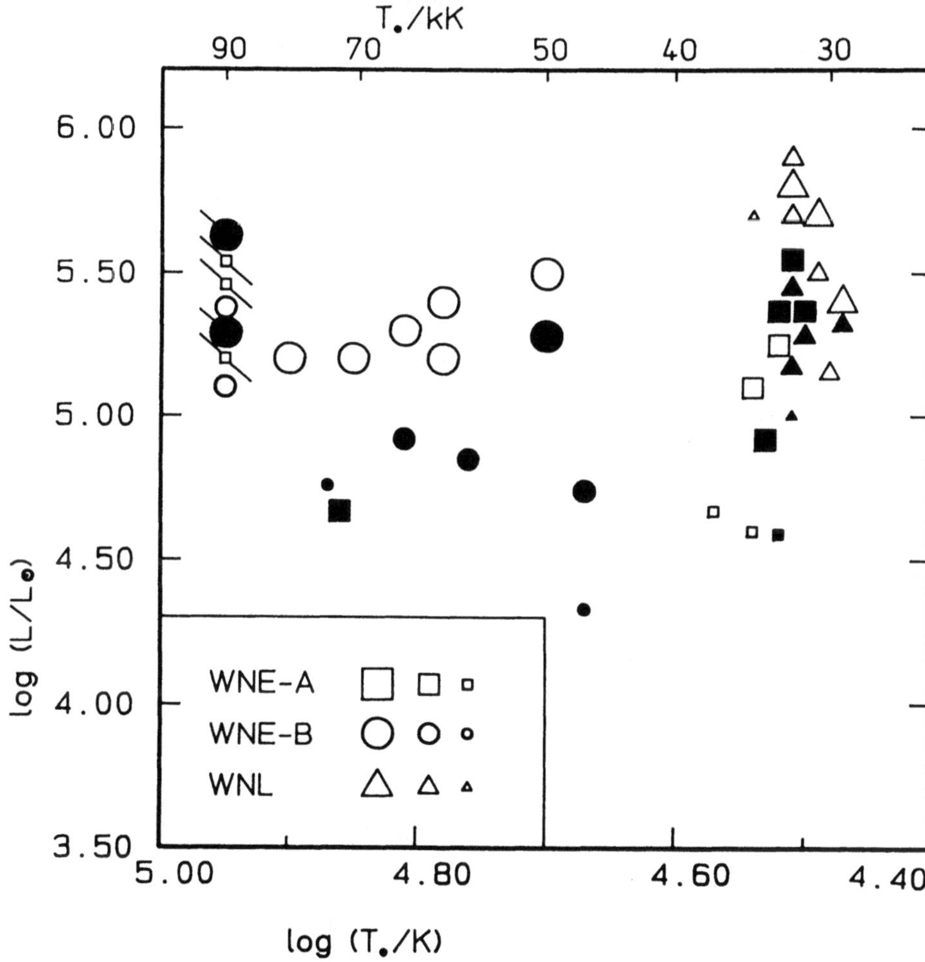

Fig. 2. Positions of WN stars in the HR diagram, as obtained by analyses with pure-helium models. LMC stars (filled symbols) tend to smaller luminosities than galactic WN stars (open symbols). Different subtypes are distinguished by different symbols, as explained in the inlet. The different sizes of the symbols indicate the mass loss rate, depending on whether log $\{\dot{M}/(M_\odot/yr)\}$ is larger than -4.2 (large symbols), smaller than -4.8 (small symbols) or in between (medium size).

From the theoretical models we learned that models with the same ratio $R_*/\dot{M}^{2/3}$ show very similar spectra. Consequently, the unique determination of stellar radius and mass-loss rate requires the knowledge of one distance-dependent quantity, e.g. the absolute visual magnitude M_v. Fortunately the distances of many WR stars are known from cluster or association memberships, but for individual stars this assignment may be in error.

For WR stars in the Large Magellanic Cloud the absolute magnitudes are more reliable, because their uniform distance is known and the interstellar reddening is generally small. Moreover, it will be interesting to compare their parameters with the galactic members. Therefore we extended our analysis of WN stars to the LMC (Koesterke *et al.*, in preparation). Spectra were obtained with the ESO 3.6m telescope and EFOSC. The analysis of the helium lines is performed in the same way and with the same grid of models as used for the galactic stars. Thus the comparison should be unbiased from systematic errors of the models. The - still somewhat preliminary - results for the LMC stars are also indicated in the HR diagram (Fig. 2, filled symbols). The comparison with the galactic WN stars reveals that, on the average, the LMC stars lie at lower luminosities. Strikingly different is the position of the early-type WN stars with strong lines. Among our sample we find five such WNE-B stars in the LMC with luminosities below 10^5 $L_\odot$ which have no galactic counterparts. But, of course, we must be cautious with statistical conclusions because our program stars are selected arbitrarily.

4. The Eclipsing Binary V444 Cygni (WN5 + O6?)

The binary system V444 Cygni provides a crucial test of our WR models. We might compare the results from a spectral analysis with those obtained by analyzing the eclipse light curve.

4.1. Spectral Analysis

For that purpose we use an observation at phase 0.24, i.e. when both components appear at their largest separation and any eclipse effects are minimized. We assume that then the spectrum can be modelled by simply adding the fluxes from both stars.

A WR model is specified by the four parameters T_*, R_*, $\dot{M}$ and v_∞ (cf. Sect. 2). The terminal velocity is easily read off from the line fits which require v_∞ = 1500 km/s. As mentioned in Sect. 3, only the ratio $R_*/\dot{M}^{2/3}$ can be found by a spectral analysis, as long as the distance is not introduced. For the mass-loss rate of V444 Cyg we adopt $10^{-4.9}$ $M_\odot$/yr for reasons concerning the light curve (cf. Sect. 4.2). Thus T_* and R_* remain as the free parameters which have to be determined by the spectral analysis.

A complication arises additionally, compared to the analysis of a single star, because it is not known which fraction of the continuum flux is contributed by either of the binary components. Thus we must introduce a further free parameter, which we define as the ratio of the continuum fluxes (WR:O) at the v band (5160 Å). Given that "brightness ratio", the observed line profiles can be scaled to the WR continuum alone. Then our usual analysis is performed and yields R_* and T_* of the WR component. The achieved agreement between synthetic and observed helium line profiles is of similar quality as we are used from single stars.

Table I. Parameters of V444 Cygni from Spectral Analysis

M_v - ratio	WR star T_*/kK	WR star $R_*/R_\odot$	WR star $\log L/L_\odot$	O star $R/R_\odot$	Remarks
1:1	35	6.0	4.7	7.2	bad fit of He I 5876
1:2	40	3.5	4.4	7.3	
1:4	60	1.4	4.4	7.3	preliminary result

For the O star we take the continuum flux from a non-LTE model with T_{eff} = 30 kK, log g = 4.0, and solar helium abundance. The radius of the O star then follows from the adopted brightness ratio. The observed continuous spectrum is matched excellently by the composite models, when the standard reddening law is adopted with E_{B-V} = 0.9 mag. For a brightness ratio of, e.g., 1:2, the absolute magnitude M_v of the WR component results as -3.1 mag, leading to a distance of the system of 780 pc.

Table I displays the results as they depend on the adopted M_v ratio. Tuning of the brightness ratio thus yields a whole manifold of solutions, but note that the resulting WR luminosity is always nearly the same.

4.2. Light Curve Analysis

In order to synthesize the eclipse light curve, we adopt the following model. The O star revolves round the WR star still within the outer regions of its extended atmosphere. The O star shadows a distinct volume of the WR atmosphere which thus cannot contribute to the total flux seen by the observer. On the other hand, additional radiation emerges from the surface of the O star. Its contribution to the total flux is more or less diminished by absorption when the rays pass through those parts of the WR atmosphere which lie between the O star and the observer.

The WR atmosphere is given by our usual models. Thus any disturbance of the spherical symmetry is neglected. Neither the influence of the O star radiation on the population numbers, nor the scattered O star photons are accounted for. The intensity emitted by the O star is taken from non-LTE continuum models, accounting for the limb-darkening.

The orbit of V444 Cygni is circular with a radius $a \sin i$ = 40 $R_\odot$, as one knows from the radial velocity curve. The orbital inclination angle i is close to 78°, according to polarimetry (Robert *et al.* 1989).

A couple of free parameters describing the WR star (R_*, T_*, $\dot{M}$; v_∞ is 1500 km/s according to the line profiles) and the O star (T_{eff}, R) now can be varied until the synthetic light curve matches the observation. The mass-loss rate has a very characteristic influence, as it determines the width of the first minimum when the O star is occulted by the extended WR atmosphere. This width depends hardly on the other parameters, and thus allows to derive $\dot{M}$ nearly independently. We find $\dot{M}$ = $10^{-4.9}$ $M_\odot$/yr, by the way remarkably consistent with the observed change in the orbital period (Khalliullin 1974).

Now we still have five free parameters (temperatures and radii of both stars, and the inclination angle) in order to fit essentially two properties of the light curve, namely the depths of the two minima. It is not likely that there is a unique solution for that problem.

In order to find at least one particular solution, we arbitrarily adopt T_{eff} = 30 kK (O star) T_* = 40 kK (WR star) and i = 78°. Then we vary both stellar radii and achieve a good fit of the observed light curve at all wavelengths (Fig. 3) with R_* = 6.5 $R_\odot$ (WR star) and R = 6.7 $R_\odot$ (O star). But we can fit the light curve with different sets of parameters as well. E.g., an increase of the inclination i and a corresponding decrease of both radii leads to the same good agreement with observation. For T_* = 60 kK (WR star) we also find a light-curve solution. Thus the basic result is that we can reproduce the observed light curve of V444 Cygni with our model, but the solution is not unique.

4.3. Comparison

The particular light-curve solution shown in Fig. 3 is based on the arbitrary choice of T_* = 40 kK and yielded a WR radius of R_* = 6.5 $R_\odot$. This may be compared with the results from the spectral analysis (Sect. 4.1). The particular solution with the same T_* (brightness ratio 1:2) implies a WR radius of only 3.5 $R_\odot$ (cf. Table I). Thus we have not fitted both, the spectrum and the light curve, with exactly the same model. However, note that we have compared only two *particular* solutions so far, and we expect that there exist other solutions which simultaneously satisfy both observational constraints. This point has still to be investigated.

Recall that Cherepashchuk *et al.* (1984) performed a light curve analysis of V444 Cygni and obtained a solution with a "hot" WR star (90 kK effective temperature, related to the radius where τ = 1). We suspect that this is another *particular* solution which is not unique. Note that our solutions agree with Cherepashchuk *et al.* (1984) in the mass-loss rate of $10^{-4.9}$ $M_\odot$/yr, and in the implied distance of the system of only about 1 kpc, thus placing V444 Cygni in the foreground of the open cluster Berkeley 86 (1.72 kpc, Forbes 1981) instead of a membership as suspected, e.g., by Lundström and Stenholm (1984).

5. Hydrogen Abundances in WN Stars

The most comprehensive study about hydrogen abundances in WN stars was performed by Conti *et al.* (1983). Their results confirmed that WN atmospheres consist predominantly of helium, while hydrogen is more or less depleted or even absent within the detection limit. A histogram (Fig. 12 in Conti *et al.* 1983) illustrates that there is no clear correlation between hydrogen detection and spectral subtype: stars with hydrogen and stars without detectable hydrogen are found among both, the WNE ("early") as well as the WNL ("late").

This picture drastically changes when the hydrogen detection is not correlated with the subtype, but instead with the stellar parameters as obtained by our spectral analyses. Fig. 4 shows a clear separation between the two groups: the "cool" stars with $T_* \approx 35$ kK exhibit hydrogen, while the "hot" stars (T_* > 60 kK) do not. WR136 is intermediate in temperature and in hydrogen abundance. The only exception, WR3, is a peculiar object in many aspects. Note that two stars classified as "early" subtypes (WR10: WN4.5-A; WR128: WN4-A) are actually "cool" and show hydrogen.

Conti *et al.* (1983) derived the hydrogen abundances from a "semiquantitative" study of the Balmer-Pickering decrement. We now repeat this analysis with our model calculations in order to obtain more precise results (Dünnebeil *et al.*, in preparation). We

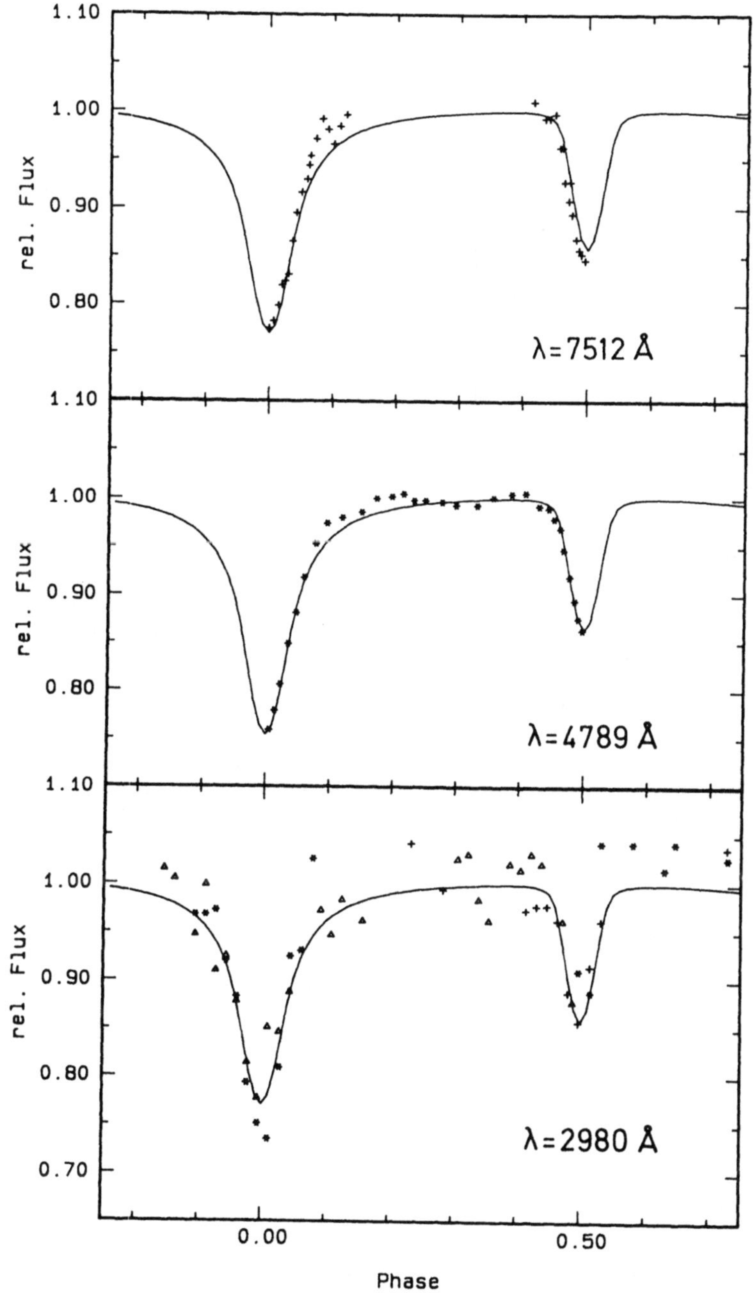

Fig. 3. Light curve of V444 Cygni. Discrete symbols denote observational data taken from Cherepashchuk and Khalliullin (1975) (7512 Å - averaged data from several periods), Cherepashchuk and Khalliullin (1973) (4789 Å - averaged data) and Cherepashchuk *et al.* (1984) (2980 Å - individual measurements). The synthetic light curve (continuous line) results from a WR model with $R_* = 6.5\ R_\odot$, $T_* = 40$ kK, $\dot{M} = 10^{-4.9}\ M_\odot/yr$, an O star with $T_{eff} = 30$ kK, $R = 6.7\ R_\odot$, and an orbital inclination of $i = 78°$.

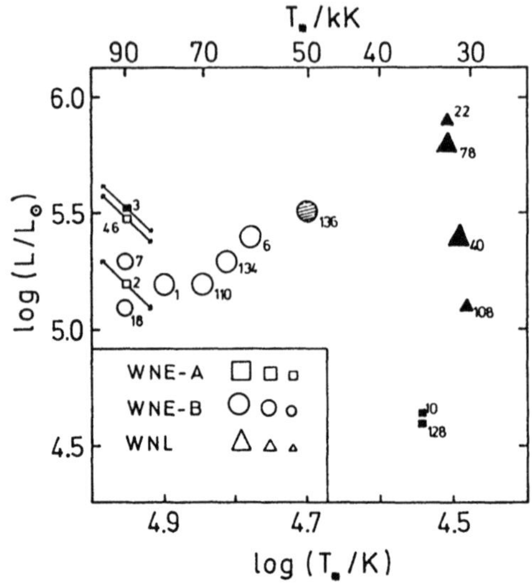

Fig. 4. Hydrogen in WN stars. Open symbols denote stars without hydrogen, filled symbols stars with hydrogen detected (according to Conti *et al.* 1983). The hatched symbol indicates a case where Conti *et al.* could not detect hydrogen, while we found a very small abundance (β_H = 6%). The individual stars are identified by their WR number. The stellar parameters (L, T.) are from our spectral analyses (Schmutz *et al.* 1989, partly improved from "fine analyses"). Different subtypes are distinguished by different symbols, as explained in the inlet. The different size of the symbols indicates the mass-loss rate: $\log\{\dot{M}/(M_\odot/yr)\} > -4.2$ (large symbols), < -4.8 (small symbols), or in between (medium size).

even extended our model code for a fully consistent treatment of line blends, in the radiation transfer as well as in the radiative rates of the statistical equilibrium. These improvements are responsible for some very minor changes of the results, compared to Hamann *et al.* (1990).

Table II. Fine Analyses Including Hydrogen

	WR 40	WR 128	WR 22*	WR 136
Subtype	WN8	WN4-A	WN7abs	WN6
T_*/kK	31.2	37	32	51
$R_*/R_\odot$	18.	5.3	30	6.0
$\log\{\dot{M}/(M_\odot\, yr^{-1})\}$	-4.05	-5.0	-4.35	-3.85
v_∞/(km/s)	1000	2000	1500	1700
$\log L/L_\odot$	5.4	4.7	5.9	5.5
β_H [%]	16 ± 5	20 ± 10	≈ 50	6 ± 3

* preliminary results for WR22

When performing the "fine analysis" of the helium spectrum, we give special weight to the fit of those He II Pickering lines which are not blended by hydrogen. Furthermore, the theoretical profiles of those He II lines which coincide with hydrogen Balmer lines are compared with the observation. The best fit yields the stellar parameters (R_*, T_*, $\dot{M}$, v_∞) and the hydrogen mass fraction, β_H. Fig. 5 gives an example of the line fits finally achieved.

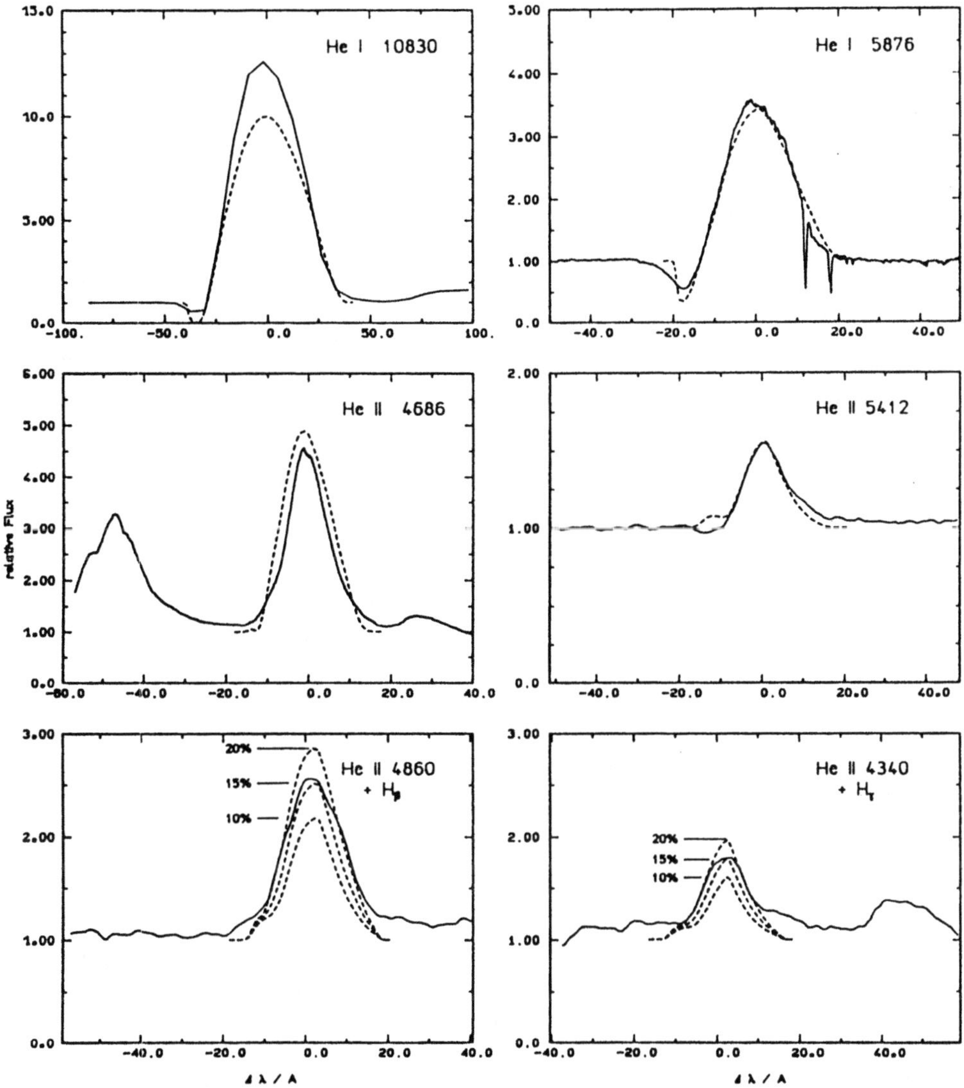

Fig. 5. Line fit of WR40 (alias HD96548). Observations obtained at ESO are compared with theoretical profiles for the model specified by $R_* = 18.0\ R_\odot$, $T_* = 31.2$ kK, $\dot{M} = 10^{-4.05}\ M_\odot/yr$, $v_\infty = 1000$ km/s. For the He II blends with H_β and H_γ (bottom) the theoretical profiles are given for three different hydrogen abundances (labels: hydrogen mass fractions), indicating that the adequate value is 0.16 ± 0.05.

Up to now we have completed four fine analyses of the described kind (cf. Table II). In one case (WR22) the hydrogen abundance nearly reaches the solar value. For the other two "cool" stars ($T_* < 35$ kK) the hydrogen mass fraction is of the order of $\beta_H = 20\%$. WR136, which is considerably hotter ($T_* \approx 50$ kK), has only 6% hydrogen, and thus appears as a link to the "hot" WN stars in which hydrogen is not detectable (cf. Fig. 4).

6. WR Models with Nitrogen and Carbon

The non-LTE models for WR atmospheres reach their next stage of sophistication when not only helium (and possibly hydrogen), but also the most important "metals" are taken into account. Hillier (1988) was the first who introduced nitrogen into his calculations, and meanwhile he proceeded to carbon in order to synthesize WC spectra (Hillier, 1989).

Our recent models now also include nitrogen, represented by a model atom with 90 levels and about 1500 lines, 350 of them treated explicitely. Low-temperature dielectronic recombination via hundreds of transitions is allowed. Together with the nitrogen we introduced an improved temperature structure which now fully accounts for radiative equilibrium, instead of the grey LTE approximation applied so far.

Test calculations with these improved models (Hamann and Wessolowski 1989), setting the nitrogen abundance to 1.5% by mass, show no significant reaction of the helium spectrum. Thus the previous results obtained from models with pure helium and "grey" temperature structure remain valid, as far as WN subtypes are concerned.

As our next step we try to reproduce in detail the nitrogen line spectrum of one WN star (WR 136; Wessolowski *et al.*, in preparation). From preliminary results we conclude that some general agreement can be achieved, while remaining discrepancies can be attributed to problems with the very complex nitrogen model atom.

Possibly, nitrogen is slightly higher ionized than predicted on the basis of radiative equilibrium. In a separate study (Baum *et al.*, in preparation) we investigate whether this "superionization" can be consistently explained by the action of X-rays, which have been observed from a couple of WR stars and which are probably caused by shocks in unstable parts of the accelerated outflow.

Calculations for WC stars which account for a complex carbon model atom are also available now in our group (Leuenhagen *et al.*, in preparation). A first analysis of WR111 confirms the results of Hillier (1989), especially concerning the carbon mass fraction of about one half. As the carbon opacities control the radiative transfer, the analysis of WC spectra with pure-helium models (Schmutz *et al.*, 1989) could only be preliminary. The improved models with carbon now yield, e.g., for WR111 a stellar temperature of about 60 kK, instead of 35 kK obtained previously. Therefore we will refrain from discussing the parameters of WC stars until we have analyzed more of them with the adequate models.

7. Conclusions: the Evolutionary Status of Wolf-Rayet Stars

The spectral analyses indicate that most WR stars are considerably cooler and less luminous than predicted by the standard evolutionary calculations, which generate WR stars as a post-red-supergiant stage (cf. Fig. 6). The hydrogen abundances in WN stars provide a further check of the evolutionary szenario. The most luminous WNL star in our

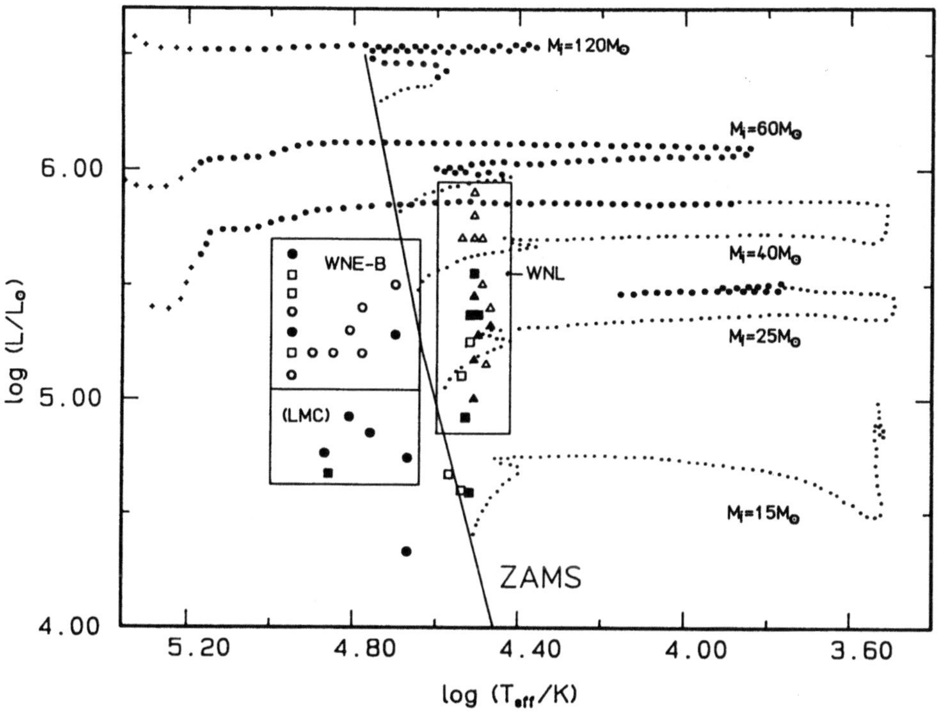

Fig. 6. Loci of WR stars in the HR diagram, as obtained from spectral analyses. Different symbols denote the subgroups WNL (△), WNE-A (□), and WNE-B (○), while open or full symbols indicate whether the stars belong to our Galaxy or the LMC, respectively. The overall location of the WNL and the WNE-B subgroups is additionally emphasized by large boxes. Superimposed are evolutionary tracks (from Maeder and Meynet, 1987) and the predicted surface abundances: ··· "normal", ••• WN-type, + + + WC-type.

sample (WR22) has nearly solar composition, implying that this star is just leaving the main-sequence band and has not yet evolved to the red. For the two other "cool" WN stars ($T_* \approx 35$ kK), irrespective whether classified as WNL or WNE-B, we find typical hydrogen mass fractions of about 20%, just as predicted for the phase of rapid blueward post-RSG evolution. The observed disappearance of hydrogen towards higher stellar temperatures also agrees qualitatively with the theoretical tracks.

The agreement with the evolutionary calculations ends, however, when the stellar parameters are considered. The observed luminosities are smaller than predicted by more than a factor of ten. Very small luminosities are determined for a couple of WNE-B stars in the LMC, possibly indicating that under low-metallicity conditions even less massive stars can reach the WN phase, compared to our Galaxy. Recall that the ratio between the number of WN stars and WC stars also strongly differs between the Galaxy and the LMC.

Summarizing, we state that we can perform now a quantitative analysis of Wolf-Rayet spectra. For a large number of stars we already obtained stellar parameters and atmospheric abundances, and further results are to be expected soon. This empirical basis strongly constrains any evolutionary explanations. Hopefully it will help us to a better understanding of the evolution of massive stars.

References

Cherepashchuk, A.M., Khalliullin, K.F.: 1973, *Soviet Astr. - AJ* **17**, 330
Cherepashchuk, A.M., Khalliullin, K.F.: 1975, *Soviet Astr. - AJ* **19**, 727
Cherepashchuk, A.M., Eaton, J.A., Khalliullin, K.F.: 1984, *Ap. J.* **281**, 774
Conti, P.S., Leep, E.M., Perry, D.N.: 1983, *Ap. J.* **268**, 228
Forbes, D.: 1981, *Pub. A.S.P.* **93**, 441
Hamann, W.-R.: 1985, *Astr. Ap.* **148**, 364
Hamann, W.-R.: 1986, *Astr. Ap.* **160**, 347
Hamann, W.-R.: 1987, in *Numerical Radiative Transfer*, ed. W. Kalkofen, Cambridge University Press, p. 35
Hamann, W.-R., Schmutz, W.: 1987, *Astr. Ap.* **174**, 173
Hamann, W.-R., Schmutz, W., Wessolowski, U.: 1988, *Astr. Ap.* **194**, 190
Hamann, W.-R., Wessolowski, U.: 1990, *Astr. Ap.* **227**, 171
Hamann, W.-R., Wessolowski, U., Schwarz, E., Dünnebeil, G.: in *Properties of Hot Luminous Stars: Boulder-Munich Workshop*, ed. C. D. Garmany, The A.S.P. Conference Series, Vol. 7, 1990 (in press)
Hillier, D.J.: 1988, *Ap. J.* **327**, 822
Hillier, D.J.: 1989, *Ap. J.* **347**, 392
Howarth, I.D., Phillips, A.P.: 1986, *M.N.R.A.S.* **222**, 809
Khalliullin, K.F.: 1974, *Soviet Astr. - AJ* **18**, 229
Lundström, I., Stenholm, B.: 1984, *Astr. Ap. Suppl.* **58**, 163
Maeder, A., Meynet, G.: 1987, *Astr. Ap.* **182**, 243
Robert, C., Moffat, A.F.J., Bastien, P., Drissen, L., St.-Louis, N.: 1989, *Ap. J.* **347**, 1034
Schmutz, W., Hamann, W.-R., Wessolowski, U.: 1989, *Astr. Ap.* **210**, 236
Wessolowski, U., Schmutz, W., Hamann, W.-R.: 1988, *Astr. Ap.* **194**, 160

The Transition of Supergiant CS Matter fom Cool Winds to Coronae – New Insights with X Aur Binary Systems

K.-P. Schröder

Hamburger Sternwarte, Gojenbergsweg 112,
D-2050 Hamburg 80, Fed. Rep. of Germany

I. Summary

The age of astronomical satellites gave us access to X-rays and UV spectroscopy. New challenging questions for astronomy have been raised, one of which is the transition from hot, corona-like matter to cool winds and chromospheres. Such a transition occurs in a region of the HR diagram which extends as a band from G super giants to K giants.

After an introduction to the topic from the point of view of single star observations, the author then focusses on recent research of ζ Aur binary systems. These stars, with primaries situated in or near the region of the HR diagram in question, offer two unique advantages: - the main sequence B6 - A2 secondary star offers a reliable way of determining the system's distance and thereby the absolute magnitude of the primary (much more accurately than for single supergiants): - moving through the wind and passing behind the supergiant's chromosphere before and after eclipse, the secondary is an ideal light probe which provides us with a unique opportunity to monitor chromospheric densities and to achieve spatial resolution of the structure of the CS matter of the primary. But sophisticated diagnostic tools are required, such as two-dimensional radiative transport in the wind.

From high-resolution IUE spectra and ground based Coudé-spectroscopy (for $\lambda \lesssim 4500$ Å) we can separate the light of the hot companion from the composite spectrum of the spatially unresolvable binary. While detailed optical observations of several chromospheres and their interpretation are still in preparation, some recent results from IUE-based studies are presented. The most important one may be the direct detection of 2 'cool coronae' (HR 6902 and HR 2554) which coincide with depleted chromospheres. HR 6902 is also a crucial star for fulfilling the v < r criterion of MgII h_2 and k_2 emission but definitely not showing cool outflow at all. Earlier interpretations of this criterion are therefore questioned.

II. Introduction

Substantial progress has been achieved during the last decade in our knowledge about the circumstellar matter of giants and supergiants having spectral types between G and M, with new observational data being collected mainly by the IUE satellite but also by X-ray emperiments such as Einstein and EXOSAT. We have learned that circumstellar matter of giants and supergiants undergoes a drastic change with a transition in the HR diagram, which manifests itself by several 'dividing lines', each defined by certain observational criteria. Fig.1 (from Mullan and Stencel, 1982) illustrates the situation in the part of the HR diagram in question. A review is given by Jordan and Linsky (1987).

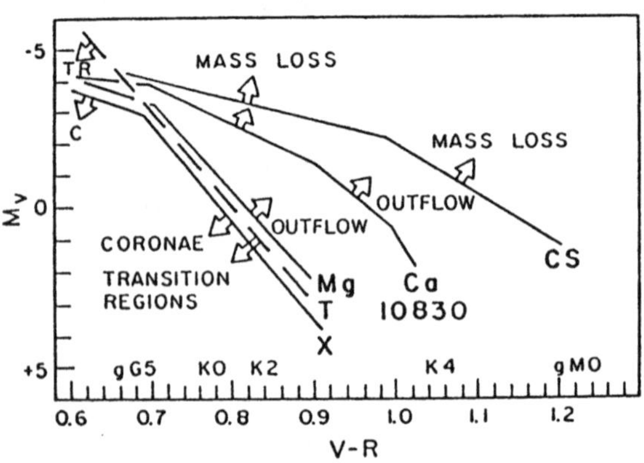

Fig.1.:
Plot of the 'diving lines' in the HR diagram, as taken from Mullan and Stencel, 1982, who suggest a sharp transition from coronae to cool winds at the 'MgII-dividing line'.

To the right of this transition we observe cool ($<10^4$K) and massive winds, while to its left there is hot ($\geq 10^5$K), corona-like CS matter. The individual dividing lines have more or less the same orientation, though their positions in the HR diagram correspond to the particular observational criterion. The left-most (in the sense of Fig.1) is the 'X-ray dividing-line'. To its left, X-ray emission, probably arising from corona-like CS matter, is observed. The same stars also show CIV, SiIV and often NV emission in IUE spectra, indicators of a hot ($\approx 10^5$K) region similar to the solar transition region. The corresponding dividing line, defined by the visibility of CIV and SiIV emission, is therefore called the 'transition-region dividing line' and is located a bit further right of the X-ray dividing line in the HR diagram. At about this same location, a gradual change in the asymmetry of the MgII h_2 and k_2 emission (spectral flux ratio of violet and red emission peaks) occurs and defines the 'MgII dividing line' (Stencel and Mullan, 1980a,b), which has so far often been interpreted as the onset of outflows of cool matter. But the example of HR 6902, which has no wind but lies clearly to the right of the MgII dividing line (showing an emission ratio of $v/r \approx 0.7....0.8$, Reimers et al., 1990) contradicts this early interpretation of the MgII $v<r$ criterion.

To the right of the 'MgII dividing line' the appearance of additional circumstellar, blue-shifted absorption features in the CaII K line defines the 'CaII dividing line' (Reimers, 1977), which is about identical with the CS-dividing line in Fig.1. While the asymmetry in the MgII absorption cores corresponds to only a fraction of the typical terminal wind velocity, the CaII K CS absorption features are found at about the full terminal wind velocity. The CaII K criterion can therefore be regarded as the strongest and most reliable evidence for the existence of a cool, massive wind, while the sequence and nature of the other dividing-lines may indicate a sort of gradual change into hot, corona-like CS matter (whatever the particular nature of CS matter of candidates in that region of the HR diagram may be). So far, there is no detailed concept of this transition, nor can we model the cool mass-loss in a clear way but have to deal rather with competing approaches !

To complicate the situation further, this region between the dividing lines is populated by so-called 'hybrid stars', which show indicators simultaneously of both hot CS matter (SiIV, CIV emission) and of cool outflow (MgII and partly CaII K CS absorption features (Hartmann et al. 1980 ; Reimers, 1982). That contradicts earlier concepts of a sharp bifurcation of the state of CS matter into <u>either</u> hot coronae (energy being put into heating) <u>or</u> cool winds (energy being put into mass loss), and again raises the question about the nature of this transition. Another important question is whether the scatter of positions in the HR diagram of a set of stars, which defines a certain dividing line, reflects reality or is the result of observational error or selection effects. To resolve that question a better-observed sample is therefore necessary. Fortunately there are a few stars in the region in question of the HR diagram which are primaries of well known ζ Aur-type eclipsing binaries. Chromospheric elicpses of the small but hot companion (B6...A3V) enable us to monitor the chromosphere and to achieve true height resolution. Chromospheric absorption features appear in the spectrum of the companion; the features are variable with phase and depend upon the particular projected height of the secondary at the time of observation. A cool wind can also be studied in detail by phase-dependent UV P Cygni-type profiles, [provided sophisticated diagnostic tools such as two-dimensional radiative transfer codes (Hempe, 1982 ; Baade, 1989, 1990) are used)]. At IUE wavelengths (1200-3200 Å) the hot companion dominates the spectrum to such an extent that the primary contribution to it can be neglected in most cases. And there are many strong resonance lines that are favourable for IUE studies of the wind. Furthermore, spectrophotometry of the hot main-sequence companions with IUE offers a good method to determine positions in the HR diagram for both components.

At blue and near-UV wavelengths a proper subtraction of the contribution of the primary has to be undertaken (Griffin, 1988), whereas the visual and red spectrum is of course completely dominated by the giant. Subtraction work in the optical region is rewarding because this spectral range is less subject to multiple blends at high column densities. It can also yield a much better s/n and resolution, given an adequate equipment such as the Calar Alto 2.2 m Coudé of the DSAZ. This region is therefore the best choice for observing deep chromospheric eclipses.

ζ Aur itself and 32 and 31 Cygni are well within the wind region in the HR diagram. But two ζ Aur-type stars, 22 Vul and HR 6902, are of particular interest owing to their position, in the HR diagram, and have already been studied in more detail, using IUE high resolution spectra. While 22 Vul shows an extended (1 R_G) neutral chromosphere and evidence of a cool wind (Schröder and Reimers, 1989), HR 6902 has only a very thin neutral chromosphere but hot (about 10^5K), ionized circumstellar matter like a corona and rather a solar-type wind instead of a cool wind (Reimers et al. 1990). That is a dramatic difference, in spite of the close proximity of the two stars in the HR diagram, and introduces the idea to use ζ Aur systems not only for studies of individual chromospheres and winds but also to probe the transition to hot CS matter. For this purpose, Hünsch (1990) extended the samll sample by studying IUE spectra of 15 binaries, which are similar to ζ Aur systems physically, but in the main do not happen to show eclipses. Those binaries, however, offer crucial tests of the existence of a cool wind and exact determination of their positions in the HR diagram, since they offer most of the advantages of true ζ Aur. systems. In addition, recent results of a study of IUE spectra of several ζ Aur. systems during chromospheric eclipse (Schröder, 1990) are presented.

III. Wind, chromospheres and coronae in ζ Aur type binaries

As already pointed out, the application of sophisticated diagnostic tools is required to make use of the advantages offered by ζ Aur type binaries. The main complication is the offset of the photon source (the secondary) from the symmetry-centre (the primary) of velocity-field and density distribution of the wind. K. Hempe (1982) introduced a generalized two-level Sobolev approximation (for the source function, the formal integral is solved correctly by the Hempe code). Later, a superior integral-operator method was developed by R. Baade which has been described in detail recently (Baade, 1989, 1990). ζ Aur. systems are well detached ($a = 5\text{-}50\ R_*$ $P = 0.5\text{-}20$ yrs) and interaction of CS matter with the secondary can be spectroscopically distinguished from ordinary wind. The wind is typically accelerated between $2...10\ R_*$.

Windline profiles from a ζ Aur system are observed in the prominent resonance lines (Fe II UV mult. 1,2 ...; SiII UV mult. 2,3; ...), where fortunately most secondaries do not show prominent photospheric features. These profiles are principally of the P-Cygni type but undergo a strong variation with phase (i.e. position of the secondary in the wind). Fig.2 (from Baade, 1989) demonstrates how the absorption apparently increases when the secondary is in the backward part of the wind and the line of sight passes through higher column densities. During chromospheric eclipse this effect pronounces absorption so much that many new, intrinsically weaker lines appear which may then well be treated as pure absorption lines.

As far as the wind is concerned a turbulent stochastic velocity parameter V_{sto} and the wind velocity can be determined separately: the absorption width in phases with the secondary in front of the primary is formed nearly by v_{sto} alone, while in backward positions, v_{wind} adds twice to the width because we see through both the approaching and the receding wind matter. Given a sample of phases well distributed around the orbit, the values of v_{sto}, v_{wind} (described by v_∞ and β, which parametrizes the acceleration) and $\dot{M}$

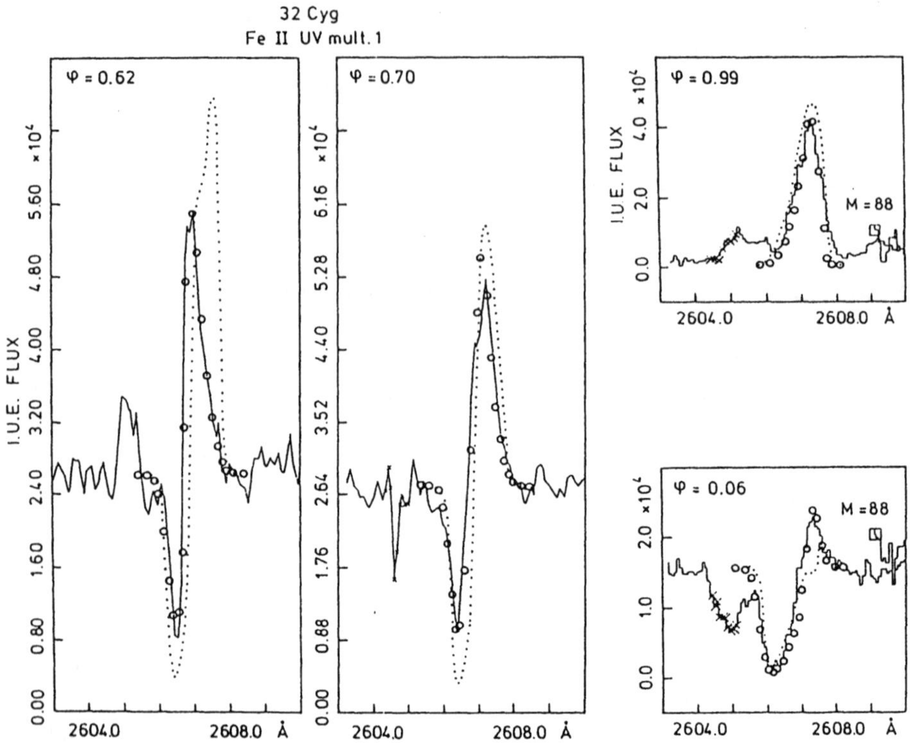

Fig.2:
Resonance wind lines from ζ Aur type binaries. IUE observations are compared to theoretical models made by the Hempe code (.......) and superior Baade-code (o o o o o); from Baade, 1989. Photons from the orbiting B stare are scattered, forming phase depending P Cygni wind-profiles.

are clearly determined by fitting the wind line profiles of all phases with the same set of these parameters of the Hempe and Baade codes. The derived massloss rates and wind velocities are therefore much more accurate than any such values from single-star observation. This 'binary technique' is only limited by intrinsic deviations from a continuous, spherically symmetric outflow. Some former simplifications with the Hempe code have been avoided by Baade's new approach.

By application of the 'binary technique' to a sample of 15 candidates, which are physically similar to ζ Aur systems but in the main do not happen to show eclipses, Hünsch (1990) presents clear evidence for the presence or absence of a cool wind. By modelling

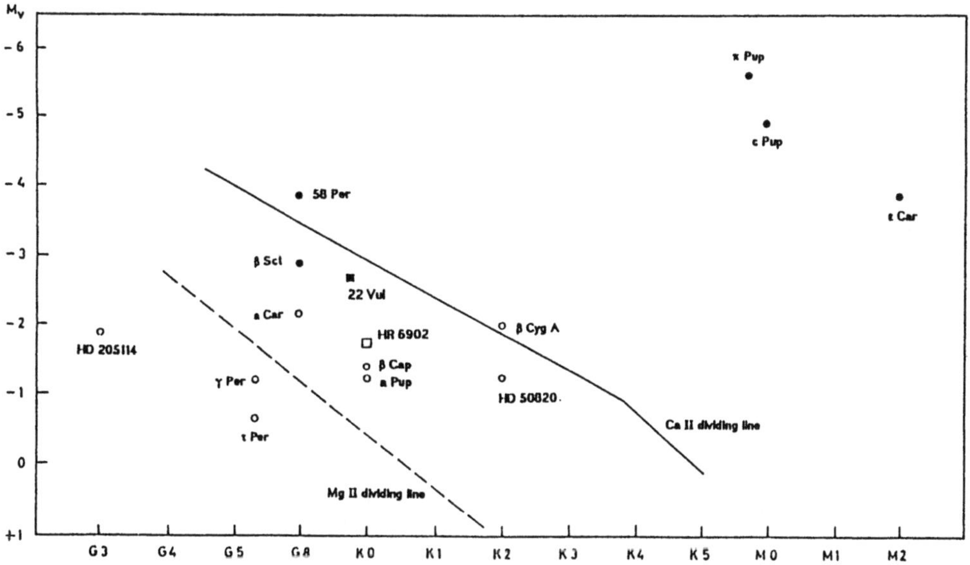

Fig.3:
15 binaries are plotted into the HR diagram according to the positions of the primaries (Hünsch, 1990). Detection of a wind with the binary technique' (• = yes, o = no) is indicated and compared to the position of the MgII- and the 'CaII-dividing line'.

the IUE fluxes of the main-sequence secondaries by Kurucz models, T_{eff} and angular diameter θ can be determined. By further applying a $R(T_{eff})$ relation and UBV colour, he determines the distance d, M_v and $(B-V)^G$ of the primary, resulting in quite accurate positions in the HR diagram. The result is demonstrated in Fig.3. A transition is indicated from 'cool wind' to 'no cool wind' near to the 'CaII K-dividing line' (= CS line Fig.1), but clearly not as far to the left as the 'MgII- dividing-line'.

There are 3 stars - 22 Vul, HR 6902, HR 2554 - at crucial positions in the diagram as according to Fig.3, which do show eclipses, and such offer a study of their chromospheres. They are presently included in a campaign of optical Coudé-spectroscopy (Griffin et al., 1990, paper 1) . It is already possible to determine principal properties of these chromospheres, from IUE observations made during chromospheric and partial eclipses. They can be compared in respect of their positions in the HR diagram (Schröder, 1990). Fig.4 gives those positions in the relevant part of the HR diagram, using a $(B-V)_o$-scale. Fig.5 compares the different FeII column densities as obtained from curve-of-growth techniques, treating unblended lines without wind-components from V II and FeII in pure absorption. Details are given by Schröder, 1990.

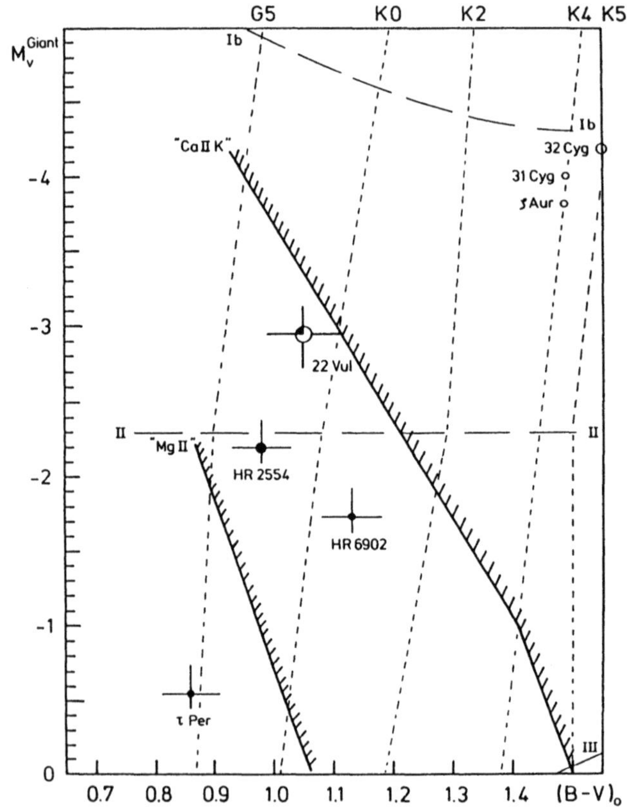

Fig.4:
7 ζ Aur systems are plotted into the HR diagram according to the positions of the primaries (Schröder, 1990). HR 2554, HR 6902 and 22 Vul are in that same region where the onset of cool winds is observed.

22 Vul has a neutral, extended cool chromosphere and at least a partly neutral wind (Schröder, Reimers, 1989). 22 Vul therefore shows the highest FeII column densities in Fig.5. Very little FeI is observed (10^{-3}...10^{-4} of FeII); furthermore' photoionization is not capable of producing FeIII or V III as long as a neutral chromosphere shields its matter from any Lyman-photons. V II, FeII densities of 22 Vul therefore correspond closely to total V, Fe-densities and can, via assumed solar abundances also be converted into total H-densities.

In contrast to 22 Vul, HR 2554 shows Fe II column densities which are less by ~4 orders of magnitude τ Per, situated more to the left in the HR diagram and not included in Fig.5, does not show clear evidence for Fe II absorption at all, except perhaps for the innermost ($h<4\cdot10^5$ km) layers (Schröder, 1990).

In intermediate position in the HR diagram, with respect to the orientation of the 'dividing lines' HR 6902 starts with FeII column densities close to those in 22 Vul.

According to Reimers et al. (1990), those layers are almost neutral ($n_e \ll n_H$ as estimated from the Fe I: II density ratio/Fe total densities and from a simplified Fe-ionization equilibrium). Then the chromospheric column densities of HR 6902 appear to fall suddenly to the level of those in HR 2554 (for heights h $2.5\cdot10^6$ km). For those

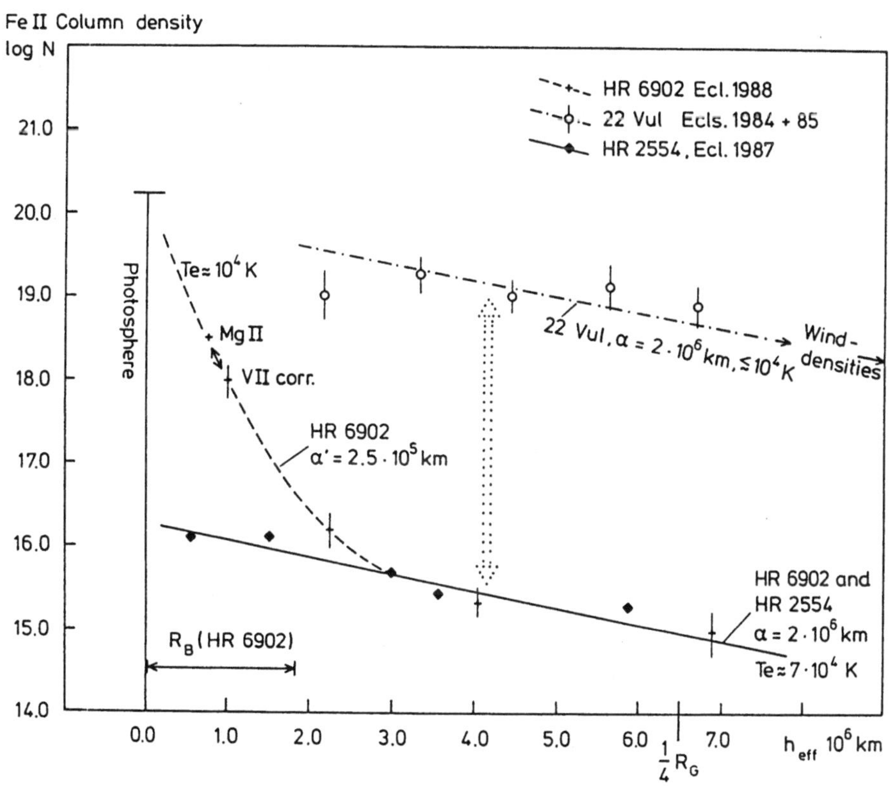

Fig.5:
FeII column densities of HR 2554, HR 6902, and 22 Vul (from IUE observations during chromospheric eclipse are compared on a height scale. The dramachange has to be compared to the individual positions of those stars in the HR diagram with respect to the 'dividing lines'. It marks the change of ionisation and temperature of the circumstellar matter.

heights, Reimers et al. deduce $T_e \approx 6...7 \cdot 10^4$ K from the observed CIV:SiIV CS absorption ratio (Reimers et al., 1990). A more detailed study of this absorption, so far detected up to a height of ≈ 0.8 R_* from a new series of SWP spectra, is in preparation. Not surprisingly, HR 2554 also shows this absorption close to eclipse. However, it must not be confused with absorption from an accretion wake , which is completely different in its phase behaviour and is not focused on phases near eclipse. We therefore may conclude the existence for both stars of circumstellar matter as hot as the solar transition region (and density approximately scaled down by the ratio in gravities) but as extended as a corona (→ 'cool corona') and which may well form a hot wind. The sudden fall of Fe II column densities in HR 6902 mark the change of the innermost cool neutral chromospheric matter to ionized and much hotter circumstellar matter. This is optically thin for Lyman photons which, together with collisional rates, produce a large Fe III : Fe II ionization ratio (in contrast to cool, neutral chromospheres, where Fe II dominates).

IV) Discussion

These observations give for the first time some direct insight into the structure and nature of circumstellar matter and its transition from hot to cool regions in this particularly interesting region of the HR diagram. Unfortunately, the samples are too small to define a new dividing line. Furthermore, the presence of the companion partly prevents the application of single star criteria even during eclipse: C IV and Si IV emission is often created by local interaction of CS matter with the companion or, in cases, by resonance scattering of photons from the companion in hot CS matter. Both processes outnumber those photons which are indead generated from transition region matter by collisions, as in simple stars.

Nevertheless the exact positions in the HR diagram, combined with such unique information that ζ Aur.-type systems offer us about the structure, extent and other properties of chromospheres, wind and coronae, permit a discussion of those 'dividing lines' in a new light.

First, we can verify that the 'CaII K dividing line' (based on additional CS absorption features in the CaII K emission) is not only the most stringent criterion (and therefore at the rightmost position) but seems to fall quite close to the onset of winds as

Fig.5 indicates for the sample by Hünsch, 1990. The X-ray dividing line', which is the most stringent criterion for the onset of coronae, is also beyond any question of doubt. 'Cool coronae' of the sort in HR 6902 and HR 2554 (to the right of the 'X-ray dividing line') are not able to release powerful enough X-rays for detection by 'Einstein' and 'Exosat' detectors. Their existence therefore does not contradict the 'X-ray dividing line'. It might perhaps be possible to detect such 'cool coronae' from suitable giants, which should be nearby (to avoid strong interstellar absorption), using the soft X-ray experiment of Rosat.

It is unfortunately not possible to check whether by showing SiIV, CIV emission, these two particular stars contradict the 'transition region dividing line' since, as we pointed out already, other processes dominate this emission. They should, however, not fall into the category of 'hybrid stars', since they do not show evidence for cool winds. 22 Vul could potentially be such a candidate, but in that case evidence for a hot component could not be drawn from the spectra: interaction of the wind with the companion (bow shock and accretion wake, Che-Bohnenstengel, Reimers, 1986) forms very prominent and complex features in SiIV and CIV, such that detection of additional absorption in those lines during chromospheric eclipse is severely hindered.

In any case we may question the 'MgII-dividing line' (V<r asymmetric h_2 and k_2 emission) as a criterion for cool outflow: there are 5 candidates definitely without wind on the right hand side of it but none to its left that do show wind. Of these 5, 2 can be investigated for their chromospheric FeII-absorption, which is detectable but very much decreased (by 4 orders of magnitude as in HR 6202 and HR 2554, if compared to 22 Vul). We may suggest that the 'MgII-dividing-line' somehow reflects the onset of an increasing column density of chromospheric ion species such as MgII, FeII, SiII, but not the onset of an extended neutral cool chromosphere. The onset of an extended cool chromosphere coincides with the onset of at least partly neutral winds, as the example of 22 Vul indicates.

Models of stars in the HR diagram at the very onset of cool winds are also crucial for interpreting the processes of the winds, e.g. approaches based on Alphèn waves (Kuin, Ahmad, 1989) and stochastic shock waves (Cuntz, 1987). They should be able to explain the precise position of the transition. Furthermore, parameters such as log g, v_∞, $\dot{M}$, V_{Sto}, T_e and n_e, which are well determined observationally, should help to constrain theory. Kuin and Ahmad

(1989) claim to reproduce the critical position of 22 Vul in that respect in their application of an Alfven-wave code. Our current optical studies of several chromospheres should give an improved empirical basis for such attempts.

For further arguing, we also do need more bright ζ Aur systems! Efforts of enlarging this exclusive club of stars have been successful during past years (to mention R.&R. Griffin and Parsons and Ake here), and are going on.

References

Baade,R., 1989: AG-Highlight-Series, Klare,G., Ed., p. 324

Baade,R., 1990: Astron.Astrophys., in press

Che-Bohnenstengel,A., Reimers,D., 1986: Astron.Astrophys.156, p.172

Cuntz,M., 1987: Astron.Astrophys.188, L5

Griffin,R.E.M., 1988: J.Roy.Astron.Soc.Canada 87, p.49

Griffin,R.E.M., Griffin,R.F., Schröder,K.-P., Reimers,D., 1990: Astron.Astrophys., in press

Hartmann,L., Dupree,A.K., Raymond,J.C., 1982: Astrophys.J.252, p.214

Hempe,K., 1982: Astron.Astrophys.115, 133

Hünsch,M., 1990, Astron.Astrophys., submitted

Jordan,C., Linsky,J.L., 1987: 'Exploring the universe with the IUE satellite', Kondo,Y., editor, p.259

Kuin,N.P.M., Ahmad,I.A., 1989: Astrophys.J., 344, 856

Mullan,D.J., Stencel,R.E., 1982: 'Advances in UV astronomy: four years of IUE research', NASA CP-2238, p.235

Reimers,D., 1977: Astron.Astrophys.57, p.395

Reimers,D., 1982: Astron.Astrophys.107, p.292

Reimers,D., Baade,R., Schröder,K.-P., 1990: Astron.Astrophys. 227, p.133

Schröder,K.-P., Reimers,D., 1989: Astron.Astrophys.208, p.223

Schröder,K.-P., 1990: Astron.Astrophys., in press

Dust Driven Mass Loss in the HRD

C. Dominik

Institut für Astronomie und Astrophysik, TU Berlin,
Sekr. PN 8-1, D-1000 Berlin 12, Fed. Rep. of Germany

1 Introduction

The upper right corner of the Hertzsprung Russel diagram (HRD) is populated by luminous red giant stars. The most prominent properties of such stars are the presence of small solid particles in their extended atmospheres and the existence of a cool slow massive wind. The dust particles are observable via the pronounced infrared excesses of the objects (e.g. Merril and Stein, 1976). The observation of molecular lines in the radio range shows that the wind is carrying very high mass loss rates (up to $10^{-4} M_{\odot}/yr$) at wind velocities around 20 km/s (e.g. Knapp and Morris 1985). These mass loss rates are so high that the corresponding time scales ($M_*/\dot{M}$) are comparable or smaller than the typical time scales for the evolution of the stars on the asymptotic giant branch. Modelling the atmospheres of dust forming red giants is, therefore, not only of interest for a better understanding of the observations of such objects. It is also important for the understanding of the mechanism of wind formation and for a consistent calculation of mass loss rates which are needed for evolutionary calculations.

As a step toward this goal we have constructed stationary models of dust forming winds which allow for a calculation of the mass loss rate.

Figure 1 shows a sketch of the most important processes within the dust forming shell of a red giant. The giant at the center of the figure may be characterized by $L_* \simeq 10^4\ L_{\odot}$, $T_* = 2000 \ldots 2500\text{K}$, $M_* \simeq 1\ M_{\odot}$. The photosphere of the star is the origin of a stellar wind carrying material to the outer parts of the envelope. This levitation of the atmosphere may be supported by the dissipation of sound waves, shock waves or Alfven waves (see Gail, 1990, this volume) but it may also be due to the usual gas pressure gradient only. This is the case in the stationary models discussed in this paper where we neglegt the existence of such waves. For the effects of an additional atmosphere levitation on stationary models see Sedlmayr (1989) and Gauger *et al.* (1990).

In the outflowing cooling gas, a rich chemistry evolves. Due to the UV radiation field from the photosphere of the star and due to the velocity field of the wind, the chemical composition of the gas may be far from the equilibrium concentrations of the species at the given gas temperature. Therefore in general we have to treat a complex non-equilibrium chemistry.

When the temperature of the cooling gas becomes sufficiently low, dust formation may take place. Since this process is very sensitive to the local conditions in the gas (as molecular densities, temperature), the hydrodynamic and

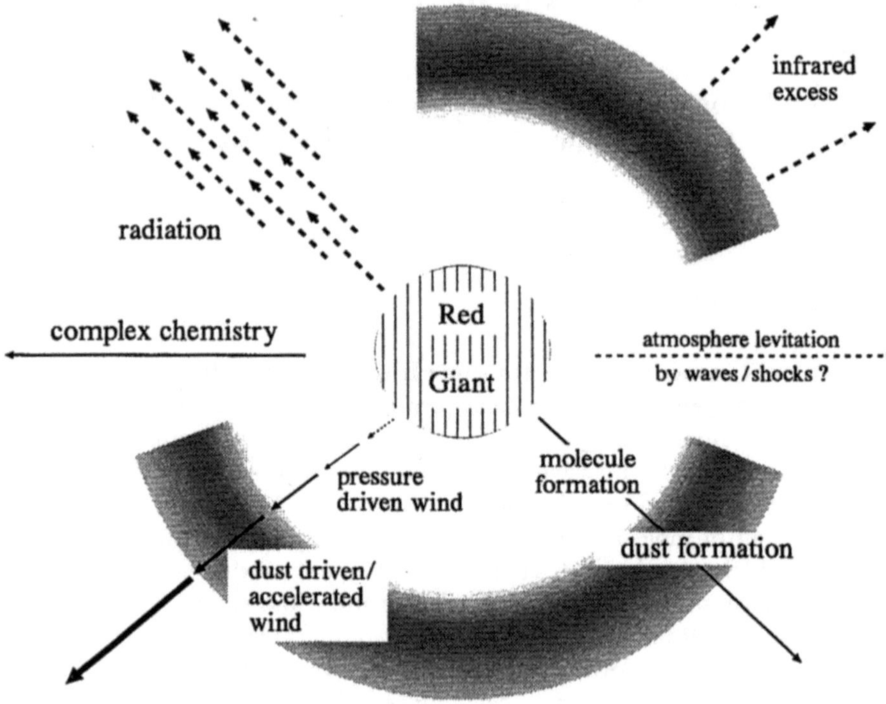

Fig. 1. Sketch of the most important processes in the dust forming envelope of a non pulsating red giant star with massive mass loss. Adopted with modifications from Sedlmayr (1989).

thermodynamic structure of the wind determines the formation and growth of dust particles. However, once dust has formed, its existence has considerable influence on the structure of the red giant envelope: First of all dust is an efficient energy transmitter. Dust grains absorb short wavelength photons and emit the energy at larger wavelengths corresponding to their internal temperature of 500...1000 K. In this way, the infrared excess of the object is produced. Furthermore, the dust transfers the momentum of the absorbed and scattered photons to the gas (via grain–gas collisions). In this way, the dust determines the hydrodynamic and thermodynamic structure of the envelope and, thus, the conditions for the *formation* of dust. Therefore, a successful modeling of those dust shells can be expected only if the *interaction* between dust formation, and the shell structure is treated.

2 The model

A wind model where radiation pressure on dust grains is the dominating hydrodynamic force which drives the wind through the sonic point is called a *dust driven wind* model. Such models including a detailed treatment of dust formation have been calculated by Kozasa *et al.* (1984) for the case of M stars and by Gail and Sedlmayr (1987a) for the case of C stars. We are using the set of equations given by Gail and Sedlmayr (1987a). It is describing a stationary spherical symmetric wind and consists of the following parts (for details see Gail and Sedlmayr (1987a)):

A) hydrodynamic equations
These are the equation of continuity and the equation for the momentum balance in the gas. The forces considered are gas pressure gradient, gravitation and radiation pressure on dust grains. Other forces due to the dissipation of all kinds of waves are neglected. Since those forces act in the same direction as the radiative acceleration, our models should rather be considered as a lower limit for the real winds from such stars. The *stellar mass* M_*, the *luminosity* L_* and the *mass loss rate* $\dot{M}$ are the stellar parameters that enter into these equations.

B) radiative transfer
If time dependent processes like shock waves are neglected, the absorption and emission of radiation are the dominating heating and cooling processes in the envelope of a red giant. Then, the temperature stratification may be obtained from the assumption of radiative equilibrium. Gail and Sedlmayr (1987a) used the grey Eddington approximation for the spherical case which has been given by Lucy (1976). From a solution of the radiative equilibrium one obtaines the temperature at every point in the shell. Furthermore, an *effective temperature* T_* is defined. T_* is close to the temperature of the photosphere of the star and must not be confused with the effective temperature of the dust shell. The HR-diagrams shown in this paper use this T_* as abscissa and, therefore, are diagrams for the central star inside rather than for the dust shell itself. However, T_* is not exactly the same as in plane parallel models (which is an usual assumption in calculations of stellar evolution) or in models without an extended dust shell.
T_* is a result of the solution of the radiative transfer. However, we may want to fix T_* as a parameter of the model in order to obtain another parameter as the result of the model calculation. It is most advantageous to choose $\dot{M}$ for this procedure. Then, it is possible to obtain the mass loss rate selfconsistently for given L_*, T_* and M_*.

C) chemical composition of the gas phase
Though the presence of non-equilibrium effects in the gas is likely (Goeres *et al.*, 1988), we have chosen a simple equilibrium chemistry in order to obtain the molecular composition of the gas. This assumption considerably reduces the expenditure for the model calculation. Since in a carbon star

envelope the molecules containing carbon (basically CO, C_2H_2, C_2H, C_3H, C_2, C_3) are most important for the formation of dust, we have reduced our chemistry to these molecules. The parameters entering into this block are the element abundances, especially the ratio ϵ_C/ϵ_O of carbon to oxygen.

D) dust formation and growth
The formation and growth of dust grains is calculated by means of the heteromolecular homogeneous nucleation theory (Gail and Sedlmayr, 1988). Only the formation of pure carbon grains has been considered. For details see Gail and Sedlmayr (1987a).

In the above mentioned work, only three models had been calculated. We have extended this work to a large range of stellar parameters in order to cover the relevant area in the HRD and different masses and chemical abundances. The results of this work will be published in detail elsewhere (Dominik *et al.*, 1990). In the present contribution, we will concentrate on one special point: the selfconsistently determined mass loss rate.

3 The mass loss rate

A dust driven wind solution does not exist for all possible sets of parameters. First of all, the star has to be sufficiently cool and extended in order to allow for dust formation in its outermost layers. Furthermore, radiation pressure on dust grains can only be the dominating hydrodynamic force if both the luminosity of the star and the absorption coefficient of the dust (which is proportional to the fraction of condensable material that actually has been converted to grains) are high. Therefore, there exists a minimum luminosity of the star at about $10^3 L_\odot$ if we assume complete condensation of all condensable material (Gail and Sedlmayr, 1987b). Since dust formation is effective only for high gas densities (which are proportional to the mass loss rate), there also exists a minimum mass loss rate for the wind to be dust driven (Gail and Sedlmayr, 1987b). If the condensation is not complete (due to low gas densities), the minimum luminosity is shifted to higher values.

In Fig. 2a we have plotted a dividing line for purely dust driven winds in the HRD which results from numerical calculations. It is located around $10^4 L_\odot$. Above this line purely dust driven winds are possible, below this line they are not. Note that the precise location of this line is dependent on the mass of the star and on the element abundances (for more details see Dominik *et al.*, 1990).

The selfconsistently determined mass loss rates that result from the model calculations are plotted in Fig. 3 as contour lines of constant mass loss rate which are, of course, present only within the allowed area. The minimum mass loss rate is about $10^{-7} M_\odot$/yr. Mass loss rates as high as $10^{-4} M_\odot$/yr can be reached for sufficiently high luminosities. The value of the mass loss rate increases with increasing luminosity and decreasing effective temperature of the star. This is interesting for the investigation of stellar evolution of AGB stars.

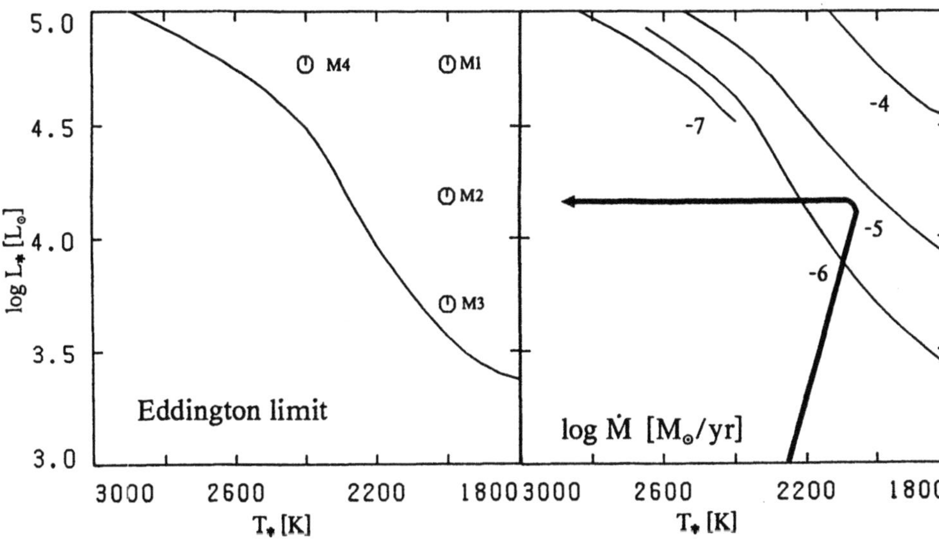

Fig. 2. a: HR-diagram with the dividing line for dust driven winds. Above the line, purely dust driven winds are possible, below they are not. The line is valid for a carbon star with $M_* = 1M_\odot$ and $\epsilon_C/\epsilon_O = 2$. The circles M1 ... M4 mark four selected sites in the HRD where the dependence of the mass loss rate upon M_* and ϵ_C/ϵ_O has been investigated (see Figs. 3, 4, 5). **b**: HR-diagram with contour lines of constant mass loss rate. The curves are labelled by $\log \dot{M}[M_\odot/\mathrm{yr}]$. This diagram is valid for a carbon star with $M_* = 1M_\odot$ and $\epsilon_C/\epsilon_O = 2$. The bold line shows a scematic evolutionary track of a AGB star

The bold line in Fig. 2b shows a schematic evolutionary track of such a star along the Hayashi line up to the tip of the AGB and at constant luminosity toward higher effective temperatures (e.g. Schönberner, 1980. Note that the effective temperature of the model is defined differently in our calculations than in stellar evolution calculations and, therefore, the low temperatures at the sketched track are not necessarily in contradiction to the usual evolutionary tracks). It can be seen that the mass loss rate will steadily increase as the evolution of the star progresses until the star evolves to a central star of a planetary nebula. This already suggests that dust driven winds might be an important mass loss mechanism during the transition from AGB stars to planetary nebulae.

Figure 3 shows the dependence of the mass loss rate upon the stellar mass. In order to construct this diagram, we have chosen four pairs of (L_*, T_*), i.e. four points in the HR-diagram. These points are marked in Fig. 2a and have been chosen in order to cover the area in the HRD where dust driven winds are possible. For these (L_*, T_*) we have varied the stellar mass M_* from a lower boundary of $0.5M_\odot$ (which seems to be a good value of the mass of the white dwarf inside the red giant) up to the highest value where a dust driven

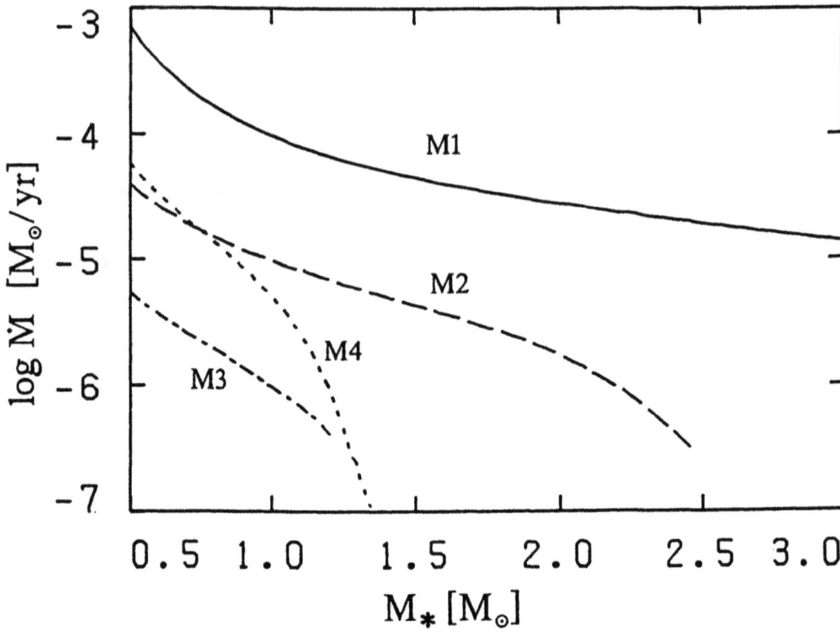

Fig. 3. Dependence of the mass loss rate upon the stellar mass at the four sites in the HRD which are marked in Fig. 2a. $\epsilon_C/\epsilon_O = 2$ for all models.

wind solution is possible. At this mass, the dividing line in the HRD meets the respective point (L_*, T_*). An exception has been made for the model M1. Here we have stopped the calculations at $M_* = 3M_\odot$, although the limit is higher. It can be obtained directly from the figure that $\dot{M}$ increases dramatically with decreasing mass. Halving M_* may increase the mass loss rate by a factor 3 to 100!

Though it is not yet possible to follow the evolution of a AGB star with a treatment of dust driven mass loss, the results of the last section give rise to the following interesting estimate: Let us assume that a low mass star ($M_* \approx 1 \ldots 3M_\odot$) evolves along the asymptotic giant branch. We do not consider the mass loss of the star until its evolutionary track crosses the dividing line for the possibility of purely dust driven winds (note again that the corresponding luminosity depends on both mass and effective temperature of the star). Let us assume that, from that moment on, the star does not change its luminosity and effective temperature. This can be interpreted as the assumption that the typical time scale for the mass loss is smaller than the time scales for the evolution of the stars interior. Then any change of the mass loss rate is due to the decreasing stellar mass only. The time evolution of the stellar mass $M_*(t)$ may then be obtained by simply integrating the corresponding $\dot{M}(M_*)$ curve in Fig. 3.

The results of this calculation are plotted in Fig. 4. It shows that for all four models the mass loss in fact is a selfaccelerating process. All models loose their

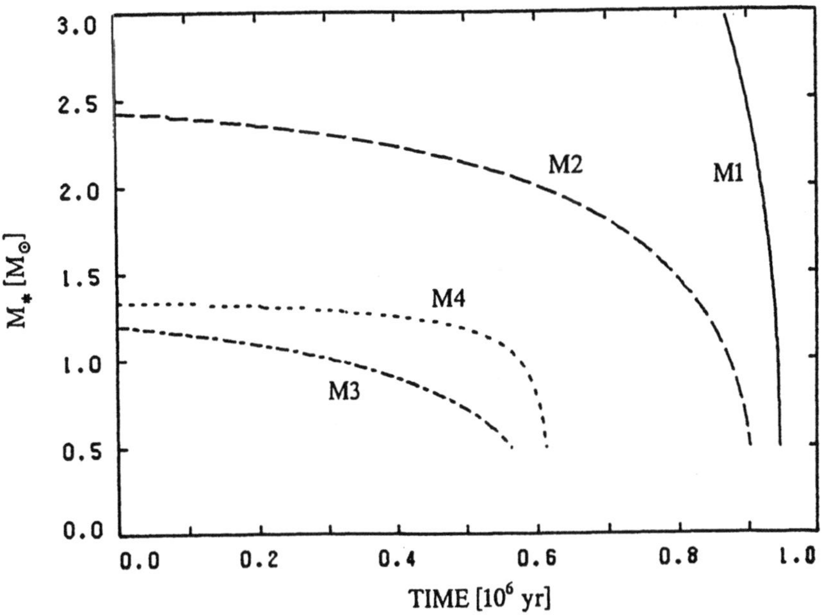

Fig. 4. Theoretical time evolution of the stellar mass with the assumption of constant L_* and T_*. (L_*, T_*) is given by the marks in Fig. 2, respectively. For M2, M3, M4, $t = 0$ is the onset of a dust driven wind when the star meets the dividing line for the possibility of a purely dust driven wind. M1 has been shifted arbitrarily in time.

envelope mass within a typical timescale of 10^6 yrs. However, a large part of the mass is lost in much shorter times of about 10^5 yrs and faster at the end of the evolution before the white dwarf at the core becomes visible. The models M1, M2, M4 which have luminosities above $10^4 L_\odot$ can reach very high mass loss rates. Their "evolution" looks very much like a mass loss instability and all three models loose at least $0.5 M_\odot$ during the last 10^4 yrs. This behaviour is very close to the specification of the so called "superwind" given by Renzini (1981) in order to explain the formation of a planetary nebula during the late phases of AGB evolution.

Therefore, it may again be suggested that dust driven wind may be a candidate for a mechanism driving this transition.

4 Mass loss rates during the evolution of low mass stars

For calculations of the evolution of low mass stars on the AGB it might be useful to construct a fit formula for mass loss rates due to dust driven winds as a function of the stellar parameters. In the last section, we showed the

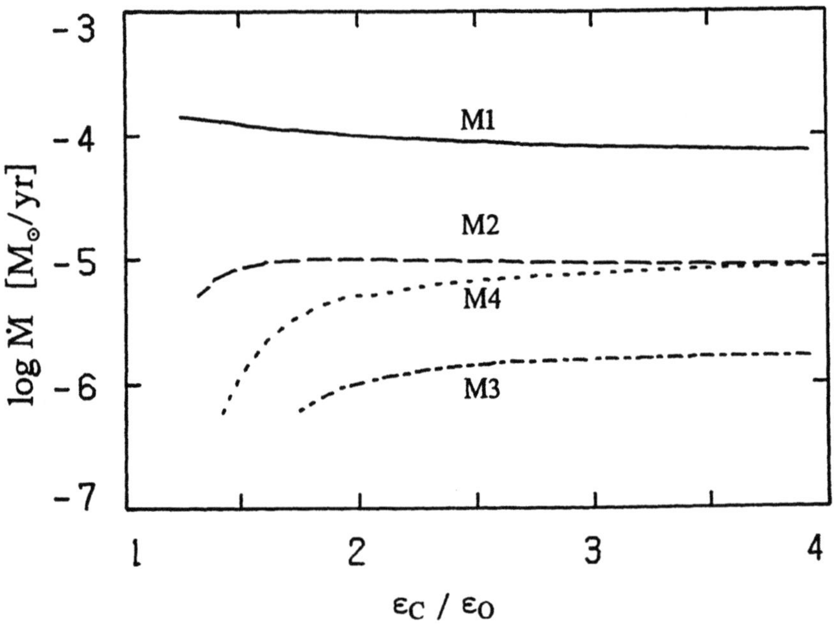

Fig. 5. Dependence of the mass loss rate upon ϵ_C/ϵ_O at the four sites in the HRD marked in Fig. 2a. $M_* = 1M_\odot$ for all models.

dependence of $\dot{M}$ upon L_*, T_* and M_*, respectively. The dependence upon the abundance of the dust forming elements for our special case is shown in Fig. 5.

For values of ϵ_C/ϵ_O very close to unity, almost all carbon is locked up in CO and only little carbon dust may condense which does not suffice to drive a wind. However, if ϵ_C/ϵ_O exceeds a value of about 1.4, the mass loss rate hardly depends on the precise value of ϵ_C/ϵ_O. Thus, the dust driven mass loss rate does not depend critically on the amount of dust formed in the envelope. If the radiative acceleration becomes sufficiently large to drive a wind, the mass loss rate is essentially a function of L_*, M_* and T_*. The mass loss rates of our models may be fitted by

$$\dot{M}[M_\odot/\mathrm{yr}] = 10^{-3.7-0.93M_*[M_\odot]} \cdot \left(\log L_*[L_\odot] - \frac{T_*}{600\mathrm{K}}\right)^{2.59M_*[M_\odot]+1.71}$$

which is correct within a factor of five for mass loss rates larger than $10^{-6} M_\odot/\mathrm{yr}$ and, therefore, covers the region of interest. It should be noted that this formula has been constuced C stars where only the formation of carbon grains has been considered. Adding other dust species should rather increase the resulting mass loss rates.

Concerning the use of this formula in calculations of stellar evolution we like to emphasize that it is necessary to use the correct definition of T_* (similar to that we used in our calculations). This is important since the fit formula is very sensitive to the value of T_*. However,a consistent treatment requires an

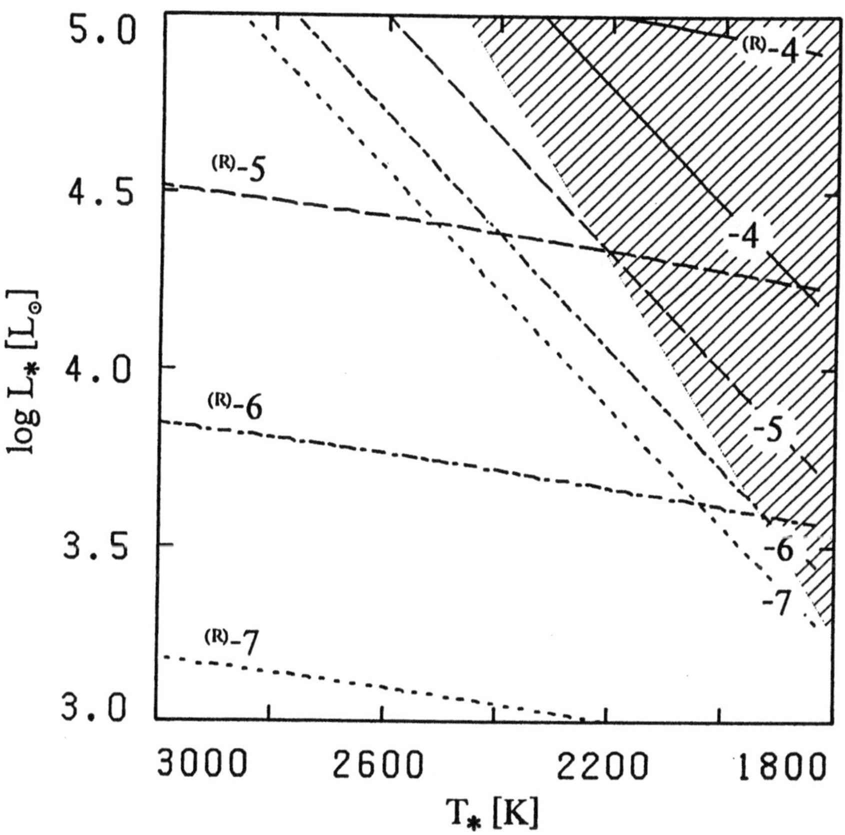

Fig. 6. Comparison of the mass loss rates derived from the Reimers formula (see text) and from our fit formula. This diagram is valid for $M_* = 1M_\odot$ and $\epsilon_C/\epsilon_O = 2$. The curves are labelled by log $\dot{M}[M_\odot/\mathrm{yr}]$. A leading (R) marks the mass loss rates according to the Reimers formula. Within the hatched area, the dust driven mass loss rates are larger than those from the Reimers formula.

evaluation of the whole outer layer integration in stellar evolution calculations with the dynamical equations. For a discussion see Dominik *et al.* (1990).

Calculations of stellar evolution including mass loss have been carried out by several authors (e.g. Harpaz and Kovetz, 1981; Schönberner, 1981; Boothroyd and Sackman, 1988; Maeder and Meynet, 1988). In all cases, the mass loss rate has been calculated from the so called *Reimers formula* (Reimers, 1975) which is given by

$$\dot{M}[M_\odot/\mathrm{yr}] = 5.5 \cdot 10^{-13} \frac{L_*}{L_\odot} \frac{R_*}{R_\odot} \frac{M_\odot}{M_*}$$

(Kudritzki and Reimers, 1978). This is an empirical formula which has been constructed for the mass loss rates of K and early M giants. However, it does certainly not apply for the dust driven mass losses discussed in this paper. This can be seen from Fig. 6 where we have plotted the mass loss rates resulting from our formula and from the Reimers formula, respectively. The most important difference is the steep temperature dependence of dust driven mass loss rates. Due to this fact, purely dust driven winds are possible only for quite cool stars. Another important difference is the fact that the lines of constant mass loss rate for dust driven winds are much closer together. Therefore, if the limit line for the onset of dust driven winds is reached, the corresponding mass loss rate rapidly exceeds the values resulting from the Reimers formula.

Given this results we think that the calculations of AGB evolution can considerable be improved by a consistent treatment of dust formation and the induced mass loss rates.

References

Boothroyd, A.I., Sackmann, I.J.: 1988, *Astrophys. J.* **328**, 653

Dominik, C., Gail, H.-P., Sedlmayr, E., Winters, J. M.: 1990, *Astron. Astrophys.*, submitted

Gauger, A.,Gail, H.-P., Sedlmayr. E.: 1990, *Astron. Ges. Abstr. Ser.* **4**, 13

Gail, H.-P., Sedlmayr. E.: 1987a (Paper III), *Astron. Astrophys.* **171**, 197

Gail, H.-P., Sedlmayr. E.: 1987b, *Astron. Astrophys.* **177**, 186

Gail, H.-P., Sedlmayr. E.: 1988 (Paper IV), *Astron. Astrophys.* **206**, 153

Goeres, A., Gail, H.-P., Henkel, R.,Sedlmayr. E.: 1988, in *Rev. in Modern Astron.* **2**, p.231, ed. G. Klare, Springer Berlin, Heidelberg

Harpaz, A., Kovetz, A.: 1981, *Astron. Astrophys.* **93**, 200

Knapp, G.R., Morris, M.: 1985, *Astrophys. J.* **292**, 640

Kozasa, T., Hasegawa, H., Seki, J.: 1984, *Astrophys. Space Sci.* **98**, 61

Kudritzky, R.P., Reimers, D.: 1978 *Astron. Astrophys.* **70**, 227

Lucy, L.B.: 1976, *Astrophys. J.* **205**, 482

Maeder, A., Meynet, G.: 1988, *Astron. Astrophys.* **210**, 155

Merril, K.M., Stein, W.A.: 1976, *Publ. Astron. Soc. Pac.* **88**, 285

Reimers, D.: 1975, in *Problems in Stellar Atmospheres and Envelopes*, p. 229, Eds. B. Baschek, W.H. Kegel, G. Traving, Springer, Berlin

Renzini, A.: 1981, in *Physical Processes in Red Giants*, eds. Iben and A. Renzini, Reidel, Dordrecht

Schönberner, D.: 1981, *Astron. Astrophys.* **103**, 119

Sedlmayr,E. : 1989 in *From Miras To Planetary Nebulae*, Montpellier, eds. M.O. Mennessier and A. Omont, Editions Frontières, in press

The Close Circumstellar Environment of Young Stellar Objects

T. Montmerle

Service d'Astrophysique, Centre d'Etudes Nucléaires de Saclay, F-91191 Gif-sur-Yvette Cedex, France

Abstract. Young stellar objects ("YSOs") are characterized by a wide variety of activity phenomena and a strong interaction with their environment, on a wide range of spatial scales. I review the solar-like activity in T Tauri stars, and the X-ray and radio [cm] evidence for extended magnetic structures ($\lesssim 10\, R_*$). I then describe the close circumstellar material ($\lesssim$ 1000 AU) recently associated with a fraction of these stars and with objects embedded in molecular clouds: circumstellar accretion disks and their associated boundary layers, ionized and neutral winds. Recent mm data, in conjunction with IR data, suggest that accretion disks have complex structures, which evolve on timescales ~ a few 10^5 to ~ a few 10^7 yrs, and are probably the result of various gravitational instabilities. A tentative unified picture emerges, linking magnetic fields, accretion disks, and mass-loss phenomena up to the much larger spatial scales of molecular outflows (several 10^4 AU). The possible relevance of this new picture for the origin of the solar system is pointed out.

1. What is a "Young Stellar Object"?

Thanks mainly to the advent of telescopes operating at "penetrating" wavelengths (X-rays, IR, mm, cm), rapid progress has been accomplished in recent years in our understanding of the earliest stages of star formation. Indeed, very young or forming stars are embedded inside molecular clouds, and it is only at the above wavelengths that one is able to see them. However, when these objects are too embedded to be visible in the optical domain, it is difficult to characterize them in the same way as optically visible stars (absence of HR diagram classification in many cases, for instance). On the other hand, the very fact that these stars are embedded in molecular clouds testifies of their youth, typically less than a few 10^7 yrs. In addition, as we shall see, many stars are surrounded by a complex circumstellar environment: magnetic fields, dust shells or disks, ionized or neutral winds, molecular ouflows, etc., and therefore do not appear as "stars" in the conventional sense. At the cost of some vagueness, it has become customary in recent years to call the various sources close to, or embedded in, molecular clouds, by the generic name of "young stellar objects" (YSOs, for short).

In most cases, YSOs can be shown to have moderate masses (≤ a few $M_\odot$) and temperatures (≤ 6000 K). Their luminosities, however, can be quite high (up to several 10 $L_\odot$), placing them generally high above the main sequence: in that sense, the expressions "pre-main sequence objects" ("PMS objects") and "YSOs" are synonymous. But other YSOs have higher masses (10-20 $M_\odot$, for instance), and are almost always found near or on the ZAMS. (This difference is of course connected with the much shorter evolutionary timescale for objects of larger masses.) In what follows, we will discuss the *low-mass* objects (≤ 2 $M_\odot$), which are much more numerous, and therefore better known as a class, than higher-mass objects (see, e.g., Lada 1987 for a discussion of high-mass stars in general, and Catala 1989 for a discussion of Herbig Ae-Be stars).

The most well-known (and historically oldest-known) low-mass PMS objects are the T Tauri stars. Very good reviews on this subject have recently appeared (Bertout 1989, Appenzeller and Mundt 1989). I refer the interested reader to these for details, but summarize here their relevant properties.

T Tauri stars, first characterized by Joy, have been known since ca. 1945. They were discovered in the vicinity of dark clouds (mainly in Taurus) thanks to their peculiar emission-line spectra, which made them conspicuous for instance in Hα objective-prism surveys. In the last few years, however, mainly as a result of X-ray observations by the *Einstein* satellite, these stars have been divided into two broad classes: the "classical" T Tauri stars ("CTTS"), and the "weak-line T Tauri stars" ("WTTS").

CTTS constitute the vast majority of the originally cataloged TTS (Herbig and Rao 1972; for a revised list, see Herbig and Bell 1988). Their spectra display strong optical and UV emission lines. The equivalent width of Hα is > 5 Å, and may reach 200 Å or more in "extreme" TTS. The underlying spectrum is that of a late-type photosphere (K7, typically), with respect to which there is a large UV and IR excess. The amount of UV excess is correlated with the strength of the emission lines, and so does, to a lesser extent, the near-IR part of the IR excess. By contrast, the WTTS show very weak, if any, emission lines (Hα equivalent width < 5Å), no UV excess, and little or no IR excess. Their most remarkable feature is their comparatively large X-ray emission, which has made their discovery possible. However, there appears to be no statistically significant difference, in the existing data base, between their X-ray properties and those of CTTS.

YSOs too deeply embedded to be visible in the optical and X-rays (A_V > a few magnitudes) generally remain visible at longer wavelengths, and have been characterized by their near- to mid-IR spectral distribution (Lada 1987, 1988). They have been broken down in "Class I" (very strong IR excess, with a λF_λ spectrum ascending with wavelength), "Class II" (IR excess comparable to that of CTTS, i.e., flat or moderately descending spectrum), and "Class III" (no IR excess, i.e., pure blackbody spectrum). Current theories link these IR spectra with evolutionary stages, in terms of decreasing amount of warm dust, from the protostar stage (Class I), in which the central stellar condensation is surrounded by a dust shell, to the dustless, advanced PMS stage (Class III) (e.g., Shu, Adams, and Lizano 1987). We shall see that this scheme is being modified by new results in the radio mm range.

Perhaps some of the most intriguing characteristics of YSOs are their intense activity, and a strong interaction with their environment, in the form of mass loss (stellar winds, collimated jets, molecular outflows), accretion from a circumstellar disk, and solar-like magnetic activity. In this paper, we will be mainly concerned with the question of the "close" circumstellar environment of YSOs, i.e., within a radius ~ 1000 AU. This includes stellar magnetic fields (§ 1 and 2), stellar winds (§4), circumstellar disks (§ 4 and 5), but excludes a detailed discussion of large-scale phenomena like jets or bipolar flows, which extend up to a fraction of a pc, or of the related "Herbig-Haro objects". (The interested reader is referred respectively to Lada 1985, Mundt 1988, and Reipurth 1989 for reviews on these topics.)

2. Solar-like activity in low-mass PMS stars

2.1. X-ray evidence

The bulk of the data we now have comes from observations using the *Einstein* satellite, operating in the imaging mode between ~ 0.4 to ~ 4 keV (for reviews, see, e.g., Feigelson 1987, Feigelson, Giampapa, and Vrba 1990). Most of the observed regions were nearby dark clouds, well known to undergo active star formation and containing many T Tauri stars: ρ Ophiuchi (Montmerle et al. 1983), Taurus-Auriga (Gahm 1981, Feigelson et al. 1988, Walter et al. 1988), Chameleon (Feigelson and Kriss 1989), all at distances ~ 160 pc, and Orion (Ku, Righini-Cohen, and Simon 1982, Caillault and Zoonematkermani 1989, Strom et al. 1990) at ~ 450 pc. (Note that ρ Oph and Orion in addition contain stars of earlier spectral types, the earliest being B3 and O7, respectively, themselves X-ray emitters.)

All these observations revealed dozens of new sources, including many more PMS stars than were previously discovered by applying the usual criteria based on the presence of strong emission lines, notably Hα. Although a definite figure cannot be given for lack of completeness in the various surveys, some authors (e.g., Walter et al. 1988) argue that these new, X-ray discovered stars may be up to 10 times more numerous than the older, traditional population of T Tauri stars. Initially called "naked" TTS because it was thought that the absence of Hα (equivalent width < 5 Å) implied the absence of circumstellar material (see below, § 4.4), these stars are now preferably called more empirically "weak-line" TTS. These newly found "WTTS" therefore make up a new, and in fact dominant, population of YSOs.

Where do the X-rays come from? When a good X-ray spectrum of a PMS star is available (which is not always the case at the level of sensitivity of *Einstein* because in general the X-ray flux is weak and too few counts are recorded), it is consistent with bremsstrahlung emission from a hot ($kT \sim 1$ keV) plasma, significantly absorbed along the line of sight ($N_H \lesssim$ a few 10^{21} cm^{-2}, or $A_V \lesssim 10$, in agreement with the optical data when available). The typical luminosities are in the range $\langle L_X \rangle \sim 10^{31-32}$ erg.s^{-1}.

But more significant results can be gained from the study of time variability. Much evidence has accumulated in favor of the existence of large X-ray flares (e.g., Feigelson and DeCampli 1981, Walter and Kuhi 1984, Montmerle et al. 1983, 1984), which dominate the X-ray emission. Within the still modest precision of the data, these flares are solar-like, but their luminosities are enhanced by factors up to 10^6 with respect to the Sun. They are also enhanced with respect to flares on dMe "flare stars" (see discussion in Montmerle et al. 1983), and quite similar to flares observed on RS CVn close binary systems (e.g., Mutel et al. 1987). However, strong flaring activity takes place only during ~ 5 % of the time, on average (Gahm 1988).

2.2. Radio evidence

Starting from the solar-flare analogy of the X-ray emission of TTS, it is a logical next step to look for radio emission in the cm (GHz) band, as observed on the Sun; in this case, the emission mechanism is expected to be non-thermal and associated with energetic electrons and magnetic fields. However, there is also the possibility that the emission mechanism may be thermal and associated with circumstellar ionized gas, which makes radio observations doubly interesting.

Taking advantage of the increased sensitivity and the imaging capability in the cm range offered by the "Very Large Array" interferometer in New Mexico, a number of workers have done radio observations following two approaches. In the first approach, known YSOs, selected according to a variety of criteria, are the targets of pointed observations: optically visible TTS found in early catalogs (e.g. Bieging, Cohen, and Schwartz 1984), sources of molecular outflows (Snell and Bally 1986; Lada 1985). In the second approach, a large region is subject to an unbiased survey, without any prior selection of targets.

The first approach gives evidence that ~ 10% of stars now known as CTTS, or molecular outflow sources, are radio emitters at a flux density sensitivity of ~ 0.5 mJy. This emission is of thermal origin (free-free emission), and confirms the existence of intense ionized stellar winds, already suspected to be present on the basis of some Hα and NaD (e.g., Mundt 1984) or [OIII] (Edwards et al. 1987) emission line profiles. According to the standard model (e.g., Panagia and Felli 1975), the mass-loss rates are typically $\dot{M} \sim 10^{-8} M_\odot$ yr^{-1} for an assumed wind temperature $T_w \sim 10^4$ K, and using velocities $v_w \sim$ 200-300 km. s^{-1} indicated by the P Cygni profiles of optical lines. The size of the emitting region is $R \sim 1000 R_*$, which, at the distance of the nearest molecular clouds, is resolvable by the VLA.

The second approach makes no assumption about the nature of the putative radio emitters, and has revealed the existence of hitherto unknown objects, in ρ Oph (André, Montmerle, and Feigelson 1987), CrA (Brown 1987), and Orion (Trapezium region, Garay, Moran, and Reid 1987, Churchwell et al. 1987). Several stellar radio sources had been previously found in X-rays, but many were discovered to be associated with near-IR sources deeply embedded in the clouds, so that even the X-rays, if present, would be absorbed. In other words, there is evidence for yet another population of YSOs, but the number of radio-emitting objects found down to a flux density of ~ 1 mJy is small, on the order of 10 % of known near-IR sources (K-magnitude limit typically 12). However, as the sensitivity increases, the number of radio-emitting objects also increases, as does the fraction of identifications with near-IR sources, which reaches ~ 30 % when the minimum detectable flux density is $\lesssim$ 0.1 mJy (Leous et al. 1990).

As in X-rays, long or repeated observations proved crucial in understanding the nature of the radio emission: strong variability was noticed as early as 1985, with the discovery of the first radio flare in a PMS star (Feigelson and Montmerle 1985), and was later established to be rather frequent among PMS radio emitters (Stine et al. 1988). For lack of adequate coverage, the typical timescale for variability is unknown, but variability in flux of factors of at least 2 have been found in a few hours, and up to 10 between observations separated by a few months (e.g., Cohen and Bieging 1986). Such timescales can only be explained in terms of a nonthermal emission mechanism, because the interpretation in terms of an ionized wind implies large emission sizes and thus long timescales, $\gtrsim$ years. (Note that the commonly used criteria based on the spectral index α, where $S_\nu \propto \nu^\alpha$, namely that if $\alpha > 0$ the emission is thermal and if $\alpha < 0$ the emission is nonthermal, does not reliably indicate the nature of the emission mechanism, contrary to what is often found in the literature, but simply indicates the opacity [resp. optically thick, and optically thin]; for details, see André 1987.) A remarkable characteristic is that the majority of the new non-thermal sources emitting above ~ 1 mJy do not exhibit an IR excess and are therefore of the WTTS type. (Sources detected in the radio with a lower flux have in most cases been observed at only one wavelength, and their nature is therefore uncertain at present; see discussion in Leous et al. 1990.) At this point, one can speak of "solar-like" activity only loosely: we will be able to be more specific only after quantitative interpretations.

2.3. Optical and UV evidence

T Tauri stars have long been known to be optically variable, having been put in the category of "flare stars" in the early fifties by such pioneers as Haro and Ambartsumian. Analysis of the time structure of these flares (Worden et al. 1981) has shown that they are distributed according to a power law, smaller flares being more frequent than larger ones. The optical flares are solar-like, but enhanced several thousand times with respect to solar flares.

Variability is also present in the near-IR and near-UV bands (e.g., Gahm 1988), but the flare activity is best observed in the near-UV band, since the contrast with the essentially red photosphere is enhanced. According to a recent study, there is some difference in the time distribution of UV flares between CTTS and WTTS (Gahm 1990).

Another relevant feature is the existence of starspots. In many TTS (CTTS and WTTS alike), multicolor photometry (typically UBVRI) has revealed a rotational modulation of the light. This modulation can be modelled, in most cases, by a single large starspot (filling factor up to ~ 20 % of the stellar surface), cooler than the photosphere by a few 100 K (see, e.g., Bouvier and Bertout 1989, Bouvier 1990). In addition to confirming the existence of solar-like activity, these observations allow to determine the rotation period of the TTS, a key parameter both for activity-related aspects and for their early evolution.

3. Magnetic nature of the activity

3.1. The solar analogy

The fact that the activity described above appears to be solar-like is already an indication that it is basically magnetic in origin. However, "solar-like" and "magnetic" are not necessarily synonymous, as magnetic configurations different (in time and space) from the solar ones may exist.

We have seen above (§ 2.1) that the X-rays were emitted by bremsstrahlung of a ~ 10^7 K opticaly thin plasma. It is therefore necessary to confine this plasma at least for the duration of a flare (up to a few hours), and, on the Sun, this is done by magnetic loops, to first approximation dipolar. Equating the gaseous and magnetic field pressures yields a surface magnetic field B_* ~ 1 kG, i.e., of the same order than in sunspots, and, by inference, starspots in general. The main difference between TTS and the Sun, however, lies in the size H_X of the loops, which is much larger: typically, H_X is a large fraction of the star's radius (which is typically R_* ~ 2 $R_\odot$), and may be even larger, reaching ~ 2 R_* (see, for instance, the "superflare" described in Montmerle et al. 1983).

As regards the non-thermal radio-emitting YSOs, the emission mechanism is likely to be gyrosynchroton radiation from ~ MeV electrons spiraling in magnetic fields ~ 1 G; at least, quantitative models exist for a few well-documented cases (in the ρ Oph cloud: DoAr21, a late-type PMS star, see André 1987 and S1, a very young ZAMS B3 star, André et al. 1988; Taurus: V410 Tau, Bieging and Cohen 1989), based on non-homogeneous emission models featuring dipolar magnetic loops, and developed for the Sun (e.g., Klein and Chiuderi-Drago 1987). The magnetic field at the stellar surface is in general $B_{*,\mathrm{rad}}$ ~ a few 100 G, consistent with the values $B_{*,X}$ deduced from the X-rays. In the case of S1, however, a higher value is required (≳ 10 kG), about the same as prevails in the well-known magnetic B stars, and this star is surrounded by a dipolar magnetosphere. The implication is that these objects must possess extended magnetic structures, the size of which H_{rad} may extend up to ~10 R_* (Montmerle and André 1988; see Fig. 1), which is significantly larger than the size H_X deduced from the X-rays. The validity of this approach has recently been demonstrated by the VLBI observations of the two stars DoAr21 and S1, in which the size of the magnetic structures could be determined and found to be consistent with model predictions (André et al. 1990b).

Altogether, it turns out that X-rays and radio emission give values for the surface magnetic field which are solar or nearly solar. These values are only indirect, and direct determinations (such as those obtained by Saar 1988 on main

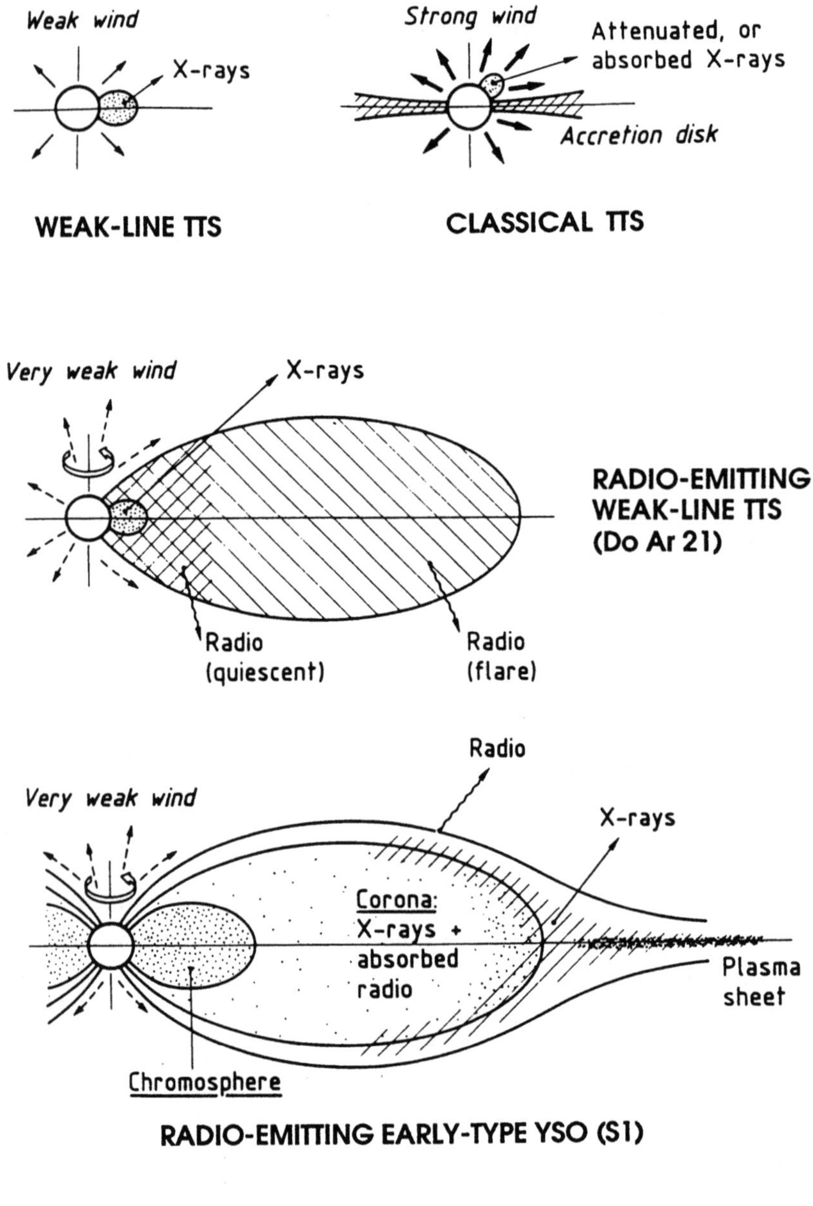

Fig. 1. *Sketch (to scale) of the magnetic structures of weak-line and classical T Tauri stars, as deduced from their X-ray and radio (cm) properties (updated from Montmerle and André 1988).*

sequence stars) would be highly desirable on T Tauri stars, but their faintness makes the measurements very difficult. The most important difference with the Sun concerns the size of the magnetic structures, found (*directly* in the case of VLBI observations) to be very large, up to several stellar radii.

3.2. Origin of magnetism

As on the Sun, the origin of the magnetism is widely thought to be the dynamo mechanism, in which magnetic fields are created by the combined action of convection and differential rotation (the so-called "$\alpha\omega$" mechanism; see Gilman 1983). Unfortunately, dynamo theory is not developed well enough in stars to lead to any quantitative prediction. One simply expects some correlation between the level of magnetic activity and rotation (itself correlated somehow with differential rotation), both being defined in a broad sense. Different authors use different parameters to quantify the "magnetic activity": it can be the X-ray luminosity, or chromospheric activity tracers such as the MgII H + K or CaII h + k line fluxes (e.g., Rutten and Schrijver 1987, Bouvier 1990). Luckily, these different parameters seem to be well correlated with each other, over 3-5 orders of magnitude (so that variability is a comparatively small factor). "Rotation", on the other hand, is also ill defined: it can be the rotation velocity (v sin$i$), the rotation period (with an embarassing stellar radius as a multiplying factor between the two), or even the so-called "Rossby number" (see, e.g., Hartmann and Noyes 1987), which is a mixture of rotation and convection (and equal to the ratio of the observed rotation period and of the computed convective turnover time). As it turns out, however, all combinations tend to yield the same rather conforting trend, namely a decrease of the "magnetic activity" with "rotation", as expected from dynamo theory, but little more than that can be extracted from the available data. (For a more extensive discussion, see e.g., Montmerle 1987; on the more specific correlation between the X-ray flux and the rotation period, see Bouvier 1990).

4. Non-solar activity and circumstellar material

4.1 Mass-loss phenomena

As mentioned above, there are several indications that strong stellar winds are associated with CTTS. The standard interpretation is that they are high-temperature, fully ionized winds, possibly accelerated by Alfvén waves (Lago 1984, Hartmann, Edwards and Avrett 1982). Interestingly, in this view the winds could themselves also be considered as "solar-like", because the ratio $\dot{M}(\mathrm{CTTS})/\dot{M}_{\odot}$ is of the same order ($\sim 10^6$) as that of the corresponding X-ray luminosities, and acceleration by Alfvén waves is one of the proposed mechanisms for the solar wind. However, it is very difficult to account in particular for the strength of Hα in such winds (see DeCampli 1981), unless one makes particular assumptions on the expansion velocity field (Hartmann et al. 1990).

A new approach suggests that these winds are in fact *cold* ($T_w \sim$ a few 10^3 K) and only partially ionized (Natta and Giovanardi 1989; see the review by Natta 1989). The actual mass-loss rate would then be the radio rate multiplied by a factor 1/[ionization degree], i.e., by at least an order of magnitude. Although the cold wind model depends on some unmeasured parameters, there is some direct observational support for the existence of "neutral" extended winds with the detection of H_2 moving at several tens of km.s^{-1} near a few well-known embedded objects, like L1551 (Lizano et al. 1989, Natta 1989). The associated crossing times (= size/flow velocity) are ~ several 10^3 yrs, pointing to very young exciting sources. Additional, indirect support comes from the consideration of the energetics of CO outflows. Indeed, these large-scale phenomena, extending over several 10^4 AU, imply high mass-loss rates (up to several 10^{-6} $M_{\odot}$.yr^{-1}; but see the large uncertainties on the published figures, Cabrit and Bertout 1989), much higher than those computed for fully ionized stellar winds. One can causally link molecular outflows and stellar winds if a momentum-conserving shock transfers the cold material from one to the other (Natta 1989). The driving mechanism for such high mass loss is still controversial; we will return to this question in the last section.

4.2. Accretion disks around young stellar objects

Other activity tracers, like the UV emission lines, and perhaps more conspicuously Hα (which cannot, as we have seen, be easily explained by stellar winds), are in excess by up to 2 orders of magnitude with respect to the other tracers as a function of rotation (Bouvier 1990; see Fig.2), in the case of CTTS, whereas for WTTS Hα, for instance (which has a few Å equivalent width at most), seems to be entirely of solar-like (chromospheric) origin. In other words, there is clear evidence in CTTS that an intense *non-solar* activity exists. As we shall now discuss, the current paradigm is that the reason for this does not lie in the stars themselves, but in the presence of

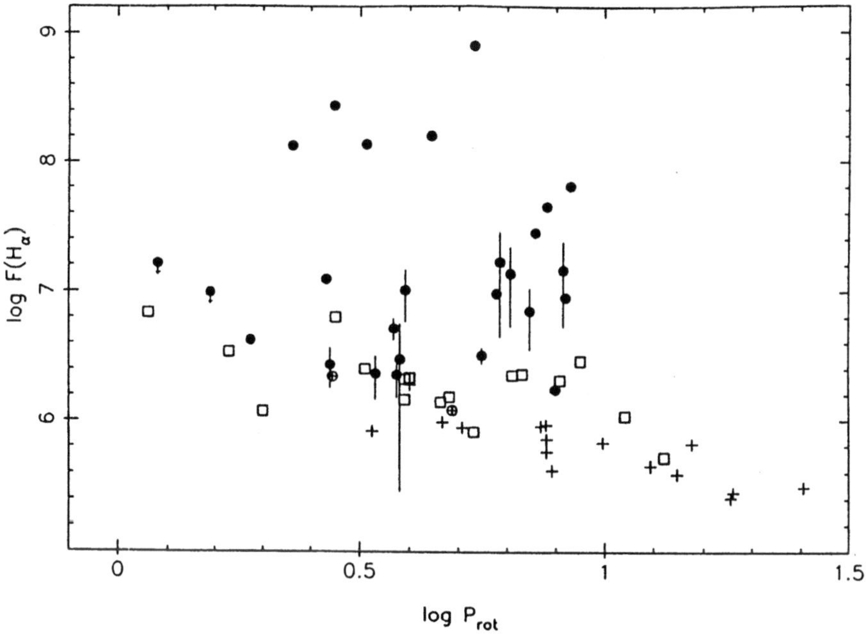

Fig. 2. *Relation between the Hα emission line flux* F(Hα) *(in erg s*$^{-1}$ *cm*$^{-2}$*) and the rotation period* P *(in days) deduced from the modulation of the stellar light by large starspots for a sample of T Tauri stars, as compared with other magnetically active stars (Bouvier 1990). The excess Hα emission of TTS over the other stars reaches 2 orders of magnitude.*
Symbols: ● = TTS, + = late-type dwarfs, ⊕ = dMe stars, □ = RCVn close binary systems.

close circumstellar material, and that there is in fact a continuum across the YSO classes (including TTS).

Let us begin with the YSOs for which the most advanced models are available, the CTTS. Current models (see Bertout, Basri, and Bouvier 1988; Hartmann and Kenyon 1987) depict a CTTS as a complex object comprizing 3 distinct components:

- a late-type star,
- a circumstellar disk,
- a boundary layer between the star and the disk.

The existence of a Keplerian, circumstellar disk of gas and dust, has first been inferred from the IR excess. The idea is that this excess comes from the existence of an additional energy source, which the star itself is unable to provide. The introduction of a disk is natural: first, the idea is inescapably present since Kant and Laplace in the late 18th century and is suggested by the coplanar orbits of the main bodies of the solar system; second, it is the logical outcome of the condensation of rotating structures like the dense "ammonia cores" (Myers et al. 1987), half of which contain IR sources thought to be protostars.

To first order the IR excess can be fitted by a continuous blackbody temperature distribution as a function of radius, implying in turn a continuous, optically thick matter distribution. This has a fundamental consequence: because the Keplerian rotation velocities are different at r and at $r+dr$, there must be some shear between successive rings, hence viscous stresses and heating. The energy radiated in the process at r must exactly balance the input of potential energy from $r+dr$ to r. But this also means that matter must slowly spiral inwards, and eventually fall onto the star. In other words, any spatially continuous disk losing energy by viscous process must be an *accretion disk.* Since viscosity is a poorly known process on a microscopic scale (especially under these astrophysical conditions), the temperatures and densities used in modelling the disks are generally taken as power laws as a function of radius, resulting in a power-law for the IR excess, which in many cases gives a satisfactory fit to the observational data. In spatially thin, optically thick viscous Keplerian disks, for instance, the temperature decreases with radius as $r^{-3/4}$, and the density as r^{-2}; the resulting near-IR spectrum λF_λ has a slope $-4/3$. The radiated IR luminosity is ~ 1/2 of the available gravitational energy, and is proportional to the accretion rate (e.g., Lynden-Bell and Pringle 1874, Pringle 1988). Typical accretion rates computed from such models are ~ a few $10^{-7}\,M_\odot\,\mathrm{yr}^{-1}$. The near-IR (~ 2 μm) traces warm material (~ 1500 K) relatively close to the star ($\lesssim$ 1 AU); longer IR wavelengths (up to the longest *IRAS* band, i.e., ~10-100 μm) trace cooler material (a few 10-100 K) and correspond to a few AU. Since these dimensions are small with respect to the overall size of the disk (which may extend up to 1000 AU or more, see below, § 5.1), and because the disk is optically thick, it is clear that the IR is essentially a temperature tracer, not a mass tracer; to trace the mass, one has to go to longer wavelengths, where, in addition, there is mounting direct evidence for the existence of extended dust disks (mm range, where they are optically thin: § 5.2).

4.3. Boundary layers

Close to the star, an additional, very important phenomenon must occur. At $r \simeq R_*$, the Keplerian velocity in the disk around a ~ 1 $M_\odot$ star is ~ 250 $km.s^{-1}$, which is also (by definition) the break-up velocity of this star. But rotation velocity measurements (like the abovementioned starspot analysis, § 2.3.) show that, contrary to early expectations, T Tauri stars (in general) are slow rotators, with equatorial velocities in the range ~ 20 $km.s^{-1}$ or even less, implying that they have very quickly lost most of their initial angular momentum. As a result, near the equator, the star rotates ~ 10 times slower than the disk! Then a *boundary layer* (a well-known phenomenon in aerodynamics) must develop, to ensure a continuous transition of the fluid velocities between the disk and the star. However, again because the viscosity is the central dissipative mechanism and is poorly known, the physical conditions within the boundary layer, its dimensions, etc. are also poorly known and are the subject of active research (see, e.g., Basri and Bertout 1989; for basic principles, see, e.g., Frank, King, and Raine 1985). The condition of energy balance of the accretion disk shows that the other ~ 1/2 of the gravitational energy lost during the inward motion of the material must be radiated at the level of the boundary layer. The way in which this energy is radiated will of course depend on the opacity. Several arguments lead to think that it must be optically thick (Bertout et al. 1988); some of the emission, like the Balmer jump, must however arise from optically thin regions (Basri and Bertout 1989). The temperatures reached are, much like in HII regions (balance between *in situ* heating and cooling), ~ 8000 K. The bulk of the radiation will therefore belong to the optical and near-UV domains, and this is why it is believed that the observed emission lines (and in particular Hα) and the UV excess originate in the boundary layer. No definitive calculations on this point exist yet, because these require a knowledge of the physical conditions within the boundary layer which we do not have at present. But this explains qualitatively nicely the correlation between the IR and UV excesses and the strength of the emission lines. In some models (see Bertout 1989) the combined contribution to the overall spectrum of the boundary layer (UV range) and the disk (IR range) is so high that it can completely dominate the photospheric contribution (optical range): from the point of view of spectral emission properties, the star is then "smothered" by the disk.

4.4. What about WTTS?

At the other extreme, the implication is that because they have no UV or IR excess, and little, if any, Hα emission, the WTTS should not be surrounded by circumstellar material, but simply represent evolved, active, solar-like PMS stars. However, many WTTS, in particular those found by André et al. (1987) in the ρ Oph cloud, must be very young because of their location in the core of the cloud, although, strictly speaking, no age can be attributed to them because most are not seen in the optical and hence cannot be placed on an HR diagram. But they are visible in the near-IR (e.g., Wilking, Lada, and Young 1989), and, as mentioned above, they show no excess in this range. According to the IR classification, they are "Class III" objects. So we are faced with a problem: how is it that very young stars, still embedded in cloud cores, apparently do not show evidence for circumstellar material?

A related problem exists more generally when considering optically visible CTTS and WTTS as a whole. Indeed, when put on an HR diagram, these two classes of PMS stars appear mixed (see discussion below, § 5.3), whereas, according to "standard" evolutionary models (e.g., Adams, Lada, and Shu 1987), CTTS, being surrounded by accretion disks, should be younger, and WTTS, deprived of such disks, should represent a more advanced evolutionary stage.

The probable answer to this problem lies in the fact that the "standard" scenario of early stellar evolution is based on IR data, which, as mentioned above, traces only warm material comparatively close to the star ($\lesssim$ 1 AU). Outer ($\gtrsim$ 10 AU), cold (<< 100 K) material is visible only in the mm range, and recent work in this range has given new insight into the presence of circumstellar matter around young stars, and into the possible role of magnetic fields (see Montmerle and André 1989).

5. New developments: mm data and evolutionary implications

5.1. Observational results

The first evidence for a flattened molecular structure around a YSO has been the detection by Beckwith et al. (1986) of strong (~ 900 mJy) mm emission from the extreme TTS HL Tau, both in continuum and in the ^{12}CO (1→0) line. Because these observations were performed with an interferometer (at Owens Valley, California) yielding an angular resolution of ~ 6", the fact that the continuum source was unresolved implied a size < 1000 AU, whereas the resolved ~ 30" size of the elongated ^{12}CO source implied a size ~ 4000 AU. Later ^{13}CO observations with the same instrument led Sargent and Beckwith (1987) to the conclusion that this structure was consistent with a circumstellar disk of gas and dust in Keplerian rotation.

Since then, more sensitive mm studies of several tens of T Tauri stars have been reported, as new, large single-dish telescopes came into operation. With the James Clerk Maxwell telescope in Hawaii, operating in atmospheric windows from 0.35 to 1.1 mm, Weintraub, Sandell, and Duncan (1989) observed a sample of 16 PMS stars, including 10 CTTSs of which 9 were detected and one, undetected, WTTS, while Adams, Emerson, and Fuller (1990) selected a sample of 13 CTTSs (11 in Taurus-Auriga and 2 in the ρ Oph cloud), of which 7 were detected. Beckwith et al. (1990) used the IRAM 30-m radio telescope near Granada, Spain, equipped with the MPIfR bolometer operating at 1.3 mm, to study a much larger sample of 86 PMS stars in Taurus-Auriga, including 53 CTTS and 21 WTTS: 53 % of the CTTS, and 29% of the WTTS were detected. Millimeter studies of low-luminosity YSOs more deeply embedded than T Tauri stars are more scarce, and have been mainly dedicated to individual cases so far (e.g., Chandler et al. 1990, Keene and Masson 1990). Two mm surveys of cold *IRAS* sources (Wilking et al. 1989, Walker, Adams, and Lada 1990) are now available, but they are essentially limited to luminous (i.e., ≥ 100-1000 $L_{\odot}$) YSOs. However, André et al. (1990c) have independently undertaken a study similar to that of Beckwith et al. (1990) in the ρ Oph core using the IRAM 30-m telescope, but selecting a sample of embedded sources spanning a broader range of IR properties, including Class I (6 objects), Class II and neighbouring IR spectra (9 objects), Class III (3 objects), with a very good sensitivity (down to an r.m.s. of ~ 1 mJy). Out of 18 such objects, 14 were detected, belonging to all classes and spanning a large range in fluxes, from ~ 30 to ~ 400 mJy. In addition, the sample included two exceptional objects exciting molecular outflows and believed to be extremely young (a few 10^3 yrs), IRAS16293 (Walker et al. 1988) and VLA1623 (André et al. 1990a); both are very bright at mm wavelengths (resp. 5 and 0.9 Jy).

While the high detection rate of Weintraub et al. (1989) and Adams et al. (1990) was not unexpected, because there was already from IR data strong independent evidence for circumstellar material around their program stars, the results of Beckwith et al. (1990) are more surprising. Indeed, many WTTS turn out to be surrounded by cold material, whereas there are indications that no or little warm material is present. On the other hand, André et al. (1990c) find a lack of obvious correlation between the mm emission and near-IR excess of the objects in the

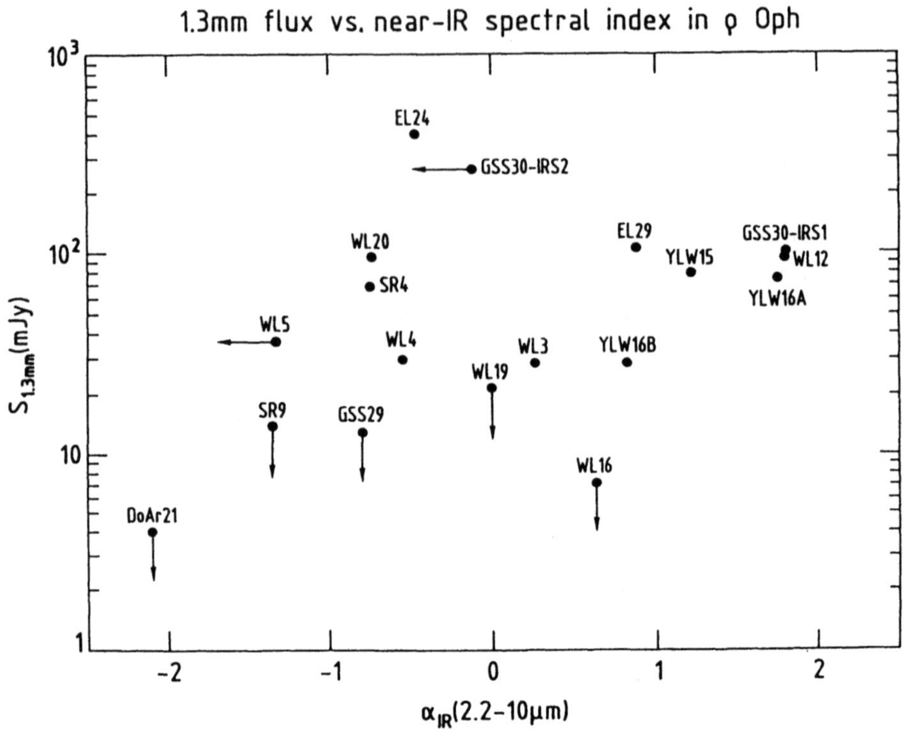

Fig. 3. *Scatter diagram of the 1.3 mm flux density vs. the near-IR spectral index (slope of the λF_λ spectrum between 2.2 μm and 10 μm) for a selected sample of sources in the ρ Oph cloud (André et al. 1990c). Class I, Class II, and Class III IR sources lie to the right, center, and left parts of this diagram, respectively. There is no obvious correlation between the mm flux (which traces optically thin, cold dust, at several AU from the central star) and the near-IR spectral energy distribution (which traces the optically thick, warm material within 1 AU), suggesting the existence of "gaps" in the accretion disks.*

sample. In particular, as shown on Fig. 3, heavily obscured (Class I) objects do not appear brighter at 1.3 mm than CTTS (Class II) in general, and conversely a few objects without near-IR excess (Class III) do show a definite mm emission. Although the samples of Beckwith et al. (1990) and André et al. (1990c) are qualitatively different, the two results are quite consistent with each other. In the framework of the disk interpretation of the emission of YSOs at long wavelengths, this means that, in many cases, these disks are not continuous. More precisely, in the extreme cases, WTTS with substantial mm emission (~ 30 % of the objects) would be surrounded by hollow, "45 r.p.m." disks, since they have no UV excess nor significant Hα emission, hence no boundary layer, implying that there is simply no physical contact between the star and the disk: this is confirmed by the absence of near-IR emission, i.e., of warm material close to the star. Conversely, CTTS with weak or absent mm emission (~ 50 %) would be surrounded by "compact" disks, at most a few 10 AU in size. Anyway, since none of the detected sources is resolved within the ~ 12" resolution of the 30–m observations, the size of such disks must be < 1000 AU.

5.2. Properties of the emitting material

A quantitative interpretation of continuum mm results is delicate. There is no doubt that the emission at mm wavelengths is due to optically thin cold dust (see the detailed discussion in André et al. 1990c), but the dust emissivity properties (and associated opacity properties) at these wavelengths are still poorly known, in particular because the dust grain size, shape, and composition are uncertain (see, e.g., Emerson 1988, and also the discussion in Beckwith et al. 1990). For $\lambda >> 100\ \mu m$, it is customary to take a power-law representation of the emissivity as a function of wavelength, with an index $\beta \geq 0$ (i.e., spectrum proportional to the Planck function at temperature T and to $\lambda^{-\beta}$). The interstellar value of β is ~ 2; lower values of β correspond to large grains, of size > 100 μm. The shape of the grains may itself affect the value of β: for instance, Wright (1987) finds $0.6 < \beta < 1.4$ for fractal aggregates of conducting spheres.

In principle, then, with enough measured spectral points in the mm range, one can determine both T and β, and deduce the total disk mass M_D (H_2 + dust, assuming a standard dust-to-gas ratio of 100) from the measured flux. In practice, however, this is possible only as long as the $\lambda^{-\beta}$ emissivity approximation holds, and this may not be true below ≈ 1 mm. In particular, it is impossible to reliably use the ~100 μm *IRAS* point (if available) in conjunction with the single wavelength presently available with the IRAM bolometer. If one nevertheless takes a temperature of ~ 50-100 K as typical of continuum mm sources, there seems to be a tendency for β to vary from object to object, from low values (at face value even compatible with 0) for CTTS, to high values (> 2) in a few cases. Beckwith et al. (1990) find that β is on average ≤ 1 in their sample, i.e., significantly smaller than the interstellar value and pointing to the existence of large dust grains. Taking as representative values $\beta = 1$ and $T = 50$ K, one then finds M_D in the range ~ 0.001 to ~ 0.5 $M_\odot$, which, interestingly, encompasses the estimated mass of the primitive solar nebula, ~ 0.005 - 0.1 $M_\odot$ (Cameron 1988).

5.3. Evolutionary consequences

The consequences of the new picture presented above are important for the evolution of low-mass YSOs. Indeed, it is now well established that WTTS and CTTS coexist in large regions of the HR diagram, corresponding to ages from ~ 3×10^5 yrs to ~ 3×10^7 yrs (Strom et al. 1989; see also Skrutskie et al. 1989, and Walter et al. 1988). One may argue that, precisely because of the presence of the disks, the positions of CTTS in the HR diagram are uncertain (Hartmann and Kenyon 1990); still, this cannot seriously affect the fact that the range of ages is very large. As Strom et al. (1989) point out, on average, CTTS tend to be younger than WTTS, but some WTTS are very young ($< 3 \times 10^5$ yrs; the non-thermal radio-emitting WTTS or Class III sources of ρ Oph, as discussed above, are also probably very young), and some CTTS are very old ($> 3 \times 10^7$ yrs), as are stars surrounded by conspicuous disks like β Pic (A star, $\geq 10^7$ yrs; see, e.g., Smith and Terrile 1984, Lagrange-Henri et al. 1989).

Therefore, the transition between CTTS and WTTS occurs at different times from star to star, somewhere between $< 3 \times 10^5$ and $> 3 \times 10^7$ yrs. In our framework, this is not because of the disappearance of the disk, but because of the disappearance of the boundary layer. The time scale for this transition is short, estimated to be $< 10^6$ yrs, compatible with the age of the younger CTTS and small compared to the age of the older WTTS (Skrutskie et al. 1989).

This means that the disk evolution of YSOs is in fact decoupled from their stellar evolution. Comparable stars (that is, identically located on the HR diagram) have different disks: "accretion" disks (which may be "compact": CTTS), or more or less "hollow" ones (WTTS). At least to a first approximation, a star of a given initial mass evolves according to standard laws of stellar structure, but some additional factors make the disks evolve differently for a given star, some quickly, some slowly, with a comparatively abrupt transition between CTTS and WTTS. This a complex subject, because the problem of disk evolution *per se* is still in its infancy. In addition to the fact that, after all, what we call a "disk" is in fact still very poorly known observationally, probably important factors which must be incorporated in any theory include:

– spatial evolution: appearance of a central cavity, perhaps even of rings, etc., or disappearance of the outer regions; change in thickness;

– dust grain evolution and possible condensation or evaporation phenomena (changing β index), presumably (but not necessarily) leading to planet formation;

– influence of mass loss (likely also from the disk itself, § 6) and possible sporadic bipolar flow phases, or intense winds, as is the case for FU Orionis events (in which the mass-loss rate is typically 100 times stronger than in CTTS: Herbig 1989, Kenyon and Hartmann 1989);

– stellar or *in situ* generated magnetic fields;

– appropriate initial conditions (in particular initial angular momentum).

Of course, binarity (Zinnecker 1989) may also play an important role on disk evolution as a result of tidal effects, in the case of a sufficiently tight system.

5.4. Disk stability and magnetic fields

One recent avenue is that of the intrinsic stability of the accretion disks around single stars, and some promising theoretical developments have recently appeared. Adams, Ruden, and Shu (1989) and Tagger et al. (1990) have shown that spatially thin disks (scaleheight << size) are subject to efficient dynamical instabilities ("swing amplification" phenomenon). Supplementing the initial work of Papaloizou and Pringle (1987) by including magnetic fields (here assumed to be locally perpendicular to the disk) in addition to gravitation, Tagger et al. (1990) have shown that, depending on the respective strength of the magnetic field and the local gravity, several instabilities are possible. (The magnetic field may be of stellar origin, or produced locally by some kind of dynamo resulting from differential rotation.)

If the magnetic field dominates gravity, these instabilities may result in a cavity in the inner regions of the disk and thus explain the existence of hollow disks, although it is premature to say whether they will explain the transition from a regular accretion disk to a hollow disk and the disappearance of the boundary layer. Support for this situation may be found in the existence of large magnetic structures around the non-thermal radio sources; we also note that the widespread X-ray activity of WTTS is indicative of strong surface magnetic fields (see Montmerle and André 1988, and above, § 3). Farther out, other instabilities are possible: if the magnetic field is negligible and the self-gravity small, the hydrodynamical instabilities of Papaloizou and Pringle (1987) occur, and if self-gravity is large enough, the spiral instabilities suggested to be present in galactic disks take over (Pellat, Tagger, and Sygnet 1990). Work is in progress to determine the timescales over which such instabilities develop in circumstellar disks, but obviously such considerations leave ample room for many possible, time-dependent configurations (including spiral arms in the outer regions).

6. Concluding remarks

There is today no doubt about the presence of warm and cold circumstellar material around YSOs. Its exact morphology is in general not known, but the existing evidence supports, albeit indirectly, the idea that this material should be close to the central star in the form of a flattened, disklike structure, perhaps as small as a few 100 AU in radius or less. The disk masses are highly uncertain, but still roughly within one order of magnitude both ways of the mass of the primitive solar nebula. This recent image therefore essentially confirms what had been guessed, or speculated, over more than two centuries.

But what is perhaps more surprising is that many YSOs *do not* have continuous "disks" around them. Would we be then witnessing some fragmenation of the initial, supposedly continuous disks, in other words, the birth of planetary systems? It is certainly premature to answer this question, because we have yet to elucidate many aspects of the stability of circumstellar disks, and in particular the role of magnetic fields. In the above discussion, we have shown ample evidence of their presence on YSOs, and we have mentioned the possible role extended stellar magnetic fields may play.

However, magnetic fields probably also play a key role in another important and related aspect of the circumstellar environment of YSOs, namely the mass loss associated with CTTS or embedded molecular flow sources. An assumption common to all existing models is that open magnetic field lines confine the outflowing material in a bipolar fashion; we have seen that a number of arguments point to the possibility that the stellar wind is in fact mostly neutral even close to the star, and becomes molecular farther out. Since the implied mass-loss rate is very high (possibly up to a few $10^{-6}\,M_{\odot}\,\mathrm{yr}^{-1}$), it is very difficult to account for it in terms of a purely stellar mechanism. For instance, Alfvén waves act only on the ionized part of the wind, but likely cannot, even through friction between ions and neutral atoms, lift the remaining neutral part. This is why current models draw the mass lost through the wind from the accretion disk itself, much in the way the disk is also invoked to explain the extra UV and IR luminosity. In other words, we are led to the concept that *accretion drives mass loss*! The general scheme is that accretion is equatorial (disk), and mass ejection is polar (jets) or, at least, confined to a relatively small angle to the polar axis of the star. And what about the energetics? In many cases (see, e.g., Lada 1985), molecular outflows have a kinetic "luminosity" which approaches, or even exceeds, that of the exciting source. Again, the trick is to tap this energy from the ultimate reservoir, namely gravity, by using the rotation of the star, or of the disk.

In one school of thought (Shu et al. 1988), the central star is strongly magnetized and rotates near break-up; the disk material flows inwards, touches the stellar equator at the critical, "X point", and then "rebounds" in open magnetic field lines bending away from the disk (so-called "X-celerator" mechanism). Because this magnetic field is anchored to the star, the ex-disk material is dragged along the lines by the centrifugal force and follows their direction, increasingly parallel to the rotation axis. Obviously, given the fact that no YSO has yet been observed to rotate close to the break-up velocity, this model could be applied only to the youngest stellar objects, before their initial angular momentum is lost. Deeply embedded sources with outflows having very short dynamical timescales (like IRAS 16293 and VLA 1623 in ρ Oph) might be in this case. However, none of these sources show at present evidence for magnetic fields such as would be indicated by non-thermal radio emission.

A second school of thought assumes that the accretion disk is itself entirely responsible for the mass loss: the disk "evaporates", its material flowing along magnetic field lines this time anchored to the disk (Uchida and Shibata 1985, Pudritz and Norman 1986, Kwan and Tademaru 1988). Here again the centrifugal force plays a role, but so does the shear and spiraling of the magnetic field lines, following the shear of the disk material. The bending of the field lines in more efficient in this case (they even join at some distance from the star along the polar axis), and it seems that the collimated jets are easier to explain in this framework. In addition, the central star plays no other role than to set up a gravitational potential well holding the accretion disk together, thus removing the stellar rotation velocity constraint. Other arguments, for instance based on forbidden optical lines (see Edwards et al. 1987) tend to support a mass loss directly from the disk, at least in the case of CTTS.

An interesting consequence is that, in the X-celerator mechanism, a physical contact is required between the star and the disk (which is not a boundary layer, since the star and the disk are assumed to rotate at the same velocity). Thus, irrespective of the stellar rotation, this mechanism could not exist for WTTS surrounded by a hollow disk, whereas the "evaporation" mechanism could conceivably still work: a possible test would be to look for evidence of cold outflows from these stars.

Whatever the (significant) differences between these models, a unified picture of YSOs tends to emerge. At an early stage, the initial protostar, at the center of a rotating spheroidal shell of gas and dust, evolves into a star surrounded by a (possibly, but not necessarily massive) accretion disk. Then, by means of the rotation of either the star or the disk, and because of the presence of stellar and/or disk magnetic fields, an intense, mostly neutral wind is accelerated and collimated. This wind gives rise to bipolar jets and molecular outflows farther away, and disperses what is left of the protostellar shell. Probably because of this intense magnetic field-driven mass loss, the central star slows down quickly, and a boundary layer appears, giving rise to a CTTS if optically visible. Under the influence of these magnetic fields, gravitational instabilities occur, and at some point the disk breaks, forming for instance rings or spiral arms. If only the outer part breaks, we have CTTS without mm emission; if, on the contrary, a central cavity forms, the boundary layer disappears, leading to a WTTS surrounded by cold, mm emitting material. This transition may take place at any time (depending on as yet unidentified conditions, as discussed in § 5.3) between ~ a few 10^5 yrs and ~ a few 10^7 yrs, and is rather fast (less than 10^6 yrs); the lifetime of the outer material is not known, but at some point all the remaining disk material disappears, leading to a truly "naked" TTS on its way to the main sequence. One reason for this disappearance may be, of course, the formation of planetary systems...

Admittedly, such a scheme is still qualitative, and even (in part) speculative. But it provides at least a stimulating line for future work, and the advent of many new instruments operating at many wavelengths, both on the ground and in space, as well as of improved theoretical tools (plasma physics, etc.), should help us very soon to give a better answer to the ultimate question: where does the Sun come from?

Acknowledgements. It is a pleasure to thank E. Sedlmayr, G. Klare, and the organizers of the Deutsche Astronomische Gesellschaft meeting in Berlin for their kind invitation, especially during historical times which hopefully will foster cooperation in Europe and in the world.

References

Adams F.C, Emerson J., Fuller G. 1990, *Ap.J.*, in press.

Adams F.C., Lada C.J., Shu F.H. 1987, *Ap.J.* **312**, 788.

Adams F.C., Ruden S.P., Shu F.H. 1989, *Ap.J.* **347**, 959.

André P. 1987, in *Protostars and Molecular Clouds*, ed. T. Montmerle and C. Bertout, (Saclay: CEA/Doc), p. 143.

André P., Despois D., Martin-Pintado J., Montmerle T. 1990a, *Astr.Ap.*, in press.

André P., Klein K.L., Lestrade J.-F., Phillips R. 1990b, in preparation.

André P., Montmerle T., Feigelson E.D. 1987, *Astr.J.* **93**, 1182.

André P., Montmerle T., Feigelson E.D., Steppe H. 1990c, *Astr.Ap.*, in press.

André P., Montmerle T., Stine P.C., Feigelson E.D., Klein K.L. 1988, *Ap.J.* **335**, 940.

Appenzeller I., Mundt R. 1989, *Astr.Ap.Rev.* **1**, 291.

Basri G., Bertout C. 1989, *Ap.J.* **341**, 340.

Beckwith S.V.W., Sargent A.I., Chini R.S., Güsten R. 1990, *Astr.J.*, in press.

Beckwith S., Sargent A.I., Scoville N.Z., Masson C.R., Zuckerman B., Phillips T.G. 1986, *Ap.J.* **309**, 755.

Bertout C. 1989, *Ann.Rev.Astr.Ap.* **27**, 351.

Bertout C., Basri G., Bouvier J. 1988, *Ap.J.* **330**, 350.

Bieging J.H., Cohen M. 1989, *Astr.J.* **98**, 1686.

Bieging J.H., Cohen M., Schwartz P.R. 1984, *Ap.J.* **282**, 699.

Bouvier J. 1990, *Astr.J.* **99**, 946.

Bouvier J., Bertout C. 1989, *Astr.Ap.* **211**, 99.

Brown A. 1987, *Ap.J.(Letters)* **322**, L31.

Cabrit S., Bertout C. 1990, *Ap.J.* **348**, 530.

Caillault J.P., Zoonematkermani S. 1989, *Ap.J.(Letters)* **338**, L57.

Cameron A.G.W. 1988, *Ann.Rev.Astr.Ap.* **26**, 441.

Catala C. 1989, in Proc. ESO Workshop *Low Mass Star Formation and Pre-Main Sequence Objects*, ed. B. Reipurth (Garching: ESO), p. 471.

Chandler C.J., Gear W.K., Sandell G., Hayashi S., Duncan W.D., Griffin M.J. 1990, *M.N.R.A.S.*, in press.

Churchwell E., Felli M., Wood D.O.S., Massi M. 1987, *Ap.J.* **321**, 516.

Cohen M., Bieging J.H. 1986, *Astr.J.* **92**, 1396.

DeCampli W.M. 1981, *Ap.J.* **244**, 124.

Edwards S., Cabrit S., Strom S.E., Heyer I., Strom K.M., Anderson E. 1987, *Ap.J.* **321**, 473.

Emerson J.P. 1988, in *Formation and Evolution of Low Mass Stars*, ed. A.K. Dupree and M.T.V.T. Lago, NATO ASI Series (Dordrecht: Kluwer Academic Publishers), p. 21.

Feigelson E.D. 1987, in Proc. *5th Cambridge Cool Star Workshop*, eds. J.L. Linsky and R. Stencel (Berlin: Springer), p. 455.

Feigelson E.D., DeCampli W.M. 1981, *Ap.J.(Letters)* **243**, L89.

Feigelson E.D., Giampapa M.S., Vrba F.J. 1990, in *The Sun in Time*, eds. C.P. Sonett and M.S. Giampapa (Tucson: University of Arizona Press), in press.

Feigelson E.D., Jackson J.M., Mathieu R.D., Myers P.C., Walter F.M. 1988, *Astr.J.* **94**, 1251.

Feigelson E.D., Kriss G.A. 1989, *Ap.J.* **338**, 262.

Feigelson E.D., Montmerle T. 1985, *Ap.J.(Letters)* **289**, L19.

Frank J., King A.R., Raine D.J. 1985, *Accretion Power in Astrophysics* (Cambridge: Cambridge U. Press)

Gahm G.F. 1981, *Ap.J.(Letters)* **242**, L163.

Gahm G.F. 1988, in *Formation and Evolution of Low Mass Stars*, ed. A.K. Dupree and M.T.V.T. Lago, NATO ASI Series (Dordrecht: Kluwer Academic Publishers), p. 295.

Gahm G.F. 1990, in *Flare Stars in Star Clusters, Associations, and the Solar Vicinity*, IAU Symp. 137 (Dordrecht: Kluwer Academic Publishers), in press.

Garay G., Moran J.M., Ried M.J. 1987, *Ap.J.* **314**, 535.

Gilman D.A. 1983, in *Solar and Stellar Magnetic Fields: Origin and Coronal Effects,* ed. J.O. Stenflo (Dordrecht: Reidel), p. 247.

Hartmann L., Calvet N., Avrett E.M., Loeser R. 1990, *Ap.J.* **349**, 168.

Hartmann L., Edwards S., Avrett E.M. 1982, *Ap.J.* **261**, 279.

Hartmann L., Kenyon S.J. 1988, *Ap.J.* **325**, 231.

Hartmann L., Noyes R.W. 1987, *Ann.Rev.Astr.Ap.* **25**, 271.

Herbig G.H. 1989, in Proc. ESO Workshop *Low Mass Star Formation and Pre-Main Sequence Objects,* ed. B. Reipurth (Garching: ESO), p. 233.

Herbig G.H., Bell K.R. 1988, *Lick Obs. Bull. No 1111.*

Herbig G.H., Rao N.K. 1972, *Ap.J.* **174**, 401.

Keene J., Masson C.R. 1990, *Ap.J.*, in press.

Kenyon S.J., Hartmann L. 1987, *Ap.J.* **323**, 714.

Kenyon S.J., Hartmann L. 1990, *Ap.J.* **349**, 197.

Klein K.L., Chiuderi-Drago F. 1987, *Astr.Ap.* **175**, 179.

Ku W.H.-M., Righini-Cohen G., Simon M. 1982, *Science* **215**, 61.

Kwan J., Tademaru E. 1988, *Ap.J.(Letters)* **332**, L41.

Lada C.J. 1985, *Ann.Rev.Astr.Ap.* **23**, 267.

Lada C.J. 1987, in *Star Forming Regions,* IAU Symp. 115, ed. M. Peimbert and J. Jugaku (Dordrecht: Reidel), p. 1.

Lada C.J. 1988, in *Formation and Evolution of Low Mass Stars,* ed. A.K. Dupree and M.T.V.T. Lago, NATO ASI Series (Dordrecht: Kluwer Academic Publishers), p. 93.

Lago M.T.V.T. 1984, *M.N.R.A.S.* **210**, 323.

Lagrange-Henri A.M., Beust H., Ferlet R., Vidal-Madjar A. 1989, *Astr.Ap.* **215**, L5.

Leous J.A., Feigelson E.D., André P., Montmerle T. 1990, *Ap.J.*, submitted.

Lynden-Bell D., Pringle J.E. 1974, *M.N.R.A.S.* **168**, 603.

Lizano S., Heiles C., Rodriguez L.F., Koo B.-C., Shu F.H., Hasegawa T., Hayashi S., Mirabel I.F. 1988, *Ap.J.* **328**, 763.

Montmerle T. 1987, in *Solar and Stellar Physics,* Proc. 5th Eur. Solar Meeting, ed. E.H. Schröter and M. Schüssler (Berlin: Springer), *Lect. Notes in Phys.* **292**, 117.

Montmerle T., André P. 1988, in *Formation and Evolution of Low Mass Stars,* ed. A.K. Dupree and M.T.V.T. Lago, NATO ASI Series (Dordrecht: Kluwer Academic Publishers), p. 225.

Montmerle T., André P. 1989, in Proc. ESO Workshop *Low Mass Star Formation and Pre-Main Sequence Objects,* ed. B. Reipurth (Garching: ESO), p. 407.

Montmerle T., Koch-Miramond L., Falgarone E., Grindlay J.E. 1983, *Ap.J.* **269**, 182.

Montmerle T., Koch-Miramond L., Falgarone E., Grindlay J.E. 1984, in Proc. *Very Hot Astrophysical Plasmas,* ed. L. Koch-Miramond and T. Montmerle, *Phys. Scripta* **T7**, 59.

Mundt R. 1984, *Ap.J.* **280**, 749.

Mundt R. 1988, in *Formation and Evolution of Low Mass Stars,* ed. A.K. Dupree and M.T.V.T. Lago, NATO ASI Series (Dordrecht: Kluwer Academic Publishers), p. 257.

Mutel R.L., Morris M.H., Doiron D.J., Lestrade J.-F. 1987, *Astr.J.* **93**, 1220.

Myers P.C., Fuller G.A., Mathieu R.D., Beichman C.A., Benson P.J., Schild R.E., Emerson J.P. 1987, *Ap.J.* **319**, 340.

Natta A. 1989, in Proc. ESO Workshop *Low Mass Star Formation and Pre-Main Sequence Objects,* ed. B. Reipurth (Garching: ESO), p. 365.

Natta A., Giovanardi C. 1990, *Ap.J.*, in press.

Panagia N., Felli M. 1975, *Astr.Ap.* **39**, 1.

Papaloizou J.C.B., Pringle J.E. 1987, *M.N.R.A.S.* **225**, 267.

Pellat R., Tagger M., Sygnet J.F. 1990, *Astr.Ap.* **231**, 347.

Pringle J. E. 1988, in *Formation and Evolution of Low Mass Stars,* ed. A.K. Dupree and M.T.V.T. Lago, NATO ASI Series (Dordrecht: Kluwer Academic Publishers), p. 153.

Pudritz R., Norman C. 1986, *Ap.J.* **301**, 571.

Reipurth B. 1989, in Proc. ESO Workshop *Low Mass Star Formation and Pre-Main Sequence Objects,* ed. B. Reipurth (Garching: ESO), p. 247.

Rutten R.G.M., Schrijver C.J. 1987, *Astr.Ap.* **177**, 155.

Saar S.H. 1988, *Ap.J.* **324**, 441.

Sargent A.I., Beckwith S. 1987, *Ap.J.* **323**, 294.

Shu F.H., Adams F.C., Lizano S. 1987, *Ann.Rev.Astr.Ap.* **25**, 23.

Shu F.H., Lizano S., Ruden S.P., Najita J. 1988, *Ap.J.(Letters)* **328**, L19.

Skrutskie M.F., Dutkevitch D., Strom S.E., Edwards S., Strom K.M., Shure M.A. 1989, *Astr.J.* **99**, 1187.

Smith B., Terrile R. 1984, *Science* **226**, 1421.

Snell R.L., Bally J.B. 1986, *Ap.J.* **303**, 683.

Stine P.C., Feigelson E.D., André P., Montmerle T. 1988, *Astr.J.* **96**, 1394.

Strom K.M., Strom S.E., Edwards S., Cabrit S., Skrutskie M. 1989, *Astr.J.* **97**, 1451.

Strom K.M., Strom S.E., Wilkin F.P., Carrasco L., Cruz-Gonzalez I., Recillas E., Serrano A., Seaman R.L., Stauffer J.R., Dai D., Sottile J. 1990, *Ap.J.*, in press.

Tagger M., Henriksen R.N., Sygnet J.F., Pellat R. 1990, *Ap.J.(Letters),* in press.

Uchida Y., Shibata K. 1985, *Pub.A.S. Japan* **37**, 515.

Walker C.K., Lada C.J., Young E.T., Margulis M. 1988, *Ap.J.* **332**, 335.

Walter F.M., Brown A., Mathieu R.D., Myers P.C., Vrba F.J. 1988, *Astr.J.* **96**, 297.

Walter F.M., Kuhi L.V. 1984, *Ap.J.* **284**, 194.

Weintraub D.A., Sandell G., Duncan W.D. 1989, *Ap.J.(Letters)* **340**, L69.

Wilking B.A., Lada C.J., Young E.T. 1989, *Ap.J.* **340**, 823.

Wilking B.A., Mundy L.G., Blackwell J.H., Howe J.E. 1989, *Ap.J.* **345**, 257.

Worden S.P., Schneeberger T.J., Kuhn J.R., Africano J.L. 1981, *Ap.J.* **244**, 520.

Wright E.L. 1987, *Ap.J.* **320**, 818.

Zinnecker H. 1989, in Proc. ESO Workshop *Low Mass Star Formation and Pre-Main Sequence Objects,* ed. B. Reipurth (Garching: ESO), p. 447.

Magnetized Disk-Winds and the Origin of Bipolar Outflows

M. Camenzind

Landessternwarte Königstuhl, D-6900 Heidelberg, Fed. Rep. of Germany

Summary. Interstellar magnetic fields play certainly a distinct role in star formation and in galactic molecular disks. These magnetic fields are, however, also a key ingredient for accretion disks around young stellar objects and in the nuclei of galaxies. In these objects we find a strong observational link between accretion disks and bipolar outflows in the form of winds and jets.

In turbulent disks, the magnetic fields evolve according to the induction equation of mean field electrodynamics. In case of axisymmetric configurations, this leads to a coupled system of partial differential equations for the poloidal magnetic flux and the toroidal magnetic field including the effects of advection, spatial diffusion, differential rotation and helicity of the turbulence. In geometrically thin accretion disks, the lowest axisymmetric dynamo mode which can grow exponentially until equipartition with turbulence, is the quadrupolar field. Dipolar fields are found to decay exponentially.

The magnetic fields exterior to the disk have the structure of a rotating force-free magnetosphere. The shape of the poloidal field lines follows here from the solution of the relativistic Grad-Schlüter-Shafranov equation. The velocity and density in open magnetospheres is then essentially determined by the structure of the magnetic flux tubes. The luminosity of the wind is given by the magnetic luminosity of the disk surface. Magnetized winds ejected from the surface of the disk will be collimated by magnetic effects on scales typically larger than the light cylinder radius for these objects. The VLBI-jets of Quasars have therefore a radius of a fraction of a light year and reach a highly relativistic motion due to the extremely strong magnetization in the corona of the accretion disk. Protostellar jets have moderate velocities and jet radii of a few hundred AU.

1. Introduction

The various activities shown by young stellar objects (YSOs), some cataclysmic variables and LMXBs, and in particular by the nuclei of Quasars, Seyfert galaxies and radio galaxies are commonly attributed to the presence of large accretion disks around the central mass condensations in these objects. There is a great similarity in the physics involved in these various objects, since the structure of steady accretion disks has a deep scaling behaviour in the sense that temperature, density and pressure in the disk scale appropriately with mass and accretion rate.

Magnetic fields play certainly some distinct role in star formation and galactic disks. But the great majority of accretion disk theory is still gas dynamical rather than magnetohydrodynamical. In contrast, our experience with turbulent space plasmas suggests that magnetic fields are of primary importance also for the theory of accretion disks. In this review talk, we briefly summarize the observational link between the existence of accretion disks and of bipolar outflows in the same type of objects. Large scale magnetic fields built up by the underlying accretion disk will launch a pair of bipolar outflows (or jets) normal to the disk. This is a natural explanation for the observed link between disks and bipolar outflows. The conditions for a possible amplification of magnetic fields in a disk are therefore crucial for the origin of bipolar outflows. We show in particular that dynamos acting in the boundary layer between the accretion disk and the central object are the most likely origin for these outflows and that the winds are driven away by strong toroidal magnetic fields in the boundary layer. Magnetic forces will collimate the outflows on scales much larger than the scale of the boundary layers, disk-winds are collimated into disk-jets. We show at the end that the optical jets observed in T Tauri stars are very similar to the parsec-scale radio jets of Quasars observed on the VLBI-scale – except for their velocities which only depend on the strength of the magnetization of the jet-plasma.

The importance of magnetic fields for accretion disks and the formation of jets has been discussed in various reviews which include Blandford (1989), Camenzind (1988, 1989a, b), Coroniti (1984), Königl (1989), Pudritz (1989), Zeldovich, Ruzmaikin and Sokoloff (1983). The simulations for axisymmetric dynamo action shown in the following are results from our 2D dynamo code MAGDISK with details given in Camenzind and Lesch (1990).

2. Observational Link between Accretion Disk and Bipolar Outflows

Accretion disks are present in various astrophysical systems: protostellar systems, cataclysmic variables, binary systems containing neutron stars and in the centers of active galaxies (Seyfert galaxies, Quasars and Radio Galaxies). A great fraction of these systems also show bipolar outflows with varying wind velocities and mass-loss. In this talk we discuss a completely general scenario for the origin of accretion-driven outflows in protostellar systems and AGNs. It turns out in fact that protostellar systems and the centers of galaxies have many properties in common.

2.1 Protostellar Systems

There is now great evidence that low-mass YSOs ($M_* \leq 2$ M$_\odot$) are surrounded by circumstellar disks with masses $M_D \simeq 0.001 - 1.0$ M$_\odot$ on scales $\simeq 10^2 - 10^3$ AU (Beckwith et al., 1990). These disks extend in some systems even downward to the stellar surface (Bertout, 1989), but Beckwith et al. (1990) found in general an emission gap between the stellar surface and the inner edge of the disk. Observations on YSOs provide evidence for the simultaneous presence of accretion disks and energetic outflows in these sources and they point to a relationship between the outflows and the possible role of magnetic fields in disks and outflows (for many details see Königl, 1989). The relevant cornerstones of the observations can be summarized as follows (Appenzeller and Mundt, 1989; Bertout, 1989). YSOs are frequently found to drive energetic, bipolar outflows which sweep up the ambient gas into expanding shells. These sources also show collimated high-velocity ($v_j \simeq 200$ - 1000 km/s) ionized jets (see Mundt, 1988) with mass-loss rates of $10^{-7} - 10^{-10}$ M$_\odot$ yr^{-1}. There are also neutral winds present in these sources with higher mass-outflow rates, but with lower velocities, $v_w \simeq 10 - 50$ km s^{-1}. Since L_*/c, where L_* is the bolometric luminosity, is typically much smaller than the estimated momentum discharge in the wind, this indicates that the outflow is not driven by radiation pressure. There is also direct evidence for a link between circumstellar disks and energetic outflows in YSOs. Naked T Tauri stars (WTTS) are very similar to classical T Tauri stars (CTTSs) except they do not show any infrared excess and UV-excess, which are signatures for efficient accretion from some circumstellar disks. They also lack the strong low-ionization and forbidden emission lines as well as the P Cygni line profiles which are the indicator for energetic outflows (Appenzeller and Mundt, 1989). This simultaneous absence of evidence for inner disk accretion and strong winds in WTTSs points to a likely connection between inner disks and disk-winds. The properties of CTTSs suggest therefore that the outflows are powered by accretion. There is also evidence that this association between accretion disks and energetic outflows extends to higher luminosities as well (Strom, Strom and Edwards, 1988; Poetzel et al., 1989; Hessman et al., 1990). Circumstellar disks also affect the forbidden lines in T Tauri stars which often show a systematic blue shift (Appenzeller, 1989; Edwards et al., 1989). These line shifts are generally attributed to obscuration in the receding part of TT winds by the disk and therefore to a predominance of emission from the blueshifted part of the flow.

The energetic winds emanating from young stellar objects manifest themselves as high-velocity molecular flows, and in the form of strongly collimated optical jets (Lada, 1985; Fukui, 1989; Mundt et al., 1989). The mechanical luminosities in these flows are in the range of 0.002 - 0.1 of the luminosity L_* of the underlying star. The collimation of the molecular flows may occur within 0.02 pc of the central source. On the other hand, the optical jets are highly collimated (with opening angles of $5^0 - 10^0$) and appear to be collimated at radii ≤ 100 AU from the central star. Since the momentum flux exceeds L_*/c by factors of 100 - 1000, the flow is not driven by radiation pressure. Nowadays, the idea is generally adopted that magnetic forces are involved in the driving mechanism (Hartmann and McGregor, 1982; Draine, 1983; Pudritz and Norman, 1986; Shu et al., 1988). These magnetic fields themselves can provide also collimation of the flow (Blandford and Payne, 1982).

That these outflows originate from magnetic effects in disks follows from various observational facts. Polarization measurements revealed that the axis of the optical jets as well as of the swept-up molecular lobes are aligned within $\simeq 30^o$ of the ambient field direction for bipolar outflow sources (Appenzeller, 1989). This suggests that the magnetic field of molecular clouds influences the orientation of the circumstellar disks. These fields can therefore not be neglected in the accretion process, since differential rotation and advection will restructure and amplify the seed fields.

2.2 Accretion Disks in AGN

Direct evidence for the existence of accretion disks in AGN follows from the presence of a strong optical-UVX excess in the continuum of these sources (the so-called "Big Blue Bump", Malkan, 1989; Sun and Malkan, 1989; Ulrich, 1989). This bump cannot be fitted in terms of a single temperature blackbody (Laor and Netzer, 1989). The innermost part of these disks is somewhat hotter than the temperature expected from the boundary layer in the protostellar systems, $T_{max} \simeq 30'000 - 100'000$ K. Since the temperature of standard disks scales as $T_D(R) \propto R^{-3/4}$, the temperature of disks in AGN also drops below the ionization limit for about 100 inner radii. The outer parts of the disks in AGN are then also relatively cool and consist of dusty matter similar to the disks in protostellar systems (Fig. 1).

Antonucci and Miller (1985) demonstrated that a torus of obscuring material completely hides a type 1 Seyfert nucleus from direct view in the case of the type 2 Seyfert NGC 1068. This model has been confirmed recently by X-ray observations (Elvis and Lawrence, 1988; Koyama, 1989). The torus of NGC 1068 is thick enough to block the 2 – 10 keV X-rays from direct view, implying an absorbing column density of 10^{25} atoms cm^{-2}. These central condensations of molecular gas are considered to be the source of accretion

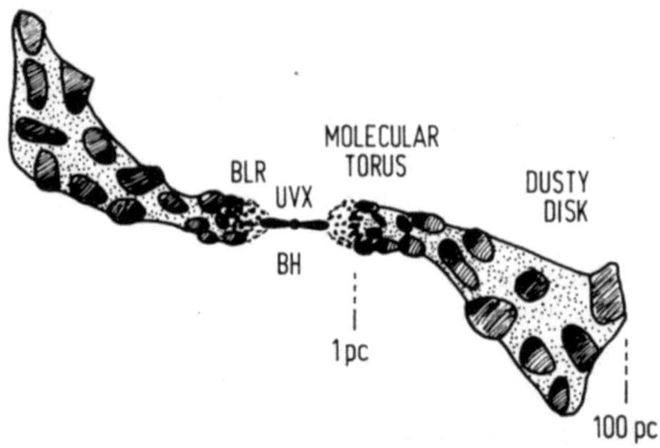

Fig. 1. Schematic view of the disk in AGN. The hot inner disk around a supermassive black hole (BH) produces the "big blue bump" (UVX), the broad emission line clouds (BLR) originate from the cooler middle part of the disk, and the infrared bump of radio-quiet objects may be emitted by dust in the outer part of the molecular torus. This molecular gas is also the fuel for the inner accretion disk.

fuel for the nucleus or for starburst activity (Fig. 1). It is still unclear how this gas is assembled on the parsec-scale from the much larger kiloparsec-scale (Krolik and Begelman, 1988). In this model, the classification of Seyfert types is determined by the angle of the line of sight to the torus: type 1 Seyferts are those viewed face-on, type 2 Seyferts are those viewed edge-on. A similar classification should hold for radio-weak quasars (Kawara et al., 1990). Barthel (1989) has also proposed that the distinction between radio galaxies and radio-loud quasars is based on the angle of the line of sight to the torus: quasars are those sources seen face-on, radio-galaxies are those viewed edge-on. In the torus models of Krolik and Begelman (1988) and Sanders et al. (1989), molecular clouds are destroyed at $\simeq 1$ pc from the center, where the grains evaporate, producing in this way the famous 3.5 μm bump (the evaporation temperature of silicate grains is $\simeq 1500$ K). When the dust sublimates, the molecular gas loses its primary opacity and cooling agent, the temperature rises above $\simeq 3000$ K, and most molecules are destroyed. The observed H_2 emission of type 1 Seyferts and Quasars (Kawara et al., 1990) would come from the inner region of the molecular torus, while CO emission has its origin in the outer parts of molecular gas concentrations. The infrared bump of radio-quiet objects is then the result of dust emission from the outer parts of the disk (Sanders et al., 1989). Part of the infrared bump could however be due to non-thermal emission from magnetized coronae around the inner accretion disk (see later on). The angular sizes of these two emission regions differ by a factor of $\geq 10^4$.

Apart from the continuum emission of the disks in AGN, there is also direct kinematic evidence for rotation in a few objects. These objects show the classical double-peaked, broad emission lines which are considered to be characteristic for Keplerian rotation. The prime candidates are 3C 390.3 (Oke, 1987; Perez et al., 1988) and Arp 102B (Halpern and Filippenko, 1988; Chen et al., 1989). Double-peaked difference spectra were also found in the Seyfert galaxy Akn 120 (Alloin et al., 1988) and in NGC 5548 (Stirpe et al., 1988). Standard thin disks do, however, not release sufficient gravitational energy locally at the radii where the Keplerian velocity corresponds to the observed width of the emission lines (typically at $\simeq$ 1000 Schwarzschild radii). Some extra source of energy is therefore needed (Collin-Souffrin, 1987).

Similar to protostellar systems, a subclass of AGN shows extremely energetic outflows extending even to scales beyond the outer edge of the galaxy in the form of strongly collimated radio jets (Bridle and Perley, 1984; Perley, 1989). The luminosities of these objects is an appreciable fraction of the luminosity of the underlying central object, and the flow speeds in these jets is at least semi-relativistic with a preponderance of extremely relativistic flows in superluminal sources, $\gamma_{\rm jet} \geq 2 - 20$ (Porcas, 1988; Zensus, 1989). As shown by VLBI observations, the collimation of these jets must occur on scales in general less than a fraction of a light year from the central source. It is generally accepted that magnetic forces are involved in the driving mechanism (Blandford and Payne, 1982; Blandford, 1989; Camenzind, 1986a, 1988, 1989b), and that the magnetic fields will also provide the collimation of the flow, since huge currents are in general involved in the jet flows (Benford, 1978, 1987; Camenzind, 1987b, 1989b; Heyvaerts and Norman, 1989; Lesch et al., 1989).

Magnetic fields are therefore very important for the understanding of the origin of bipolar outflows from disk-systems. Concerning the structure of these magnetic fields, there are two main questions. First, we would like to understand the origin of the fields in accretion disks and to find out the difference between sources with strong jets (such as Quasars) and sources with weak jets (QSOs and Seyfert galaxies). Secondly, once we know the structure of the magnetic fields in the disk, we have to elaborate the structure of the winds driven by these magnetic fields and to understand the collimation and acceleration mechanism.

3. The Origin of Magnetic Fields in Accretion Disks

The interstellar matter which accretes onto protostellar objects and supermassive objects in the centers of galaxies carries along magnetic fields. The evolution and influence of these magnetic fields on the structure of the disk are completely neglected in the theory of standard accretion disks. The evolution of magnetic fields in the disk depends on the nature of the electrical

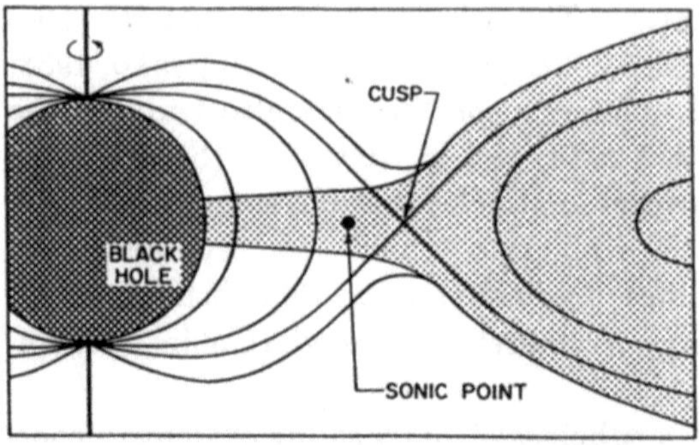

Fig. 2. A meridional cross-section through the boundary layer of a rotating black hole. The equipotential surfaces around a Kerr black hole show a cusp-like structure near the marginal stable orbit. The self-crossing surface forms a kind of Roche lobe. If the accretion disk overflows this Roche lobe, matter is accreted onto the horizon supersonically (Abramowicz and Zurek, 1981). The frame-dragging effect requires the plasma to corotate with the black hole near the horizon.

conductivity in the disk. At high temperatures Ohmic resistivity dominates in the disk, while in weakly ionized molecular gas ambipolar diffusion is an important process (the drift of ions and electrons, to which the magnetic field is frozen, relative to the neutrals in a weakly ionized medium). In a fully turbulent disk, the effective conductivity is however much lower due to turbulent processes.

3.1 Boundary Layers

Consider a steady accretion disk with accretion rate $\dot{M}$ around a central object of mass M and radius R_*. Throughout most of the disk the angular velocity of the disk material is Keplerian, $\Omega = \Omega_K(R)$. Between the central object and the disk we find in general a boundary layer (BL), in which the angular velocity varies between $\Omega_K(R_*)$ and Ω_*. For a black hole, $\Omega_* = \Omega_H$ is the angular velocity of the black hole ($\Omega_H = 0$ for a Schwarzschild black hole, $\Omega_H = a/(r_H^2 + a^2)$ for a Kerr black hole with angular momentum a). In purely hydrodynamical models, the width ΔR of the BL is small, $\Delta R \simeq H^2/R$, when H is the disk-thickness (Kley, 1989). For rotating black holes, this width is much broader, extending to a few Schwarzschild radii (Fig. 2), since there are no stable Keplerian orbits between the horizon and the marginally stable orbit (Breuer, 1975). Relativity, however, requires the accreting plasma to corotate with the black hole near the horizon, quite similar to accretion onto a non-collapsed object. The only difference is that the plasma flows supersonically through the horizon (Abramowicz and Zurek, 1981).

The energy generated in a Newtonian boundary layer is

$$L_{\mathrm{BL}} \simeq \frac{GM\,\dot{M}}{2R_*}\left(1-\frac{\Omega_*}{\Omega_K}\right)^2 \tag{1}$$

with $L_{\mathrm{BL}} \simeq L_D$ for the disk-luminosity when $\Omega_* \ll \Omega_K(R_*)$,

$$L_D \simeq \frac{GM\,\dot{M}}{2R_*}\,. \tag{2}$$

This is half of the available accretion luminosity L_{acc}

$$L_{\mathrm{acc}} = \frac{GM\,\dot{M}}{R_*}\,. \tag{3}$$

The boundary layer around black holes must be treated relativistically. This is a completely unsolved problem.

The discussion of BLs must be based on some assumption about the viscosity. The standard picture of viscosity in disks relies on some kind of turbulent viscosity with turbulent eddies of maximal size $L_T \simeq H$ and turbulent speeds $v_T \leq c_S$, where c_S is the local sound speed. This gives rise to the Shakura and Sunyaev (1973) α-parametrization for the kinematic viscosity

$$\nu_T \simeq \alpha\, c_S\, H \simeq \alpha\, \Omega\, H^2(R) \quad , \quad \alpha = \frac{L_T}{H}\frac{v_T}{c_S} \leq 1\,. \tag{4}$$

In the boundary layer itself the length scales become typically smaller. Here one needs some formula for ν_T which takes the reduced length scales into account (Tylenda, 1981). The α-formula also applies for magneto-hydrodynamic turbulence, $v_A \leq c_S$, when $v_A = \sqrt{B^2/4\pi\rho}$ is the Alfvén speed in the disk.

3.2 Axisymmetric Dynamos in Disks

Fully turbulent disks are ideal set-ups for magnetic dynamos, quite similar to the dynamo action in the disks of spiral galaxies (Parker, 1971; Ruzmaikin et al., 1988; Krause et al., 1990). The essential ingredients for $\alpha\Omega$-dynamos are strong shear plus a density gradient perpendicular to the shear. Any small seed field which is initially present will decay exponentially or will be amplified by the dynamo until it reaches a maximum value with $v_A \simeq c_S$, i.e. $\alpha \simeq 1$. This would indicate that the magnetic contribution to the viscosity is dominant.

In turbulent disks, the magnetic fields evolve according to the induction equation of mean field electrodynamics (Moffatt, 1978; Krause and Rädler, 1980; Pudritz, 1981)

$$\partial_t \boldsymbol{B} + (\boldsymbol{u}\cdot\nabla)\,\boldsymbol{B} = (\boldsymbol{B}\cdot\nabla)\boldsymbol{u} - (\nabla\cdot\boldsymbol{u})\,\boldsymbol{B} + \nabla\wedge\left\{\mathcal{E} - \eta_T\,\nabla\wedge\boldsymbol{B}\right\}. \tag{5}$$

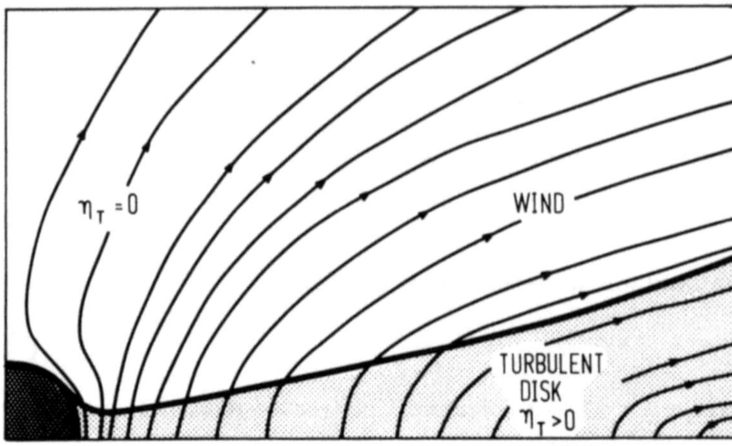

Fig. 3. **The general set-up of a fully turbulent accretion disk around some collapsed or non-collapsed object. Strong differential rotation and turbulence in the disk will generate large-scale magnetic fields which build up a rotating magnetosphere around the disk. Plasma from the disk can flow into this magnetosphere and initiate strong outflows.**

$\boldsymbol{B}$ is the mean magnetic field, and $\boldsymbol{u}$ the mean velocity field in the disk, $\boldsymbol{u} = (u_R, R\Omega, 0)$. η_T is the total magnetic diffusivity including the microscopic one and the turbulent diffusivity. $\mathcal{E} = \alpha_D\,\mathbf{B}$ represents the electromotive force induced by correlations between turbulent motions and turbulent magnetic fields. For axisymmetric fields, the above equation can be written as a coupled system for the poloidal magnetic flux $\Psi(t,R,z)$ and the poloidal current flux $T(t,R,z) = R\,B_\phi(t,R,z)$ (Camenzind, 1990b)

$$B_R = -\frac{1}{R}\frac{\partial\Psi}{\partial z} \quad , \quad B_z = \frac{1}{R}\frac{\partial\Psi}{\partial R}\,, \tag{6}$$

$$\frac{\partial\Psi}{\partial t} + u_R\frac{\partial\Psi}{\partial R} - \eta_T\,R^2\nabla\cdot\left\{\frac{1}{R^2}\,\nabla\Psi\right\} = \alpha_D\,T\,, \tag{7}$$

$$\frac{\partial T}{\partial t} + u_R\frac{\partial T}{\partial R} - R^2\,\nabla\cdot\left\{\frac{\eta_T}{R^2}\,\nabla T\right\} + T\,\frac{R\,\partial(u_R/R)}{\partial R} = R^2\,\mathbf{B}_p\cdot\nabla\Omega$$
$$+R^2\,\nabla\cdot\left\{\frac{\alpha_D}{R^2}\nabla\Psi\right\}. \tag{8}$$

These equations show the redistribution of the fluxes Ψ and T over the effect of spatial diffusion and radial advection. The shear $R\nabla\Omega$ will convert the poloidal field into a toroidal one. The fluxes Ψ and T are regenerated by helicity $\alpha_D(R,z)$ in the turbulent fluid (Ruzmaikin et al., 1988; Rüdiger, 1990)

$$\alpha_D = L_T^2\,\Omega\,\bar{\alpha}(z)/H(R)\,. \tag{9}$$

Since differential rotation is extremely strong in boundary layers, where the rotation is adjusted to the rotation of the central object, we expect strong amplification of magnetic fields in such boundary layers. Then also the Parker instability occurs which allows flux to escape from the disk and reconnection to take place. If viscosity is of magnetic origin, the free shear energy in the disk is converted into magnetic energy at the rate (per unit disk area)

$$Q^{+} \simeq \nu_T \, \Sigma \, (R\Omega')^2 \simeq \frac{B^2}{8\pi} v_A \quad , \quad v_A \simeq \alpha H \Omega \, . \tag{10}$$

In equilibrium, this flux must be destroyed or removed from the disk in a kind of magnetized wind.

The solutions of the dynamo equations are in general not stationary. For this reason, we developed the time-dependent simulation code MAGDISK, based on finite elements, for the solution of the coupled system (7) and (8) (Camenzind, 1990b; Camenzind and Lesch, 1990) and for given structure functions $\eta_T(R,z)$, $\Omega(R,z)$, $\alpha_D(R,z)$ and $u_R(R)$. When the dynamo equations are normalized for the inner edge of the accretion disk ($R \rightarrow R/R_{\rm in}$, $z \rightarrow z/R_{\rm in}$, $t \rightarrow t/t_D$, $t_D = R_{\rm in}^2/\eta_{T,\rm in}$, $u_R = -u_{\rm in}\, u(R/R_{\rm in})$, $\alpha_D = \alpha_{D,\rm in}\, \bar{\alpha}_D$)

$$\frac{\partial \Psi}{\partial t} - R_m \, u \, \frac{\partial \Psi}{\partial R} - \bar{\eta}_T \, R^2 \nabla \cdot \left\{ \frac{1}{R^2} \nabla \Psi \right\} = R_\alpha \, \bar{\alpha}_D \, T \, , \tag{11}$$

$$\frac{\partial T}{\partial t} - R_m \, u \, \frac{\partial T}{\partial R} - R^2 \, \nabla \cdot \left\{ \frac{\bar{\eta}_T}{R^2} \nabla T \right\} - R_m \, T \, \frac{R \partial (u/R)}{\partial R} = R_\Omega \, R^2 \, \mathbf{B}_p \cdot \nabla \bar{\Omega}$$
$$+ R_\alpha \, R^2 \, \nabla \cdot \left\{ \frac{\bar{\alpha}_D}{R^2} \nabla \Psi \right\} , \tag{12}$$

the strength of the various terms in these dynamo equations are measured by means of the corresponding dynamo numbers

$$R_\Omega = \frac{R_{\rm in}^2 \, \Omega_{\rm in}}{\eta_{T,\rm in}} = \frac{R^2 \, \Omega}{\alpha \, P_m H^2(R) \Omega} = \frac{1}{\alpha P_m} \left(\frac{H_{\rm in}}{R_{\rm in}} \right)^{-2} , \tag{13}$$

$$R_\alpha = \frac{R_{\rm in} \, \alpha_{D,\rm in}}{\eta_{T,\rm in}} = R_\Omega \left(\frac{H_{\rm in}}{R_{\rm in}} \right) \left(\frac{L_T}{H_{\rm in}} \right)^2 , \tag{14}$$

$$R_m = \frac{R_{\rm in} \, u_{R,\rm in}}{\eta_{T,\rm in}} = \frac{1}{P_m} = \alpha \, R_\Omega \left(\frac{H_{\rm in}}{R_{\rm in}} \right)^2 . \tag{15}$$

α is the parameter of standard accretion disks (Eq. (4)) and P_m the magnetic Prandtl number, $\eta_T = P_m \, \nu_T$, $P_m \simeq 1$. For geometrically thin accretion disks, the dynamo numbers have the following relations (characteristic for $\alpha\Omega$-dynamos)

$$R_m < R_\alpha < R_\Omega \, . \tag{16}$$

In geometrically thick disks, differential rotation and helicity would be of the same order of magnitude, $R_\Omega \simeq R_\alpha \simeq R_m/\alpha$ ($\alpha^2 \Omega$-dynamos).

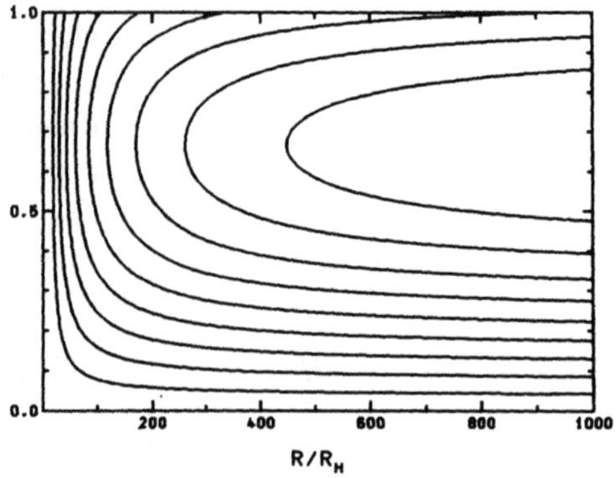

Fig. 4a. Poloidal field lines for stationary solutions in geometrically thin accretion disks without dynamo effect, $R_\alpha = 0$, $R_\Omega = 600$, $R_m = 200$, R_H: horizon radius, $\zeta = z/H$ (Khanna and Camenzind, 1990).

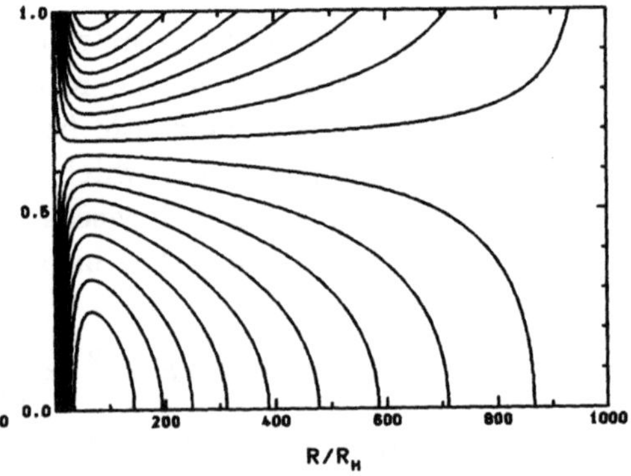

Fig. 4b. The distribution of the poloidal currents for the solutions, $RB_\phi = const$ (in units of Gauss cm).

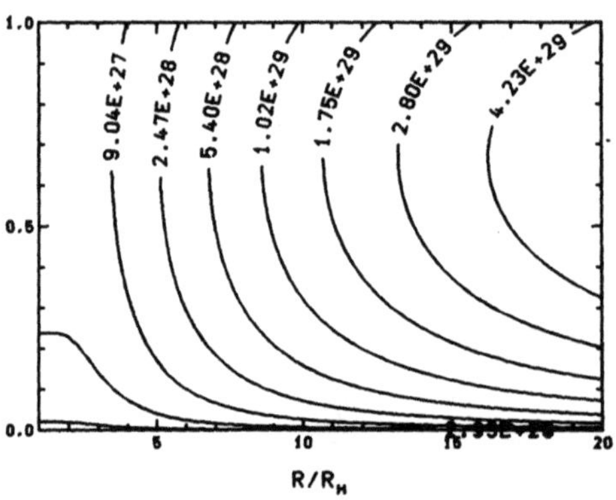

Fig. 4c. Inner part of the poloidal field structure shown in Fig. 4a (Ψ in units of Gauss cm^2).

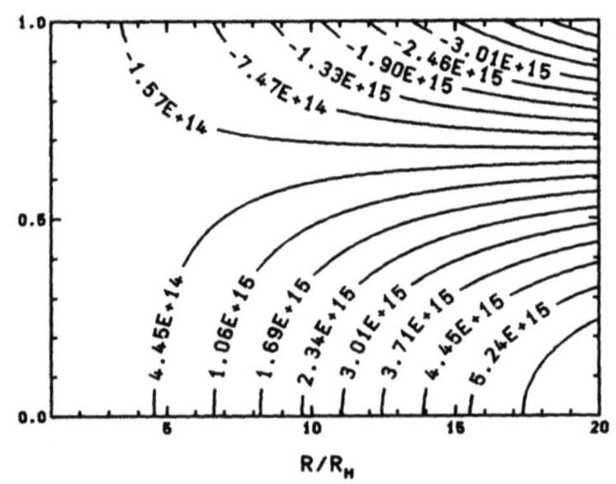

Fig. 4d. The inner part of the poloidal current structure of Fig. 4b (RB_ϕ in units of Gauss cm).

3.3 Stationary Solutions

When the α-effect is neglected in a disk, $\alpha_D = 0$ throughout the disk, the strong differential rotation amplifies the toroidal field. The poloidal flux Ψ is however only redistributed in the disk by advection and diffusion (there is no source term for Ψ). In this case, stationary solutions are generic and have been discussed by Lovelace et al. (1987) and Königl (1989). In Fig. 4 we show solutions for $R_\alpha = 0$ in the thin disk approximation for a R-dependent magnetic diffusivity $\eta_T(R)$ around supermassive black holes (Khanna and Camenzind, 1990). In this case, accretion from the molecular torus advects a total magnetic flux $\Psi_\infty (= 5.7 \times 10^{30}$ Gauss cm^2) inwards and builds up a global quadrupolar field geometry which can also close to the horizon (Fig. 4c). The negative radial magnetic field near the equatorial plane produces there a positive toroidal field (Fig. 4b), and the positive radial magnetic field in the upper part of the disk is converted into a negative toroidal field (Fig. 4d). The maximum of B_ϕ occures near 30 R_H. The total current driven by differential rotation is $\simeq 10^{17}$ Ampère corresponding to a maximal toroidal field of $\simeq 10$ Gauss.

This example of a thin disk solution of the stationary dynamo equations shows that the magnetic field has not a self-similar structure (as assumed a priori e.g. by Königl (1989)). It also demonstrates that advection and diffusion alone cannot produce a flux distribution which is highly peaked near the inner edge of the disk (compare with the true dynamo solution in Fig. 6). For $R_\alpha = 0$, one can also construct similar solutions for dipolar configurations.

3.4 Disk-Corona Systems

The stationary solutions discussed in the previous section are in general not stable against the inclusion of the dynamo effect (i.e. for a non-vanishing α_D). From the experience with spherical dynamos we know that in systems with strong differential rotation ($R_\Omega \gg R_\alpha$) axisymmetric quadrupole modes have the highest probability to be excited, while for $\alpha^2\Omega$-dynamos also axisymmetric dipolar modes can grow (Brandenburg et al., 1990). In geometrically thin disks we expect therefore that dipolar modes would decay, but quadrupolar modes to grow under certain conditions (Brandenburg et al., 1990). The behaviour of the disk dynamo depends essentially on the two dynamo numbers R_Ω and R_α (Fig. 5). For each value of R_Ω we find a corresponding critical value for R_α, where the system is stationary. For values of R_α higher than this critical value, the magnetic fields grow exponentially, but they would decay exponentially for values smaller than the critical one. The solid line in Fig. 5 represents this critical curve for axisymmetric quadrupolar modes, the dahsed line for dipolar modes. The exact shape of this critical line is not yet known for accretion disks.

Using our time-dependent solver MAGDISK for the system (11) and (12) (Camenzind and Lesch, 1990), we found one critical point for $\alpha P_m = 1/3$ and $H_{\rm in}/R_{\rm in} = 0.1$ ($R_\Omega = 300$, $R_\alpha = 30$ and $R_m = 3$) which is denoted as a star in Fig. 5. Since for disks, $R_\alpha = \sqrt{R_\Omega/\alpha P_m}$, disk models with fixed values of αP_m would move along the thin solid lines in Fig. 5, when $H_{\rm in}$ is varied. Large scale magnetic fields would not exist in the region below the critical line.

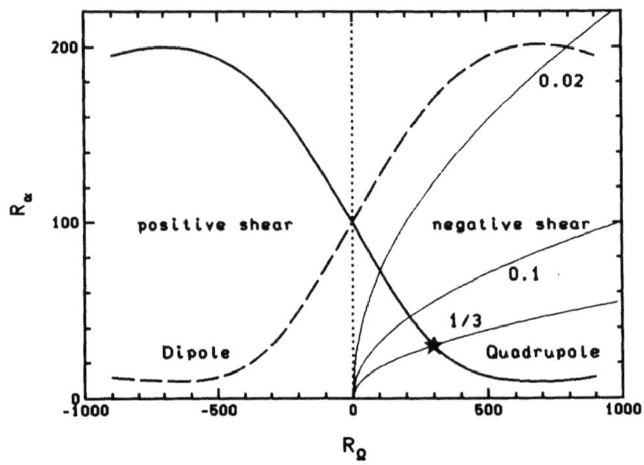

Fig. 5. The dynamo number plane (R_Ω, R_α) for axisymmetric modes in accretion disks. $R_\Omega > 0$ means a negative shear, $R\partial_R \Omega < 0$. Axisymmetric modes can only grow above the solid line for the corresponding mode. The exact critical dynamo numbers are not yet known for turbulent accretion disks. The star marks one point we found in our simulations. Quadrupolar modes grow exponentially above the solid line, dipolar modes above the dashed line. Accretion disks move along the thin solid lines for given αP_m, when the disk height H is varied.

In Fig. 6 we show the solutions of the dynamo equations for a disk-corona system, where the turbulent diffusivity η_T vanishes in the corona. For $R_\Omega = 300$, $R_\alpha = 30$ and $R_m = 3$, the quadrupolar field is marginally stationary and grows exponentially for higher dynamo numbers (Fig. 5), but would decay exponentially for much lower dynamo numbers. Differential rotation in the corona also excites a positive toroidal field there which grows to the same strength as the negative toroidal field in the disk itself. We find that the global behaviour of the magnetic fields depends largely on the z-dependence of η_T, quite similar to the behaviour we found for galactic dynamos (Camenzind and Lesch, 1990).

For supercritical dynamo numbers, seed fields in the disk are rapidly amplified and grow in general exponentially, until equipartition with the turbulence is reached. Realistic solutions of the dynamo equations must include therefore the feedback of the magnetic fields on the evolution of the turbulence in the disk and corona and the feedback on rotation and radial motion (Heyvaerts and Priest, 1989). A configuration with $B_\phi \gg B_p$ is in general unstable to magnetic buoyancy. The entire azimuthal flux tube is lifted up above the disk and now forms a low-β hot corona ($\beta = P_g/P_m$) (see also Shibata et al., 1989; Matsumoto et al., 1990). This interaction between the disk-corona and the disk itself will saturate the exponential growth of the disk-fields. For extremely high dynamo numbers, more complicated magnetic field distributions with various nodes can be excited. As shown by Eq. (13), this will occur for extremely thin disks or for low α-values.

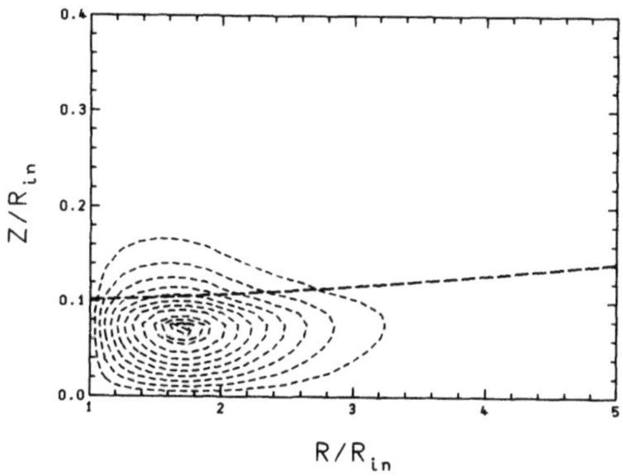

Fig. 6a. Poloidal field distribution in geometrically thin disks with parameters $\Omega_* = 0.8\Omega_K(R_*)$ for the dynamo numbers $R_\Omega = 300$, $R_\alpha = 50$, $R_m = 3$, $t = 1.5t_D$, $H_{\rm in}/R_{\rm in} = 0.1$.

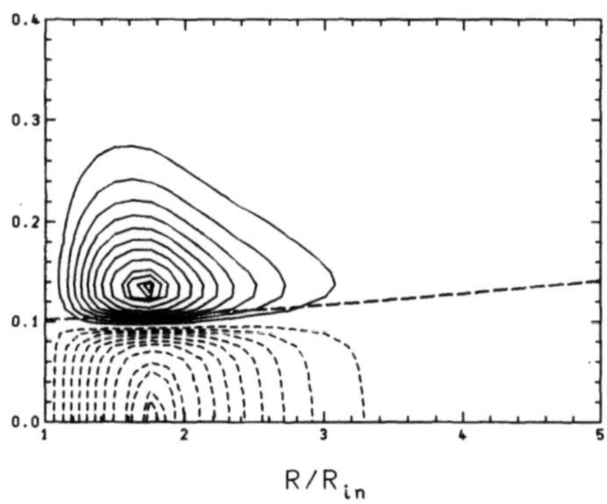

Fig. 6b. The toroidal field distribution for Fig. 6a. The dashed line marks the scale height of the disk as used in the simulation.

4. The Structure of Axisymmetric Magnetized Outflows

Dynamo action in the disk forms a rotating magnetosphere which will be deformed by outflowing corona plasma. The poloidal plasma flow must, therefore, be included in the description of the exterior solutions. Exterior to the disk the magnetic structure has to be modeled as a wind magnetosphere. The rapid rotation of inner disks in AGN forces us to use general relativistic MHD on a Kerr background (Camenzind, 1986a, b, 1987, 1989a, b; Haehnelt and Camenzind, 1990). The following treatment of the problem, however, neglects the influence of gravity. Since we are mainly interested in the equilibrium configurations for these wind magnetospheres, we can use a stationary approach. It is well known that in this case the plasma has to flow parallel to the magnetic field, i.e. the poloidal velocity $\mathbf{u}_p$ is proportional to the poloidal magnetic field $\mathbf{B}_p$

$$\mathbf{u}_p = \frac{\eta}{n}\mathbf{B}_p \quad , \quad u^\phi = \left(\frac{\eta}{n}B_\phi + R\Omega^F\right) . \tag{17}$$

n is the particle density in the flow and η is constant along the magnetic flow lines ($\mathbf{B}\cdot\nabla\eta = 0$). This relation means that we know the flow direction, once we know the structure of the magnetosphere. With these quantities we define the special relativistic Mach number M (Camenzind, 1986a)

$$M^2 = \frac{4\pi\mu\, n\, u_p^2}{B_p^2} \quad , \quad \mu = \frac{\rho + P}{n} \simeq m_p c^2 , \tag{18}$$

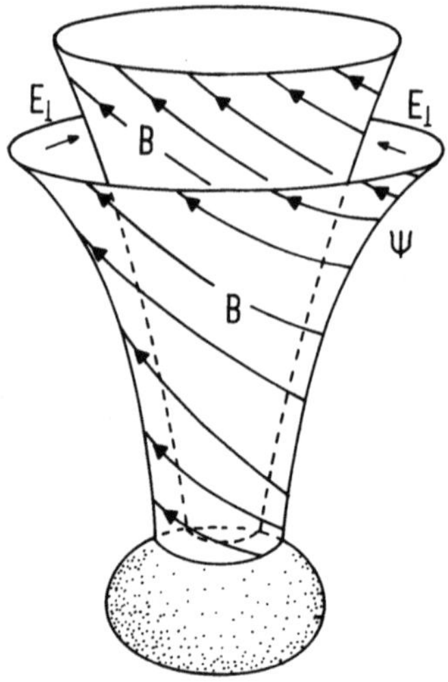

Fig. 7. The structure of relativistic magnetized winds. The wind ejected by rotating objects consists of a nested family of magnetic surfaces. Due to the rotation, toroidal fields are also produced. This rotation of the field lines provides an electric field perpendicular to the magnetic surfaces. The wind plasma is flowing along these magnetic surfaces

which tells us, whether the flow is strongly magnetized ($M^2 \ll 1$) or weakly magnetized ($M^2 \geq 1$). For low velocities near the disk surface, the magnetic field controls the flow, but far away from the disk the flow becomes super-Alfvénic.

4.1 The Concept of Magnetic Surfaces

The essential ingredient for the global structure of axisymmetric hydromagnetic flows is the concept of magnetic surfaces. Due to axisymmetry, hydromagnetic flows consist of a family of nested magnetic surfaces $\Psi = const$, where the quantity $2\pi\Psi$ describes the magnetic flux enclosed by a cross section, which cuts the corresponding magnetic surface and the rotational axis (see Fig. 7). In the high conductivity limit, plasma is forced to flow along these surfaces, and these surfaces are also electric potential surfaces, i.e. the electric field $\mathbf{E}$ is orthogonal to the magnetic surface

$$\mathbf{E} = -\frac{R\Omega^F}{c}\,\mathbf{e}_\phi \wedge \mathbf{B}_p = -\frac{\Omega^F}{c}\,\nabla\Psi\,. \tag{19}$$

The magnetic surfaces have an additional property: they rotate with the angular frequency $\Omega^F(\Psi)$ which is constant along the surface, but varies from surface to surface. Its value is given by the rotation of the magnetic field at its foot point in the disk. This angular velocity defines a light cylinder radius for each field line, $R_L = c/\Omega^F$.

The magnetosphere also carries poloidal currents $\mathbf{j}_p$ which follow from the toroidal magnetic field B_ϕ

$$\mathbf{j}_p = \frac{c}{4\pi R}\,\nabla(RB_\phi)\wedge \mathbf{e}_\phi\,. \tag{20}$$

The toroidal field itself follows from conservation equations ($x = R/R_L$)

$$RB_\phi = -4\pi\,\frac{\eta(\Psi)E(\Psi)}{\Omega^F}\,\frac{x_A^2 - x^2}{1 - M^2 - x^2}\,, \tag{21}$$

where $E(\Psi)$ represents the total energy per particle carried along the wind

$$E(\Psi) = \mu\gamma - \frac{R\,\Omega^F\,B_\phi\,B_p}{4\pi\,nu_p}\,. \tag{22}$$

The first term is the total mass-energy (γ the Lorentz factor of the flow) and the second term represents the magnetic energy in the flow, which vanishes for non-rotating configurations. Similarly, $L(\Psi)$ is the total angular momentum carried in the wind

$$L(\Psi) = \frac{\mu}{c^2}\gamma\,R^2\,\Omega - \frac{R\,B_\phi\,B_p}{4\pi\,nu_p}\,. \tag{23}$$

From expression (21) for B_ϕ we see that the Alfvén point of the flow is in the relativistic case not at $M^2 = 1$ as in Newtonian MHD ($R = R_A$, $x = x_A$), but slightly changed by

$$M_A^2 = 1 - \Omega^F\,L(\Psi)/E(\Psi) \leq 1\,. \tag{24}$$

When the flow is strongly magnetized, $M_A^2 \ll 1$ and therefore $\Omega^F L/E \to 1$, the Alfvén point moves towards the light cylinder, $R_A \to R_L$, and RB_ϕ is then only a function of the magnetic surface

$$RB_\phi = -\frac{2}{c}\,I(\Psi) = -4\pi\frac{\eta(\Psi)\,E(\Psi)}{\Omega^F} = -4\pi\,\eta(\Psi)\,L(\Psi)\,, \tag{25}$$

i.e. in this case the angular momentum distribution determines the current distribution $I(\Psi)$ in the magnetosphere. As a consequence, we find $\mathbf{j}_p \parallel \mathbf{B}_p$,

$$\mathbf{j}_p = -\frac{dI}{2\pi\,d\Psi}\,\mathbf{B}_p\,. \tag{26}$$

In summarizing, we find that magnetized coronae and winds consist of a nested family of magnetic surfaces, the plasma must flow along these surfaces and the poloidal currents also flow along the surfaces, when the flow is strongly magnetized. The next question is therefore to find out the form of these surfaces.

4.2 The Grad-Schlüter-Shafranov Equation (GSS)

One of the essential points of the computation of self-consistent MHD disk-winds is the particular form of the toroidal current j_ϕ, which determines the structure of the wind-magnetosphere via Ampère's equation

$$\nabla \wedge \mathbf{B}_p = \frac{4\pi}{c}\, j_\phi \mathbf{e}_T \quad , \quad \mathbf{B}_p = \frac{1}{R}(\nabla \Psi \wedge \mathbf{e}_T)\,. \tag{27}$$

With the introduction of the magnetic flux function Ψ, which we now use as our stream function, this equation can be transformed into a divergence equation of the form

$$R\nabla \cdot \left\{ \frac{1}{R^2} \nabla \Psi \right\} = -\frac{4\pi}{c}\, j_\phi\,. \tag{28}$$

Plasma confinement in the laboratory and in solar filaments is based on this equation, known as the Grad-Schlüter-Shafranov (GSS) equation. When plasma is moving along the magnetic flux surfaces, the general form of the current j_ϕ follows from a force balance perpendicular to the flux surfaces. For this purpose one considers the relativistic Lorentz force

$$n(\mathbf{u}\cdot\nabla)(\mu\mathbf{u}) + \nabla P = \rho_e\, \mathbf{E} + \frac{1}{c}(\mathbf{j} \wedge \mathbf{B})\,. \tag{29}$$

The projection of this equation perpendicular to the magnetic flux surfaces provides then the current j_ϕ (a derivation is given in Camenzind, 1987a)

$$\begin{aligned} \frac{1}{c} B_p j_\phi (1 - M^2 - x^2) = {} & -\frac{R B_\phi}{4\pi R^2} \nabla_n (R B_\phi) + \rho_{\mathrm{GJ}} E_n \\ & - \mu\gamma n \{\nabla_n \gamma - \Omega \nabla_n l\} - \nabla_n P + \mu\eta B_p^2 \nabla_n (\eta/n)\,. \end{aligned} \tag{30}$$

$\mathbf{n}$ is the normal unit vector of the magnetic surfaces. In the *force-free limit*, where plasma inertia is neglected, $\nabla_n P = 0 = \nabla_n \gamma = \nabla_n l$, we obtain

$$B_p j_\phi (1 - x^2) = -c \frac{R B_\phi}{4\pi R^2} \nabla_n (R B_\phi) + c\rho_{\mathrm{GJ}} E_n\,, \tag{31}$$

so that the current is given as (ρ_{GJ} is the Golreich-Julian density)

$$j_\phi (1 - x^2) = 4c \frac{I(\Psi)}{4\pi R} I'(\Psi) + c\rho_{\mathrm{GJ}} \frac{E_n}{B_p}\,. \tag{32}$$

It is then easy to show that the GSS-equation (28) can be written in the suggestive form of

$$R\nabla \cdot \left\{ \frac{1 - x^2}{R^2} \nabla \Psi \right\} = -\frac{4}{R} I I'\,, \tag{33}$$

which is the *classical pulsar equation* originally derived by Scharlemann and Wagoner (1973). This equation shows that the poloidal current distribution $I(\Psi)$ also determines the effective source for the poloidal fields.

The general GSS equation is a highly non-linear equation, since all the source terms have a complicated dependence on Ψ itself. In particular, $I(0) = 0 = I(\Psi_{\rm disk})$, when the total current flowing in the magnetosphere vanishes. Let us normalize the coordinates in units of light cylinder radii, the flux Ψ in units of the total flux Ψ_D produced by the disk, and the current in terms of its maximal value, $I = I_{\max}\bar{I}$. Then the general GSS equation including inertia effects can be written in a dimensionless form

$$x\nabla \cdot \left\{ \frac{1 - M^2 - x^2}{x^2} \nabla\Psi \right\} + \frac{g_I}{x} \bar{I}\bar{I}' = -\frac{1}{\sigma_0^2} \bar{j}^{\rm inert} , \qquad (34)$$

where the quantity

$$\sigma_0 = \frac{\Psi_D}{4\pi\eta_0 mcR_L^2} \simeq \frac{\Psi_D^2}{2\dot{N}_{\rm wind}\, mc\, R_L^2} \qquad (35)$$

acts as the inverse of a coupling constant for the inertial current $\bar{j}^{\rm inert}$. The source function $\bar{j}^{\rm inert}$ contains all contributions to the toroidal current density j_ϕ due to pressure and inertial effects in the force-balance equation (30). g_I is the coupling constant for the effects of the poloidal currents

$$g_I = \frac{4\,I_{\max}^2\, R_L^2}{c^2\, \Psi_D^2} \simeq 0.04 \left(\frac{I_{\max}}{10^{18}\,\rm Amp} \right)^2 \left(\frac{R_L}{10^{15}\,\rm cm} \right)^2 \left(\frac{\Psi_D}{10^{33}\,\rm G\,cm^2} \right)^{-2} . \qquad (36)$$

When the plasma density is small in the magnetosphere, we find $\sigma_0 \to \infty$, $M^2 \to 0$. In this case, the current distribution $I(\Psi)$ alone determines the form of the magnetic surfaces. This limit holds for AGN ($M > 10^8$ $M_\odot$), where

$$\sigma_0^{\rm AGN} \simeq 10\, \Psi_{D,33}^2\, \dot{M}_{w,-1}^{-1}\, R_{L,15}^{-2} , \qquad (37)$$

for a mass flux $\dot{M}_{\rm wind} \simeq 0.1$ $M_\odot$/yr in the wind and a disk magnetic flux $\Psi_D \simeq 10^{33}$ Gauss cm^2. This term is, however, dominant for stellar winds

$$\sigma_0^* \simeq 10^{-11}\, \Psi_{D,25}^2\, \dot{M}_{w,-8}^{-1}\, R_{L,15}^{-2} . \qquad (38)$$

The computation of wind magnetospheres for AGN can therefore be handled in the force-free limit (33), while those for proto-stellar winds must be done with the full GSS equation (34). In the past few years we developed methods to solve the force-free GSS equation (Camenzind, 1987a)

$$x\nabla \cdot \left\{ \frac{1 - x^2}{x^2} \nabla\Psi \right\} + \frac{g_I}{x} \bar{I}\bar{I}' = 0 , \qquad (39)$$

including also the effects of the gravitational background of a rotating Kerr geometry (Haehnelt and Camenzind, 1990). Unfortunately, there is

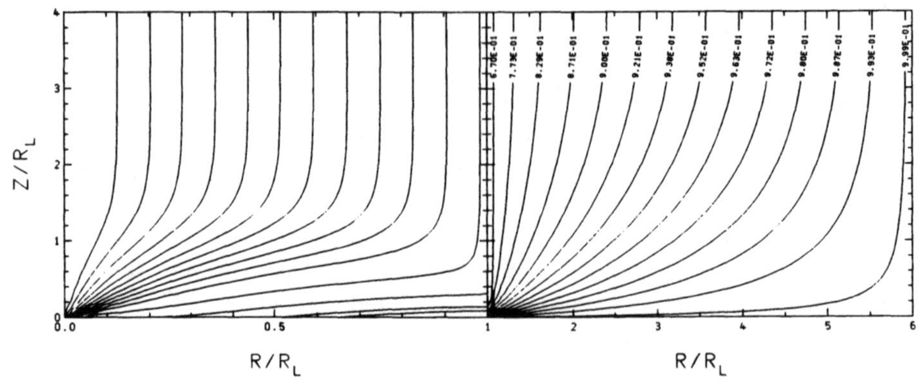

Fig. 8. The structure of the poloidal magnetic flux surfaces of a rotating magnetosphere produced by the accretion disk around a black hole. The outflow is collimated due to the strong pressure in galactic centers and the pinching force of the current-carrying magnetosphere.

not much known about analytic solutions of this equation, except for linear $I(\Psi)$ (Michel, 1982; Lovelace et al., 1987). We solved this equation by fixing the form of $\bar{I}(\Psi)$, and increasing slowly the coupling constant g_I. For low coupling constant, $g_I \ll 10^{-2}$, the magnetosphere has a universal form with field lines perpendicularly crossing the light cylinder (Figs. 8, 9). The solution exterior to the light cylinder is then completely determined by the flux distribution along the light cylinder and the boundary conditions at large distances. In the core of a galaxy, the escaping wind plasma will encounter a finite pressure $P_{\rm ext} \simeq 10^{-2}$ dyn cm^{-2}, which enforces pressure equilibrium of the wind with this ambient medium. The wind is therefore collimated in the direction of the rotational axis outside the light cylinder with a radius $R_{\rm jet}$ given by perpendicular pressure equilibrium (Fig. 8). The same holds for disk-winds in molecular clouds, but here the collimation already starts beyond the Alfvén radius $R_A \simeq 10\,R_{\rm in} \simeq 1$ AU (see Sec. 5.2).

4.3 Magnetized Winds

The terminal speed v_∞ of a cold MHD wind which flows along a given magnetic surface can readily been derived from the energy equation (see e.g. Michel, 1969)

$$v_\infty \simeq V_m = \left(\frac{\Omega_*^2\,\Psi_*^2}{\dot{M}_w}\right)^{1/3} = \left(\frac{\Psi_*^2\,c^2}{\dot{M}_w\,R_L^2}\right)^{1/3} . \qquad (40)$$

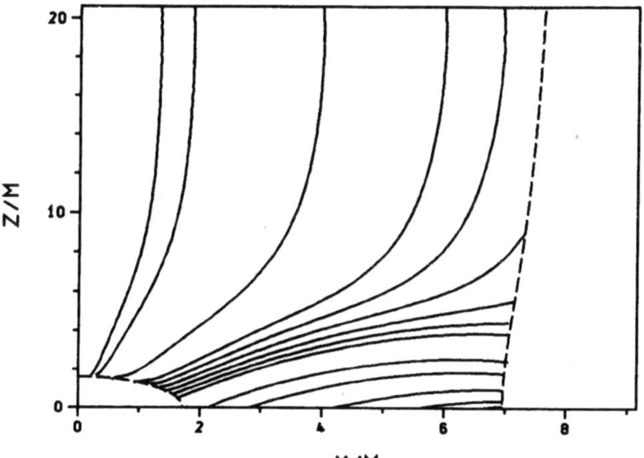

Fig. 9. Distribution of the dipolar poloidal flux around a rotating black hole with angular momentum $a = 0.8\,M_H$ (Haehnelt and Camenzind, 1990). This solution includes the gravitational effects of a rotating Kerr black hole. $M = GM_H/c^2$, the dashed lines denote the position of the inner and outer "light cylinder".

Ψ_* is the poloidal magnetic flux enclosed by the axisymmetric flux-tube at its foot point on the disk, $\dot{M}_w$ the mass-flux in the wind and R_L the corresponding light cylinder radius. In general, this relation will be scaled by some factor because the exact value of the terminal speed will depend on the particular form of the flux-tube.

In the relativistic case, the poloidal velocity of the plasma streaming along a given flux tube in the rotating magnetosphere follows from the normalisation of the 4-velocity, $u^\alpha\, u_\alpha = 1$. Since the Lorentz factor γ and the specific angular momentum l are determined by the conservation laws for the total energy and angular momentum, the solution of the poloidal velocity defined as $u_p^2 = -u^A u_A$ ($A = 1, 2$) follows from the expressions discussed in the last section

$$u_p^2 + 1 = \gamma^2\,\left(1 - l^2/R^2\right)\,. \tag{41}$$

The above equation for u_p is still implicit, since the Mach number M also contains the poloidal velocity. By using the definition of η, $\eta/n = u_p/B_p$, $B_p^2 = -B^A B_A$, $x = R/R_L$, we can, however, decompose this Mach number in a suitable way as

$$M^2 = 4\pi m\eta c^2\,\frac{\mu}{mc^2}\,\frac{\eta}{n} = u_p\,x^2\,\frac{\Phi(R)}{\sigma_*(\Psi)}\,, \tag{42}$$

with the following parameters

$$\Phi(R) = \frac{B_{p,D}R_D^2}{B_p R^2}\,, \tag{43}$$

$$\sigma_*(\Psi) = \frac{B_{D,\mathrm{in}} R_D^2}{4\pi\eta(\Psi) mc R_L^2} \simeq \sigma_0 \,. \tag{44}$$

With all these parameters, the above implicit equation for u_p can be transformed to a polynomial of degree 16 for a polytropic index $\Gamma = 5/3$, when the quantity $z_p = u_p^{1/3}$ is used instead of u_p (Camenzind and Endler, 1990)

$$\sum_{n=0}^{16} A_n\left(\frac{R}{R_L}, E, L, \frac{P_{\mathrm{in}}}{n_{\mathrm{in}} mc^2}; \frac{\Phi(R)}{\sigma_*}; g_{tt}, g_{t\phi}, g_{\phi\phi}\right) z_p^n = 0. \tag{45}$$

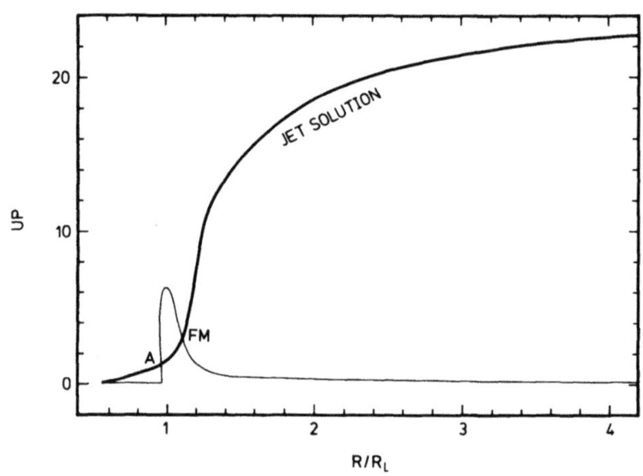

Fig. 10. The various solutions of the wind equation for a flux-tube of the accretion disk (Fig. 8). A: Alfvén point, FM: fast magnetosonic point, $u_p = \gamma v_p$.

The solutions of this hot wind equation depend on the initial parameters R_{in}, R_L, $B_{p,\mathrm{in}} R_{\mathrm{in}}^2$, $u_{p,\mathrm{in}}$, $P_{\mathrm{in}}/n_{\mathrm{in}} mc^2$, and the particle injection law $\eta(\Psi) = d\dot{N}/d\Psi$. Of central importance for the acceleration of the plasma is the form of the flux tube function Φ, defined by equation (43). Φ is constant for a monopole geometry, but has a complicated form for an open flux tube in the rapidly rotating magnetosphere. The equation for the poloidal velocity (or for the Lorentz factor) has two additional critical points, the slow and fast magnetosonic points. The parameters L and E are now constrained by the requirement that the physical wind solution passes through the Alfvén point and the fast magnetosonic point.

In Fig. 10 we show the various branches of solutions of the hot wind equation for a finite initial pressure in the background of a Kerr black hole with a flux tube given by the solution shown in Fig. 8. The injection of plasma occurs near the radius of maximum energy production in the disk. The physical wind solution starts with finite velocity near the slow magnetosonic point (near the injection point), crosses an unphysical branch at the Alfvén point near the light cylinder and a second time at the fast magnetosonic point, and reaches finally a constant outflow velocity. Since the inertia of the plasma is included in our treatment, the light cylinder is no critical point of the wind equation. Without inclusion of gravity, the slow magnetosonic point does not appear in the solutions.

4.4 Energetics of Outflows

Magnetized disk winds carry in general two forms of energies, on the one hand kinetic energy from the acceleration and on the other hand also magnetic energy in form of the Poynting flux. The initial magnetic energy flow can be estimated from the expression for the Poynting flux $\mathbf{P}_p$, which follows from the expression for the total energy carried away, Eq. (22),

$$\mathbf{P}_p = \frac{c}{4\pi}(-B_\phi)\frac{R}{R_L}\mathbf{B}_p\,. \tag{46}$$

If we integrate this flux over the entire disk surface, we obtain the initial magnetic luminosity L_m of the flow

$$L_m \simeq \frac{c}{2}(-B_\phi)_D\frac{R_{\rm in}}{R_L}\Psi_{\rm disk} \simeq \frac{1}{c}\,\Omega^F\,I_{\rm max}\,\Psi_{\rm disk} \simeq I_{\rm max}\,\Psi_{\rm disk}/R_L\,. \tag{47}$$

For *protostellar objects* we get then

$$L_m \simeq 0.1\,L_\odot\,(-B_\phi)_D\,\Psi_{D,25}\,, \tag{48}$$

when the toroidal field in the disk is in units of Gauss and the total magnetic flux of the disk Ψ_D in units of 10^{25} Gauss cm^2. This magnetic luminosity will be partly transformed into kinetic energy of the outflowing wind, so that finally $L_w \le L_m$. This estimate conforms nicely with the kinetic luminosities derived for the ionized jets in protostellar objects (Mundt et al., 1987). This magnetic luminosity has to be compared with the accretion luminosity $L_{\rm acc}$

$$L_{\rm acc} = \frac{1}{2}\dot{M}\frac{GM}{R_{\rm in}} \simeq 1.0\,L_\odot\,\dot{M}_{-7}\,\frac{3R_\odot}{R_{\rm in}}\,. \tag{49}$$

Similarly, we can estimate the magnetic luminosity of supermassive accretion disks in AGNs. Here we find toroidal fields typically of the order of a few kilo-Gauss and magnetic fluxes not exceeding 10^{34} Gauss cm^2. This corresponds to a total magnetic luminosity above the disk surface of

$$L_m \simeq 10^{47}\,{\rm erg\,s^{-1}}\,(-B_\phi)_{D,4}\,\Psi_{D,34}\,, \tag{50}$$

which is remarkably close to the accretion luminosity of strong Quasars.

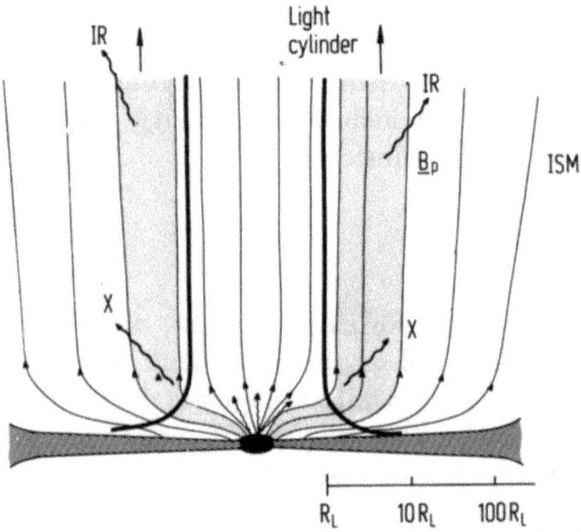

Fig. 11. The parsec-scale structure of magnetized jets in Quasars (Courvoisier and Camenzind, 1989). The central accretion disk carries a rotating magnetosphere which is strongly deformed by the presence of the light cylinder. The escaping disk-wind material is collimated outside the light cylinder by pinching forces. Stationary synchrotron emission (IR) occurs either near the light cylinder or the outer edge of the jet. The hot wind material can efficiently cool by Comptonization of the UV-flux from the inner disk and produce the hard X-ray emission (X).

5. Quasar and Protostellar Jets

The ionized winds emanating from the central boundary layers in AGN and protostars will be collimated on scales typically larger than the light cylinder radius of these objects. Disk-winds are therefore accelerated to disk-jets.

5.1 Quasar Jets

The parsec-scale jets of Quasars such as 3C 120 (Walker et al., 1987), 3C 273 (Cohen et al., 1987), or 3C 345 (Biretta et al., 1986) are visible in the synchrotron light in the GHz-range. At frequencies below 5 GHz, VLBI images of 3C 273 show a slightly curved jet which is 50 mas long (for z = 0.158 1 mas corresponds to 3.6 pc for H_0 = 50 km s^{-1} Mpc^{-1}) and has a complex brightness distribution. In this jet, superluminal components follow a common trajectory which is guided by the magnetic structure of the underlying relativistic plasma jet. With the higher resolution of millimeter-VLBI the motion of these superluminal knots can be traced even closer to the core (Krichbaum et al., 1990).

Relativistic jets in Quasars are formed in the rapidly rotating magnetosphere of accretion disks (Blandford and Payne, 1982; Lovelace et al.,

1987; Camenzind, 1986a, 1987b, 1988, 1989b). The accretion disk around a supermassive black hole creates a strong magnetosphere with typical field strengths of the order of a few kilo-Gauss at the surface of the disk (Figs. 9, 11). Normal plasma is then injected from the surface of the inner disk into this magnetosphere and is accelerated along the open magnetic flux tubes to final Lorentz factors $\gamma_{\rm jet} \simeq \sigma_* \simeq 10$, where σ_* is Michel's magnetization parameter defined in Eq. (44). If the flux tubes contain an extremely thin plasma, even higher Lorentz factors can be achieved. The collimation of the flux tubes towards the rotational axis of the disk occurs outside the light cylinder through the magnetic pinch effect and finite pressure effects in the center of a galaxy. The resulting jet radius is $\simeq (10 - 100)\ R_L$, and this inner region of the escaping jet is the origin of the self-absorbed synchrotron core emission (Krichbaum et al., 1990). The inner jet is also surrounded by a kind of magnetic cocoon formed through the interaction of the magnetized jet with the ambient hot interstellar matter. The expanding VLBI knots are then most probably the result of resistive instabilities which occur near the edge of the jet (for further discussion, see Courvoisier and Camenzind, 1989). The stability of these relativistic magnetized jets will depend on the total current carried along (Appl, 1990). There is also not much known about the propagation of these jets through the interstellar medium of a galaxy. Some numerical work has been done quite recently only on the propagation of non-relativistic magnetized jets (Kössl et al., 1990; Lind et al., 1989). Besides relativity, the essential unknown ingredient for such calculations is the finite resistivity of the magnetized jet material.

5.2 Protostellar Jets

We have argued that the most likely site of the origin of a hydromagnetic wind from an accreting protostar is the boundary layer between the inner accretion disk and the protostar (see also Pringle, 1989). Observationally, protostars are known to have rotation rates well below the break-up speed, $\Omega_*/\Omega_K \simeq 0.1$; the jet-source DG Tau has $v_* \sin i \simeq 22$ km s^{-1}, while $v_K(R_* \simeq 3R_\odot) \simeq 260$ km s^{-1}, i.e. $\omega = \Omega_*/\Omega_K(R_*) \simeq 1/5$ (Hartmann and Stauffer, 1989). This indicates that despite the high mass accretion rate the star has not been spun up to break-up rotation.

In the following we discuss a new model for the origin of the winds, the ionized jets and the low rotation rates observed in young stellar objects (Montmerle and André, 1988; Camenzind, 1990c). The essential assumption relies on the existence of a magnetized protostar (with fields of the order of a few hundred Gauss) and the existence of cool disks on the scale from 0.03 AU to ≥ 100 AU. This is the range of a few stellar radii to thousands of stellar radii. Provided the central protostar carries a magnetosphere with a more or less dipolar structure, there will be an equilibrium surface between this magnetosphere and the accretion disk, determined by pressure balance.

The pressure in a standard accretion disk follows from the expression (see e.g. Straumann, 1986)

$$P(R) = \frac{1}{I_{N+1}} \frac{\dot{M}}{4\pi\, \alpha\, R^2} \left(\frac{GM_*}{R}\right)^{1/2} \left(\frac{H}{R}\right)^{-1} . \tag{51}$$

α is the viscosity parameter of standard accretion disks and $H(R)$ the height of the disk. Pressure equilibrium with a dipolar magnetosphere is then obtained at the radius $R_{\rm in}$

$$R_{\rm in} = 2.4\, R_* \left(\frac{\alpha\, I_{N+1}}{2}\right)^{2/7} \left(\frac{B_{*,3}\, R_*}{3R_\odot}\right)^{4/7} \left(\frac{H}{\dot{M}_{-7}\, R}\right)^{2/7} \left(\frac{R_* c^2}{10^6\, GM_*}\right)^{1/7} . \tag{52}$$

I_{N+1} is a structure parameter of order unity (Straumann, 1986). For accretion rates in the range of $\dot{M} \simeq 10^{-7}$ $M_\odot$ yr^{-1}, the inner edge of such a disk is typically at a few stellar radii, provided the stellar magnetic field $B_* \simeq 10^3$ Gauss. The value for this inner edge of the accretion disk is only approximate, since the structure of the stellar magnetosphere is not exactly dipolar and the pressure in the disk can deviate from the standard Keplerian approximations. When the accretion rate increases considerably, the disk is driven towards the stellar surface, and the magnetic gap between stellar surface and the inner accretion disk completely disappears. When the accretion rate decreases, the magnetic gap will grow.

This kind of interaction between magnetic fields and accretion disks is extensively discussed in the case of accretion disks around magnetized neutron stars (Lamb, 1989). Since the boundary layer is highly turbulent, the magnetic field enters the disk on the diffusion time-scale $t_D = R_{\rm in}^2/\eta_T$, where $\eta_T \simeq P_m\, \nu_T$ and ν_T is the turbulent kinematic viscosity in the boundary layer

$$\nu_T \simeq \alpha\, \Omega\, H(R)^2 \simeq 10^{15}\, \alpha\ \mathrm{cm}^2\, \mathrm{s}^{-1} . \tag{53}$$

With these expressions we obtain a diffusion time scale given by the rotation period $P_{\rm in} \simeq$ days at the inner edge of the disk

$$t_D \simeq \frac{P_{\rm in}}{2\pi} \frac{1}{\alpha\, P_m} \left(\frac{H}{R}\right)^{-2} \simeq 30\, P_{\rm in} \left(\frac{10\, H}{R}\right)^{-2} \frac{1}{\alpha\, P_m} . \tag{54}$$

This is the fundamental time-scale for the evolution of fluctuations in the boundary layer. It essentially depends on the α-parameter and on the magnetic Prandtl number P_m.

The fast rotation in the disk generates electric radial fields in the disk, which drive a poloidal current system in the magnetosphere. These currents produce toroidal magnetic fields, which exert a momentum on the central star. The magnetic structure of this star-disk-system must then be calculated self-consistently including all the currents driven by the rotation. In a first approximation we neglect the influence of the plasma in the corona

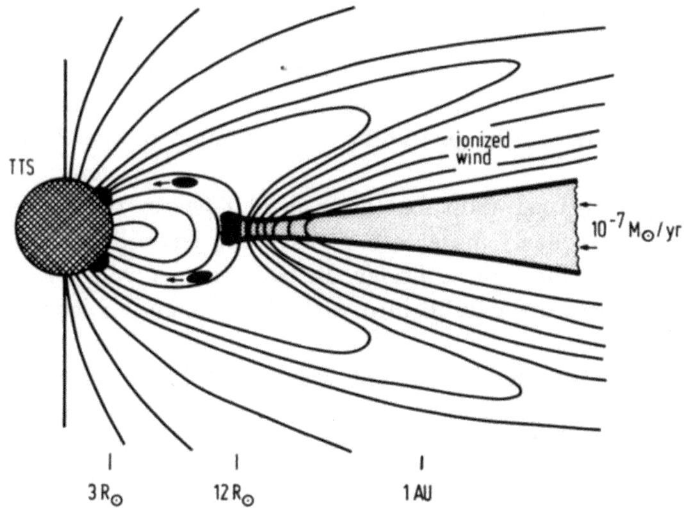

Fig. 12. The interaction between the rotating magnetosphere of the central star and the accretion disk in TTSs. The azimuthal motion of the plasma in the disk generates a radial electric field in the disk that drives the field aligned currents within the magnetosphere and cross-field currents in the disk and the corona of the star. The resulting magnetic torque density RB_ϕ is then negative in the upper hemisphere and positive in the lower hemisphere and acts to spin down the star. A strong wind is initiated along the open field lines emanating from the disk.

of the disk and the star. This means we can treat the problem in the force-free approximation. We can, however, not neglect the rapid rotation of the magnetosphere, since the light cylinder R_L is not at infinity

$$R_L = \frac{c}{\Omega_*} = \frac{c}{\omega \Omega_K(R_*)} = \frac{1}{\omega} R_* \left(\frac{R_* c^2}{GM_*}\right)^{1/2} \simeq 10^4\, R_*/10\omega \simeq 10^{15}\ \mathrm{cm}/5\omega\,. \tag{55}$$

The light cylinder radius of T Tauri stars is typically at 100 AU, and the open part of the stellar magnetosphere is then confined inside the light cylinder radius. Since the outer part of the magnetosphere can only be closed outside the light cylinder, plasma flowing away along these field lines will open up the field structure with the result that now fields enter from the light cylinder and cross the disk (Fig. 12). Field lines emanating from beyond the corotation radius

$$R_{\mathrm{cor}} = \left(\frac{GM_*}{\Omega_*^2}\right)^{1/3} = R_* \left(\frac{R_L}{R_*}\right)^{2/3} \left(\frac{GM_*}{c^2\, R_*}\right)^{1/3} \simeq 4.6\, R_* \left(\frac{1}{10\omega}\right)^{2/3} \tag{56}$$

will carry wind plasma driven outwards centrifugally. In this domain, the force-free approximation is no longer correct and the plasma inertia should be included. Apart from this effect, we obtain a plasma carrying magnetosphere along the surface of the disk which is closed beyond the light cylinder with field lines from inside the light cylinder. In this way, the field structure is globally closed, though the disk-wind blows up a magnetic bubble into the ambient medium. As is shown by Fig. 13, the disk-wind cannot be collimated to a radius smaller than the light cylinder - in fact the light cylinder is also here the natural scale for the jet-radius quite similar to the AGN jets.

Mundt et al. (1987) report jet radii for the ionized flows in the range of 10^{15} cm - 10^{16} cm. This is nicely in agreement with the jet radius being upto a factor 10 larger than the light cylinder. In fact, the jet will try to expand transversely until pressure equilibrium with the ambient molecular cloud is reached (see later on).

These magnetospheres also carry a current system. T Tauri stars are found to be in a kind of equilibrium state, where the angular momentum gain by mass accretion, $\dot{M}\sqrt{GM\,R_{\rm in}}$, is balanced by angular momentum loss through currents flowing from the star to the disk

$$\frac{dJ_*}{dt} = -\frac{1}{4\pi}\int_{\rm disk} RB_\phi\,\mathbf{B}_p\cdot d\mathbf{S} = -\frac{1}{c}\int I(\Psi)\,d\Psi \simeq -\frac{1}{c}\,I_{\rm max}\,\Psi_D\,. \qquad (57)$$

In the rotational equilibrium state, we need a current flow of the order of

$$I_{\rm max} = \frac{c^2}{\Psi_D}\,\dot{M}\,R_{\rm in}\sqrt{\frac{GM_*}{c^2\,R_{\rm in}}} \simeq 10^{14}\,{\rm Amp}\,\dot{M}_{-7}\,\Psi_{D,25}^{-1}\left(\frac{R_{\rm in}}{5R_*}\right)^{1/2}, \qquad (58)$$

corresponding to toroidal fields $\simeq 100$ Gauss at the surface of the star, or $\simeq 10$ Gauss at the inner edge of the disk. Such field strengths can be carried by the inner part of the disk. From pressure equilibrium between magnetic pressure and the disk pressure given by Eq. (51) we get the maximally possible disk field strength

$$B_\phi \simeq 42\,{\rm Gauss}\,\dot{M}_{-7}^{1/2}\,\alpha^{-1/2}\left(\frac{R_{\rm in}}{5R_*}\right)^{-5/4}. \qquad (59)$$

These fields are built up in the disk by differential rotation and dynamo effects (as discussed in Sec. 3). This rotational equilibrium which enforces $R_{\rm in} \simeq R_{\rm cor}$ solves the long-standing problem why accreting T Tauri stars are not rotating much faster. The observed slow rotation is especially surprising in view of the evidence for accretion from circumstellar disks in the pre-main sequence evolution. Accretion occurs at $\simeq 10^{-7}\,M_\odot\,{\rm yr}^{-1}$ over a typical age of 10^6 years. In the absence of angular momentum loss, this would spin up a 1 $M_\odot$ star to about half the break-up speed (Hartmann and Stauffer, 1989). The magnetic coupling between star and disk provides then a natural explanation for the observed rotational velocities. The above model is also different from the one proposed by Shu et al. (1988), where

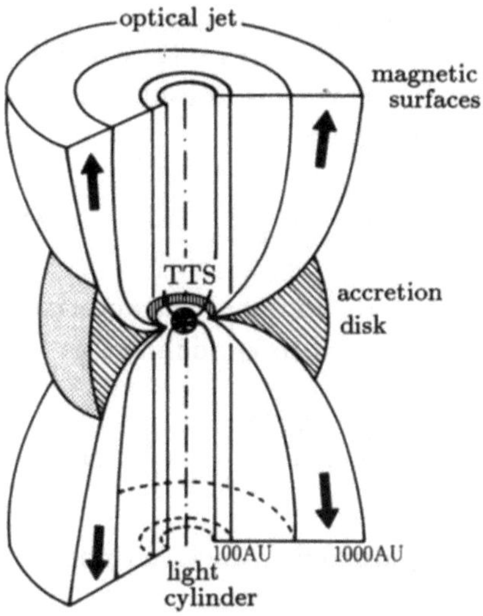

Fig. 13. The structure of magnetized T Tauri jets driven by the rapidly rotating magnetosphere. The magnetized disk-wind is collimated outside the light cylinder ($\simeq$ 50 AU, or less for younger objects) into bipolar jets with a typical radius of 500 AU. This is also the site of the origin of the forbidden lines in TTSs. The entire jet consists of a family of nested rotating magnetic surfaces which guide the plasma flow (arrows). Inside the light cylinder, the magnetic surfaces are closed to the stellar surface. They carry the return current

pre-main sequence stars rotate nearly at break-up because of high accretion from the disk, and bipolar outflows are driven away by this rapid rotation.

The disk-winds driven away by the rapid rotation of the magnetosphere are collimated outside the light cylinder of the central star into bipolar jets with a typical radius of 500 AU. Mundt et al. (1987, 1990) find a jet radius of a few 10^{15} cm in their sample, where the radius of the jet is estimated from the diameter of the optical knots. These jets consist of a family of nested magnetic surfaces, where the ionized wind flows between the surfaces (Fig. 13). The magnetic field is predominantly toroidal in this region with a field strength of $\simeq 1$ milli-Gauss, corresponding to a current of 10^{14} Ampère. The terminal speed v_∞ in the jet follows from Eq. (40)

$$v_\infty \simeq V_m = 521\,\mathrm{km\,s^{-1}}\, \Psi_{*,25}^{2/3}\, R_{L,15}^{-2/3}\, \dot{M}_{w,-8}^{-1/3} \,. \tag{60}$$

This estimate is in nice agreement with the observed velocities in the optical jets of TTSs, $v_j \simeq 200 - 1000$ km s^{-1} (Mundt et al., 1990).

The mass flux in the jet, $\dot{M}_j$, and the kinetic luminosity L_j can be determined observationally, if the jet radius R_j, jet velocity v_p and density n_j in the jet are known (Mundt et al., 1987, 1990)

$$\dot{M}_j = 3.0 \times 10^{-9}\, M_\odot\, yr^{-1}\, R_{j,15.5}^2\, n_{j,2}\, v_{p,7.5}\,, \tag{61}$$

$$L_j = 8 \times 10^{-3}\, L_\odot\, \dot{M}_{j,-9}\, v_{p,7.5}^2 \,. \tag{62}$$

The thermal pressure in the jet is at least one order of magnitude lower than the magnetic pressure ($n_j \leq 10^2\ \mathrm{cm}^{-3}$, the temperature $T_j \leq 10^4$ K, and therefore for the sound speed $c_S \simeq 10$ km/s). Since the sound speed in the jet is smaller than the magnetosonic speed in the jet,

$$v_{MS} = \sqrt{B^2/4\pi m_p n_j} \simeq 220\,\mathrm{km\,s^{-1}}\,B_{-3}\,n_{j,2}^{-1/2}\,, \tag{63}$$

strong protostellar jets are magnetically dominated, and the characteristic speed excited in the jet is the fast magnetosonic speed v_{FM}, defined as

$$v_{FM} = \frac{1}{\sqrt{2}}\left((v_{MS}^2 + c_S^2) + \sqrt{(v_{MS}^2 + c_S^2)^2 - 4\,c_S^2\,v_A^2}\right)^{1/2} \simeq c_S + v_{MS}\,. \tag{64}$$

Plasma flows therefore in the jet with a magnetosonic Mach number of $M_{FM} \simeq 1 - 5$ given by

$$M_{FM}^2 = \frac{4\pi m_p n_j v_p^2}{B^2} \simeq \frac{4\,\dot{M}_j\,v_p}{B^2\,R_j^2} \simeq \frac{\dot{M}_j\,c^2\,v_p}{I_{\max}^2} \simeq 1.6\,\dot{M}_{j,-8}\,v_{p,7.5}\,I_{14}^{-2}\,. \tag{65}$$

Whether the outflows will always reach super-magnetosonic speeds depends largely on the global topology of the magnetic surfaces.

The exact behaviour of the jet material only follows from self-consistent solutions of the Grad-Shafranov equation. All existing calculations for the stellar case have used quite strong simplifications in solving the transfield equation. Chan and Henriksen (1980), Blandford and Payne (1982) and Königl (1989) assumed self-similarity, Pudritz and Norman (1983) some special scaling with radius and Sakurai (1985, 1987) solved the transfield equation numerically only for a simplified initial monopole type geometry.

6. Conclusions

In this review, a unified discussion of hydromagnetic aspects of accretion disks in protostellar systems and the central region of galaxies has been given. These systems have been associated with collimated outflows which are most probably mediated by large scale poloidal magnetic fields that thread the central part of the disks. Jet generation can then be seen as a natural consequence of the magnetic activity of accretion disks. There are some extremely important issues which should be solved within the next years. The first is to investigate the dynamo problem for accretion disks including the backreaction of the magnetic fields on the evolution of the disk. Secondly, the large scale topology of the magnetic fields excited in accretion disks has to be elaborated – dipolar fields behave differently from quadrupolar fields. Thirdly, stationary solutions of MHD wind magnetospheres should be enlarged to include also time-dependent evolution of non-axisymmetric flows.

Acknowledgements. The author would like to thank Prof. I. Appenzeller and Dr. T. Montmerle for stimulating discussions on the subject of T Tauri stars and to Dr. H. Lesch, S. Appl, M. Haehnelt and R. Khanna for general enlightening comments on the properties of magnetized plasmas. This work is supported by the Deutsche Forschungsgemeinschaft (SFB 328).

References

Abramowicz, M.A., Zurek, W.H.: 1981, *Astrophys. J.* **246**, 314
Alloin, D., Boisson, C., Pelat, D.: 1987, *Astron. Astrophys.* **200**, 17
Antonucci, R.R.J., Miller, J.S.: 1985, *Astrophys. J.* **297**, 621
Appenzeller, I.: 1989, in *Modeling the Stellar Environment*, ed. P. Delache (Editions Frontières, Paris), p. 47
Appenzeller, I., Mundt, R.: 1989, *Astron. Astrophys. Review* **1**, 291
Appl, S.: 1990, *Ph. D. thesis*, University of Heidelberg
Barthel, P.: 1989, *Astrophys. J.* **336**, 606
Beckwith, S.V.W., Sargent, A.I., Chini, R., Güsten, R.: 1990, *Astron. J.* **99**, 924
Benford, G.: 1978, *Mon. Not. Roy. Astron. Soc.* **183**, 29
Benford, G.: 1987, in *Astrophysical Jets and their Engines*, ed. W. Kundt, NATO Advanced Science Institute **208**, Reidel (Dordrecht), p. 197
Bertout, C.: 1989, *Ann. Rev. Astron. Astrophys.* **27**, 351
Biretta, J.A., Moore, R.L., Cohen, M.H.: 1986, *Astrophys. J.* **308**, 93
Blandford, R.D.: 1989, in *Theory of Accretion Disks*, eds. F. Meyer, W.J. Duschl, J. Frank and E. Meyer-Hofmeister, Kluwer (Dordrecht), p. 35
Blandford, R.D., Payne, D.G.: 1982, *Mon. Not. Roy. Astron. Soc.* **199**, 883
Brandenburg, A., Tuominen, I., Krause, F.: 1990, *Geophys. Astrophys. Fluid Dyn.* **50**, 95
Breuer, R.A.: 1975, *Lecture Notes in Physics* (Springer, Berlin) **44**
Bridle, A.H., Perley, R.: 1984, *Ann. Rev. Astron. Astrophys.* **22**, 319
Camenzind, M.: 1986a, *Astron. Astrophys.* **156**, 137
Camenzind, M.: 1986b, *Astron. Astrophys.* **162**, 32
Camenzind, M.: 1987a, *Astron. Astrophys.* **184**, 341
Camenzind, M.: 1987b, in *Interstellar Magnetic Fields*, eds. R. Beck and R. Gräve, Springer (Berlin), p. 229
Camenzind, M.: 1988, in *High Energy Astrophysics*, ed. G. Börner, Springer Proceedings, Springer (Berlin), p. 118
Camenzind, M.: 1989a, in *Theory of Accretion Disks*, eds. F. Meyer, W.J. Duschl, J. Frank and E. Meyer-Hofmeister, Kluwer (Dordrecht), p. 59
Camenzind, M.: 1989b, in *Accretion Disks and Magnetic Fields in Astrophysics*, ed. G. Belvedere, Kluwer (Dordrecht), p. 129
Camenzind, M.: 1990a, in *IAU Symp. 140: Galactic and Intergalactic Magnetic Fields*, eds. R. Beck, P.P. Kronberg, R. Wielebinski, Reidel (Dordrecht), p. 413
Camenzind, M.: 1990b, Habilitation, Univ. Heidelberg
Camenzind, M.: 1990c, in preparation
Camenzind, M., Endler, M.: 1990, in preparation
Camenzind, M., Lesch, H.: 1990, in preparation

Chan, K.L., Henriksen, R.N.: 1982, *Astrophys. J.* **241**, 534
Chen, K., Halpern, J.P., Filippenko, A.V.: 1989, *Astrophys. J.* **339**, 742
Cohen, M.H., Zensus, J.A., Biretta, J.A., Comoretto, G., Kaufmann, P., Abraham, Z.: 1987, *Astrophys. J.* **315**, L89
Collin-Souffrin, S.: 1987, *Astron. Astrophys.* **179**, 60
Coroniti, F.V.: 1985, in *Unsteady Current Systems in Astrophysics, IAU Symp.* **107**, eds. M.R. Kundu and G.D. Holman, Reidel (Dordrecht), p. 453
Courvoisier, T.J.-L., Camenzind, M.: 1989, *Astron. Astrophys.* **224**, 10
Draine, B.T.: 1983, *Astrophys. J.* **270**, 519
Edwards, S., Cabrit, S., Ghandour, L.O., Strom, S.E.: 1989, in *ESO Workshop on Low Mass Star Formation and Pre-Main Sequence Objects*, ed. Bo Reipurth, ESO Proceedings **33**, 385
Elvis, M., Lawrence, A.: 1988, *Astrophys. J.* **331**, 161
Fukui, Y.: 1989, in *ESO Workshop on Low Mass Star Formation and Pre-Main Sequence Objects*, ed. Bo Reipurth, ESO Proceedings **33**, 95
Haehnelt, M., Camenzind, M.: 1990, in preparation
Halpern,J.P., Filippenko, A.V.: 1988, *Nature* **331**, 46
Hartmann, L., Stauffer, J.R.: 1989, *Astron. J.* **97**, 873
Hartmann, L., MacGregor, K.B.: 1982, *Astrophys. J.* **259**, 180
Hessman, F.V., Eislöffel, J., Mundt, R., Hartmann, L., Herbst, W., Krautter, J.: 1990, submitted to *Astrophys. J.*
Heyvaerts, J., Norman, C.A.: 1989, *Astrophys. J.* **347**, 1055
Heyvaerts, J., Priest, E.R.: 1989, *Astron. Astrophys.* **216**, 230
Kawara, K., Nishida, M., Gregory, B.: 1990, *Astrophys. J.* **352**, 433
Khanna, R., Camenzind, M.: 1990, in preparation
Kley, W.: 1989, *Astron. Astrophys.* **208**, 98
Kley, W.: 1990, this volume
Königl, A.: 1990, *Astrophys. J.* **342**, 208
Kössl, D., Müller, E., Hillebrandt, W.: 1989, *Astron. Astrophys.* **229**, 378
Koyama, K.: 1989, in *IAU Symposium 134, Active Galactic Nuclei*, eds. D.E. Osterbrock and J.S. Miller, Kluwer (Dordrecht), p. 167
Krause, F., Meinel, R., Elstner, D., Rüdiger, G.: 1990, in *IAU Symp. 140: Galactic and Intergalactic Magnetic Fields*, eds. R. Beck, P.P. Kronberg, R. Wielebinski, Reidel (Dordrecht), p. 97
Krause, F., Rädler, K.-H.: 1980, *Mean Field Magnetohydrodynamics and Dynamo Theory*, Pergamon Press (Oxford)
Krichbaum, T.P. et al.: 1990, MPIfR preprint, submitted to *Astron. Astrophys.*
Krolik, J.H., Begelman, M.C.: 1988, *Astrophys. J.* **329**, 702
Lada, C.J.: 1985, *Ann. Rev. Astron. Astrophys.* **23**, 267
Lamb, F.K.: 1989, in *Timing Neutron Stars*, eds. H. Ögelman and E.P.J. van den Heuvel, Kluwer (Dordrecht), p. 649
Laor, A., Netzer, H.: 1989, *Mon. Not. Roy. astron. Soc.* **238**, 897; **242**, 560
Lesch, H., Appl, S., Camenzind, M.: 1989, *Astron. Astrophys.* **225**, 341
Lind, K.R., Payne, D.G., Meier, D.L., Blandford, R.D.: 1989, *Astrophys. J.* **344**, 89
Lovelace, R.V.E., Wang, J.C.L., Sulkanen, M.E.: 1987, *Astrophys. J.* **315**, 504
Malkan, M.: 1989, in *Theory of Accretion Disks*, eds. F. Meyer, W.J. Duschl, J. Frank and E. Meyer-Hofmeister, Kluwer (Dordrecht), p. 19
Matsumoto, R., Horiuchi, T., Hanawa, T., Shibata, K.: 1990, *Astrophys. J.* **356**, 259
Michel, F.C.: 1969, *Astrophys. J.* **157**, 1183
Michel, F.C.: 1982, *Rev. Mod. Phys.* **54**, 1
Moffatt, H.K.: 1978, *Magnetic Field Generation in Electrically Conducting Fluids*, Cambridge Univ. Press (Cambridge)
Montmerle, T., André, P.: 1988, in *Formation and Evolution of Low Mass Stars*, eds. A.K. Dupree and M.T.V.T. Lago (Kluwer, Dordrecht), p. 225
Mundt, R.: 1988, in *Formation and Evolution of Low Mass Stars*, eds. A.K. Dupree and M.T.V. Lago (Kluwer, Dordrecht), p. 257
Mundt, R., Brugel, E.W., Bührke, T.: 1987, *Astrophys. J.* **319**, 275
Mundt, R., Ray, T.P., Bührke, T., Raga, A.C., Solf, J.: 1990, *Astron. Astrophys.* **232**, 37

Parker, E.N.: 1971, *Astrophys. J.* **163**, 255
Perley, R.: 1989, in *Hot Spots in Extragalactic Radio Sources*, eds. K. Meisenheimer and H.-J. Röser, Lecture Notes in Physics (Springer, Berlin) **327**, 1
Poetzel, R., Mundt, R., Ray, T.P.: 1989, *Astron. Astrophys.* **224**, L13
Porcas, R.W.: 1987, in *Superluminal Radio Sources*, Eds. J.A. Zensus and T.J. Pearson, Cambridge University Press (Cambridge), p. 12
Pringle, J.: 1989, *Mon. Not. Roy. Astron. Soc.* **236**, 107
Pudritz, R.E.: 1981, *Mon. Not. R. astr. Soc.* **195**, 881; 897
Pudritz, R.: 1989, in *Galactic and Extragalactic Star Formation*, eds. R.E. Pudritz and M. Fich (Kluwer, Dordrecht), p. 135
Pudritz, R.E., Norman, C.A.: 1983, *Astrophys. J.* **274**, 677
Pudritz, R.E., Norman, C.A.: 1986, *Astrophys. J.* **301**, 571
Rüdiger, G.: 1990, *Geophys. Astrophys. Fluid Dynamics* **50**, 53
Ruzmaikin, A.A., Shukurov, A.M., Sokoloff, D.D.: 1988, *Magnetic Fields of Galaxies*, Kluwer (Dordrecht)
Sakurai, T.: 1985, *Astron. Astrophys.* **152**, 121
Sakurai, T.: 1987, *Publ. astron. Soc. Japan* **39**, 821
Sanders, D.B., Phinney, E.S., Neugebauer, G., Soifer, B.T., Matthews, K.: 1989, *Astrophys. J.* **347**, 29
Scharleman, E.T., Wagoner, R.V.: 1973, *Astrophys. J.* **182**, 951
Shakura, N.I., Sunyaev, R.A.: 1973, *Astron. Astrophys.* **24**, 337
Shibata, K., Tajima, T., Matsumoto, R., Horiuchi, T., Hanawa, T., Rosner, R., Uchida, Y.: 1989, *Astrophys. J.* **338**, 471
Shu, F.H., Lizano, S., Adams, F.C.: 1987, *Ann. Rev. Astron. Astrophys.* **25**, 23
Shu, F.H., Lizano, S., Ruden, S.P., Najita, L.: 1988, *Astrophys. J. Letters* **328**, L19
Stirpe, G.M., deBruyn, A.G., von Groningen, E.: 1988, *Astron. Astrophys.* **200**, 9
Straumann, N.: 1986, *General Relativity and Relativistic Astrophysics*, Texts and Monographs in Physics, Springer (Berlin)
Strom, S.E., Strom, K.M., Edwards, S.: 1988, in *Proc. NATO Advanced Study Institute on Galactic and Extragalactic Star Formation*, eds. R.E. Pudritz and M. Fich (Kluwer, Dordrecht), p. 53
Sun, W.-H., Malkan, M.: 1989, *Astrophys. J.* **346**, 68
Tylenda, R.: 1981, *Acta Astron.* **31**, 267
Ulrich, M.-H.: 1989, in *Theory of Accretion Disks*, eds. F. Meyer, W.J. Duschl, J. Frank and E. Meyer-Hofmeister, Kluwer (Dordrecht), p. 3
Walker, R.C., Benson, J.M., Unwin, S.C.: 1987, *Astrophys. J.* **316**, 546
Zel'dovich, Ya.B., Ruzmaikin, A.A., Sokoloff, D.D.: 1983, *Magnetic Fields in Astrophysics*, Pergamon Press (Oxford)
Zensus, J.A.: 1989, in *BL Lac Objects*, eds. L. Maraschi, T. Maccacaro, M.-H. Ulrich, Lecture Notes in Physics (Springer, Berlin) **334**, 3

Bipolar Nebulae Driven by the Winds of Young Stars

H.J. Staude and Th. Neckel

Max-Planck-Institut für Astronomie, Königstuhl 17,
D-6900 Heidelberg, Fed. Rep. of Germany

Summary. We review the observed properties of young bipolar nebulae. S106 and Cep A are discussed in detail, three recently discovered objects (GN 20.18.3, NS14 and L810) are presented. The characteristics of young bipolar nebulae are then compared to those of young stellar objects (YSOs) observed as cold luminous IRAS sources associated with dense molecular cores and bipolar outflows. It is concluded that they belong to the same early evolutionary stage.

1. Introduction

During the last decade, three classes of objects have emerged from systematic studies of star forming regions in the far infrared and radio domain: the "dense cores" in molecular clouds, with masses $M \approx 0.5...760\ M_\odot$ and densities $n_H = 2 \cdot 10^3 ... 2 \cdot 10^5\ cm^{-3}$, the cold and luminous IRAS sources embedded in molecular clouds, with luminosities $L \leq 4 \cdot 10^5\ L_\odot$ and temperatures $T \leq 300$ K, and the bipolar molecular outflows. There is now ample observational evidence that these three phenomena are tightly related:

1) About 2/3 of the dense cores are associated with IRAS sources, thus implying that, once the conditions for formation of a dense core are fulfilled, also an embedded IRAS source will form within a comparable timespan [6].

2) Within a given star forming region, bipolar molecular outflows are associated with the most luminous and coolest IRAS sources, that is with those surrounded by the densest and largest amounts of circumstellar matter [20]. This indicates that the bipolar molecular outflow sets in early in the stage of PMS evolution represented by the most deeply embedded IRAS sources.

3) Conversely, about one half of the luminous embedded IRAS sources have bipolar molecular outflows [53]. Thus the outflow phase is not substantially shorter than the lifetime of the YSOs as embedded IRAS sources, which is estimated to be $< 4x10^5$ a.

About 200 objects are presently known, which share these characteristics. Most of them cannot be observed at optical or near IR wavelengths. We will argue below that the young bipolar nebulae systematically studied since the late seventies, of which now about a dozen are known, represent the fortunate exceptions to this rule. But first we will describe the two most thoroughly studied young bipolar nebulae, S106 and Cep A, to illustrate the kind of informations which are provided by optical, infrared and radio studies, and present three recently discovered objects of this class.

2. S106

Figure 1 shows an I image of this bipolar nebula, whose overall size is 30"x100" or 0.1 pc x 0.3 pc at the distance d = 600 pc. The two sharply defined bright lobes are separated by a dark equatorial lane, in which the star exciting the nebula is embedded. Its spectral type is O9.5V-B0III, the visual extinction reaches A_V = 21 mag in front of the star and drops to A_V = 5...8 mag towards the poles of the nebula [12], [56].

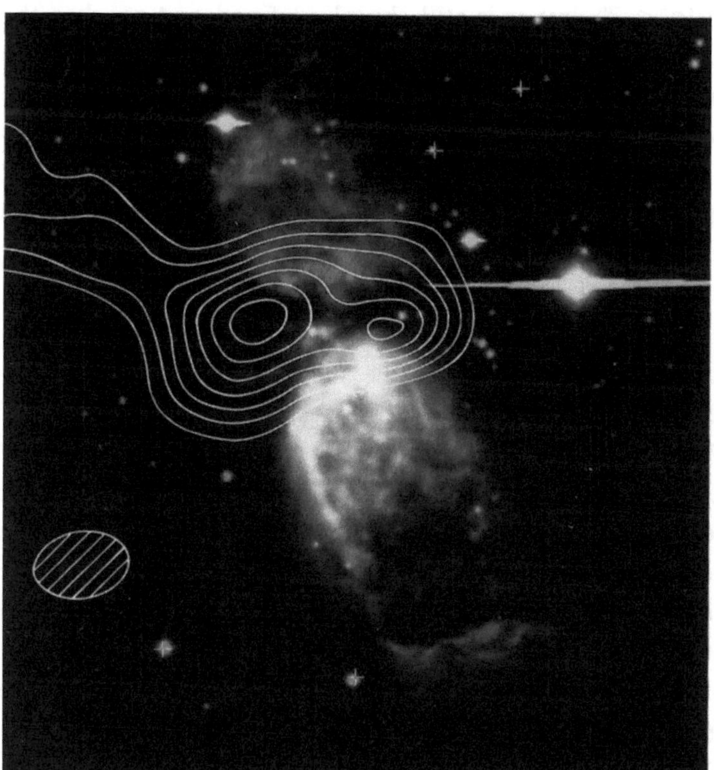

Figure 1: CCD image in I of the bipolar nebula S106, taken at the prime focus of the 3.5 m telescope on Calar Alto by K. Birkle and U. Hopp. The superimposed isophotes show the HCN distribution [7].

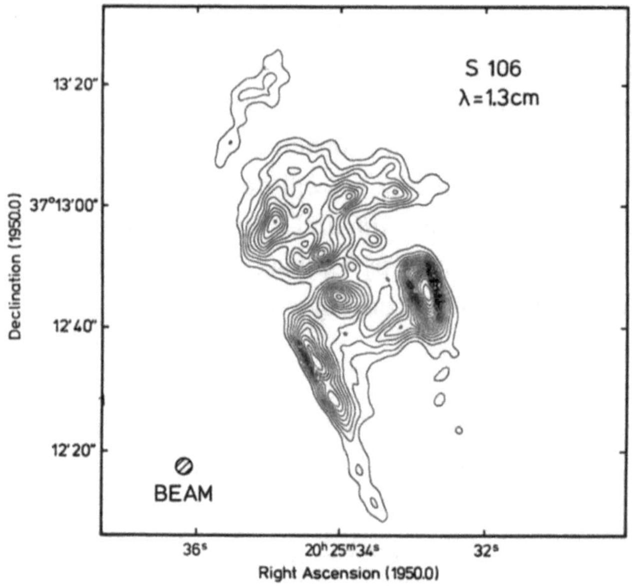

Figure 2: VLA map of S106 at λ 1.3 cm [17].

The distribution of the ionized gas unaltered by extinction is shown by the VLA map at λ = 1.3 cm in Figure 2 [17]. The gap between the ionized lobes now shrinks to < 2", but it is still sharply defined and reaches the immediate surroundings of the central star. The star appears in the radio map as a faint unresolved source (Ø < 0".5) which emits only about 0.1% of the total radio continuum. The shape of its spectrum, $S_\lambda \sim \lambda^{-0.7}$, is characteristic of a thick, expanding ionized shell. The fit of the model by Felli and Panagia [16] to the optical, infrared and radio data, including the resolved Brγ emission line [52], yields

$$\dot{M} = 10^{-6}\ M_\odot/a, \quad Ø = 30\ AU,$$

$$v_{exp} = 200\ km/s, \quad L_{kin} = 5\ L_\odot \approx 10^{-3}\ L \quad ,$$

where L_* is the bolometric luminosity of the star. The expanding circumstellar shell is nearly optically thick. It absorbs 75% of the stellar UV photons, only 25% are available to ionize the lobes of the bipolar nebula.

But the VLA map impressively demonstrates that the shell is not spherically symmetric. Near the equatorial plane it totally absorbs the ionizing stellar radiation, thereby allowing the neutral gap to reach the immediate surroundings of the star. Also in the middle IR (10-20 μm) within the equatorial belt no sign of interaction is observed between UV radiation and neutral gas or dust, while the photons are free to escape towards the poles and to ionize the extended HII region.

These facts lead us to the *ad hoc* hypothesis: the parameter $\dot{M}/v_{exp}$, which determines the optical depth of the shell, increases from the pole to the equator by a factor > 2, e.g. due to a decrease of v_{exp} towards the equator from its maximal value at the poles. Such a variation of the wind velocity with latitude has been observed by Jones and Herbig [26] in the cometary reflection nebula NGC 2261. Here, the P Cygni profile of the shell emission can be measured in the reflected light at various distances from R Mon, the illuminating star. The derived wind velocity v_{exp} varies from 100 km/s at the equator to 250 km/s at high latitudes and smoothly joins the transversal velocity of the Herbig-Haro object HH39, which moves away from the star with the wind along the polar axis in the plane of the sky at 280 km/s. This observation strongly supports the notion that the stellar wind has got its bipolar structure already in the immediate surroundings of the star. In this picture, the jets of young stars simply mark the direction of maximum velocity in a bipolar stellar wind fanning out into the entire lobes. In fact, a bipolar jet emanates from R Mon in both directions along the polar axis of the cometary nebula [8]. The search for shock excited emission along the axis of symmetry of bipolar nebulae has been successful several times, as for instance in the Boomerang and the PV Cephei nebula [49].

The high extinction in front of the central star and within the dark lane of S106 already hints at the existence of a dense equatorial belt of gas and dust. The search for an extended, neutral equatorial torus in S106 first became succesful when Bieging [7] presented a high resolution HCN map of S106 (Fig. 1). In the light of the emission lines of other molecules ([28], [2]) the existence of such a thick torus could not be confirmed. But the strongest argument in favour of its reality comes from the 1.3 mm continuum map obtained with the 30 m MRT [38] which almost exactly the same structure as in the NH_3 emission apears. The authors derive a mass $M_{H} \geq 20$ $M_{\odot}$ at T = 18 K. The ratio τ (1.3 mm)/A_V turns out to be strongly abnormal: the dust grains evidently are much larger in the torus than in the ISM. The discrepancy between observations of different molecular species thus appears to be due to depletion of molecules by accretion onto the dust grains within the dense equatorial torus.

A dramatic insight into the structure of the southern lobe of S106 is provided by the Coudé spectra taken by Solf and Carsenty [55]. The Hα line is split into two components with a total width of 220 km/s: the clumpy and turbulent ionized gas flows towards the observer along the bell shaped walls of the lobe at a velocity which agrees well with the wind velocity derived at

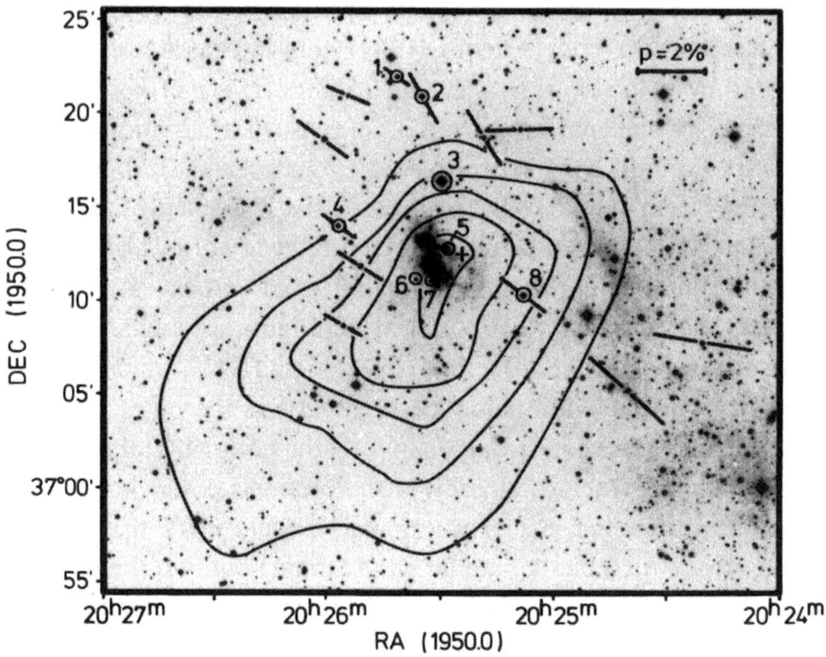

Figure 3: S106 in its parent molecular cloud [37]. Polarimetry of field stars shining through the cloud from [56].

the central star. The southern lobe is essentially empty, while no line splitting is observed in the northern lobe: here the gas is flowing into the densest parts of the molecular cloud in which the new star was formed. Fig. 3 shows this cloud by its CO emission [37], permeated by a magnetic field oriented parallel to the polar axis of S106 [56].

3. Cepheus A

This bipolar nebula is not visible in the optical. It was discovered at $\lambda = 2.2$ μm in a search for the source exciting the complex HH object GGD37 ([32], Fig. 4). Surface polarimetry (Fig. 5) demonstrates that the extended nebulosity is a reflection nebula illuminated by a source unobservable even at 3.6 μm, which is located near to its SE border. The analysis of 2.2 μm and 3.6 μm maps reveals a structure similar to that of the southern lobe of S106: in its central part it is nearly empty, while near to the illuminating star the extinction rises to $A_V \approx 75$ mag. The illuminating star is a cool and luminous IRAS source ($T = 90$ K, $L = 2 \cdot 10^4$ $L_\odot$). The radio continuum VLA map brings up a chain of compact sources (Fig. 6), all of which have flat spectra, except the one coinciding with the exciting star: its radio spectrum ($\alpha = 0.6$)

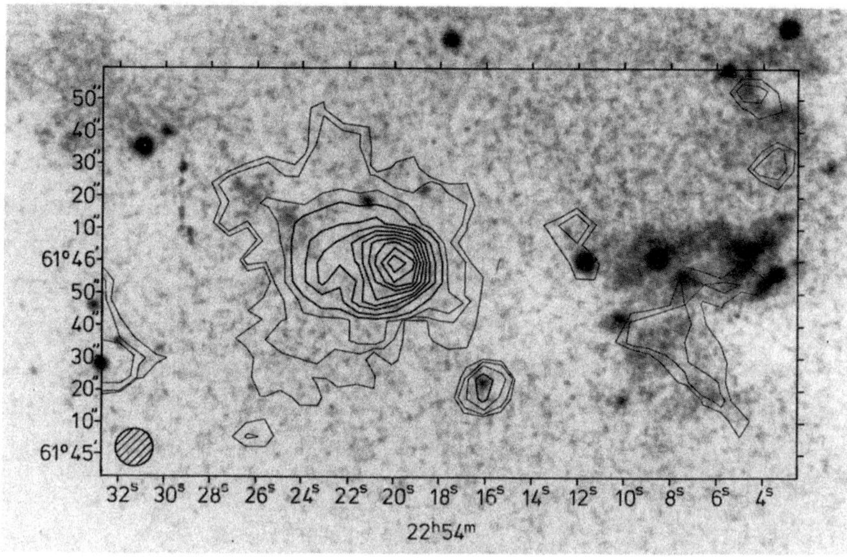

Figure 4: 2.2 μm map of Cep A, superimposed on POSS R [32].

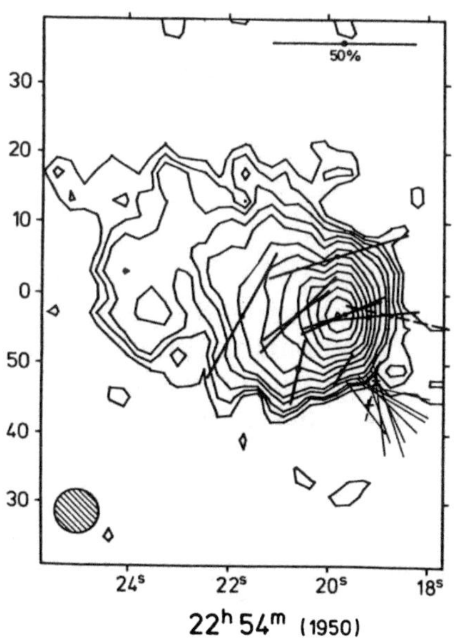

Figure 5: Polarimetry of the IR nebula Cep A at λ = 3.6 μm [32].

again can be modeled by a thick expanding ionized envelope which completely surrounds a B0.5 star [25]. Although the central star has a spectral type similar to that in S106, there still is no way out for ionizing photons; the other HII regions are excited by the wind from the central star: this requires $\dot{M} = 10^{-5}\ M_\odot/a$ and $v_{exp} = 400$ km/s [58].

Very similar values are obtained assuming the bipolar CO outflow (Fig. 6) to be driven by a stellar wind: $\dot{M} = 4 \cdot 10^{-6}\ M_\odot/a$, $v_{exp} = 400$ km/s [34]. This picture is corroborated by the extended HH emission detected on the deep Hα image (Fig. 6). Here in the NE a well formed bow shock appears. Lenzen [31] determined its proper motion, as well as that of various components of GGD37 in the west: taken together with the radial velocities [32] again velocities $v_{exp} = 300$-450 km/s away from the central star are derived.

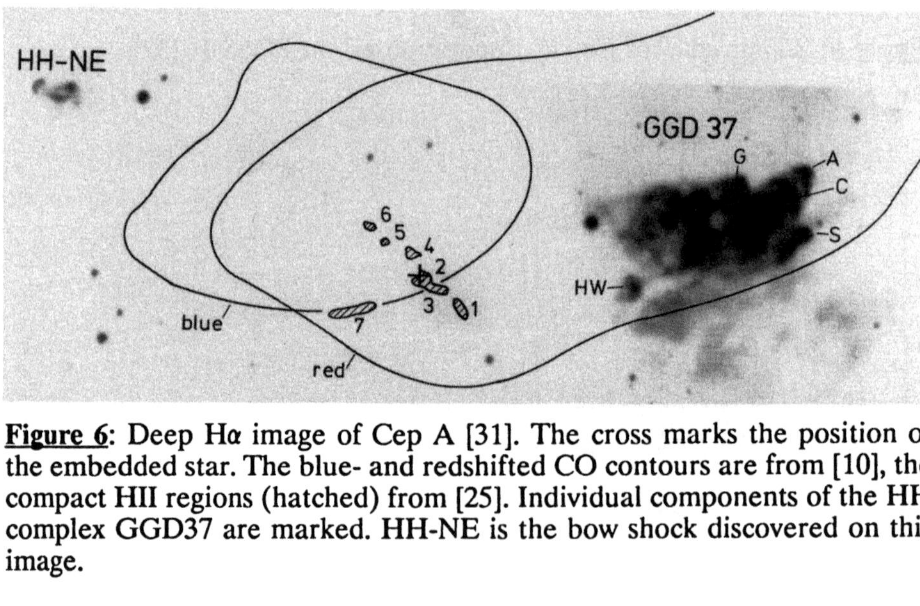

Figure 6: Deep Hα image of Cep A [31]. The cross marks the position of the embedded star. The blue- and redshifted CO contours are from [10], the compact HII regions (hatched) from [25]. Individual components of the HH complex GGD37 are marked. HH-NE is the bow shock discovered on this image.

The most direct view of this wind driven bipolar nebula is provided by the mosaic of 28 CCD images in the light of the 2.12 μm line emitted by the shock excited H_2 molecule [29]. Here the eastern lobe appears in its full extent, the powerful bow shock in the NE now turns out to be located on its axis of symmetry, as expected; the outflow follows the direction of the local magnetic field [24]. The western outflow is heavily disturbed - apparently it is deflected to the NE by a massive condensation within the molecular cloud.

The central stars of S106 and Cep A have similar winds. But the envelope of Cep A is still completely optically thick for ionizing photons - therefore no extended HII region has developed yet, while the stellar wind can manifest itself in the bipolar CO outflow and in the fast moving HH objects: characteristic features, which otherwise can only be observed near later type, not ionizing stars.

4. GN 20.18.3

This compact reflection nebula in Cygnus [45] is associated with a dense dust filament ($n_H = 6 \cdot 10^4$ cm^{-3}, M = 250 $M_\odot$). On the deep I image in Fig. 7a the exciting star becomes visible at the apex of the sharply defined cometary nebula; the surface polarimetry proves it to be the sole source illuminating the nebula. From its R and I magnitudes and its IRAS luminosity ($L \approx 100\ L_\odot$) we estimate its mass to be M = 3-5 $M_\odot$. Thus its spectral type is B7-B9 (if it has already reached the main sequence) and its visual extinction is A_V = 11-20 mag, far more than the extinction in front or within the nebula (as follows from its high surface polarization) - this is an indirect but strong hint at the presence of a thick circumstellar torus. On the Hα image (Fig. 7b) a bright HH object appears, which is located upon the axis of symmetry of the cometary nebula.

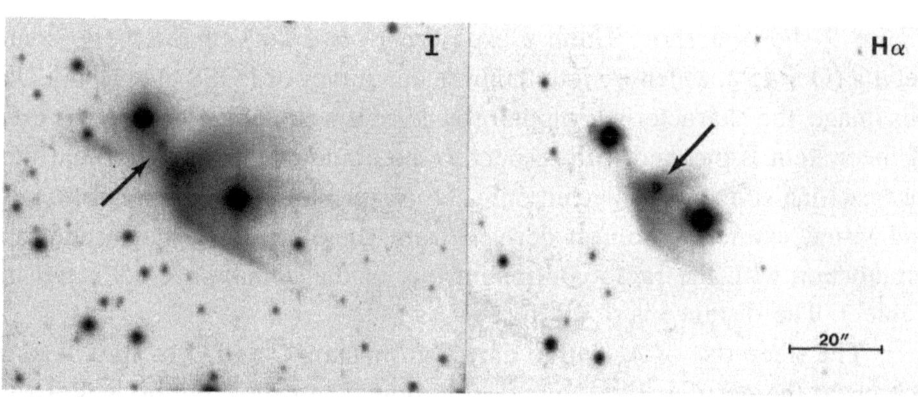

Figure 7: Left - I image of GN 20.18.3. The arrow points at the exciting star at the apex of the nebula. Right - Hα image of GN 20.18.3. The arrow points at the HH knot. The two brighter stars are in the foreground [45].

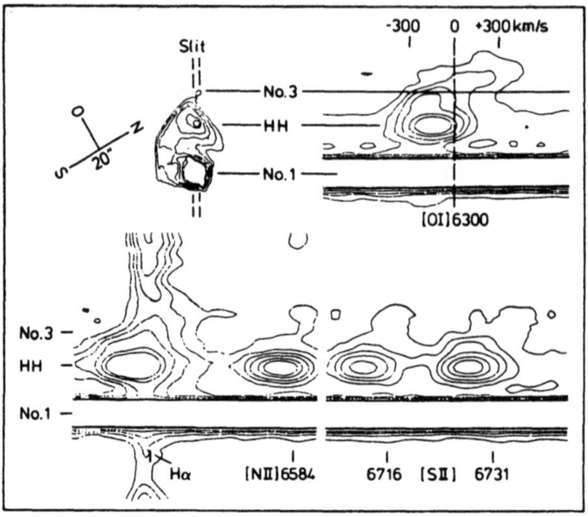

Figure 8: Spectrogram of GN 20.18.3 showing the [OI] 8446, Hα, [NII] and [SII] emission lines. Top left - on the red isophotes of the nebula the position of the exciting star (No. 3), the HH object, the field star in the SW (No. 1) and the slit position are indicated.

The spectrum of the shock excited emission lines (Fig. 8) reveals radial velocities of up to -300 km/s at the HH object. At the other side of the exciting star the direction of the outflow is reversed: at +300 km/s the jet pierces into the dark cloud, where due to the high density ($n > 10^5$ cm^{-3}) the [SII] and [NII] emission is suppressed. Thus also in this case the outflow is intrinsically bipolar. As in the case of S106, only the lobe directed towards the observer has fully developed, after having disrupted the outer layers of the molecular cloud.

5. NS 14

Figure 9 shows a short Gunn z exposure ($\lambda \approx 1$ μm) of this very small nebula (Ø ≈ 15"), which we found during our survey of POSS plates [44]. On this image, the characteristic bipolar structure is well defined. The polar axis of the system is inclined with respect to the plane of the sky, such that the *four* exciting stars are projected onto the eastern lobe. Their spectral types and visual extinction values, derived from their R and I magnitudes in conjunction with the radio continuum flux of the nebula, are collected in Table 1. The distance is d ≈ 2 kpc.

The steep rise of A_V within only 3" from stars C and D to stars A and B hints at the peculiar local dust distribution. The true situation is illustrated by the 1.3 mm map in Fig. 10, obtained with the 30 m MRT [48]. Coinciding with the optical obscuration lane bisecting the nebula and centered exactly upon star A, we see a bar or torus resolved only in N-S direction (beam diameter = 11"), containing 350 $M_\odot$ of dense, neutral

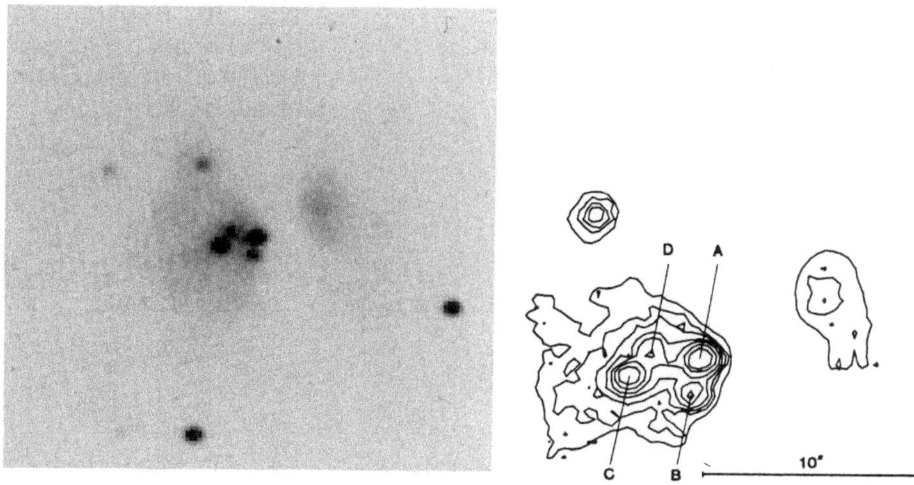

Figure 9: Left - Gunn z CCD image of NS14. Right - on the isophotes constructed from this image the letters A-D identify the four exciting stars, see also Table 1.

Table 1: Spectral types and visual extinction values of the "Trapezium" stars embedded in NS14

Star	Type	A_V
A	B0.5	15.8
B	B1	17.1
C	B9	8.3
D	A5	7.7

material at T ≈ 10 K. The VLA map at λ = 1.3 cm (Fukui, priv. comm.) shows an extremely compact (diameter 3" or 0.03 pc), clumpy HII region, whose brightest, unresolved knot also coincides with star A. Our longslit spectra of the Hα, [NII] and [SII] lines show, that the HII region is expanding at v = 15 km/s out of the dense molecular torus. This leads to an age of the HII region of only 2000 years.

The surface polarimetry of NS14 [51] shows the pattern typical for bipolar nebulae: high degrees of polarization in the lobes, oriented centro-symmetrically around the "Trapezium", and lower polarization in the dark lane, oriented parallel to the equator. From this Scarrott et al. infer the

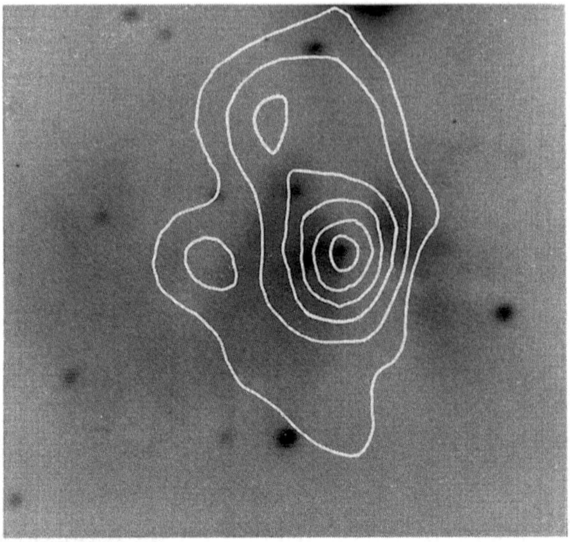

Figure 10: Isophotes of the 1300 μm emission superposed on the Gunn z image of NS14.

presence of a toroidal magnetic field aligning the dust particles within the neutral torus, which give rise to polarization by selective absorption (Davis-Greenstein mechanism). But this interpretation is at variance with our observation that stars A and C are essentially unpolarized. The surface polarization is due to pure scattering: single scattering in the lobes, multiple scattering in the equatorial belt, as confirmed also by recent model calculations due to Bastien and Ménard [3], who developed further the basic Elsässer-Staude model [15] by including second order Mie scattering. There is no need for the assumption of a toroidal magnetic field to explain the observed polarization pattern.

The H_2CO map obtained at Effelsberg (Fig. 11) shows that the Trapezium was born within a dense core belonging to a filamentary molecular cloud oriented N-S like the torus; 350 $M_\odot$ are concentrated within the torus, only 500 $M_\odot$ in the entire dense core, and 1900 $M_\odot$ in the whole filament. Fukui [19] has discovered a bipolar outflow centered on NS14. In the second, northern dense core we have also found violent signs of multiple star formation.

6. Lynds 810

After ESO 210-6A, this is the second Bok globule in which active star formation has been observed [43]. Fig. 12 shows L 810 on POSS R, with the NH_3 and H_2CO maps obtained at Effelsberg. Its physical parameters (n, M, T) do not differ from those of a typical dense core. In its centre a weak,

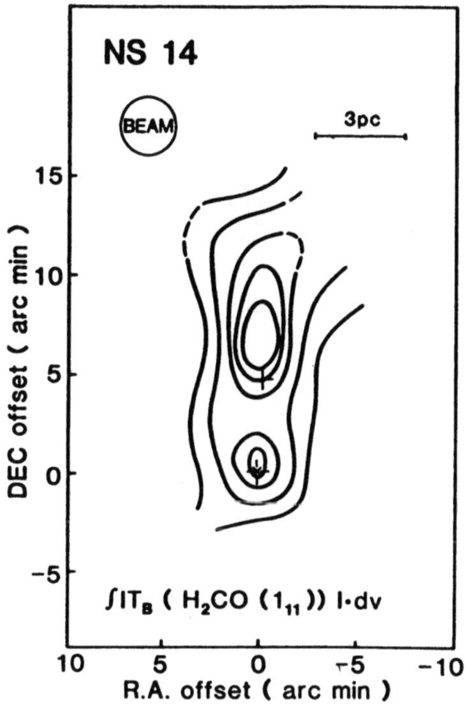

Figure 11: Map of integrated H_2CO line temperature towards the dark cloud associated with NS14. Crosses mark the positions of IRAS point sources. NS14 coincides with the southern one [48].

filamentary reflection nebula points to the embedded H_2O maser. On the deep R image in Fig. 13 [47] this filament breaks up into two cometary nebulae, both oriented roughly E-W, emanating from the embedded stars No. 7 and 18. Their spectral types and extinction values derived from CCD BVRI photometry are A5, A_V = 7.0 mag and G2, A_V = 5.1 mag respectively. Both stars have broad Hα emission lines with strong blueshifted absorption components at -60 and -110 km/s, as typical for T Tauri stars. Three more stars, No. 17, 8 and 16, are associated with similar cometary nebulosities, all oriented roughly E-W. Altogether we found 15 stars embedded in the globule.

The surface polarimetry of the reflection nebula (Fig. 14) shows that star No. 7 is its main illuminating source - despite its heavy extinction towards the observer. This situation recalls the case of L1551-IRS5 [30], where the hidden star illuminates a cavity blown free by its wind, in which the jet [40, 46, 50] and the blueshifted CO flow [54] move outwards at high velocity. This interpretation of the optical data of L810 has recently been confirmed by Fukui (priv. comm.) with the discovery of a well defined bipolar CO outflow in L810, whose blueshifted lobe coincides with the reflection nebula.

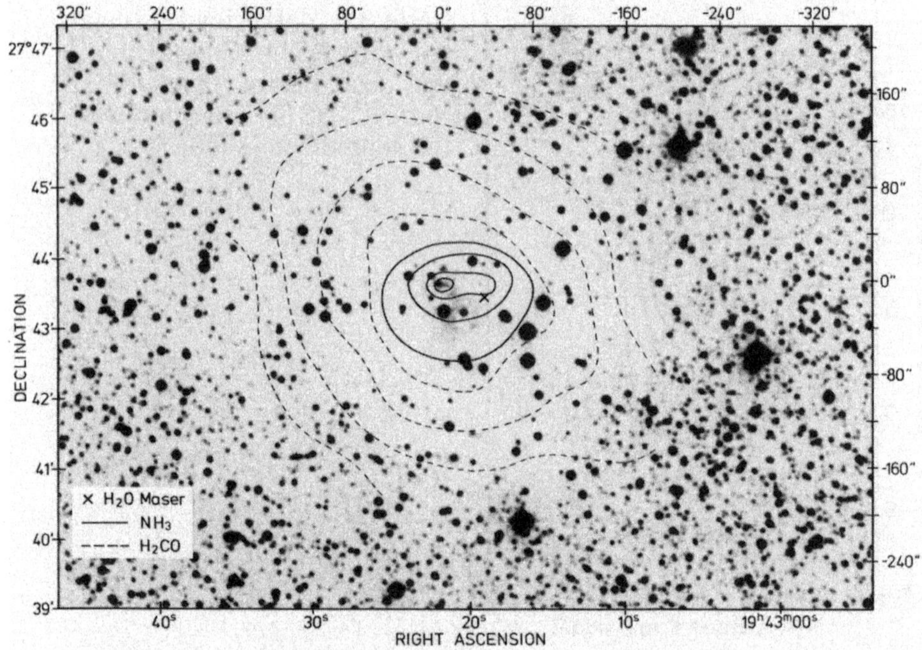

Figure 12: The Bok globule L 810 on POSS R with H_2CO and NH_3 contours. The cross marks the position of an H_2O maser. Note the bright filament elongated westwards at the centre [43].

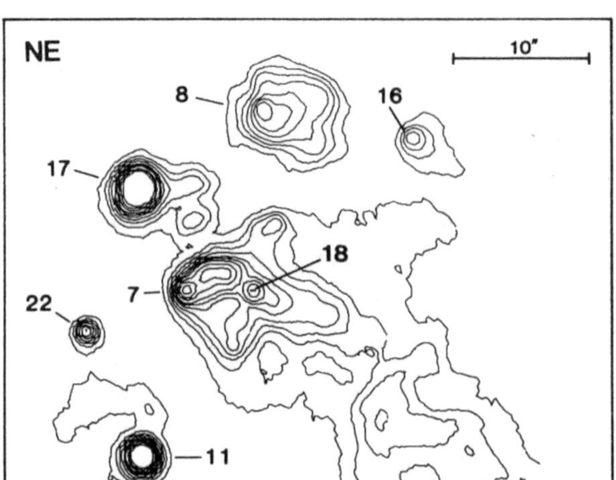

Figure 13: Gunn r image of the core region of L 810. The bright filament breaks up into two components emanating from stars No. 7 and 18. Also the embedded stars No. 17, 8 and 16 are associated with cometary nebulosities oriented approximately westwards [47].

In L810 two preferential directions are observed. The individual cometary nebulae in its core are oriented E-W, while the larger scale bipolar CO outflow is oriented N-S. The tangential component of the magnetic field in the outer parts of L810 has been worked out by Hodapp [23] - we can imagine that it continues in E-W direction in the core and determines the orientation of the outflows prevailing there. But the CO outflow breaking through the outer layers of L810 in N-S direction evidently does not follow the magnetic field lines. Here the density gradient within the globule appears to be the dominant factor determining the orientation of the outflow.

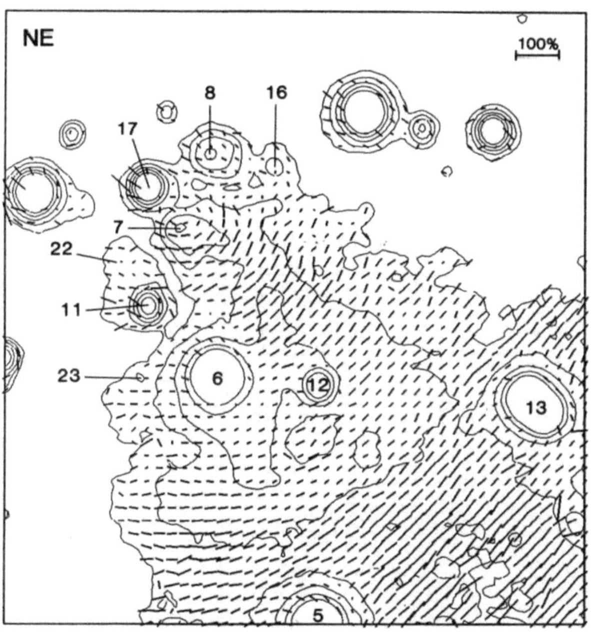

<u>Figure 14</u>: Polarimetry of the reflection nebula in L810. Star No. 7 is the main illuminating source.

<u>7. Young Bipolar Nebulae, IRAS Sources and CO Outflows</u>

In order to characterize observationally young bipolar nebulae as a stage in PMS evolution, we have assembled data related to 12 objects in Table 2. The list does not pretend to be complete, but due to the stringent requirement that deeply embedded objects should be observable in the optical or near IR range, it probably includes the majority of hitherto discovered cases.

The *central stars* have spectral types ranging from O9 to K and IRAS luminosities between 70 and 10^5 $L_{\odot}$. Compared with the sample of stars emitting jets [39], here the higher luminosities are more frequent - an obvious selection effect, since brighter stars are more likely to produce

Table 2: Observed Properties of Young Bipolar Nebulae

Bipolar Nebula	Central Star: Sp	A_V / mag	$L_{IRAS}/L_\odot$	$\dot{M}$ / $M_\odot$/a	v(wind) / km/s	Radio contin.	Lobes: Jet	HH	H	bipol. CO	Torus: pc × pc	$M/M_\odot$	Cloud Core: $M/M_\odot$	n/cm^{-3}	References
L1551-IRS 5	G - K (FU Ori)	20	30	0.8×10^{-6}	300	$\alpha \sim 0$	+	+	+	+	0.1 × 0.01	2	18	10^5	10, 27, 30, 41., 46, 50
Lk Hα 208	F0Ⅴ	0.3	50												9
R Mon	A_eB_e	4	600	1.5×10^{-8}	350	-	+	+		+	0.8 × 0.4	30			4, 5, 8, 10, 26
NS 14	B0.5 - A5	8-17	1.5×10^4			$\alpha = -0.1$				+	0.2 × <0.1	350	500	10^6	19, 44, 48, 59
Boomerang	K0Ⅲ + A0Ⅲ	3	50		100		+	+							49
NGC 6334-Ⅴ		>20	$>1\times10^5$			+			+	+					18, 22, 35, 42, 57
SRN	double pre T Tauri	7	70			-					0.05 × 0.01		220	5×10^4	13, 14, 21
L 810	A5 + G2	5-9	600		100	-	+	+		+			650	10^5	43, 47
GN 20.18.3	B7	20	100		350		+	+					250	10^5	45
S 106	O9.5	21	1.2×10^4	1.6×10^{-6}	220	$\alpha = 0.7$	-	-	+	-	0.1 × 0.02	20	320	10^4	1, 12, 17, 36, 38, 56
PV Cep	F2	5	100	2×10^{-6}	400		+	+		+			10	10^4	33, 44, 49
Cep A	B0.5	75	2×10^4	10^{-6}	400	$\alpha = 0.6$	+	+	+	+			500	10^4	10, 25, 31, 32, 58

observable reflection nebulae, while their eventually associated jets are more difficult to observe. Both samples together show that the bipolar phase is characteristic of a wide range of spectral types and luminosities (spectral types earlier than M2, luminosities $L > 0.5\ L_{\odot}$). Thus the evolution of PMS stars within a wide mass interval goes through this phase.

The high *extinction* values, taken together with the presence of bright nebulosities, clearly hint at the existence of circumstellar disks and tori. The *mass loss* rates all cluster around $10^{-6}\ M_{\odot}/a$, with the exception of R Mon, which possibly has almost reached the end of its outflow phase.

Two of the three OB stars (in S106 and in Cep A) have *radio spectra* typical of thick, ionized expanding envelopes ($\alpha \approx 0.7$). In Cep A an HII region has not yet developed, while in S106 its continuum radiation can be separated from that of the envelope. In NS 14 the HII region is already present (since only about 2000 years), but it is too compact to be separated from the contribution of an eventually present, much weaker envelope. This suggests a sequence of increasing age: Cep A - NS14 - S106.

The features characteristic for shock excitation within the bright *lobes* are found in almost all cases. In the case of S106, which is permeated by ionizing photons, only H_2 emission can be observed outside the expanding HII region.

Four massive circumstellar *tori* have been observed. Interestingly, in the most massive example (NS14) almost the entire material within the associated dense core is concentratet within the torus. Finally, the masses of the *dense cores* in this sample cover the whole range found by Benson and Myers [6].

In order to classify these sources in terms of their PMS evolution, we compare their IRAS colours and luminosities with those of the IRAS sources embedded in the star forming region L 1641. In this region, Fukui et al. [20] have performed an unbiased search for bipolar CO outflows. They found that, of all embedded sources, those associated with bipolar outflows are the coolest ones. This indicates that the molecular outflow phase takes place early in the evolution of the sources, when they are most deeply embedded in the dense cores from which they have formed. The molecular outflow apparently causes the removal of these cores. Moreover, the fact that the sources with CO outflow have higher IRAS luminosities than those without outflow might indicate that the luminosity is enhanced by accretion from the dense circumstellar matter onto a circumstellar disk.

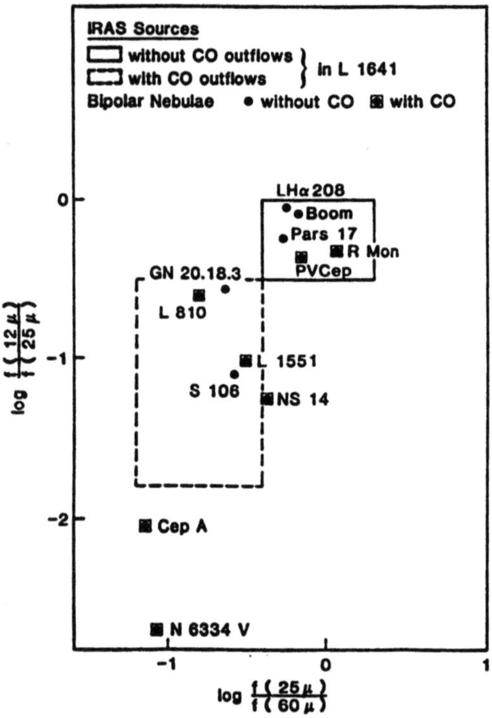

Figure 15: Two-colour diagram of embedded IRAS sources. The two boxes indicate the approximate positions of sources in L 1641 with and without associated CO outflows. The sources associated with bipolar nebulae are from Table 2.

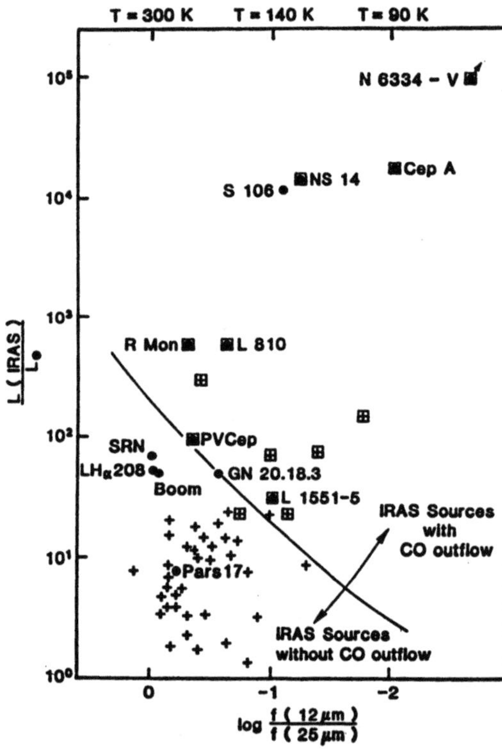

Figure 16: IRAS luminosity vs. 12 μm/25 μm flux ratio. Crosses: all sources in L1641. Filled circles: sources associated with bipolar nebulae (from Table 2). Squares mark those sources which are associated with bipolar CO outflows.

Figure 15 shows the two-colour diagram constructed from the IRAS 12 μm, 25 μm and 60 μm fluxes. The IRAS sources in L 1641 can be divided essentially into two groups: the hotter ones, displaced towards the upper right of the diagram, consist mainly of classical T Tauri stars. The cooler ones, displaced towards the lower left, are those found to be associated with bipolar molecular outflows [20]. Our sample of young bipolar nebulae spans the whole area covered by the L 1641 sources. Here again the cooler sources are those associated with bipolar CO outflows; one drastic exception is S106 - but in this case the outflowing material is ionized and cannot be detected in the molecular line emission. Cep A and NGC 6334-V, the two bipolar nebulae which are observable only in the near IR, have embedded IRAS sources cooler than any source in L 1641.

In Fig. 16 the integrated IRAS luminosities of the sources in L 1641 and of those within bipolar nebulae are plotted against their 12 μm/25 μm flux ratios. The T Tauri stars with no CO outflow are clustering in the lower left, being fainter and hotter, while the sources with molecular outflow are brighter and cooler (upper right). Again, the IRAS sources associated with bipolar nebulae exhibit similar statistical properties. The sources in Cep A and NGC 6334-V are brighter *and* cooler than any source in L1641.

In conclusion, these young bipolar nebulae represent those rare cases in which, due to the strongly asymmetric distribution of the circumstellar matter in a wide range of scales, and to a fortunate orientation of our line of sight, the youngest and most deeply embedded stellar objects and their environment become amenable to detailed studies in the optical and near infrared range.

References

1 Bally, J., Scoville, N.Z.: 1982, Astrophys. J. **255**, 497
2 Bassony, M., Scoville, N.Z., Bally, J., Claussen, M.J.: 1989, Astrophys. J. **343**, 212
3 Bastien, P., Ménard, F.: 1988, Astrophys. J. **326**, 334
4 Beckwith, S., Sargent, A.I., Scoville, N.Z., Masson, C.R., Zuckerman, B., Philipps, T.G.: 1986, Astrophys. J. **309**, 755
5 Beckwith, S., Zuckerman, B., Strutskie, M.F., Dyck, H.M.: 1984, Astrophys. J.. **287**, 793
6 Benson, P.J., Myers, P.C.: 1989, Astrophys. J. Suppl. **71**, 89
7 Bieging, J.H.: 1984, Astrophys. J. **286**, 591
8 Brugel, E.W., Mundt, R., Bührke, Th.: 1984, Astrophys. J. (Letters) **287**, L939
9 Calvet, N., Cohen, M.: 1978, Mon. Not. R.A.S. **182**, 687
10 Cantó, J., Rodriguez, L.F., Barral, J.F., Carral, P.: 1981, Astrophys. J. **244**, 102
11 Cudworth, K.M., Herbig, G.H.: 1979, Astron. J. **84**, 548
12 Eiroa, C., Elsässer, H., Lahulla, J.F.: 1979, Astron. Astrophys. **135**, 261
13 Eiroa, C., Leinert, Ch.: 1987, Astron. Astrophys. **188**, 46
14 Eiroa, C., Lenzen, R., Leinert, Ch., Hodapp, K.-W.: 1987, Astron. Astrophys. **179**, 171
15 Elsässer, H., Staude, H.J.: 1978, Astron. Astrophys. **70**, L3
16 Felli, M., Panagia, N.: 1982, Astrophys. J. **262**, 650
17 Felli, M., Staude, H.J., Reddmann, Th., Massi, M., Eiroa, C., Hefele, H., Neckel, Th., Panagia, N.: 1984, Astron. Astrophys. **135**, 261
18 Fisher, J., Joyce, R.R., Simon, M., Simon, T.: 1982, Astrophys. J. **258**, 165

19 Fukui, Y.: 1989, in: ESO Workshop on Low Mass Star Formation, Ed. B. Reipurth, p. 95
20 Fukui, Y., Iwata, T., Takaba, H., Mizuno, A., Ogawa, H., Kawabata, K., Sugitani, K.: 1989, Nature **342**, 161
21 Gomez de Castro, A.I., Eiroa, C., Lenzen, R.: 1988, Astron. Astrophys. **201**, 299
22 Harvey, P.M., Wilking, B.A.: 1984, Astrophys. J. (Letters) **280**, L19
23 Hodapp, K.-W.: 1987, Astrophys. J. **319**, 842
24 Hodapp, K.-W.: 1990, Astrophys. J. **352**, 184
25 Hughes, V.A.: 1985, Astrophys. J. **298**, 830
26 Jones, B.F., Herbig, G.H.: 1982, Astron. J. **87**, 1223
27 Kaifu, N., Suzuki, S., Hasegawa, T., Morimoto, M., Inatani, J., Nagane, K., Miyazawa, K., Chikada, Y., Kanazawa, T., Akabane, K.: 1984, Astron. Astrophys. **134**, 7
28 Kaifu, N., Hayashi, S.S.: 1987, Proc. IAU Symp. 115, Star Forming Regions, p. 369
29 Lane, A.P.: 1989, in: ESO Workshop on Low Mass Star Formation, Ed. B. Reipurth, p. 331
30 Lenzen, R.: 1987, Astron. Astrophys. **173**, 124
31 Lenzen, R.: 1988, Astron. Astrophys. **190**, 269
32 Lenzen, R., Hodapp, K.-W., Solf, J.: 1984, Astron. Astrophys. **137**, 202
33 Levreault, R.M.: 1984, Astrophys. J. **277**, 634
34 Levreault, R.M.: 1985, Technical Report No. 85-1 (University of Texas)
35 Longhram, L., McBreen, B., Fazio, G.G., Rengarajan, T.N., Maxson, C.W., Serio, S., Sciortino, S., Ray, T.P.: 1986, Astrophys. J. **303**, 629
36 Longmore, A.J., Robson, E.I., Jameson, R.F.: 1986, Mon. Not. R.A.S. **221**, 589
37 Lucas, R., Le Squéren, A.M., Kazés, J., Encrenaz, P.J.: 1978, Astron. Astrophys. **66**, 155
38 Mezger, P.G., Chini, R., Kreysa, E., Wink, J.: 1987, Astron. Astrophys. **182**, 127
39 Mundt, R., Brugel, E.W., Bührke, Th.: 1987, Astrophys. J. **319**, 275
40 Mundt, R., Fried, J.W.: 1983, Astrophys. J. (Letters), **274**, L83
41 Mundt, R., Stocke, J., Strom, S.E., Strom, K.M., Anderson, R.E.: 1985, Astrophys. J. (Letters) **297**, L41
42 Nakagawa, T., Tetsuya, N., Matsuhara, H., Okuda, H., Shibai, H., Hayashi, S.S.: 1990, Astrophys. J. **351**, 573

43 Neckel, Th., Chini, R., Güsten, R., Wink, J.E.: 1985, Astron. Astrophys. **153**, 253
44 Neckel, Th., Staude, H.J.: 1984, Astron. Astrophys. **131**, 200
45 Neckel, Th., Staude, H.J.: 1987a, Astrophys. J. (Letters) **320**, L145
46 Neckel, Th., Staude, H.J.: 1987b, Astrophys. J. (Letters) **322**, L27
47 Neckel, Th., Staude, H.J.: 1990, Astron. Astrophys. **231**, 165
48 Neckel, Th., Staude, H.J., Meisenheimer, K., Chini, R., Güsten, R.: 1989, Astron. Astrophys. **210**, 378
49 Neckel, Th., Staude, H.J., Sarcander, M., Birkle, K.: 1987, Astron. Astrophys. **175**, 231
50 Sarcander, M., Neckel, Th., Elsässer, H.: 1985, Astrophys. J. (Letters) **288**, L51
51 Scarrott, S.M., Brosch, N., Ward-Thompson, D., Warren-Smith, R.F.: 1986, Mon. Not. R.A.S. **223**, 505
52 Simon, M., Fisher, J.: 1982, Bull. Am. Astr. Soc. **14**, 925
53 Snell, R.L., Dickman, R.L., Huang, G.-L.: 1990, Astrophys. J. **352**, 139
54 Snell, R.L., Loren, R.B., Plambeck, R.L.: 1980, Astrophys. J. (Letters) **239**, L17
55 Solf, J., Carsenty, U.: 1982, Astron. Astrophys. **113**, 142
56 Staude, H.J., Lenzen, R., Dyck, H.M., Schmidt, G.D.: 1982, Astrophys. J. **255**, 95
57 Straw, S.M., Hyland, A.R., McGregor, P.J.: 1989, Astrophys. J. Suppl. **69**, 99
58 Weliachew, L., Emerson, D.T., Forvcille, T., Bieging, J.H., Wright, M.C.H., Ellder, J.: 1985, Astron. Astrophys. **153**, 139
59 White, G.J., Gee, G.: 1986, Astron. Astrophys. **156**, 301

Winds of Luminous Blue Variables

O. Stahl

Landessternwarte Königstuhl, D-6900 Heidelberg, Fed. Rep. of Germany

1. Introduction

The class of the Luminous Blue Variables (LBV's) includes the original Hubble-Sandage variables [4], the S Dor variables in the Large Magellanic Cloud (LMC) and several peculiar objects in the Galaxy (η Car, P Cyg, AG Car, HR Car etc.). The LBV's are, except for supernovae, the visually brightest stars in galaxies. Due to their high luminosities and their large light amplitude, they can be found and studied also in distant galaxies.

The stellar winds of these stars are very strong and highly variable, which makes them interesting objects for the study of stellar winds. In fact, the stellar winds are so dense, that, similar to the case of Wolf-Rayet stars, the spectral appearance of the LBV's is dominated by stellar wind lines, i.e. a static photosphere can hardly be seen in these stars.

For well known reasons, the brightest stars have traditionally been studied in the Magellanic Clouds, and we also concentrate in the following on the LBV's in the MC's. At present four LBV's have definitely been found in the LMC and none in the SMC. These four LMC LBV's are S Dor [8,6]), R 71 ([14]), R 127 ([10]) and R 110 ([12]). S Dor, although known as an LBV for a very long time, has been observed almost exclusively during relatively bright phases so far. R 127 is well observed both during minimum and maximum phase. It has the largest light amplitude of the LMC LBV's and is also the most luminous one. The stars R 71 and R 110 are less luminous and probably are close to the lower brightness limit for LBV's. In the following we first shall discuss the outburst of R 127, especially the stellar wind variations during the outburst. Then we discuss the outburst of the less luminous star R 110 and finally we describe new observations of S Dor, which shed some new light on the stellar winds of LBV's.

The following paper is based mainly on photometry obtained within the Long-Term Photometry of Variables (LTPV) program at ESO [13], which started in 1982, on data obtained with the ESO Cassegrain echelle spectrograph (CASPEC) and on spectra obtained with the International Ultraviolet Explorer (IUE) satellite.

2. The outburst of R 127

R 127 was first observed in outburst in 1982 [10]. Before the outburst, the star was classified as an Ofpe/WN9 star by Walborn [15,16]. At this time the visual magnitude was about 11.2. During the outburst, the star gradually became with V=8.8 the visually brightest star of the LMC [17]. Within this period, the spectral appearance changed from the original Ofpe/WN9 spectrum to a B-type spectrum similar to P Cyg and finally to an A-type spectrum similar to S Dor in outburst. During this time, the temperature decreased from about 30,000 K to 8,000-9,000 K. From photometry, especially in the UV, it is clear, that the bolometric luminosity did not change significantly during the outburst, i.e. the outburst in the visual is due only to a re-distribution of the flux from the UV to the optical and a corresponding increase of the stellar radius.

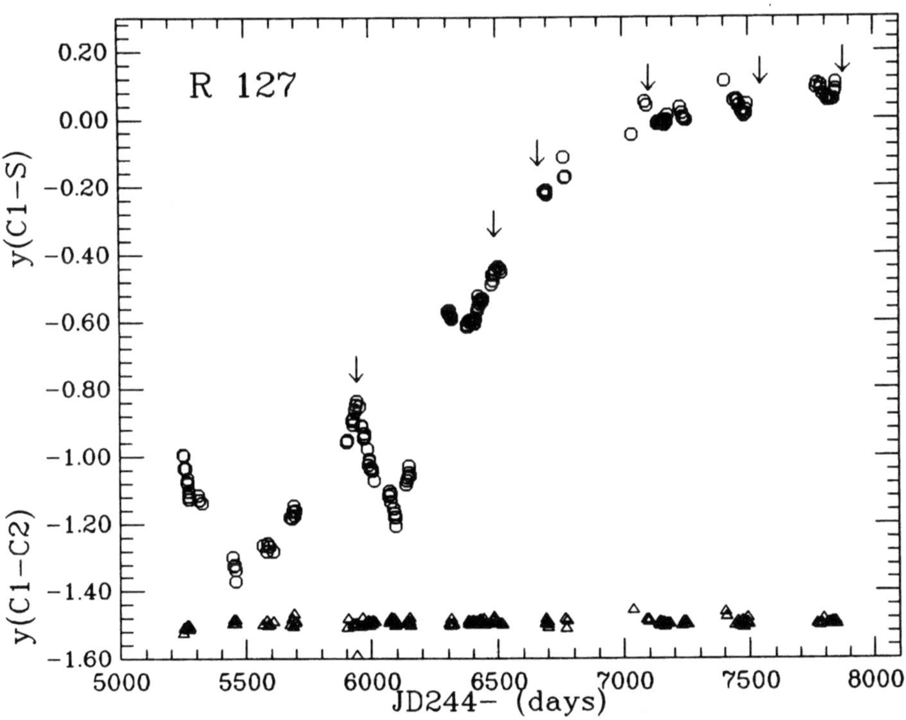

Fig. 1. Differential y-lightcurve of R 127 between 1982 and 1990 obtained within the LTPV program. The mean y-magnitudes of the comparison stars C1=HD 37772 and C2=HD 37584 are 8.89 and 8.33, respectively. Arrows indicate where CASPEC spectra habe been taken.

During the early phase of the outburst shells with discrete expansion velocities have been observed [10,17]. These shells disappeared in the later phases of the outburst. The expansion velocity of the stellar wind decreased during the

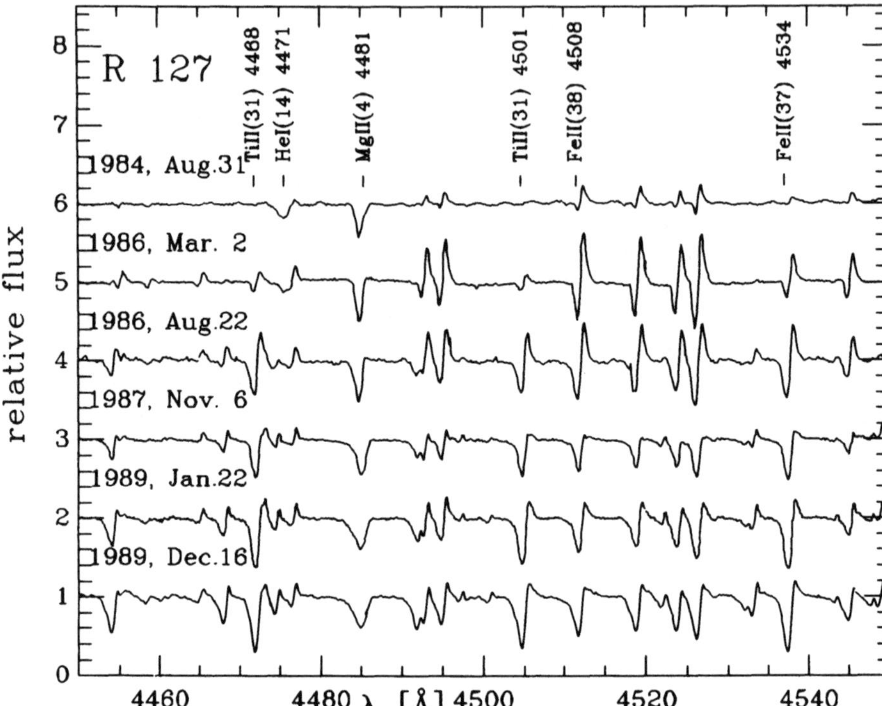

Fig. 2. The spectral evolution of R 127, obtained from CASPEC spectra, between Aug. 84 and Dec. 89, when the star brightened from 9.7 to 8.8. The HeI-lines in the spectrum disappeared during this period. FeII- and later TiII-, CrII-, etc. lines appeared strongly.

outburst by about a factor 3. This can be understood in the framework of radiation driven stellar winds, which predict that the expansion velocity scales with the escape velocity. This means, for constant bolometric luminosity, that the expansion velocity is proportional to the stellar temperature. The lightcurve and the evolution of the spectrum are shown in Figs. 1 and 2.

Due to the large range of temperature and brightness observed in R 127 during the last decade, it has become a key object for the understanding of the LBV's. The spectral similarity in minimum to the WN stars, during rise to outburst to P Cyg and during maximum to S Dor suggests a close connection between all these classes. Especially the similarity in minimum to the WN-stars clearly indicates that the LBV's are evolved stars close to the WN-phase.

From the Balmer line profiles and the IR excess of R 127, mass-loss rates can be derived. The mass-loss rate of R 127 during outburst is about $5\,10^{-5}\,M_{\odot}\,yr^{-1}$. The expansion velocity of the stellar wind is about $100\,km\,s^{-1}$. These numbers are typical for LBV's in outburst. The temperature during outburst is about 8,000-9,000 K, as derived from the spectral type and the energy distribution.

3. The subluminous LBV R 110

R 110 was first observed in a bright phase in Jan. 89. At that time, it showed an F-type spectrum and it was about $0\overset{m}{.}5$ brighter than before [12]. In Dec. 89, the spectral type was even later, see Fig. 3. Before the outburst, it was a suspected variable classified as B6 to A5 by different authors [11]. The correlated light and color changes of the star are typical for LBV's. The luminosity is at the lower limit for an LBV, but it is close to the luminosity of R 71, which is a well-established LBV. However, the star has several peculiar properties. First, its mass-loss rate during outburst is exceptionally low (see Fig. 4) and, second, its temperature is also exceptionally low (see Fig. 5). The observed amplitude of the variations of only $0\overset{m}{.}5$ is also unusual for LBV's.

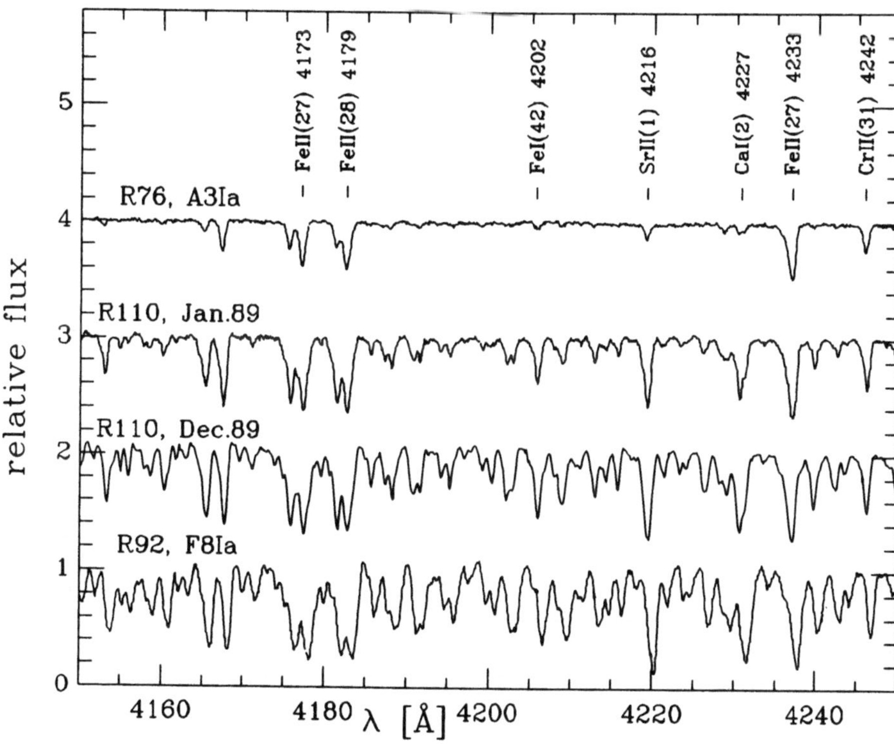

Fig. 3. Comparison of spectra of R 110 and several "normal" LMC hypergiants. By comparison with these stars R 110 can be classified as F0Ia and F5Ia in Jan. 89 and Dec. 89, respectively.

From the Hα line a mass-loss rate of the order of $10^{-6}\,M_\odot\,yr^{-1}$ [12] was derived, which is not higher than expected for a normal star of the luminosity of R 110 ($M_{bol} \approx -8.9$) and also not significantly higher than before the outburst. This is in strong contrast to classical LBV's like R 127, which have during outburst mass-loss rates about a factor of 50 higher. Classical LBV's also have during outburst higher mass-loss rates than during minimum.

From the continuum energy distribution observed in Jan. 89 from ground-based photometry and low dispersion IUE spectra, a temperature of $\lesssim 7,500$ K is derived [12]. This is in agreement with the spectral type of F0Ia. It is at the lower limit of the temperature range where radiation pressure may drive the expansion of the star. The temperature continued to decrease from Jan. 89 to Dec. 89 and probably reached a value of about 6,500 K, which is probably too low to be explained by radiation pressure alone.

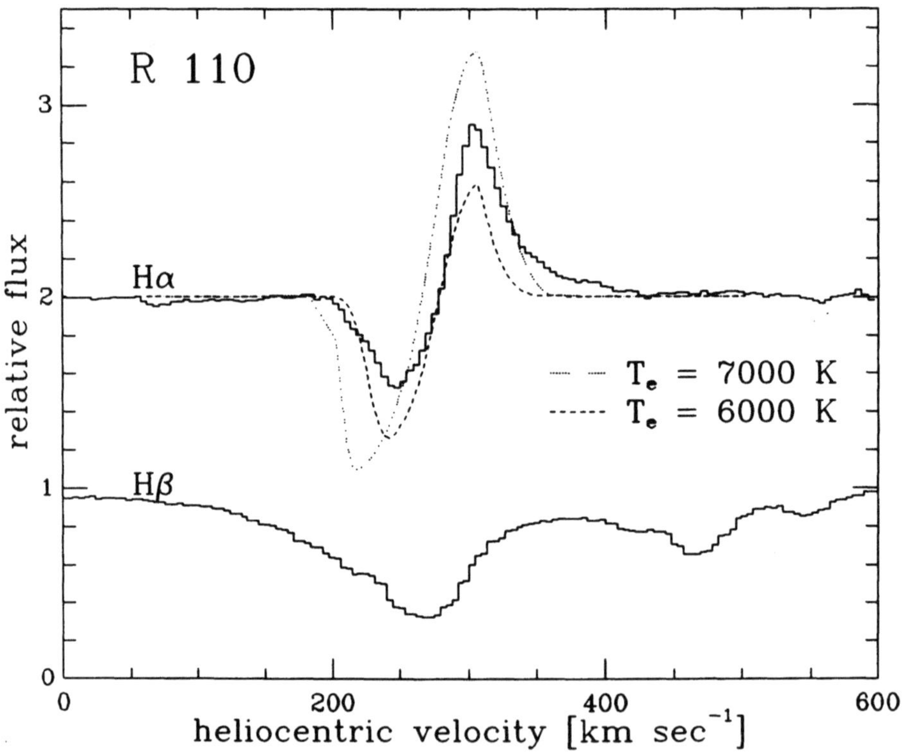

Fig. 4. The Hα and Hβ line profiles of R 110, obtained in Jan. 89. By comparison with model calculations, a mass-loss rate of about 10^{-6} $M_\odot$ yr^{-1} is derived from the Hα line.

The peculiar properties of R 110 are interesting for several reasons. The small mass-loss rate suggests that a mass-loss instability may in this case not be the trigger of the outbursts. This is important for outburst models of LBV's [1,2]. The low temperature indicates that radiation may not be the only driving force for the expansion of the star, since the radiation pressure decreases in all models drastically below a temperature of about 7,500 K.

With its low temperature and luminosity for an LBV, R 110 is located in the HRD between the LBV's and the ρ Cas stars, which have spectral types

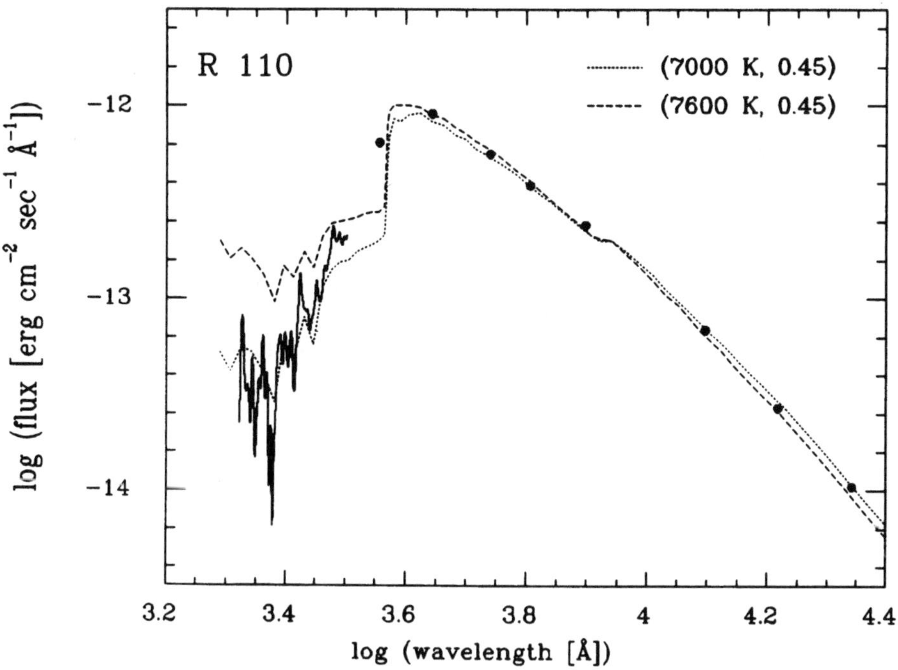

Fig. 5. The continuum energy distribution of R 110, as obtained in Jan. 89, compared with model atmosphere calculations.

between F and M. If R 110 is indeed a transition object between the LBV's and the ρ Cas variables, the two classes may be closely related, as has been suggested before [3].

The spectrum of R 110 during outburst looks so normal that it cannot be distinguished from the spectrum of a normal hypergiant. Also before outburst, i.e. in minimum of visual light, the spectrum of R 110 was not very peculiar. This property suggests that at least some apparently "normal" stars are indeed LBV's is some "stand-still" phase. Since the spectrum of R 110 is so normal, it can be analyzed by model-atmosphere techniques [12], which is much more difficult for other LBV's. From the models, temperature and surface gravity of the star can be derived. From the surface gravity, the stellar mass can be calculated to 10 $M_{\odot}$, which is surprisingly small. The small mass is in agreement with evolutionary tracks [7] only, if we assume that the star is in the post-red-supergiant phase. This is probably not the case for more luminous LBV's, since these stars are above the luminosity limit, where red supergiants are observed [5]. This, and the peculiar properties of R 110, probably means that we observe at least two distinct classes of LBV's which are in different evolutionary phases.

4. New results on S Dor

S Dor, discovered as a variable star already in 1897 [8], has for a long time been the only known LBV in the LMC. Therefore it has already been studied in considerable detail. For reference to older literature see e.g. [6].

Observations during minimum are sparse, since the star did not show a deep minimum for about 25 years. During a shallow miniumum observed about 1984, the spectral type was middle-B, with pronounced [FeII] emission lines. During maximum, S Dor shows an A-type spectrum, where almost all lines in the optical show P Cyg profiles and all lines in the UV are blue-shifted [6,9]. This type of spectrum is often described as S Dor-like.

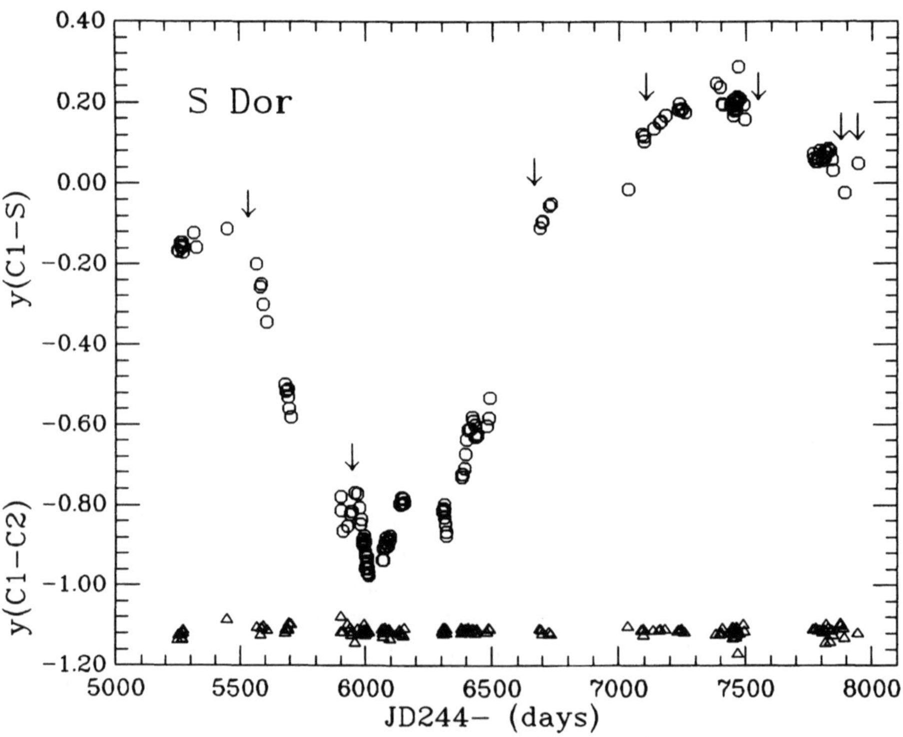

Fig. 6. Same as Fig. 1 for S Dor. The mean y-magnitudes of the comparison stars C1=HD 35293 and C2=HD 35294 are 9.25 and 8.48, respectively.

The light and spectral evolution during the last years is shown in Figs. 6 and 7. During this period, the star showed a shallow mimimum, but most of our spectra have been obtained when the star was quite bright. The spectra of S Dor and R 127 during bright phases are strikingly similar, even in small details (cf. Figs. 2 and 7).

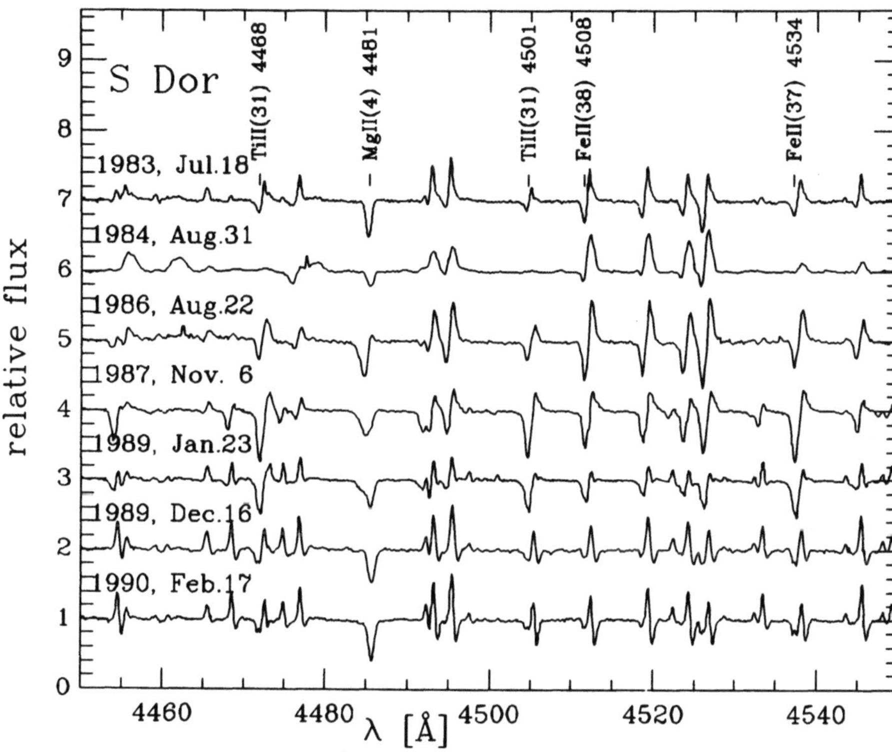

Fig. 7. The spectral evolution of S Dor, obtained from CASPEC spectra, between Jul. 83 and Feb. 90. HeI- and [FeII]-lines are present only during the minimum. The general appearance of the spectrum is strongly correlated with the light variations.

Around 1988, the star had a particularly bright phase where it reached with V=9.1 the brightest magnitude observed so far. Right after this phase, during a period of slow decline, is now shows a very remarkable spectrum, never observed so far for S Dor or for any other LBV [18]. As discussed earlier, LBV's normally show very strong stellar wind effects in most lines, at least during outburst. However, now S Dor shows strong inverse P Cyg profiles, i.e. absorption redward of the emission, in many lines, cf. Figs 7 and 8.

Faint absorptions redwards of the emission component have been observed [9] in S Dor earlier, but these have always been very faint. On our spectra taken in Dec. 89 and Feb. 90, the red component is stronger than the blue one in most lines in the optical range. Only the strong lines, like e.g. the strongest FeII-lines or the Balmer lines, still show a blue-shifted absorption component. The red-shifted absorption component was already faintly visible on spectra taken in Jan. 89. Although no spectra have been taken between Jan. 89 and Dec. 89, we assume that the phase of the inverse P Cyg components last at least one year. During this time, the red component strengthened considerably and shows so far no evidence for a decrase.

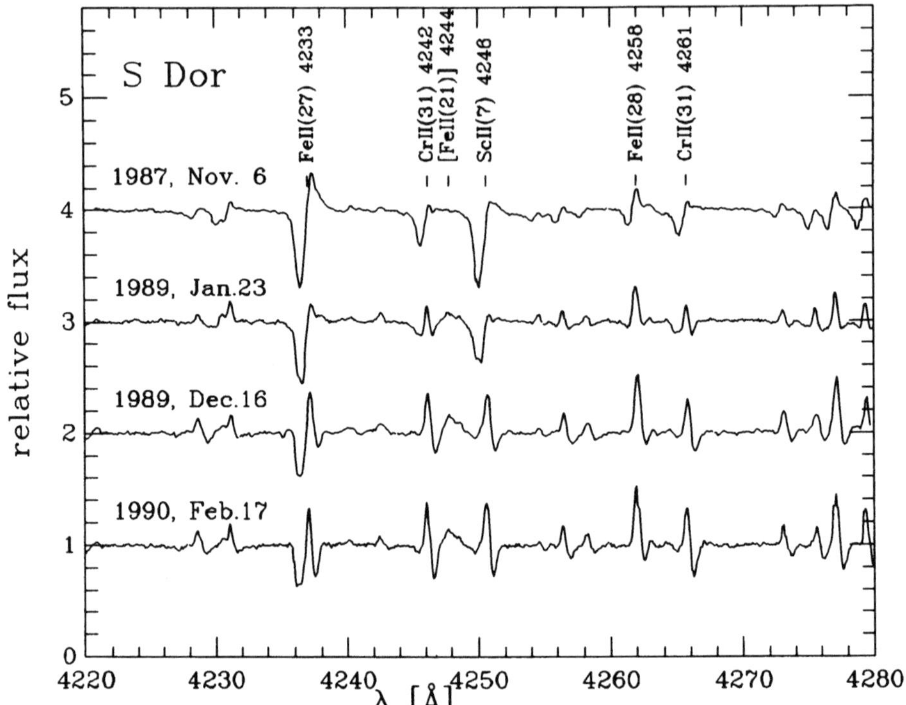

Fig. 8. The spectral evolution of S Dor between Nov. 87 and Feb. 90. During this phase the star was with V=9.1 brighter than ever before. The line profiles show strong evidence for infalling matter from Jan. 89 on, when the star started its decline from the very bright maximum. Only the stronger lines, like e.g. FeIIλ4233 continue to show the blue-shifted absorption.

The observed infall velocity (i.e. the red edge of the red absorption component compared to the systemic velocity) is about 90 $\mathrm{km\,s^{-1}}$, i.e. almost as high as the observed blue-edge velocity. The observed outflow velocity decreased during the "infall phase" from about 120 to about 70 $\mathrm{km\,s^{-1}}$. The only line in the optical spectral range which normally does not show a P Cyg emission, the MgIIλ4481 line, has a mean velocity of $+30\,\mathrm{km\,s^{-1}}$, compared to $-30\,\mathrm{km\,s^{-1}}$ in Nov. 87.

The explanation for the "infall phase" of S Dor is not easy. It is surprising, that the phase is not accompanied by strong changes of the brightness. The fact that the phase began after a very bright phase, which probably was a phase of extreme mass loss, suggests that the star expelled during this phase more mass than it could drive to infinity. The observed time scale of the "infall phase" allows to estimate the distance from where the matter is falling back. Using a velocity of 90 $\mathrm{km\,s^{-1}}$ and a time scale of one year, about 10 R_* are derived for this distance, which is surprisingly large.

5. Winds and outbursts of Luminous Blue Variables

It has been suspected for a long-time that the strong winds of the LBV's and their outbursts are related, specifically that the strong winds of the LBV's cause the outbursts via some mass-loss instability.

The observations of the LMC LBV's allow to draw several conclusions on the relation between the stellar winds of the LBV's and their outbursts. As shown in Fig. 9, the temperatures of the LBV's during outburst are all similar, about 8,000-9,000 K. This is in good agreement with the model of instabilities of radiative atmospheres, which predicts that the expansion of the star stops at a temperature of about 7,500 K [1,2]. This behaviour is due to the sudden drop of opacity at this temperature. In fact, this observation is one of the main arguments for the models.

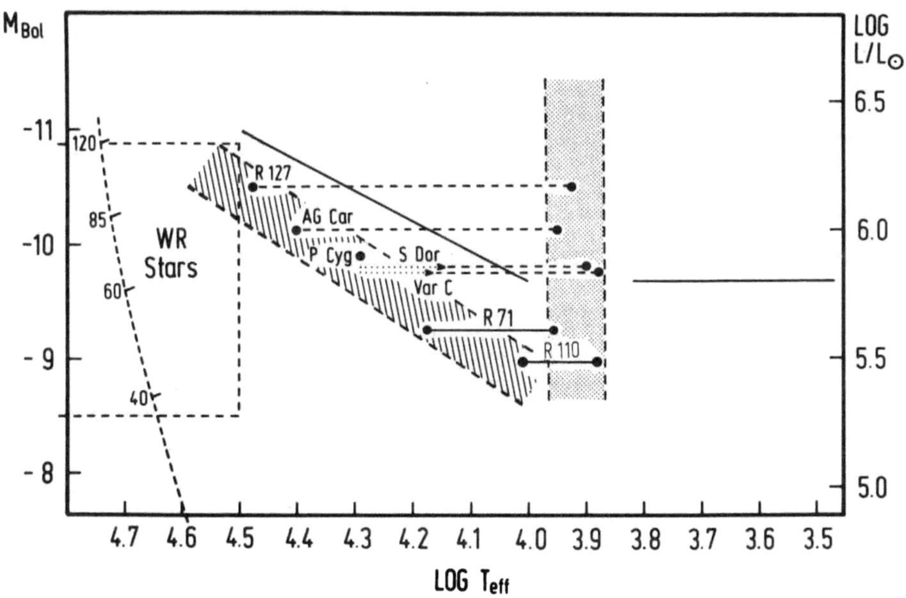

Fig. 9. The location of the LBV's in the HRD. During minima, the stars populate an inclined instability strip close to the Humphreys-Davidson limit. During maximum the temperatures are remarkably similar, with the exception of R 110.

A few recent observations cause problems for the pure radiative instability models, however:

The star R 110 is now getting too cool to be explained by the models. Its mass-loss rate is also so small, that it seems unlikely that a mass-loss instability is responsible for the outburst. Possibly, another mechanism, like e.g. turbulent pressure, takes over at lower luminosity and temperature. One might also suspect that some other mechanism leads to the expansion of the LBV's in general and the high mass-loss rates observed during outburst are not the cause of the outburst but just a consequence of the expansion of the star.

The "infall phase" observed in S Dor, without much change in brightness or temperature, is also not easily explained by purely radiative models. However, time-dependent hydrodynamic models are needed to prove or disprove the models conclusively.

References

[1] Appenzeller, I.: 1989, in *IAU Coll. No 113, Physics of Luminous Blue Variables*, eds. K. Davidson, A.F.J. Moffat, H.J.L.G.M. Lamers, Dordrecht, Kluwer, p. 195

[2] Davidson, K.: 1987, *Astrophys. J.* **317**, 760

[3] de Jager, C., van Genderen, A.M.: 1989, in *IAU Coll. No 113, Physics of Luminous Blue Variables*, eds. K. Davidson, A.F.J. Moffat, H.J.L.G.M. Lamers, Dordrecht, Kluwer, p. 127

[4] Hubble, E., Sandage, A.: 1953, *Astrophys. J.* **118**, 353

[5] Humphreys, R.M., Davidson, K.: 1979, *Astrophys. J.* **232**, 409

[6] Leitherer, C., Appenzeller, I., Klare, G., Lamers, H.J.G.L.M., Stahl, O., Waters, L.B.F.M., Wolf, B.: 1985, *Astron. Astrophys.* **153**, 168

[7] Maeder, A.: 1987, *Astron. Astrophys.* **173**, 247

[8] Pickering, E.C.: 1897, *Harvard Circ.* No. 19

[9] Stahl, O., Wolf, B.: 1982, *Astron. Astrophys.* **110**, 272

[10] Stahl, O., Wolf, B., Klare, G., Cassatella, A., Krautter, J., Persi, P., Ferrari-Toniolo, M.: 1983, *Astron. Astrophys.* **127**, 49

[11] Stahl, O., Wolf, B., de Groot, M., Leitherer, C.: 1985, *Astron. Astrophys. Suppl.* **61**, 237

[12] Stahl, O., Wolf, B., Klare, G., Jüttner, A., Cassatella, A.: 1990, *Astron. Astrophys.* **228**, 379

[13] Sterken, C.: 1983, *ESO Messenger* **33**, 10

[14] Thackeray, A.D.: 1974, *Mon. Not. Roy. Astr. Soc.* **168**, 221

[15] Walborn, N.R.: 1977, *Astrophys. J.* **215**, 53

[16] Walborn, N.R.: 1982, *Astrophys. J.* **256**, 452

[17] Wolf, B., Stahl, O., Smolinski, J., Cassatella, A.: 1988, *Astron. Astrophys. Suppl.* **74**, 239

[18] Wolf, B., Stahl, O.: 1990, *Astron. Astrophys.*, in press

The Hubble Space Telescope Before Launch: A Personal Perspective

H. Jenkner[1]

Space Telescope Science Institute[2],
3700 San Martin Drive, Baltimore, MD 21218, USA

1. Introduction

The objective of the Hubble Space Telescope project is to orbit an optical 2.4 m telescope for astronomical observations above the earth's disturbing and absorbing atmosphere. This review attempts to summarize several aspects of this exciting endeavor at a time when the main part of the mission, signified by the anticipated launch in early 1990, has not yet begun. Specifically, a brief history of the Space Telescope is given, the major characteristics of the telescope and its scientific instruments are discussed, the scientific expectations are described, and the Space Telescope Science Institute, the center for the scientific operations of this project, is introduced. As an example of the many developments which have already appeared as part of the Hubble Space Telescope program, but transgress into the direction of more general astronomical applications, the production and characteristics of the Guide Star Catalog are described as well.

2. A Brief History of the Space Telescope Project

As the launch of the Hubble Space Telescope (HST) draws near again, it is time to remember the surprisingly long and sometimes troubled history of this project. The concept of an astronomical observatory in space goes back to the early part of this century, specifically to the writings of Hermann Oberth (1923). Later, Lyman Spitzer (1946) initiated the more concrete discussions of such a project; he has contributed significantly to its existence, and followed its development ever since.

In the early 1960s, the astronomical community recommended in a report of the US National Academy of Sciences for the first time, that a large, diffraction limited telescope for operation above the earth's atmosphere be built. During the fall of 1971, the National Aeronautics and Space Administration (NASA) began a detailed study of such a project, featuring a 3 m telescope, then called the Large Space Telescope. However, it was not until 1977, after considerable political difficulties had been overcome (Hanle 1985, O'Dell 1982), that the US Congress gave approval to actually build and operate the Space Telescope in its present form. By this time, the decision had been made to enter into a partnership

[1] Affiliated to the Astrophysics Division, Space Science Department, European Space Agency.

[2] Space Telescope Science Institute (ST ScI) is operated by the Association of Universities for Research in Astronomy, Inc. (AURA), for the National Aeronautics and Space Administration (NASA).

with the European Space Agency (ESA) on this project; ESA would provide one of the Scientific Instruments, the solar arrays, as well as operations and staff support.

The scientific operation of the Space Telescope was anticipated to be carried out by an organizational entity separate from NASA. Therefore, a Request for Proposal for a Space Telescope Science Institute was issued in 1979. As the successful proposer, the Association of Universities for Research in Astronomy (AURA) began to operate the Space Telescope Science Institute (ST ScI) on the campus of the Johns Hopkins University in Baltimore, Maryland, in 1981.

At that time, the launch of the Hubble Space Telescope was planned for 1984. However, a number of delays caused the launch date to slip into 1986. The Challenger disaster, by grounding HST's launch vehicles for more than two years, led to a further considerable delay. At the time of this writing, the launch of the Hubble Space Telescope is planned for late March 1990.

3. The Telescope and its Instruments

The Hubble Space Telescope will be placed in an approximately circular orbit at about 590 km altitude by the Space Shuttle, at an inclination angle of 28.5°, and with an orbital period of approximately 95 min. Operating well above the earth's atmosphere, HST will observe the ultraviolet, visible, and infrared wavelength regions of the spectrum, achieve an increase in spatial resolution that is seven or more times better than the capability of ground-based observatories, and detect objects approximately 50 times fainter than are presently observable.

Like ground-based telescopes, the Hubble Space Telescope was designed as a general-purpose instrument, capable of using a variety of scientific instruments at its focal plane. By using the Space Shuttle for launch and to provide scientific instrument upgrading and subsystem maintenance, the useful and effective operational lifetime of the Hubble Space Telescope will be extended to more than a decade.

In the following sections, the characteristics of the HST Optical Telescope Assembly itself, as well as those of the initial set of Scientific Instruments are briefly summarized. More extensive overviews can be found in the book edited by Hall (1982), as well as in an article by Bahcall and Spitzer (1982); detailed technical information is contained in the references cited in the various sections.

3.1. The Optical Telescope Assembly

The core of the HST Optical Telescope Assembly (O'Dell 1982, Burrows 1989) is the optical system comprised of a 2.4 m, f/24, Ritchey-Chrètien Cassegrain-type telescope. The surface of the primary mirror is accurate to about 1/60 of a wave at 633 nm; at this wavelength, about 70 % of the energy emanating from a point source passes through a circle of 25 μ in the telescope's focal plane. The focal plane, with a usable field of about 28', is shared by the four axial-bay scientific instruments, and the Wide-Field/Planetary Camera as well as the three Fine Guidance Sensors in the radial bays.

The HST launch weight is approximately 11,600 kg; it is 13.1 m long and has a diameter of 4.27 m at the aft end and of 3.05 m at the forward end. Its overall appearance is dominated by the two solar arrays, provided by the European Space Agency (ESA), with a size of 2.3 m by 11.8 m when unfurled; they provide an average power of about 2400 W.

3.2. The Wide Field/Planetary Camera

The scientific objective of the Wide Field/Planetary Camera (WFPC; Westphal *et al.* 1982, Griffiths 1989) is to provide photometrically and geometrically accurate, multiband images of astronomical objects over a relatively wide field of view, with high angular resolution across a broad range of wavelengths. Its field of view is centered on the optical axis of the telescope and it therefore receives the highest quality images. The WFPC operates in two basic configurations: the wide field mode at f/12.9 provides a field of view of 2.57' × 2.57' with each 15 μ detector pixel subtending 0.10" on the sky; in the planetary camera mode at f/30, the field of view is 66" × 66", and a pixel subtends 0.043" on the sky.

In each mode, the WFPC uses 4 charge-coupled devices (CCDs) of 800 × 800 pixels each. These detectors provide a useful sensitivity in the wavelength range from 115 nm to 1100 nm. Numerous filters, polarizers, and a neutral density ("Baum") spot provide for a wide range of scientific applications.

3.3. The Faint Object Camera

The Faint Object Camera (FOC; Macchetto *et al.* 1982, Paresce 1989), one of the European Space Agency's contributions to the HST program, is a long focal ratio, photon counting device designed to take high resolution two dimensional images of areas of the sky up to 44" × 44" in size with pixel dimensions as small as 0.0072" in the 115 nm to 650 nm wavelength range. It will be the only instrument on board HST to fully utilize the unprecedented spatial resolution capabilities of the Optical Telescope Assembly.

The two prime purposes of the FOC are imagery and photometry of very faint celestial objects and imagery of bright objects at the highest possible resolution available from HST. In principle, the FOC is capable of detecting a star of blue magnitude 30 in a 10 hour exposure with a signal-to-noise ration of 5, and to resolve bright sources in the far UV up to an effective angular resolution of approximately 0.015".

Two separate and independent cameras comprise the FOC: one camera magnifies the image to an effective focal ratio of f/48, while the other magnifies to focal ratios of f/96 or f/288 (by means of a High Resolution Apodizer). The two detectors are two dimensional photon counting devices of identical design, each consisting of a three-stage image intensifier optically coupled to a television camera. The data storage capacity is adapted to an image area of 512 × 512 resolution elements. Wavelength ranges can be selected by various filters; neutral density filters, objective prisms, polarizers, and a long slit spectrographic facility are also available.

3.4. The Faint Object Spectrograph

The Faint Object Spectrograph (FOS; Harms *et al.* 1982, Ford 1985) is designed to allow spectroscopic analysis of physical conditions in faint and often extremely distant astronomical sources. The FOS has two independent optical paths which image a dispersed image of the entrance aperture onto two Digicon detectors. A filter/grating wheel allows selection of six slightly overlapping spectral regions between 110 nm and 800 nm with a resolving power of $\lambda/\Delta\lambda = 1300$, or three spectral regions with a resolving power of approximately 250, covered by two gratings and a prism. A polarizer can be used to measure linear and circular polarization over nearly the entire FOS wavelength range.

Of the two detectors, one Digicon has a bialkali photocathode on a magnesium fluoride faceplate and is sensitive to light between 110 nm and 550 nm. The second Digicon has a trialkali photocathode on a fused silica faceplate and is sensitive to light from 180 nm to 850 nm. Each detector counts photoelectron pulses in 512 independent silicon diode channels. High time resolution (up to 18 msec) and time-tagged observations are also available.

3.5. The Goddard High Resolution Spectrograph

The Goddard High Resolution Spectrograph (GHRS; Brandt *et al.* 1982, Duncan and Ebbets 1989) is basically a Czerny-Turner spectrograph, with dispersion provided by one of six gratings, mounted on a rotating carousel. The GHRS operates in the wavelength range of 105 nm to 320 nm. Three resolving powers are available: a low resolution grating with $\lambda/\Delta\lambda = 2000$ covers the wavelength range of 105 nm to 180 nm; four separate medium resolution gratings will provide spectra with a resolving power of 20,000 over the entire wavelength range; the highest spectral resolution of 100,000 is achieved in two Echelle configurations. Observations of a slightly reddened B0V star, with an exposure time of 1000 sec, a signal-to-noise ratio of 10, and in the 120 nm to 150 nnm wavelength range, can be made to limiting visual magnitudes of 19, 16, and 14, for the three respective resolving powers.

The GHRS contains two magnetically focused photon counting Digicon detectors, which differ only in the nature of their photocathodes: one detector has a CsI photocathode deposited on a LiF window, while the other has a CsTe cathode on the MgF_2 window. The photoelectrons are detected by a linear array of 500 silicon diodes; twelve additional diodes can be used for background determination, target acquisition, and monitoring the high energy charged particle environment in orbit.

3.6. The High Speed Photometer

The High Speed Photometer (HSP; Bless *et al.* 1982, White 1985) exploits the capabilities of HST by making photometric measurements over visual and ultraviolet wavelengths at rates up to 10^5 Hz and by measuring very low amplitude variability. A secondary purpose of the instrument is to measure linear polarization in the near UV.

The HSP has quite an unusual design, in that it has no moving parts; the telescope has to slew to move the target from one filter to another. In total, 23 filters for photometry, covering the wavelength range from 120 nm to 750 nm, and 4 UV filters for polarimetry are available.

There are five detectors in the HSP: four image dissector tubes and one photomultiplier tube. Two of the former have CsTe photocathodes on MgF_2 faceplates (sensitive from 120 nm to 300 nm), and two have bialkali cathodes on suprasil faceplates (sensitive from 160 nm to 700 nm; the photomultiplier features a GaAs photocathode. Three of the image dissectors are used for photometry, while one of the bialkali image dissectors is used for polarimetry.

3.7. The Fine Guidance Sensors

In addition to their operational role in pointing the telescope, the Fine Guidance Sensors (FGSs; Duncombe *et al.* 1982, Taff 1989) can also be used for astronomical observations, similar to other Scientific Instruments aboard HST.

The Fine Guidance Sensors are a system of photomultiplier tubes and white-light amplitude interferometers (Koester prisms), which are used for the fine guidance of HST. Since only two FGSs are required for precision guidance of the telescope, the third can be used to conduct astrometric observations. A precision of 0.003" in the relative positions of three stars should be possible for stars with apparent visual magnitudes in the range from 4^m to 17^m. The Fine Guidance Sensors are also expected to provide a photometric precision of 1% for stars as faint as 17^m.

The FGS is not an imaging device like a camera, but is instead an assembly of two interferometers which determine the tilt of a wavefront relative to the full HST primary mirror (the entrance pupil of the interferometer). This arrangement was chosen because the positional precision which can be achieved by the interferometer is not limited by the resolving power fo the HST, but by the variation of the interference fringe visibility.

4. Scientific Expectations

As the main technical advantages of an astronomical telescope in space are obvious, *i.e.*, its diffraction-limited resolution and the unobstructed wavelength coverage, the main driver for a project of this size and magnitude nevertheless has to be the importance of its scientific return. This was of course recognized by the astronomical community, and led to recommendations and conferences on the anticipated astronomical programs relatively early in the project; major conferences included a workshop jointly sponsored by the European Space Agency and the European Southern Observatory in Geneva (Macchetto *et al.* 1979), an IAU Colloquium in Princeton (Longair and Warner 1979), and special sessions at various IAU General Assemblies. More general summaries of these astronomical expectations can be found in various articles (Longair 1982, Longair 1985).

The first expression of the real programs, which the users of HST wanted to perform, came with the first cycle of General Observer proposals. By the deadline of early October 1988, approximately 560 proposals had been received at the Space Telescope Science Institute, requesting about 10,000 hours of observing time with HST. After the peer review process had been completed, approximately 160 proposals had been approved, with a total of only 1100 hours of observing time being available for this cycle.

Because of their importance, three main topics of astronomical research had been selected earlier, in order to ensure results in these areas soon after launch. These *key projects* are:

- the determination of the extragalactic distance scale;
- the HST medium-deep survey; and
- the quasar absorption line survey.

In general, the results from HST-related research will touch a wide variety of fields. The planetary science to be performed with HST will include the study of the atmospheres and features in planetary environments, the detailed study of planetary rings, and the search for planetary systems around nearby stars. Studies such as the high-resolution imaging of planets carried out by the Voyager missions can be regularly undertaken with HST.

Projects in stellar astrophysics will include the study of globular cluster cores, their populations, ages, and internal dynamics; studies in open clusters; stellar evolution, population, variability, and mass loss. Studies of stellar populations in nearby galaxies such as the Magellanic Clouds and the Andromeda galaxy will also be of high scientific interest.

Studies in the area of the interstellar medium may include its spatial and thermodynamic structure; the nature of the halo gas; studies of supernovae and their remnants; and areas of star formation.

Galaxy and cluster research will focus on the evolution of galaxies; the nuclei and the environment of galaxies; structure and morphology of distant galaxies and clusters; and protogalaxies.

Observations of the intergalactic medium could make significant contributions to the knowledge of the evolution of quasar absorption lines; broad absorption line systems; cooling flows in galaxies; and studies of the galactic halo gas against background quasars.

Quasistellar objects and active galactic nuclei research will probably focus on the detailed imaging and spectroscopy of nuclei and jets; the nature of host galaxies around quasars; the environment of active galaxies and quasars; and the study of gravitational lenses. Kinematic and spectroscopic studies could help to confirm massive central black holes or central stellar remnants of massive starbursts. Studies of the inner regions of active nuclei will give some information on the spatial extent and physical conditions of the narrow and broad line regions.

Finally, in the field of cosmology attention will be given to the determination of an accurate distance scale to galaxies and, therefore, a better value for the Hubble constant and related parameters.

Research with the Hubble Space Telescope will very likely touch upon a wide field of astronomical areas. As results will become available, they will undoubtedly pose more and more questions. An early motto of the HST project may well characterize this situation with regard to its scientific impact: conscious expectation of the unexpected.

5. The Space Telescope Science Institute

The Space Telescope Science Institute (ST ScI), established in 1981, located on the campus of the Johns Hopkins University in Baltimore, Maryland, and having a staff of more than 300, is broadly responsible for selecting, planning, and scheduling science observations with HST, for overseeing the operation of HST's scientific instruments, and for disseminating and archiving the science data (Giacconi 1982, Tucker 1985). ST ScI is divided into several divisions for handling the various components of this unprecedentedly complex mission.

The *Science Program Selection Office* supports the Director in establishing and implementing guidelines for the selection of the observing proposals which will comprise the HST science program, including interaction with the world-wide astronomical community and scientific advisory committees. It is also responsible for implementing and conducting the peer review selection process which judges observing proposals for scientific merit and technical feasibility.

The *Science Computing and Research Support Division* is responsible for the development of computer software and systems designed to analyze scientific data from HST. It provides software and data analysis support to ST ScI staff and HST users, operates the data archive, and provides research laboratory support. It also contains the personnel to maintain and enhance the Guide Star Catalog, as well as the related other catalogs and survey plate archives.

The *Science Programs Division* provides science instrument and calibration expertise through the instrument scientists who staff the Telescope and Instrument Branch. The Science Planning Branch establishes the long-range observing program by planning and scheduling observations to maximize spacecraft efficiency in order to gain the greatest scientific return. The User Support Branch provides coordinated support to the astronomical community, both during the proposal preparation and selection phase and during the visits to ST ScI to make the observations and reduce the data.

Comprising nearly a quarter of ST ScI's staff, the *Operations Division* carries out the daily operations of the Science Operations Ground System, such as weekly planning and scheduling, scientific "mission control" for HST observations, and monitoring of science data and engineering performance of its on-board instruments.

It is the responsibility of the *Science and Engineering Systems Division* to maintain and further develop the operational systems and software, telescope commanding procedures, and operational databases. In addition, it performs analyses of observatory systems and provides engineering support.

More than a data processing facility, the Space Telescope Science Institute is an international astronomical center where staff astronomers conduct their own research. To provide a stimulating intellectual environment, the *Academic Affairs Division* maintains the ST ScI library and organizes workshops, scientific colloquia, and weekly lectures. It also manages graduate student, post-doctoral and visiting fellow programs, and is ST ScI's inteface with the Johns Hopkins University.

The *Educational and Public Affairs Office* provides the public, news media, and educators with accurate information on Space Telescope projects and on-going research conducted at ST ScI. It also manages the Astronomy Visualization Laboratory.

In addition to the Space Telescope Science Institute, the Space Telescope European Coordinating Facility (ST-ECF), located on the premises of the European Southern Observatory (ESO) in Garching near Munich, FRG, will play a major role with regard to the Hubble Space Telescope project from the European point of view (Macchetto 1982). The ST-ECF's responsibilities include archiving, cataloging, retrieving, and disseminating HST data in Europe. In addition, it will provide coordination and information with regard to the proposal process, perform as a support center for software and data analysis needs, and serve as a centralized repository for HST-related knowledge.

In addition to the functional responsibilities of the Space Telescope Science Institute during the operational phase of the HST project, it has already provided a number of significant contributions to the astronomical community. These include, among others, the Guide Star Catalog and its related systems (see next section), the Calibration Data Base System (which includes data and software which are also useful outside their HST context), the Space Telescope Science Data Analysis System (STSDAS), developments with regard to data archiving (the Data Management Facility) and computer networking.

6. The Guide Star Catalog

One of the first responsibilities of the Space Telescope Science Institute was the development of a system to support the pointing of HST. Because this project is a prime example for pre-launch development and transgresses the line of purely project-oriented work to more general astronomical applications, and because of the involvement of the author in this project, it is described in more detail.

The Hubble Space Telescope will use off-axis guide stars to achieve its pointing performance. The selection of these guide stars uses a catalog specifically constructed for this task, the Space Telescope Guide Star Catalog (GSC), which consists of $\sim 2 \times 10^7$ stars and other objects in the 9^m to 16^m range.

6.1. Evolution of the GSC Concept

The concept of a Guide Stars Selection System (GSSS) originally envisioned for the HST by the Astrometry Science Team (Benedict 1979) featured a production scanning and astrometry system in which specific target fields would have been scanned and analyzed in response to their being incorporated into observing plans. While such a "schedule-driven" concept would have worked, the difficulty of doing such delicate work and of maintaining a suitably skilled staff for a program demanding two-shift operations over fifteen years motivated the search for a more automated approach.

This, in addition to the realization that many HST observations did not require the full 0.33" astrometric precision (*e.g.*, just placing an object within one of the camera apertures has no particular precision requirement), led to the concept of using relatively coarse (50μ) scans to construct an all-sky catalog of "rough coordinates" that could be used for scheduling those observations with minimal precision requirements and also for planning further work on fields requiring greater precision. The idea then was that improved coordinates would be obtained later by re-scanning the target area (and perhaps a few guide stars) at higher resolution to support the astrometrically most demanding observations.

The throughput requirement for building such a catalog in time for the then scheduled HST launch (early 1985 at the time of the GSSS design review) was to scan and process one plate in twelve hours. These constraints led to 50μ scans made at 200 mm/s. Further testing showed that this configuration with a suitably apodized aperture (*e.g.*, Lasker *et al.* 1986) supports the GSSS goal (0.33", guide star relative to target; 0.4^m in the V passband) in uncluttered fields but deteriorates significantly as field complexity grows. The desirability of a scan raster somewhat finer than 50μ is supported by the calculations of power spectra for typical plates (Stobie *et al.* 1984; Lasker and Kinsey 1984); these show that, while sample sizes in the range of $15 - 18\mu$ are needed to record all of the plate information, *nearly* all of it is recorded with 25μ samples.

Towards the end of 1984, this realization, combined with schedule relief due to the extension of the HST's launch date to then mid-1986, motivated the successful search for a faster PDS operating configuration, *i.e.*, less than 12 hours for a 25μ full-plate scan; this was achieved by a combination of replacing the logarithmic amplifier and of tuning the PDS's driving software, internal microcode, acceleration ramps, and mechanical dampers, thus making the construction of a catalog based on these finer scans feasible even against the constraint of the then scheduled HST launch date. For many purposes these 25μ scans effectively replace the plate; and the requirements for securely archiving the data and for making it conveniently accessible led to its transcription from magnetic tape (three 6250 bpi rolls per plate) to 12-inch optical disks (four plates per disk). These disks are currently accessible to prospective users through the Guide Star Astrometric Support Program (GASP, McLean et al. 1989); eventually they will also be incorporated into the HST data archive.

6.2. GSC Production

The catalog is based on microdensitometer scans of the UK SERC J survey in the south and a 1982 epoch "Quick V" survey made with the 48-inch Schmidt on Palomar Mountain. The image processing algorithms used to produce an inventory of classified objects from the scans are:

- the background follower, which spline-fits a surface to a grid of sky estimates covering the plate;
- the object identifier, which is based on the COSMOS algorithm (Lutz 1979);
- the blend resolver, which searches multiply-peaked objects by correlation against a library of star images;
- the astrometric centroider, which contains the Sanders-Schröder (1980) algorithm, generalized with an additional parameter, a variable exponent ("hardness") in place of the usual 2 in the Gaussian expression;
- a corresponding photometric estimator, which approximates the logarithm of the image intensity; and
- the classifier, which uses a multi-variate Bayesian approach (Duda and Hart 1973), operating on selected image features (COSMOS parameters, hardness, correlation with model diffraction spikes, etc.) to discriminate stars from other objects (*e.g.*, galaxies, unresolved blends).

The production of the GSC was carried out as a series of pipeline processes organized on a plate basis and controlled by a software executive. The plate measurements (x, y, photometric parameter, and object classification), which are the results of this sequence, are the inputs to the calibration and archive procedures described below.

Many details of these algorithms have been described in recent publications and conference proceedings (overviews in Jenkner *et al.* 1988, Lasker *et al.* 1988, and Russell *et al.* 1988; astronomical and computational methods in Jenkner 1983, Lasker 1984a and 1984b, and Russell 1986; astrometric topics in Russell and Williams 1986; and scanning techniques in Kinsey *et al.* 1984, Lasker and Kinsey 1984, and Lasker *et al.* 1986); a definitive series of publications presenting the GSC algorithms, the statistical properties of the GSC processing, and the data itself is currently being prepared for publication; preprints of these publications have been distributed with the GSC data (Lasker *et al.* 1989, Russell *et al.* 1989, and Jenkner *et al.* 1989).

6.3. Photometric and Astrometric Calibration

The photometric and astrometric data in the Hubble Space Telescope (HST) Guide Star Catalog (GSC) are based on the image processing described in section 6.2, specifically on a photometric parameter which estimates the integrated photographic intensity above a threshold as produced by the COSMOS algorithm (Lutz 1979) or a related construct for blended images, and on astrometric x, y measurements produced by a seven parameter, modified Gaussian centroider.

The reference data for the photometric data reduction are a sequence of six stars, roughly in the range from 9.0^m to 14.5^m, which were generally chosen to be near the center of each plate, near at least 2 SAOC stars, and within a one-half degree square. Each star in the sequence has photoelectric B and V magnitudes with a precision of 0.05^m. This set of approximately 9000 photometric reference stars has been published as the Guide Star Photometric Catalog (GSPC, Lasker *et al.* 1987).

The least-squares solution fits the logarithms of the photometric parameters to the reference magnitudes, generally with a second-order Chebyshev polynomial, but for a small percentage of the plates a curve of first or third order was used. The photometric sequence generally had a formal error of 0.05^m. As an independent check, the GSC magnitudes were compared to those from the $\delta = +28°$ band of sky observed by McGraw *et al.* using the CCD Transit Instrument of the University of Arizona. The results show differences of the order of 0.25^m *rms* between the two surveys. A worst case estimate of the precision of the magnitudes comes from the comparison of calculated magnitudes of stars which appear on more than one plate. The plate to plate differences may be as high as 1.0^m but are generally about 0.5^m.

The reference catalogs for the astrometric data reduction are the AGK3 in the north and the SAOC in the south. This coincides with the dividing line between the two plate surveys so that the Palomar plates are reduced with the AGK3 and most of the SRC plates are reduced using the SAOC for reference. The exception to the latter is for plates centered at $\delta = -65°$ to $-90°$, where the Cape Photographic Catalog (CPC) was used. The plate model includes ten constants in each of the ξ and η coordinates; it appears as a general third order polynomial, and only includes geometric terms, *i.e.*, no terms using magnitude or color are used.

The standard errors of unit weight of the plate solutions vary with the precision of the reference catalogs and the accuracy of the centroiding positions[3] from about 0.4" in the north to 1.5" in some areas of the south. The GSC includes a single estimate for the precision of the positions; it is based on the calculated centroiding errors and the errors in the ξ and η solutions, combined with *a priori* estimates of the systematic errors. This calculated error ranges from a few tenths near the plate center to several seconds of arc for stars in the extreme plate corners. Since the precision goal of the GSC was defined as a relative position error (because of the pointing technique used by HST), extensive investigations have been performed in this area. While the error behaviour is complex, the results seem to indicate a discontinuity with regard to the northern and southern plate material, and a magnitude effect in the sense that relative positions of stars fainter

[3]Differences in the accuracy of the position determination between the northern Quick-V and the southern SERC surveys are indicated, and are mainly believed to be due to the difference in exposure time.

than about 9^m or 10^m (the operationally relevant range for HST) are more accurate than those of brighter ones.

Advanced astrometric algorithms are currently being investigated; these will lead to removal of at least the largest astrometric errors from the GSC.

6.4. GSC Structure and Content

Access to the GSC has to be accommodated by identification as well as by celestial area. Together with the expected volume of data, this led to the formulation of a structure which would allow fast access, but would not require inordinate amounts of storage space nor specific software components.

The GSC is organized into approximately 10,000 regions; each region therefore holds a few thousand individual object entries. In order to account for the varying population of the catalog as a function of galactic latitude, the region size is not fixed, but varies, so that the number of objects per region is nearly constant.

Each object in the GSC carries a 10-digit number as identification. The first five digits encode the region number; the last five specify the number of the star within the particular region. Since only a few thousand objects are contained in each region (leaving some room for future expansion), this identification scheme is noncontiguous, *i.e.*, there are many 10-digit numbers which do not correspond to an entry in the catalog.

Each region is thus characterized by its center (reference) position and extent, as well as by the number of objects it contains. (Back-pointers to the individual plates from which objects appear in the region as well as several organizational items are available in addition.) For each entry the following information is contained in the GSC:

- Identification (GSC Number).
- Position (α, δ).
- Position standard deviation (*one* per position).
- Magnitude band descriptor.
- Magnitude and magnitude standard deviation.
- Object classification (star versus non-star).
- Source plate identification.
- Flag for multiple entries.

Note that object data derived from different plates (*i.e.*, for objects appearing on more than one plate) are stored separately, but with the *same* identification.

The catalog data themselves, as well as statistical information and user information in the form of text, have been made available in computer-readable form. Since the GSC on magnetic tape would require more than 20 rolls (even at high density and large blocking), the associated effort with regard to tape duplication is prohibitive for mass distribution. Therefore, compact optical disks (CD-ROMs) in standard ISO 9660 format, containing

FITS (Wells *et al.* 1981, Greisen and Harten 1981, Grosbøl *et al.* 1988, Harten *et al.* 1988) files, have been chosen for distribution.

The first version of the GSC, with an issue date of 1 June 1989, has been published as a set of two CD-ROMs, containing 599 and 590 MByte of data, respectively. Both contain introductory text and supporting tables; the GSC data (in the form of 9537 region files) are divided at $\delta = -7.5°$. This first version of the GSC contains a total of 25,126,027 entries for 18,819,291 objects, 15,169,873 of which are classified as stars.

6.5. Plans for GSC Enhancements

Near-term goals for improvements and maintenance of the GSC include the inclusion of bright objects from special plates and the incorporation and cross-referencing of existing astronomical catalogs, the removal of spurious objects around bright stars, and astrometric recalibration using newly developed techniques (Taff, Lattanzi and Bucciarelli, 1989).

Improvements in the photometry will require improved image processing as well as the extension of the existing reference stars both with regard to magnitude (toward fainter magnitudes) and to spatial coverage (location of sequences not only in the plate centers). A program to provide these photometric reference data is currently under way at ST ScI.

As new sky surveys become available, they will be scanned with increased resolution (with pixel sizes of 15μ to 18μ). Special emphasis will be given to improved image processing algorithms, yielding more consistent astrometric and photometric behaviour, as well as substantially more accurate object classification. As a pilot project, a catalog of galaxies (initially to about 17^m) will be generated. With the availability of these new surveys, color and proper motion information can be added to the GSC data. Finally, an extension towards fainter magnitudes will increase the size of the catalog by two orders of magnitude.

From the organizational (data base) point of view, enhanced catalog structure is anticipated for the future, which will, however, retain the basic GSC nomenclature and region scheme. This new structure is based on the need to:

- preserve raw data (image features) obtained directly from the underlying plates;
- perform re-calibration of selected plates without affecting the rest of the catalog;
- provide corrections in an orderly and trackable way; and
- provide cross-references to other catalogs.

To achieve this goal, it is expected that the main part of the catalog will contain object names, image features, and other information, directly derived from the plates, but without any additional calibration. A second part will contain these constructs, *e.g.*, plate solutions and photometric transformations. Addenda and error corrections will be carried in a third part. Cross-references to other catalogs (as well as the catalog data themselves) will also be added to the new structure. The various pieces of information related to one object would be combined by a software module — the catalog server — upon demand, *i.e.*, upon a request by a user for data from the catalog. This enhancement would allow the efficient incorporation of new survey data, and also of other catalog data as they become available in the community.

Acknowledgements

The author wishes to express his gratitude to the Astronomische Gesellschaft for the opportunity to present this review paper at their scientific meeting in Graz. Many helpful suggestions by Barry M. Lasker with regard to the manuscript, as well as supporting materials from the ST ScI Educational and Public Affairs Office are gratefully acknowledged.

Due to the very nature and size of the Guide Star Selection System project, which contained – as an integral part – all the efforts leading to the generation of the Guide Star Catalog, a large team of scientists, software professionals, engineers, operators, and (last but not least) managers, was required to cover all the different areas of expertise needed. Therefore, the results presented here reflect the efforts of all the past and present staff members of the Space Telescope Science Institute who have been and still are involved in this endeavor, as well as the contributions from a number of distinguished colleagues in the international astronomical community. Many of the main contributors to the Guide Star Catalog and the related projects can be identified from the author lists of the various references.

The Guide Star Catalog is partially based on data obtained at Palomar Observatory, operated by the California Institute of Technology; at the U.K. Schmidt Telescope, operated by the U.K. Science and Engineering Research Council and by the Anglo-Australian Observatory; at the Cerro Tololo Inter-American Observatory and the Sacramento Peak Observatory, both operated by the Association of Universities for Research in Astronomy, Inc., under contract to the National Science Foundation; and at the Mount Lemmon Observatory, operated by the University of Arizona.

References

BAHCALL, J. N., SPITZER, L.: 1982, "The Space Telescope", *Sci. Amer.* **247**, No. 1, 40.

BENEDICT, G. F.: 1979, "Space Telescope Astrometry Team, Guide Star Selection System, Specifications and Design", Univ. of Texas, Austin.

BLESS, R. C. *et al.*: 1982, "The High Speed Photometer for the Space Telescope", in: The Space Telescope Observatory (D. N. B. Hall, Ed.), *NASA CP-2244*, 106.

BRANDT, J. C. *et al.*: 1982, "The High Resolution Spectrograph for the Space Telescope", in: The Space Telescope Observatory (D. N. B. Hall, Ed.), *NASA CP-2244*, 76.

BURROWS, C.: 1989, "Hubble Space Telescope Optical Telescope Assembly Handbook", Version 1.0, *Space Telescope Science Institute.*

DUDA, R. O., HART, P. E.: 1973, *Pattern Classification and Scene Analysis,* Wiley, New York.

DUNCAN, D. K., EBBETS, D.: 1989, "Hubble Space Telescope Goddard High Resolution Spectrograph Instrument Handbook", Version 2.0, *Space Telescope Science Institute.*

DUNCOMBE, R. L., BENEDICT, G. F., HEMENWAY, P. D., JEFFERYS, W. H., SHELUS, P. D.: 1982, "Astrometric Observations with the Space Telescope", in: The Space Telescope Observatory (D. N. B. Hall, Ed.), *NASA CP-2244*, 114.

FORD, H.: 1985, "Hubble Space Telescope Faint Object Spectrograph Instrument Handbook", Version 1.0, *Space Telescope Science Institute.*

GIACCONI, R.: 1982, "Science Operations with Space Telescope", in: The Space Telescope Observatory (D. N. B. Hall, Ed.), *NASA CP-2244*, 1.

GILLESPIE, B. A., HARVEL, C. A., JENKNER, H., LASKER, B. M., RUSSELL, J. L.: 1984, "The Space Telescope Guide Star Selection System", *Bull. AAS,* **16**, 498.

GREISEN, E. W., HARTEN, R. H.: 1981, "An Extension of FITS for Groups of Small Arrays of Data", *Astron. Astrophys. Suppl. Ser.* **44**, 371.

GRIFFITHS, R.: 1989, "Hubble Space Telescope Wide Field and Planetary Camera Instrument Handbook", Version 2.0, *Space Telescope Science Institute.*

GROSBØL, P., HARTEN, R. H., GREISEN, E. W., WELLS, D. C.: 1988, "Generalized Extensions and Blocking Factors for FITS", *Astron. Astrophys. Suppl. Ser.* **73**, 359.

HALL, D. N. B. (Ed.): 1982, "The Space Telescope Observatory", *NASA CP-2244*.

HANLE, P. A.: 1985, "Astronomers, Congress, and the Large Space Telescope", *Sky and Tel.* **69**, No. 4, 300.

HARMS, R. J. *et al.*: 1982, "Astronomical Capabilities of the Faint Object Spectrograph on Space Telescope", in: The Space Telescope Observatory (D. N. B. Hall, Ed.), *NASA CP-2244*, 55.

HARTEN, R. H., GROSBØL, P., GREISEN, E. W., WELLS, D. C.: 1988, "The FITS Tables Extension", *Astron. Astrophys. Suppl. Ser.* **73**, 365.

JENKNER, H.: 1983, "Astronomical and Statistical Algorithms Used in the Space Telescope Guide Star Selection System", in: *Proc. Statistical Methods in Astronomy,* ESA SP-201, 65.

JENKNER, H.: 1984, "Principles of Astronomical Catalog and Data Base Construction and their Application towards the Space Telescope Guide Star Selection System", in: *Proc. International Course on Data Handling in Astronomy and Astrophysics*, ed. B. Hauck and G. Sedmak, Memorie Societa Astronomica Italiana.

JENKNER, H., DAMASHEK, M., GARNAVICH, P. M., LASKER, B. M., RUSSELL, J. L.: 1984, "Guide Star Catalog Construction", *Bull. AAS,* **16**, 498.

JENKNER, H., GILLESPIE, B. A., HARRISON, J. J., LASKER, B. M., MCLEAN, B. J., POLLIZZI, J. A., ROSENTHAL, D. A., RUSSELL, J. L., STURCH, C. R., WILSON, D. S.: 1985, "The Guide Star Catalog – Structure and Publication", *Bull. AAS,* **17**, 549.

JENKNER H., RUSSELL J. L., LASKER B. M.: 1988, "The Guide Star Catalog III. Structure and Publication, Status and Plans", in: Mapping the Sky – Past Heritage and Future Directions, *IAU Symp.* No. 133 (S. Debarbat, J. A. Eddy, H. K. Eichhorn, and A. R. Upgren, Eds.).

JENKNER, H., LASKER, B. M., STURCH, C. R., MCLEAN, B. J., SHARA, M. M., RUSSELL, J. L.: 1989, "The Guide Star Catalog. III: Production, Database Organization, and Population Statistics", *ST ScI Preprint Series*, No. **363**, 81.

KINSEY, J. H., DENMAN, R. J., EVZEROV, A.: 1984, "Laser Interferometer Enhanced Positional Precision of a Scanning Microdensitometer for Astronomical Plate Scanning", *Rev. Sci. Instr.,* **55**, 1960.

LASKER, B. M.: 1984a, "Schmidt Astronomy and the Space Telescope", in: *Astronomy with Schmidt-Type Telescopes (Proc. IAU Colloquium No. 78),* ed. M. Capaccioli.

LASKER, B. M.: 1984b, "The Guide Star Selection System for the Space Telescope: Long-Term Activities", in: *Science from Measuring Machines (Proc. Royal Greenwich Obs. Workshop),* ed. I. N. Reid and P. C. Hewett.

LASKER, B. M., DAMASHEK, M., GARNAVICH, P. M.: 1986, "A Shaped Aperture for Two-Dimensional Microdensitometry", *Publ. A.S.P.*, **98**, 125.

LASKER, B. M., KINSEY, J. H.: 1984, "PDS Sampling", *Bull. AAS,* **16**, 498.

LASKER, B. M., STURCH, C. R., LOPEZ, C., MALLAMA, A. D., MCLAUGHLIN, S. F., RUSSELL, J. L., WIŚNIEWSKI, W. Z., GILLESPIE, B. A., JENKNER, H., SICILIANO, E. D., KENNY, D., BAUMERT, J. H., GOLDBERG, A. M., HENRY, G. W., KEMPER, E., SIEGEL, M. J.: 1988, "The Guide Star Photometric Catalog. I.", *Ap. J. Suppl.* **68**, No. 1, 1.

LASKER B. M., JENKNER H., RUSSELL J. L.: 1988, "The Guide Star Catalog I. Overview, History, and Prospective", in: Mapping the Sky – Past Heritage and Future Directions, *IAU Symp.* No. 133 (S. Debarbat, J. A. Eddy, H. K. Eichhorn, and A. R. Upgren, Eds.).

LASKER, B. M., STURCH, C. R., MCLEAN, B. J., RUSSELL, J. L., JENKNER, H., SHARA, M. M.: 1989, "The Guide Star Catalog. I: Astronomical Foundations and Image Processing", *ST ScI Preprint Series*, No. **363**, 1.

LONGAIR, M. S., WARNER, J. W. (Eds.): 1979, "Scientific Research with the Space Telescope", IAU Colloquium No. 54, *NASA CP-2111.*

LONGAIR, M. S.: 1982, "Reflections on the Space Telescope", in: The Space Telescope Observatory (D. N. B. Hall, Ed.), *NASA CP-2244*, 121.

LONGAIR, M. S.: 1985, "The Scientific Challenge of Space Telescope", *Sky and Tel.* **69**, No. 4, 306.

LUTZ, R. K.: 1979, "An Algorithm for the Real Time Analysis of Digitized Images", *Computer Journal,* **23**, 262.

MACCHETTO, F., PACINI, F., TARENGHI, M. (Eds.): 1979, "Astronomical Uses of the Space Telescope", ESA/ESO Workshop Proceedings.

MACCHETTO, F.: 1982, "European Astronomy and the Space Telescope: The Space Telescope European Coordinating Facility", in: The Space Telescope Observatory (D. N. B. Hall, Ed.), *NASA CP-2244*, 16.

MACCHETTO, F. *et al.*: 1982, "The Faint Object Camera", in: The Space Telescope Observatory (D. N. B. Hall, Ed.), *NASA CP-2244*, 40.

MCLEAN, B. M., *et al.*: 1989, "The Guide Star Catalog. IV: User Interfaces, Utilities, and Astronomical Applications", *in preparation.*

OBERTH, H.: 1923, "Die Rakete zu den Planetenräumen", *R. Oldenbourg, Munich.*

O'DELL, C. R.: 1982, "The Space Telescope Observatory", in: The Space Telescope Observatory (D. N. B. Hall, Ed.), *NASA CP-2244*, 20.

PARESCE, F.: 1985, "Hubble Space Telescope Faint Object Camera Instrument Handbook", Version 1.0, *Space Telescope Science Institute.*

RUSSELL, J. L.: 1986, "The Guide Star Selection System and the Guide Star Catalog for Space Telescope", in: *Astrometric Techniques (Proc. IAU Colloquium No. 109),* ed. H. Eichhorn and R. J. Leacock, 721.

RUSSELL, J. L., EGRET, D.: 1986, "Hipparcos – The Tycho Input Catalog", in: *Highlights of Astronomy,* ed. J.-P. Swings, 713.

RUSSELL, J. L., WILLIAMS, C. A.: 1986, "Testing Schmidt Plates for Astrometric Purposes", in: *Astrometric Techniques (Proc. IAU Colloquium No. 109),* ed. H. Eichhorn and R. J. Leacock, 721.

RUSSELL J. L., LASKER B. M., JENKNER H.: 1988, "The Guide Star Catalog II. Astrometric and Photometric Algorithms and Precision", in: Mapping the Sky – Past Heritage and Future Directions, *IAU Symp.* No. 133 (S. Debarbat, J. A. Eddy, H. K. Eichhorn, and A. R. Upgren, Eds.).

RUSSELL, J. L., LASKER, B. M., MCLEAN, B. J., STURCH, C. R., JENKNER, H.: 1989, "The Guide Star Catalog. II: Photometric and Astrometric Models and Solutions", *ST ScI Preprint Series*, No. **363**, 45.

SANDERS, W. L., SCHRÖDER, R.: 1980, "Membership in the Open Cluster NGC 6494", *Astron. Astrophys.*, **88**, 102.

SPITZER, L.: 1946, "Astronomical Advantages of an Extraterrestrial Observatory", Project RAND Report, *Douglas Aircraft Corp.*

STOBIE, R. S., OKAMURA, S., DAVENHALL, A. C., MACGILLIVRAY, H. T.: 1984, "Power Spectrum Analysis of Astronomical Photographs Digitized with Small Apertures", in: *Astronomical Photography 1984 (Occasional Reports Roy. Obs. Edinburgh)*, ed. M. E. Sim and K. Ishida.

STURCH, C., LASKER, B. M., MALLAMA, A., MCLAUGHLIN, S., RUSSELL, J. L., SIEGEL, M., WISNIEWSKI, W., LOPEZ, C., HENRY, G.: 1984, "The GSSS Photometry Catalog", *Bull. A.A.S.*, **16**, 499.

TAFF, L. G.: 1989, "Hubble Space Telescope Fine Guidance Sensors Instrument Handbook", Version 2.0, *Space Telescope Science Institute.*

TAFF, L. G., LATTANZI, M., BUCCIARELLI, B.: 1989, "Two Successful Techniques for Schmidt Astrometry", *Ap. J.*, in press.

TUCKER, W.: 1985, "The Space Telescope Science Institute", *Sky and Tel.* **69**, No. 4, 295.

WELLS, D. C., GREISEN, E. W., HARTEN, R.H.: 1981, "FITS – A Flexible Image Transport System", *Astron. Astrophys. Suppl. Ser.* **44**, 363.

WESTPHAL, J. A. *et al.*: 1982, "The Wide-Field/Planetary Camera", in: The Space Telescope Observatory (D. N. B. Hall, Ed.), *NASA CP-2244*, 28.

WHITE, R. L.: 1985, "Hubble Space Telescope High Speed Photometer Instrument Handbook", Version 1.0, *Space Telescope Science Institute.*

Helioseismic Measurements of the Solar Internal Rotation

J. Christensen-Dalsgaard

Astronomisk Institut, Aarhus Universitet, DK-8000 Aarhus C, Denmark

Fluctuations of the Interstellar Medium

1 Introduction.

The rotation of the solar interior plays a major role in understanding the dynamics of the solar convection zone, and the generation of the solar magnetic field. Furthermore, it is possible that spin-down from an initial rapid rotation in the early phases of solar history might have led to hydrodynamical instabilities that could have caused mixing and hence affected the evolution of the Sun. Finally, it has been speculated that parts of the Sun may rotate so rapidly as to cause a centrifugal distortion, and hence a perturbation of the gravitational field outside the Sun, of sufficient magnitude to have a significant effect on the motion of Mercury; this could invalidate tests of the theory of general relativity based on the precession of the perihelium of Mercury's orbit. However, before the advent of helioseismology, essentially nothing was known about the rotation beneath the surface of the Sun.

As discussed by Stix (these proceedings) detailed observations have been made of the rotation of the solar surface, and its variation with latitude. Somewhat different values are obtained from Doppler observations of the surface plasma motion, and from observations of the advection of surface magnetic features. A typical result for the plasma rotation rate is

$$\frac{\Omega_p}{2\pi} = 452 - 49\mu^2 - 84\mu^4 \quad \text{nHz}, \tag{1.1}$$

where $\mu = \cos\theta$ (Snodgrass, 1984), whereas a typical rotation law for magnetic features is

$$\frac{\Omega_m}{2\pi} = 462 - 74\mu^2 - 53\mu^4 \quad \text{nHz} \tag{1.2}$$

(Snodgrass 1983). It should also be noted that Stenflo (1989) found a very different behaviour, with substantially smaller variation between equator and pole, from a correlation analysis of large-scale magnetic patterns. These differences may be a hint that the angular velocity varies with depth, although more quantitative estimates are difficult (see Schüssler 1987).

It is obviously impossible to carry out realistic numerical modelling of the complicated hydrodynamical processes that determine the rotation of the solar interior.

Simplified (although still quite complex) models of the solar convection zone suggest that rotation should be largely constant on cylinders, *i.e.* a function mainly of the distance to the rotation axis. Also, models of the solar spin-down indicate that the Sun could contain a rather large region in the radiative interior rotating at several times the surface rate; however, a major uncertainty in such models are the mechanisms that transport angular momentum from the radiative region to the convection zone. For example, even a weak magnetic field may be sufficient to cause an essentially complete redistribution of angular momentum, leading to uniform rotation, within the lifetime of the Sun.

The principal purpose of this review is to demonstrate that it has been possibly through observation of solar oscillations to obtain a rather detailed picture of the solar internal rotation, and to present the main results. In section 2 I give a very brief description of the properties of solar oscillations, and discuss the effect of rotation on the frequencies. Section 3 gives an overview of the techniques used to analyse the observed frequencies, and section 4 discusses techniques to invert the observed data to determine the angular velocity as a function of position within the Sun. In section 5 I present the results of the observations, and the inferences about rotation that have been made from them. Finally section 6 discusses the results and, in the light of them, addresses the questions of the origin and effects of the rotation.

2 Rotation and solar oscillations.

2.1 Properties of solar oscillations.

To provide a background for the discussion of the effects of rotation, it is useful to review very briefly some important properties of solar oscillations. More extensive presentations can be found, for example, in the reviews by Deubner & Gough (1984), Christensen-Dalsgaard, Gough & Toomre (1985) and Christensen-Dalsgaard & Berthomieu (1990).

Small-amplitude oscillations of a spherical star can be separated into normal modes, each of which has a harmonic dependence on time, and depends on the spherical coordinates θ and φ (co-latitude and longitude) as a spherical harmonic. Thus the displacement for a single mode can be written

$$\delta \boldsymbol{r}(r,\theta,\varphi,t) = \mathrm{Re}\left\{\left[\xi_r(r) Y_\ell^m \boldsymbol{a}_r + \xi_h(r)\left[\frac{\partial Y_\ell^m}{\partial\theta}\boldsymbol{a}_\theta + \frac{1}{\sin\theta}\frac{\partial Y_\ell^m}{\partial\varphi}\boldsymbol{a}_\varphi\right]\right] e^{-i\omega t}\right\}, \tag{2.1}$$

where $\boldsymbol{a}_r$, $\boldsymbol{a}_\theta$ and $\boldsymbol{a}_\varphi$ are unit vectors in the r, θ and φ directions. Here $Y_\ell^m(\theta,\varphi) = c_{\ell m} P_\ell^m(\cos\theta) e^{im\varphi}$ is a spherical harmonic, P_ℓ^m being a Legendre function and $c_{\ell m}$ a normalization constant defined such that the integral of $|Y_\ell^m|^2$ over the sphere is unity. The variation of the displacement with the distance r from the centre is determined by the *eigenfunctions* $\xi_r(r)$ and $\xi_h(r)$. The mode is characterized by three wavenumbers: n is the *radial order* which, roughly, gives the number of zeros in ξ_r; ℓ is the *degree*, and m the *azimuthal order* of the mode. The degree of the mode is related to its horizontal wavenumber k_h and wavelength λ at radius r by

$$k_h = \frac{2\pi}{\lambda} = \frac{L}{r}, \tag{2.2}$$

where $L = \sqrt{\ell(\ell+1)}$. As is seen from equation (2.1), m, which is in the range $-\ell$ to ℓ, is twice the number of zeros around the equator.

We neglect processes that damp or excite the oscillations. Then ω is real, and it follows from equation (2.1) that the dependence of the displacement (or, more generally, any quantity associated with the mode) on longitude and time is given by

$$\cos(m\varphi - \omega t). \tag{2.3}$$

Thus for $m = 0$ the mode is a standing wave. Modes with $m > 0$, which are normally called *prograde*, propagate in longitude in the direction of increasing φ, whereas *retrograde* modes, with $m < 0$, propagate in the direction of decreasing φ. When ω is real, the eigenfunctions ξ_r and ξ_h may also be chosen to be real.

In addition to the *angular frequency* ω, which is used in equation (2.1), the *cyclic frequency* $\nu = \omega/(2\pi) = 1/P$, is commonly used, particularly in discussions of observed frequencies; here P is the oscillation period.

In general the frequency $\omega = \omega_{n\ell m}$ depends on all three wavenumbers. However if rotation or other departures from spherical symmetry are ignored, $\omega_{n\ell m}$ does not depend on m. This follows from the fact that in this case there is no preferred axis in the star; since m depends on the choice of coordinate axis, the physics of the oscillations, and hence their frequencies, must be independent of m. For slow rotation a modal description as in equation (2.1) is still possible, provided that the rotation axis is chosen as coordinate axis.

The possible modes of oscillation can be separated into three classes: the p, the f and the g modes. The p modes are essentially standing acoustic waves; in solar models they have cyclic frequencies between about 250 and 5000 μHz. The f modes are surface gravity waves, whereas the g modes are standing internal gravity waves; g mode frequencies are below about 450 μHz in normal solar models. In the Sun the only modes that have been definitely observed are the so-called five minute oscillations, with frequencies between about 1500 and 5000 μHz; these correspond to p modes or, at high degree, also f modes. There have also been reports of oscillations with periods of several hours; if caused by normal modes of the Sun, these must be g modes, but the identification or even the solar origin of these oscillations is still in some doubt.

Because of their high radial order or degree the p modes observed on the Sun can be described with reasonable accuracy by means of asymptotic theory. For the present purpose the main result of such an analysis is that p modes are trapped outside a *turning point*, the location $r = r_t$ of which is determined by

$$\frac{c(r_t)}{r_t} = \frac{\omega}{L}, \tag{2.4}$$

where c is the adiabatic sound speed. This corresponds to a point of total internal reflection; for $r < r_t$ the mode decays exponentially. At the surface the wave is reflected (provided the frequency does not exceed about 5500 μHz) by the steep density gradient. Thus the wave propagates in a series of "bounces" between the surface and the turning point. A mode of oscillation, which is a standing wave, is formed as an interference pattern between such bouncing waves. At low degree r_t as determined from equation (2.3) is small; the modes penetrate essentially to the solar centre, and hence their frequencies carry information about the entire Sun. However, the amplitudes are largest near the solar surface (see also Figure 1), and hence some care is required to extract that part of the information which derives from the deep solar interior.

The g modes are trapped beneath the solar convection zone, and have large amplitudes in the solar core. This makes them potentially very valuable as probes of conditions in the deep solar interior.

2.2 Rotational splitting.

We assume that the rotation is sufficiently slow to allow us to neglect the rotational distortion of the star, or more generally terms of order Ω^2, where Ω is the angular velocity. Then the structure of the star is unchanged, and the effects of rotation can be treated as a small perturbation in the equations describing the oscillations.

That rotation causes a splitting of the frequencies can be seen from a purely geometrical argument. Assume the angular velocity Ω to be uniform, and consider an oscillation with a frequency ω_0, independent of m, in the frame rotating with the star. I introduce a coordinate system in this frame, with coordinates (r',θ',φ') which are related to the coordinates (r,θ,φ) in an inertial frame through

$$(r', \theta', \varphi') = (r, \theta, \varphi - \Omega t). \tag{2.5}$$

Since in the rotating frame the perturbations depend on φ' and t as $\cos(m\varphi' - \omega_0 t)$ (*cf.* equation (2.3)), the dependence in the inertial frame is $\cos(m\varphi - \omega_m t)$, where

$$\omega_m = \omega_0 + m\Omega. \tag{2.6}$$

Thus an observer in the inertial frame finds that the frequency is split uniformly according to m.

This description is obviously incomplete. Even in the case of uniform rotation the effects of the Coriolis force must be taken into account in the rotating frame, causing a contribution to the frequency splitting (Cowling & Newing 1949; Ledoux 1949). Furthermore, in general the angular velocity is a function $\Omega(r,\theta)$ of position. Nevertheless, as shown below, the effect of the Coriolis force is often small, and equation (2.6) is approximately correct if Ω is replaced by a suitable average of the position-dependent angular velocity.

The general case was considered by Gough (1981), Hansen, Cox & van Horn (1977) and Cuypers (1980). The result is that the *rotational splitting*, *i.e.* the perturbation in the frequencies caused by rotation, can be written

$$\delta\omega_{n\ell m} = m\frac{R_{n\ell m}}{I_{n\ell m}}, \tag{2.7}$$

where

$$R_{n\ell m} = \int_0^\pi \sin\theta d\theta \int_0^R \left\{ \xi_r^2 P_\ell^m(\cos\theta)^2 + \xi_h^2 \left[\left[\frac{dP_\ell^m}{d\theta}\right]^2 + \frac{m^2}{\sin^2\theta} P_\ell^m(\cos\theta)^2 \right] \right. \tag{2.8}$$

$$\left. - 2P_\ell^m(\cos\theta)^2 \xi_r \xi_h - 2P_\ell^m(\cos\theta)\frac{dP_\ell^m}{d\theta}\frac{\cos\theta}{\sin\theta}\xi_h^2 \right\} \Omega(r,\theta)\rho(r) r^2 dr$$

and

$$I_{n\ell m} = \frac{2}{2\ell+1}\frac{(\ell+|m|)!}{(\ell-|m|)!}\int_0^R \left[\xi_r^2 + \ell(\ell+1)\xi_h^2\right]\rho(r) r^2 dr; \tag{2.9}$$

here ρ is the density.

2.3 Splitting for spherically symmetric rotation.

To proceed an explicit assumption about the variation of Ω with θ is required. For simplicity I shall first assume that Ω is independent of θ. In this case the integrals over θ in equation (2.8) only involve Legendre functions and may be evaluated analytically. The result is

$$\delta\omega_{n\ell m} = m\frac{\int_0^R \Omega(r)\left[\xi_r^2 + L^2\xi_h^2 - 2\xi_r\xi_h - \xi_h^2\right] r^2 \rho dr}{\int_0^R \left[\xi_r^2 + L^2\xi_h^2\right] r^2\rho dr}, \tag{2.10}$$

where I have introduced $L^2 \equiv \ell(\ell+1)$. It should be noticed that the integrands in equation (2.10) are given solely in terms of ξ_r, ξ_h and ℓ, and therefore are independent of m. Hence in the case of spherically symmetric rotation, the rotational splitting is proportional to m.

It is convenient to write equation (2.10) as

$$\delta\omega_{n\ell m} = m\beta_{n\ell}\int_0^R K_{n\ell}(r)\Omega(r)dr, \tag{2.11}$$

where

$$K_{n\ell} = \frac{\left[\xi_r^2 + L^2\xi_h^2 - 2\xi_r\xi_h - \xi_h^2\right]r^2\rho}{\int_0^R\left[\xi_r^2 + L^2\xi_h^2 - 2\xi_r\xi_h - \xi_h^2\right]r^2\rho dr}, \tag{2.12}$$

and

$$\beta_{n\ell} = \frac{\int_0^R\left[\xi_r^2 + L^2\xi_h^2 - 2\xi_r\xi_h - \xi_h^2\right]r^2\rho dr}{\int_0^R\left[\xi_r^2 + L^2\xi_h^2\right]r^2\rho dr}. \tag{2.13}$$

By using this definition it is ensured that the *rotational kernel* $K_{n\ell}$ is unimodular, *i.e.*

$$\int_0^R K_{n\ell}(r)dr = 1. \tag{2.14}$$

Hence for uniform rotation, where $\Omega = \Omega_s$ is constant,

$$\delta\omega_{n\ell m} = m\beta_{n\ell}\Omega_s. \tag{2.15}$$

In this case the effect of rotation is completely given by $\beta_{n\ell}$. For high-order or high-degree p modes the terms in ξ_r^2 and $L^2\xi_h^2$ dominate, and $\beta_{n\ell} \approx 1$. Thus the rotational splitting between adjacent m-values is given simply by the angular velocity, as in equation (2.6). Physically, the neglected terms in equation (2.13) do indeed arise from the Coriolis force. For high-order g modes, on the other hand, the displacement is predominantly horizontal and the terms containing ξ_r can be neglected; hence

$$\beta_{n\ell} \approx 1 - \frac{1}{L^2}. \tag{2.16}$$

In particular, the splitting of high-order g modes of degree 1 is only *half* the angular velocity.

Returning to the case where Ω depends on r, it should be noted that the integral in equation (2.11) provides a weighted average

$$\langle\Omega\rangle = \int_0^R K_{n\ell}(r)\Omega(r)dr \tag{2.17}$$

of $\Omega(r)$. For high-order p modes one can neglect the terms in ξ_h in equation (2.12) and use that, according to the asymptotic theory for p modes $\xi_r \sim (\rho c)^{-1/2}r^{-1}$ outside the turning point. The result is

$$\delta\omega_{n\ell m} \approx m\langle\Omega\rangle \approx m\frac{\int_{r_t}^R\Omega(r)\frac{dr}{c}}{\int_{r_t}^R\frac{dr}{c}}. \tag{2.18}$$

Thus in this case we obtain the intuitively appealing result that the rotational splitting is an average of the angular velocity, weighted by the sound travel time. A more accurate expression can be obtained by including the term in $L^2\xi_h^2$ and using the complete asymptotic expressions for ξ_r and ξ_h, or from ray theory (Gough 1984).

Figure 1 shows a few examples of kernels for high-order p modes. They are clearly large near the solar surface, as is also implicit in equation (2.18). Beneath the turning point the kernels get very small, but they are locally enhanced just above it. This effect arises from the term in ξ_h in equation (2.12); physically it corresponds to the fact that the waves travel approximately horizontally in this region, and hence spend a relatively long time there.

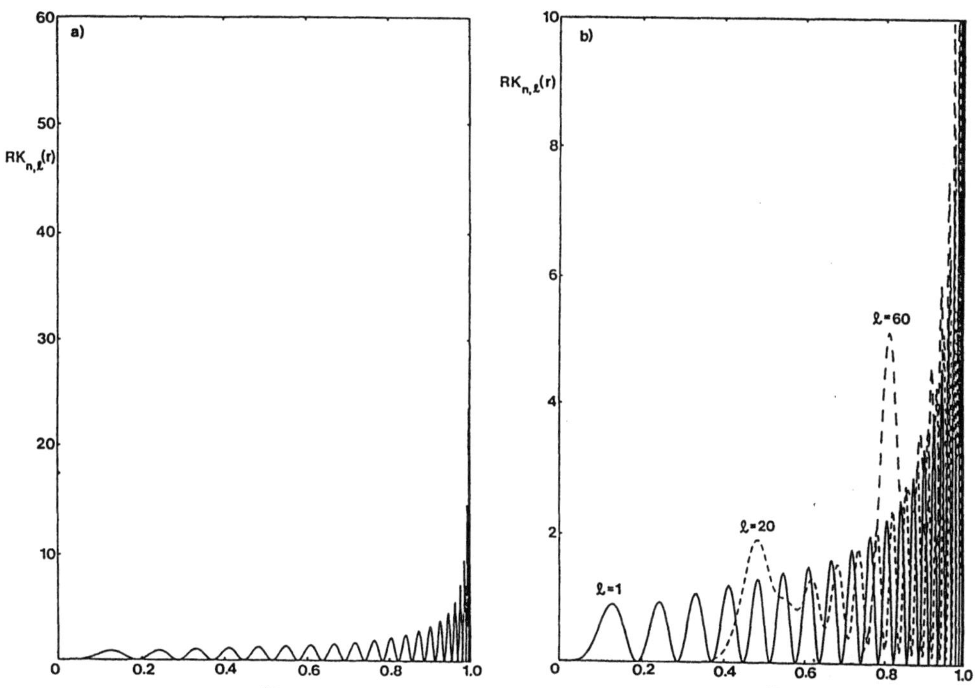

Figure 1. Kernels $K_{n\ell}$ for the frequency splitting in p modes caused by spherically symmetric rotation (*cf.* equation (2.12)). In a) is plotted $RK_{n\ell}(r)$ for a mode with $\ell = 1$, $n = 22$ and $\nu = 3233\mu$Hz. The maximum value of $RK_{n\ell}(r)$ is 57. In b) is shown the same mode, on an expanded vertical scale, (————) together with the modes $\ell = 20$, $n = 17$, $\nu = 3367\mu$Hz (- - - - - - - - - -), and $\ell = 60$, $n = 10$, $\nu = 3231\mu$Hz (— — — — — — — —). Notice that the kernels almost vanish inside the turning point radius r_t, and that there is an accumulation just outside the turning point.

A kernel for a high-order g mode is shown in Figure 2. It is evident that observations of such modes would provide sensitive measures of the rotation of the solar core. Unfortunately definite detection, let alone identification, of g modes in the Sun has yet to be made.

To get an impression of the contribution to the splitting made by different parts of the Sun, it is instructive to consider the integrated kernel

$$\int_0^r K_{n\ell}(r')dr'. \tag{2.19}$$

Examples are shown in Figure 3 for a p and a g mode. In the p-mode case only about 1 per cent of the contribution comes from the inner 10 per cent of the radius, whereas the corresponding number for the g mode is about 45 per cent. This further emphasizes the difficulty in measuring the solar core rotation with p modes, and the great potential importance of g-mode observations.

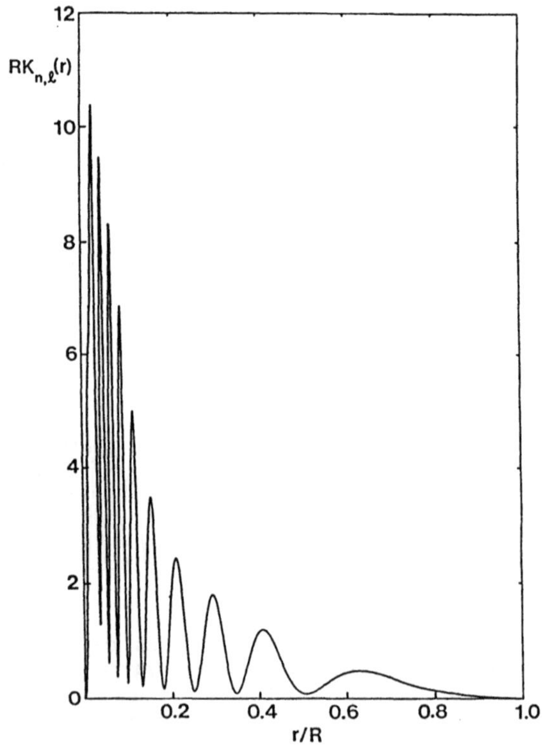

Figure 2. Kernel $K_{n\ell}$ for the frequency splitting caused by spherically symmetric rotation (*cf.* equation (2.12)), in a g mode with ℓ = 2, n = 10 and ν = 101 μHz.

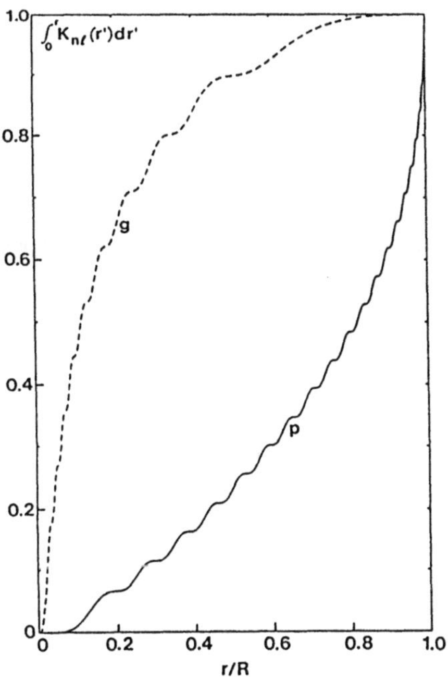

Figure 3. Integrals of kernels for spherically symmetric rotation. The solid curve is for a p mode with $\ell = 1$, $n = 21$ and $\nu = 3096$ μHz; the dashed curve is for a g mode with $\ell = 2$, $n = 10$ and $\nu = 101$ μHz.

2.4 General rotation laws.

In the general case, where Ω depends on both r and θ, the rotational splitting may be computed from equations (2.7) - (2.9), by evaluating the two-dimensional integral in equation (2.8). This integral is in general m-dependent, and so the splitting is no longer a linear function of m. To illustrate the properties of the splitting, I consider again the case of high-order p modes, keeping only the term in ξ_r^2 and approximating it as in the derivation of equation (2.18). This yields

$$\delta\omega_{n\ell m} = m\frac{\int_0^\pi \sin\theta \left[P_\ell^m(\cos\theta)\right]^2 \int_{r_t}^R \Omega(r,\theta)\frac{dr}{c}d\theta}{\int_0^\pi \sin\theta \left[P_\ell^m(\cos\theta)\right]^2 d\theta \int_{r_t}^R \frac{dr}{c}}. \tag{2.20}$$

Thus the rotational splitting is once again a weighted average of the angular velocity, but now the weight depends both on the sound travel time and on the distribution of oscillation energy with latitude, as given by $(P_\ell^m)^2$. With increasing ℓ the sectoral modes, with $\ell = |m|$, get increasingly confined towards the equator. Thus, the rotational splitting of the sectoral modes provides a measure of the solar equatorial angular velocity. For general m, P_ℓ^m is confined approximately between latitudes $\pm\sin^{-1}[(1-m^2/L^2)^{1/2}]$. The variation of the extent of the P_ℓ^m with m allows resolution of the latitudinal variation of the angular velocity, in much the same way as the variation of the depth of penetration with ω/L allows resolution of the variation with radius.

To study the splitting without making the asymptotic approximation, it is convenient to consider a parameterized representation of $\Omega(r,\theta)$. Here I use an expansion of the form

$$\Omega(r,\theta) = \sum_{s=0}^{s_{max}} \Omega_s(r)\mu^{2s}, \tag{2.21}$$

where $\mu = \cos\theta$. Then the integrals over θ can be evaluated analytically, and the rotational splitting becomes (see also Cuypers 1980)

$$\delta\omega_{n\ell m} = m \sum_{s=0}^{s_{max}} \int_0^R K_{n\ell ms}(r)\ \Omega_s(r)\ dr. \tag{2.22}$$

Here

$$K_{n\ell ms}(r) = \rho r^2\ I_{n\ell}^{-1}\Big\{\Big[(\xi_r - \xi_h)^2 + (L^2 - 2s^2 - 3s - 2)\xi_h{}^2\Big]\ Q_{\ell ms} \tag{2.23}$$

$$+\ s(2s-1)\xi_h{}^2 Q_{\ell ms-1}\Big\},$$

where

$$I_{n\ell} = \int_0^R \rho r^2\left[\xi_r{}^2 + L^2\xi_h{}^2\right]\ dr, \tag{2.24}$$

and

$$Q_{\ell ms} = \frac{2\ell+1}{2}\ \frac{(\ell-|m|)!}{(\ell+|m|)!}\int_{-1}^1 \mu^{2s}\ [P_\ell^m(\mu)]^2\ d\mu. \tag{2.25}$$

For spherically symmetric rotation, $\Omega_s = 0$ for $s > 0$. Since $Q_{\ell m0} = 1$, the kernel $K_{n\ell m0}(r) = \beta_{n\ell}K_{n\ell}(r)$ is independent of m, and the splitting is uniform in m. Thus the results of section 2.3 are recovered.

It may be shown that the factor $Q_{\ell ms}$ is a polynomial in m^2 of degree s. Hence to equation (2.22) corresponds an expansion of the rotational splitting in odd powers of m. As discussed in section 3 it is convenient, from the point of view of data analysis, to write this in terms of Legendre polynomials as

$$\delta\omega_{n\ell m} = \ell \sum_{j=0}^{s_{max}} \Delta^j\omega_{n\ell}P_{2j+1}\left[\frac{m}{\ell}\right]. \tag{2.26}$$

Here the expansion coefficients $\Delta^j\omega_{n\ell}$ of the splitting are related to the expansion coefficients Ω_s of the angular velocity through

$$\Delta^j\omega_{n\ell} = \sum_{s=j}^{s_{max}} \int_0^R K^j_{n\ell s}(r)\ \Omega_s(r)\ dr. \qquad (2.27)$$

The kernels $K^j_{n\ell s}$ are obtained from expressions similar to equation (2.23), with the $Q_{\ell ms}$ replaced by expansion coefficients of $Q_{\ell ms}$ in terms of Legendre polynomials (see Brown *et al.* 1989 for details).

In the limit of large ℓ, the terms $\xi_r{}^2$ and $L^2\xi_h{}^2$ dominate in equation (2.23). As in the case of spherically symmetric rotation, the effects of Coriolis forces are small and the rotational splitting is simply an average of $\Omega(r,\ \theta)$. Then the kernels $K^j_{n\ell s}(r)$ in equation (2.27) are approximately proportional to

$$H_{n\ell}(r) = I^{-1}_{n\ell}\left[\xi_r{}^2 + L^2\xi_h{}^2\right]\rho(r)r^2; \qquad (2.28)$$

note that this is also the expression for the kernel $K_{n\ell}$ for spherically symmetric rotation in this limit (*cf.* equation (2.12)). As shown by Brown *et al.* equations (2.27) may be approximated by

$$\begin{aligned}
\Delta^0\omega_{n\ell} &\simeq \int_0^R\left[\Omega_0(r) + \frac{1}{5}\Omega_1(r) + \frac{3}{35}\Omega_2(r) + \cdots\right]H_{n\ell}(r)dr\ ,\\
\Delta^1\omega_{n\ell} &\simeq -\frac{1}{5}\int_0^R\left[\Omega_1(r) + \frac{2}{3}\Omega_2(r) + \cdots\right]H_{n\ell}(r)dr\ , \qquad (2.29)\\
\Delta^2\omega_{n\ell} &\simeq \frac{1}{21}\int_0^R\left[\Omega_2(r) + \cdots\right]H_{n\ell}(r)dr\ .
\end{aligned}$$

These may be recombined to yield

$$\begin{aligned}
\langle\Omega_0\rangle &\simeq \Delta^0\omega_{n\ell} + \Delta^1\omega_{n\ell} + \Delta^2\omega_{n\ell},\\
\langle\Omega_1\rangle &\simeq -5\Delta^1\omega_{n\ell} + 14\Delta^2\omega_{n\ell}, \qquad (2.30)\\
\langle\Omega_2\rangle &\simeq 21\Delta^2\omega_{n\ell},
\end{aligned}$$

where the angular brackets denotes an average over r weighted with $H_{n\ell}$. From equation (2.21) one finally obtains the corresponding radial average of $\Omega(r,\theta)$ as

$$\langle\Omega(\theta)\rangle \simeq \Delta^0\omega_{n\ell} + (1 - 5\mu^2)\Delta^1\omega_{n\ell} + (1 - 14\mu^2 + 21\mu^4)\Delta^2\omega_{n\ell}, \qquad (2.31)$$

where $\mu = \cos\theta$ (Brown *et al.* 1989; see also Morrow 1988). Thus by considering different combinations of the observed expansion coefficients, one can obtain some idea about the latitudinal variation of Ω. For example, $\Delta^0\omega_{n\ell} + \Delta^1\omega_{n\ell} + \Delta^2\omega_{n\ell}$ gives a measure of the equatorial angular velocity, whereas $\Delta^0\omega_{n\ell} - 4\Delta^1\omega_{n\ell} + 8\Delta^2\omega_{n\ell}$ measures the polar value. Note also that since the depth of penetration of the modes depends on ℓ through equation (2.3), the variation of these combinations with ℓ provides an initial impression of the variation of Ω with depth.

The expansions of the angular velocity and the splitting considered here is only one among many possible pairs. Other expansions were discussed by Hill *et al.* (1986) and Durney, Hill & Goode (1988).

3 Analysis of rotational splitting observations.

Only a brief summary of the analysis techniques can be given here. A more detailed discussion of techniques for observing solar oscillations was given by Brown (1988), while Harvey (1988a) reviewed observations of rotational splitting.

Early observations of the solar five minute oscillations dealt with modes of high degree. Furthermore, the data were generally averaged in one spatial dimension. In particular, averaging in latitude largely eliminates modes with $m \neq \pm \ell$. Two-dimensional Fourier analysis of the resulting signal, in longitude and time, therefore yields spectra that are dominated by the sectoral modes, with $|m| = \ell$. From this one may determine an average rotational splitting

$$\Delta^{(sec)}\omega_{n\ell} = \frac{\omega_{n\ell\ell} - \omega_{n\ell-\ell}}{2\ell}, \tag{3.1}$$

which, according to equation (2.20) and the behaviour of P_ℓ^ℓ, is dominated by the equatorial value of Ω.

From observations of the dependence of the oscillations on both latitude and longitude, modes at all values of ℓ and m, up to the limit imposed by the resolution, can be isolated. As discussed in section 2, this permits a determination of the latitudinal variation of the angular velocity. While it would in principle be possible to analyze directly the individual splittings $\delta\omega_{n\ell m}$, in practice the data have so far been presented in terms of fits to the m-dependence of the splittings. A particularly convenient representation is to expand the splitting in terms of Legendre polynomials, as

$$\delta\omega_{n\ell m} = 2\pi\ell \sum_{k=0}^{k_{max}} a_{n\ell}^k P_k\left[\frac{m}{\ell}\right], \tag{3.2}$$

where $P_k(x)$ is a Legendre polynomial. Since the P_k are orthogonal, the coefficients $a_{n\ell}^k$ are statistically independent, as far as random errors are concerned. From equation (2.26) follows that rotation only contributes to the odd terms in this expansion. Even terms, if statistically significant, must be caused by departures from spherical symmetry in the equilibrium model, such as caused by rotational flattening, or from effects of magnetic fields which do not depend on the sign of m. (*e.g.* Dziembowski & Goode 1984; Gough & Taylor 1984; Gough & Thompson 1990a). Here we concentrate on the odd terms which, according to equation (2.26), are given by

$$a_{n\ell}^{2j+1} = \frac{1}{2\pi}\Delta^j\omega_{n\ell}, \tag{3.3}$$

and hence are related to the expansion of the angular velocity through equation (2.27).

Modes of low degree are particularly interesting, since they penetrate to the core of the Sun. Unfortunately observation of splittings for such modes is difficult.

Spatially resolved observations in general suffer from a lack of an independent stable wavelength reference. As a result, the line position averaged over the solar disk is effectively used as a reference, precluding detection of modes of the lowest degrees. Observations that do not suffer from this problem have been obtained with the resonant scattering cell technique, applied to light integrated over the solar disk. Such measurements have provided data of very high quality for modes of degree 0 - 3. An additional problem, however, is that the observations provide little discrimination between modes of different m. If, as is usually assumed, rotation is the dominant source of departure from spherical symmetry, the axis of symmetry of the oscillations coincides with the rotation axis, which is always close to the plane of the sky. In that case observations in integrated light are virtually insensitive to modes where $\ell - m$ is odd, because for these the Legendre function is antisymmetric around the equator.[1] (For a discussion of the sensitivity of such observations, see, for example, Dziembowski 1977; Hill 1978; Christensen-Dalsgaard & Gough 1982). Even so, for each value of n and ℓ the spectrum contains $\ell + 1$ peaks, with a separation which for p modes is roughly $2\langle\Omega\rangle$, where $\langle\Omega\rangle$ is the average of the angular velocity defined in equation (2.17). Thus, if the angular velocity is nowhere much larger than its surface value, the separation in cyclic frequency is only of the order of 1 μHz. To resolve this separation requires observations spanning at least 15 days (*e.g.* Luomos & Deeming 1978). Furthermore, except for modes of frequency less than about 2000 μHz, the natural line width of the peaks is greater than 1 μHz (Libbrecht 1988a). In this case random variations in mode amplitude or phase may cause fine structure in the peaks which may be mistaken for rotational splitting (*e.g.* Sørensen 1988).

4 Inversion of the rotational splitting.

Observation of the rotational splitting for a mode of oscillation provides an integral measure of the angular velocity over the region where the mode is trapped. Because of the variation of the turning point radius r_t with degree and frequency, different modes sample different parts of the Sun. Hence, the variation of the splitting with degree, say, provides an indication of the variation of Ω with r. This also indicates that it may be possible to obtain localized information about Ω: roughly speaking, the difference between splittings for modes with different r_t should be a measure of the rotation in the region between the turning points of the modes.

---

[1]By considering the effect of rotation as an amplitude modulation, van der Raay, Pallé & Roca Cortes (1986) argued that all three $\ell = 1$ modes should be visible; however they apparently overlooked the cancellation caused by the latitude variation of the $m = 0$ mode.

The expression (2.11) for the splitting caused by spherically symmetric rotation is a particularly simple example of a relation between observable properties of oscillation frequencies and properties of the solar interior. The determination of $\Omega(r)$ from the $\delta\omega_{n\ell m}$ constitutes an *inverse problem*. Such problems have a vast literature, covering their application in, for example, geophysics and radiation theory (*e.g.* Parker 1977; Craig & Brown 1986; Deepak 1977; Tarantola 1987). The application to the solar inverse problem was discussed by, for example, Gough (1978, 1985a) and Gough & Thompson (1990b). Christensen-Dalsgaard, Schou & Thompson (1990) made a systematic comparison of different inversion techniques, as applied to the problem of spherically symmetric rotation.

As background to the presentation in section 5 of results of inversions, it is instructive to consider briefly the properties of the technique of *optimally localized averages*, developed by Backus & Gilbert (1970). The inverse problem for a spherically symmetric angular velocity $\Omega(r)$ may be formulated as

$$\Delta_i = \int_0^R K_i(r)\Omega(r)dr, \tag{4.1}$$

where, for notational simplicity, I represent the pair (n,ℓ) by the single index i. Δ_i is the scaled rotational splitting $m^{-1}\beta_{n\ell}^{-1}\delta\omega_{n\ell m}$, so that the kernels K_i are normalized as in equation (2.14). The principle of the method is to construct a linear combination

$$\tilde{\Omega}(r_0) = \sum_i c_i(r_0)\Delta_i = \int_0^R K(r;\ r_0)\Omega(r)dr \tag{4.2}$$

of the observed data, where

$$K(r;\ r_0) \equiv \sum_i c_i(r_0)K_i(r). \tag{4.3}$$

The goal of the method is to choose coefficients $c_i(r_0)$ such as to make $K(r;\ r_0)$ approximate as far as possible a delta function $\delta(r - r_0)$ centred on r_0. Then $\tilde{\Omega}(r_0)$ provides an approximation to $\Omega(r_0)$. If this can be done for all r_0, an estimate of $\Omega(r)$ is obtained.

The coefficients $c(r_0)$ are determined by minimizing

$$\cos\eta\int_0^R (r - r_0)^2K(r;\ r_0)^2dr + \sin\eta\sum_{ij}E_{ij}c_ic_j, \tag{4.4}$$

where E_{ij} is the covariance matrix of the data, subject to the constraint

$$\int_0^R K(r;\ r_0)dr = 1. \tag{4.5}$$

This minimization problem leads to a set of linear equations for the c_i. The effect of the minimization is most easily understood for $\eta = 0$. Minimizing equation (4.4) subject to equation (4.5), ensures that $K(r;\ r_0)$ is large close to r_0, where the weight function $(r - r_0)^2$ is small, and small elsewhere. This is precisely the required "delta-ness" of the combined kernel. However, with no further constraints, the optimization of the combined kernel may result in numerically large coefficients of opposite sign. Hence the variance in $\tilde{\Omega}$, which can be estimated as

$$\sigma^2(\tilde{\Omega}) = \sum_{ij} E_{ij} c_i c_j, \tag{4.6}$$

would be large. The effect of the second term in equation (4.4), when $\eta > 0$, is to restrict $\sigma^2(\tilde{\Omega})$. The size of η determines the relative importance of the localization and the size of the variance in the result. Hence η must be determined to ensure a trade-off between the localization and the error, and η is generally known as *the trade-off parameter.*

To characterize the properties of the inversion, it is convenient to consider a measure of the width of the averaging kernels $K(r;r_0)$ and the *error magnification*

$$\Lambda(r_0) = \left[\sum_i c_i(r_0)^2\right]^{1/2}; \tag{4.7}$$

Λ is defined such that if the standard error $\sigma(\Delta_i)$ is the same for all the modes the standard error in the result of the inversion is

$$\sigma[\tilde{\Omega}(r_0)] = \Lambda(r_0)\sigma(\Delta_i). \tag{4.8}$$

By considering the *trade-off curve*, where Λ is plotted against the width for varying η, one may choose an appropriate value of η. It is evident that this value will depend on the properties of data, particularly the level of errors, and on the desired properties of the solution.

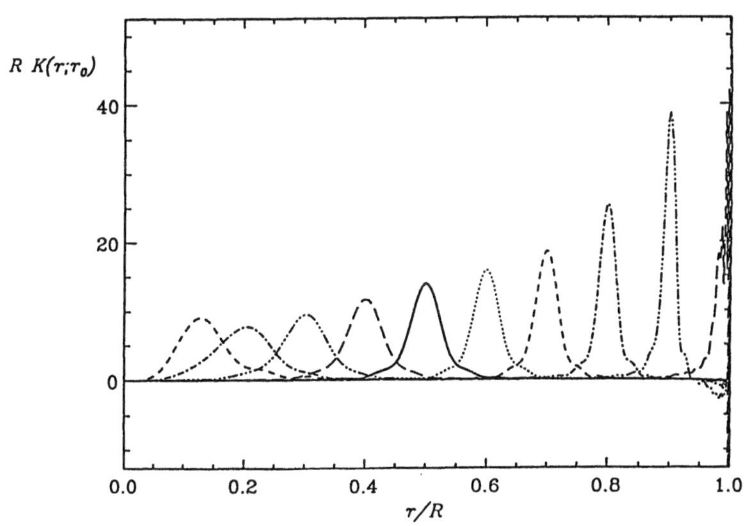

Figure 4. Averaging kernels $K(r;r_0)$ at selected radii (r_0/R = 0.1, 0.2, · · · ,1.0) for inversion by means of optimally localized averages. The kernel at r_0/R = 0.5 is shown as a bolder curve. From Christensen-Dalsgaard, Schou & Thompson (1990).

An illustration of the use of this method is provided by the results obtained by Christensen-Dalsgaard *et al.* (1990). They considered a set consisting of about 830 modes at selected degrees between 1 and 200, and frequencies between 2000 and 4000 μHz. Examples of averaging kernels $K(r;\, r_0)$ are shown in Figure 4. The trade-off parameter was chosen such that the error magnification at $r_0 = 0.5R$ was close to 1. It should be realized that the kernels entering into the combination are of the form shown in Figure 1. Thus a very large degree of cancellation has been achieved of the dominant contribution from near the surface. Nevertheless, it is obvious that the averaging kernels are only approximate realizations of delta functions; structure on a scale smaller than roughly 0.05 R is not resolved.

This limitation is inherent in any inversion method. Indeed, it is evident that from a finite set of data one can never completely resolve the function $\Omega(r)$. Thus the solution must be constrained. The constraint that is invoked in the present method, and in most other inversion methods, is that the solution be smooth. This is ensured by the representation of the solution by the averaging kernels, whose shape is determined by the minimization in equation (4.4).

A second technique that has been used for the determination of the solar internal angular velocity is the regularized least-squares method (*e.g.* Phillips 1962; Craig & Brown 1986). Here $\Omega(r)$ is parameterized, often as a piecewise constant function, and the parameters are determined through a least-squares fit to the data by minimizing the sum of the squared differences between the observed splittings and the splittings computed from the parameterized representation of Ω. To obtain a smooth solution, in the quantity to be minimized one adds a multiplum of the average of the square of Ω, or of its first or second derivative; the weight given to this term serves as a trade-off parameter, determining the balance between resolution and error for this method. Another inversion technique is spectral expansion, where Ω is approximated as a linear combination of the kernels (see for example Backus & Gilbert 1967; Gough 1985). Finally, from the asymptotic properties of p modes follows that the inverse problem can be formulated approximately as an integral equation of the Abel type, the solution to which can be written down analytically; this leads to a very simple, although approximate, technique for inversion of data consisting only of such modes (*e.g.* Gough 1984). - In these methods, also, there are parameters which determine the trade-off between resolution and error.

These methods are all *linear*, in the sense that the result of the inversion depends linearly on the data. For any such method, there exist coefficients $c_i(r_0)$ such that the solution $\tilde{\Omega}(r_0)$ may be written in terms of the observed data Δ_i as in equation (4.2), and hence, by using equation (4.1), may be expressed from the original angular velocity $\Omega(r)$ through an averaging kernel $K(r;r_0)$. It should be noticed that, apart from the choice of the trade-off parameter, the averaging kernels are independent of the data. As discussed extensively by Christensen-Dalsgaard *et al.*

(1990), a quantitative comparison of different inversion methods can be carried out in terms of the averaging kernels and the coefficients $c_i(r_0)$.

So far I have considered inversion for a function that depends on r alone. It is evidently desirable, however, to carry out inversion for a general angular velocity $\Omega(r,\theta)$ which is a function both of r and θ. As discussed in section 2.4, it is often convenient to expand $\Omega(r,\theta)$ in terms of suitable functions of θ, with coefficients Ω_s that are functions of r (*cf.* equation (2.21)). If the observed splittings are expanded as in equation (2.26), there result the relations (2.27) between the expansion coefficients of the splittings and the expansion functions $\Omega_s(r)$ of $\Omega(r,\theta)$; from these the Ω_s can be determined from r-dependent inversion, as discussed above. The practical implementation of this procedure was discussed, for example, by Korzennik *et al.* (1988) and Brown *et al.* (1989). Finally, for p modes a two-dimensional asymptotic inversion technique may be developed by noting that the dependence on P_l^m in equation (2.20) can be approximated in such a way as to lead to an integral equation of the Abel type in latitude (Kosovichev & Parchevskii 1988; Gough 1990; Gough & Thompson 1990a,b). As in the one-dimensional case the radial dependence also leads to a similar integral equation. Thus the solution to the inverse problem can once again be written down analytically.

5 Results of rotational splitting observations.

In this section I discuss observed rotational splittings and the results obtained by analysing them. No complete critical review of the observations can be attempted here. However, it might be noted that Harvey (1988a) compared a number of different observations, as did Rhodes *et al.* (1988).

5.1 The equatorial rotation rate.

As discussed in section 3, the splitting between the prograde and retrograde sectoral modes provides a measure of the angular velocity in the neighbourhood of the equatorial plane. The first successful observations of this nature were carried out by Deubner, Ulrich & Rhodes (1979). They considered modes with surface wavenumbers k_h exceeding 0.2 Mm^{-1}, corresponding to degrees greater than about 140. The splitting increased somewhat with decreasing k_h, and hence increasing depth of penetration, indicating that the angular velocity increased with depth in the upper part of the convection zone. There were indications that this trend was reversed for the most deeply penetrating modes in the data. To relate the variation in the splitting to the depth in the solar convection zone Deubner *et al.* assigned a socalled *effective depth* to each mode, essentially assuming a linear variation of Ω with depth. According to this definition, Ω increased to a depth of about 15 Mm, the maximum value of $\Omega/2\pi$ being about 50 nHz higher than the surface value. It was pointed out by Gough (1978) that such an analysis of the depth variation could be misleading. Gough demonstrated the power of inverse analysis for such data, by analysing artificial data for an assumed $\Omega(r)$; however, the analysis was not applied to the observed data.

Somewhat more extensive data, spanning up to five days, were obtained by Hill, Gough & Toomre (1984a). Hill *et al.* (1988) analyzed the data by means of optimally localized averages, to obtain the angular velocity as a function of depth. The results are shown in Figure 5, which is an average of four individual determinations. The most striking result is the increase in $\Omega/2\pi$ by about 20 nHz in the outer 2 Mm, followed by a gradual decrease. This result is at least qualitatively consistent with the earlier results obtained by Deubner *et al.*, given the uncertain definition of the depth scale in the latter analysis. It is also tempting to relate this behaviour to the difference between the Doppler and magnetic feature rotation rates (*cf.* equations (1.1) and (1.2)), on the assumption that the magnetic features are anchored at some distance beneath the visible surface. It should be noticed that the inferred surface value is about 25 nHz smaller than the value obtained from surface Doppler measurements. Hill *et al.* noted, however, that a very small uncorrected drift of the telescope would be sufficient to account for this discrepancy.

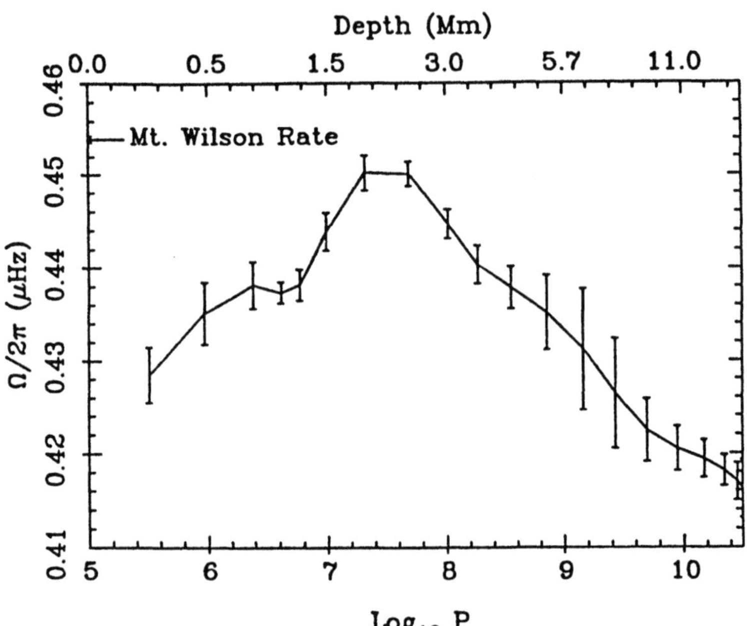

Figure 5. Equatorial rotation frequency $\Omega/2\pi$ as a function of depth (upper scale) and logarithm of pressure (lower scale). The curve was obtained as the average of four individual determinations; the bars denote the standard deviation in this average. From Hill *et al.* (1988).

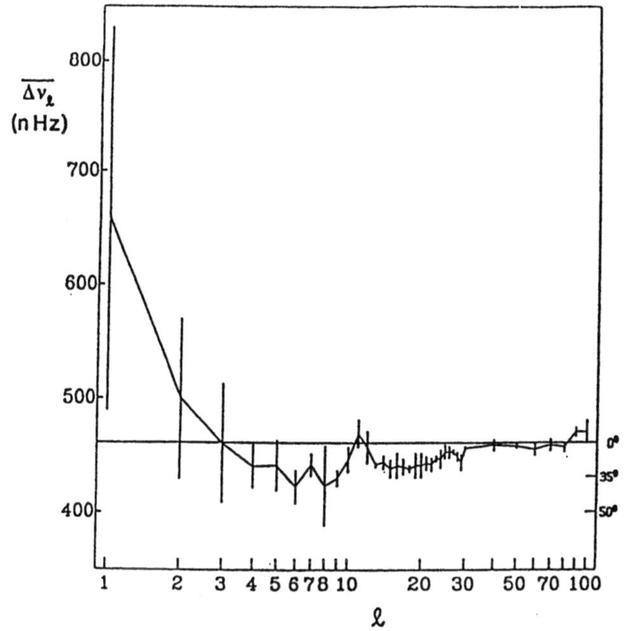

Figure 6. Splittings $\overline{\Delta\nu_\ell} = \overline{\Delta^{(sec)}\omega_{n\ell}}/2\pi$ between sectoral modes (*cf.* equation (3.1)), averaged over radial order n, as a function of the degree ℓ. The error bars represent the standard deviation of the mean. The horizontal line marks the surface equatorial angular velocity, and the surface angular velocity at latitudes 30° and 50° are also indicated at the right. From Duvall & Harvey (1984).

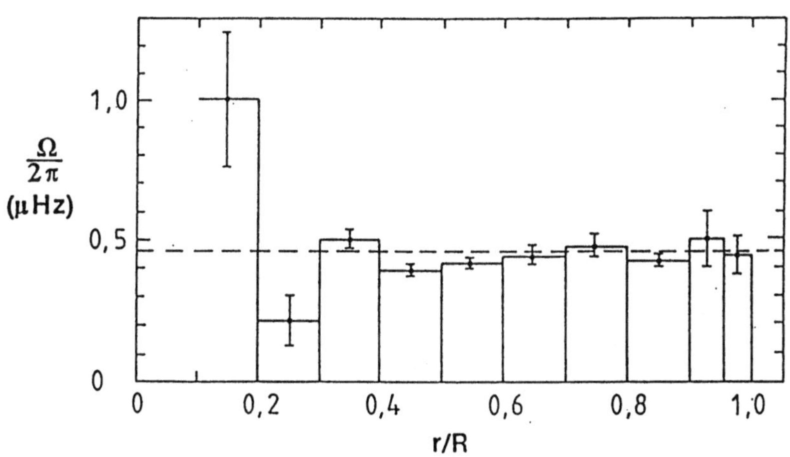

Figure 7. Equatorial rotation frequency $\Omega/2\pi$ as a function of fractional radius r/R, obtained by inverting the data shown in Figure 6. The inversion was carried out by representing Ω as a piecewise constant function and determining the constants by fitting equation (2.11) to the observed splittings. The error bars indicate the observational uncertainties. From Duvall *et al.* (1984).

Duvall & Harvey (1984) used an optical technique to obtain latitudinally averaged data covering essentially the entire solar disk. From these they were able to measure the rotational splitting for sectoral modes at degrees between 1 and 100. The results, after averaging over mode order n, are shown in Figure 6. At most degrees the splitting is slightly, but apparently significantly, below the value corresponding to the surface equatorial angular velocity. The exceptions are at the highest degrees included, in a peculiar feature at $\ell = 11$, and at $\ell = 1$ where, however, the error bars are substantial.

These data were inverted by means of several techniques by Duvall *et al.* (1984). Figure 7 shows the result of a least squares fit of a piecewise constant function to the data. As could have been expected from the splittings, Ω is below its surface value in most of the solar interior. The rapid variation at $r/R = 0.2 - 0.4$ arises principally from the local maximum in the splitting at $\ell = 11$. Finally, it is interesting that the central part of the Sun is found to rotate at roughly twice the surface rate; this, however, is almost entirely a result of the larger splitting at $\ell = 1$, and hence is subject to considerable uncertainty. I return to the rotation of the core in section 5.3 below.

If accepted, the behaviour of the angular velocity in the solar core may have far-reaching implications for the dynamics of the solar interior (Gough 1985b). The observations of Brown (1985) gave some support to the behaviour of the splitting found by Duvall & Harvey (1984); on the other hand Libbrecht (1986) found no maximum at $\ell = 11$. Thus a definite conclusion on the reality of this feature must await further observations.

Finally I note that Hill (1987) presented a summary of determinations of the equatorial rotation rate in the convection zone.

5.2 The dependence of rotation on latitude and depth.

As discussed in section 2, information about the latitude variation of the angular velocity can be obtained from observation of the m-dependence of the rotational splitting. The first to present such data was Brown (1985) (see also Brown 1986 for a more detailed description of the observing and analysis technique). From two-dimensional velocity images of the Sun he could isolate individual spherical harmonics[2] through suitable spatial filtering, and hence obtain spectra with contributions from only a small number of (ℓ, m) pairs. Rather than analyzing completely the resulting large number of spectra to determine individual frequencies $\nu_{n\ell m}$, Brown employed a correlation technique, which effectively averages the splitting over mode order n. These averages were finally fitted by cubics in m/ℓ, the cubic

---

[2]In practice it is impossible to isolate completely a single spherical harmonic from data of this nature, since at most half of the solar surface is observed. Thus the filtered spectra contain contributions from a range of typically ± 2 in ℓ and m.

term providing a measure of the variation of angular velocity with latitude. The results showed that for $15 \le \ell \le 40$ the cubic term was substantially smaller than the cubic term that would result if the surface differential rotation, as in equation (1.1) or (1.2), persisted throughout the Sun. This suggests that the variation of Ω with latitude is smaller in the solar interior than on the surface. This conclusion was confirmed by Christensen-Dalsgaard (1986), who compared the observations with rotational splittings computed for a variety of rotation laws.

A more extensive dataset of this nature was obtained by Brown & Morrow (1987). They again considered averages of the splittings over n, fitting the results to an expression of the form given in equation (3.2) to determine coefficients $\overline{a_\ell^k}$, $k = 1, \cdots, 5$, the coefficient $\overline{a_\ell^5}$ being only marginally statistically significant (*cf.*

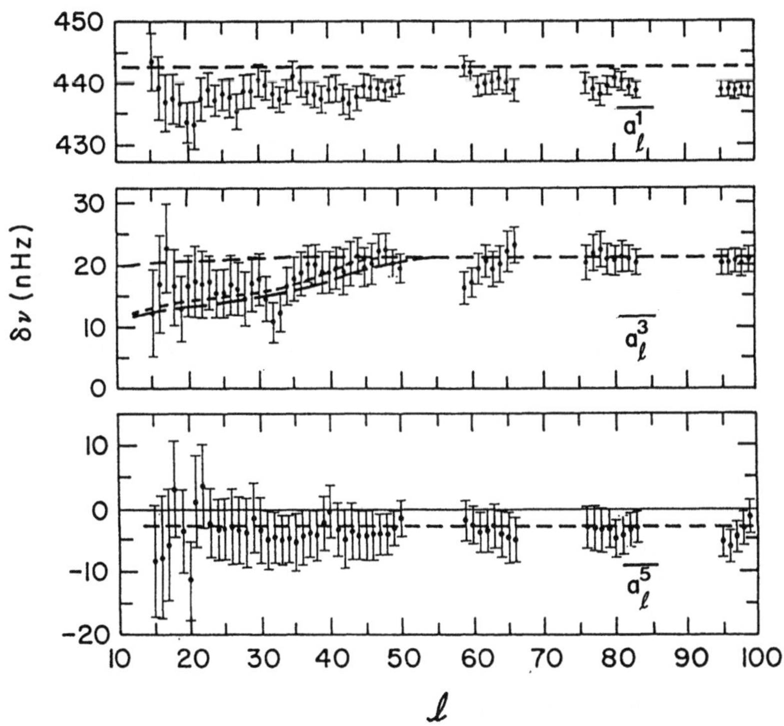

Figure 8. Expansion coefficients for frequency splitting, averaged over mode order n. The dashed lines indicate the values obtained by assuming that the surface differential rotation extended throughout the Sun. In the panel for $\overline{a_\ell^3}$ the dotted and long-dashed lines show the result of assuming radius-independent rotation, and rotation on cylinders, respectively, for $r > 0.73\ R$, with continuous match to the observed surface angular velocity; inside this radius the angular velocity was assumed to be constant, and equal to the surface equatorial value. From Brown & Morrow (1987).

Brown *et al.* 1989). The resulting odd coefficients are shown in Figure 8. It might be noticed, in particular, that $\overline{a_\ell^3}$ is significantly below the value corresponding to the surface rotation for $\ell \lesssim 50$. This again indicates that the latitude variation of Ω decreases with increasing depth.

Brown *et al.* (1989) carried out a detailed analysis of these data, both by comparing the observations with the splitting resulting from different rotation laws, and by applying different inversion techniques. The analyses gave broadly speaking similar results: the equatorial angular velocity decreases slightly, and the polar angular velocity increases substantially, with depth in the convection zone, and beneath the convection zone rotation may be independent of latitude. It should be emphasized, however, that the resolution of the inversions was relatively poor. Thus the extent of the transition region from surface-like to latitude-independent rotation could only be restricted to be less than about 0.2 R.

Analyses of the Duvall & Harvey (1984) and the Brown & Morrow (1987) data were also presented by Kosovichev (1988), who obtained results similar to those described above.

The correlation technique used by Brown (1985) and by Brown & Morrow (1987) is not ideal. Because of the dependence of the turning point radius on frequency (*cf.* equation (2.3)) the average over n mixes modes that extend to different depths, and hence sense the rotation differently. This may make a significant contribution to the scatter in the observed splitting coefficients (see also Harvey 1988a).

Extensive sets of velocity images were analyzed by Libbrecht (1986, 1989). In contrast to Brown & Morrow, Libbrecht determined rotational splitting coefficients in the fit in equation (3.2) for individual n and ℓ. Furthermore, his longest time series extended over 100 days. As a result the scatter in the measured splitting coefficients was considerably smaller than in the previously discussed data; however, Libbrecht's results were essentially consistent with those obtained by Brown & Morrow, showing a similar decrease in $a_{n\ell}^3$ at low degree. To illustrate the dependence of the expansion coefficients on degree and frequency, Libbrecht (1989) fitted them to an expression of the form

$$a_{n\ell}^k = a_k^* + b_k^*(\nu_{n\ell} - \nu_0), \qquad (5.1)$$

with $\nu_0 = 2500\ \mu\text{Hz}$. The results are shown in Figure 9. It is evident that the a_k^* have a very well-defined dependence of ℓ, and hence on the turning-point position.

Libbrecht's data have been subjected to several independent inverse analyses. Figure 10 shows a representative result, obtained by Christensen-Dalsgaard & Schou (1988) by means of optimally localized averages. Given the range of degrees (5 - 60) included in the data, and the fact that the errors in the lowest-degree data were considerable, it was only possible to perform an inversion in the range 0.35 R to 0.8 R. Furthermore the polar angular velocity, which as indicated by equation

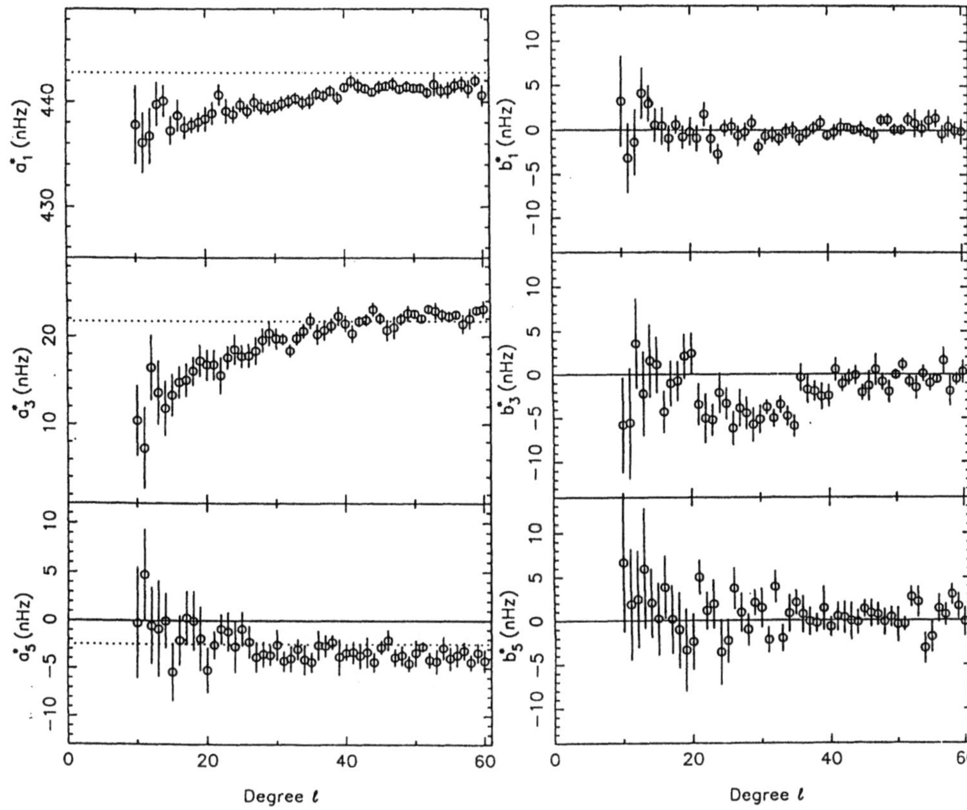

Figure 9. Splitting coefficients a_k^* and b_k^*, resulting from a linear fit of the expansion coefficients $a_{n\ell}^k$ as a function of frequency (*cf.* equation (5.1); note that the values of b_k^* assume that ν is in mHz). From Libbrecht (1989).

(2.31) is particularly sensitive to the highest-order expansion coefficients of the splitting, was relatively poorly determined. Nevertheless, the gross features of the result appear to be significant: the angular velocity is similar to the surface value in much of the convection zone, with a transition near the base of the convection zone to what may be latitude-independent rotation. It should be noted that this transition takes place within the resolution width of the inversion; thus, although with the present data the extent of the transition cannot be further constrained, the results of the inversion are consistent with it being essentially discontinuous.

These results are compared in Figure 11 with a regularized least-squares fit to the data carried out by Dziembowski & Goode (*cf.* Libbrecht 1988b). It is evident

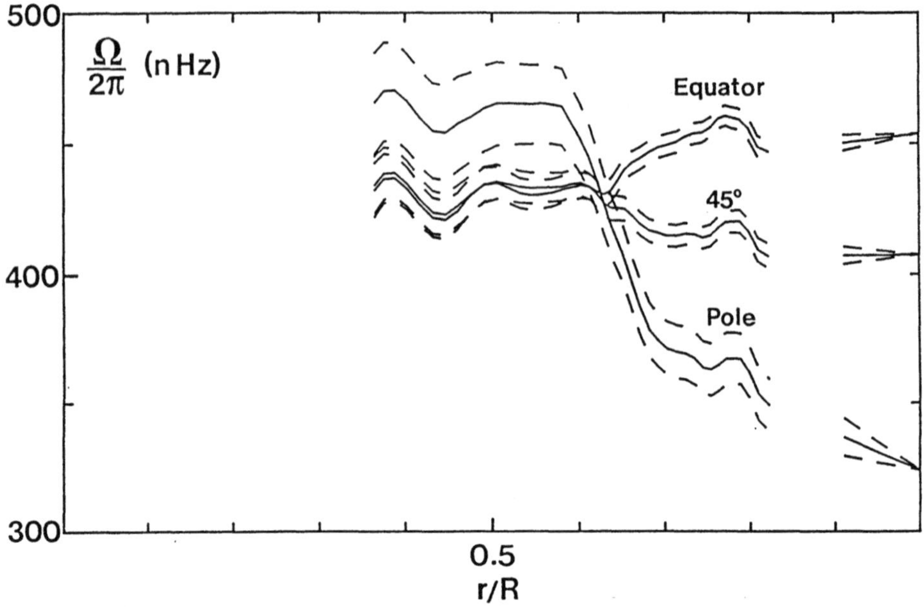

Figure 10. Rotation frequency $\Omega/2\pi$ at the equator, latitude 45° and the pole, obtained by inverting Libbrecht's (1989) observed splittings. The dashed lines correspond to one standard deviation limits. The surface values are indicated by the straight lines starting at $r/R = 1$. Note that inversion for the higher-order expansion coefficient Ω_2 (*cf.* equation (2.21)), which is particularly important for the polar angular velocity, was only possible for $r > 0.6\ R$; inside this it was essentially assumed to be constant. Thus the apparent constancy of the polar rotation frequency inside $0.6\ R$ has no significance. From Christensen-Dalsgaard & Schou (1988).

that these two independent analyses gave very similar results. Dziembowski, Goode & Libbrecht (1989) discussed the least-squares inversion in more detail, with particular emphasis on the fact that the region with a sharp radial gradient in the angular velocity is a likely site for dynamo action which may be responsible for the solar magnetic activity (*cf.* section 6). A third independent inversion of the Libbrecht data was carried out by Thompson (1990). Using a least-squares technique he obtained results very similar to those of Christensen-Dalsgaard & Schou and Dziembowski & Goode. Furthermore Thompson made an accurate estimate of the average gradient of the angular velocity in the convection zone, by means of inversion using optimally localized averages. This showed that rotation in the convection zone is almost certainly not constant on cylinders, as might otherwise have been expected from hydrodynamical simulations of convection zone dynamics (see Stix, these proceedings, for a review).

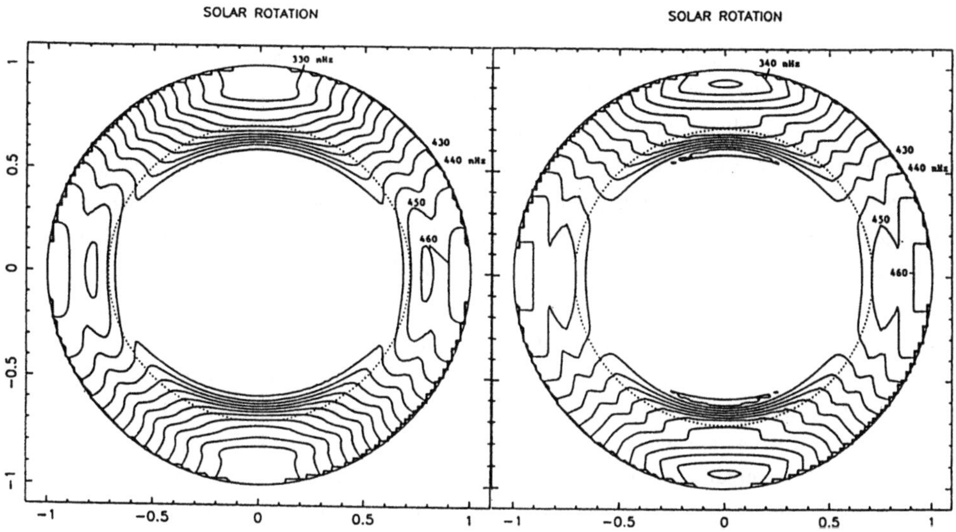

Figure 11. Contour diagrams of rotational frequency $\Omega/2\pi$ obtained by inverting Libbrecht's (1989) data. The left panel was obtained by Dziembowski & Goode using a least squares technique with first-derivative smoothing (see also Dziembowski, Goode & Libbrecht 1989). The right panel gives the results obtained by Christensen-Dalsgaard & Schou (1988) which were presented in Figure 10. From Libbrecht (1988b).

The data discussed so far were all obtained from observations of Doppler velocity. Duvall, Harvey & Pomerantz (1986) presented spatially resolved observations made in Ca II line intensities; the observations were carried out in 1981 at the South Pole. The data were analyzed by means of the cross-correlation technique of Brown (1985) and presented as averages over bins of 10 in degree. In contrast to the Doppler velocity data little variation with ℓ of the expansion coefficients was found, indicating that the surface differential rotation persists throughout most of the Sun. However, similar data obtained at the South Pole in 1987 (Jefferies *et al.* 1988), which extended to ℓ = 150, were essentially in agreement with the Doppler data at the degrees where the datasets overlapped (see also Harvey 1988a). The reason for the discrepancies with the earlier South Pole data is not clear. Since the Jefferies *et al.* data include modes at higher degree, inversion of the data, which has yet to be carried out, should enable determination of the angular velocity closer to the surface than was possible in the inversions shown in Figures 10 and 11.

Extensive sets of Doppler image observations have been obtained at the Mt Wilson Observatory. Tomczyk (1988) and Tomczyk *et al.* (1988) presented results of an analysis of the rotational splitting for part of the data, at degrees up to 120.

Results for sectoral modes, extending to higher degree, were obtained by Rhodes, Cacciani & Korzennik (1988). Comparisons, carried out by Harvey (1988a) and Rhodes *et al.* (1988), with the data obtained by Brown & Morrow (1987), Jefferies *et al.* (1988) and Libbrecht (1989) showed that for the Mt Wilson data $\overline{a_\ell^1}$ was systematically higher than for the other sets, whereas $\overline{a_\ell^3}$ was systematically lower; however, the sum $\overline{a_\ell^1} + \overline{a_\ell^3} + \overline{a_\ell^5}$, which as indicated in section 2.4 is a measure of the equatorial rotation rate, was very similar for the Mt Wilson and the other sets. The reason for these systematic differences is so far not understood.

Rhodes *et al.* (1988) combined the different independent datasets and presented the results against an average location of the kernels, obtained as the midpoint between the lower turning point and the surface. Although this at best gives a very rough indication of the variation of angular velocity with depth, the results were in qualitative agreement with the results of inversions, indicating again a decrease in the angular variation of Ω towards the base of the convection zone. Korzennik *et al.* (1988) carried out inversions of both the individual observed datasets and a combined set, by means of the spectral expansion and the optimally localized averages techniques. Only data averaged in n at each ℓ were considered. The results differed in detail between the different sets, and from the inversions carried out by Christensen-Dalsgaard & Schou (1988) and Dziembowski *et al.* (1989). However, the dominant features were again similar to those discussed previously.

5.3 Rotation of the solar core.

To measure the rotation of the solar core from p-mode data requires observation of modes of low degree. As discussed in section 3, the best observations of such modes have been made through Doppler velocity measurements in integrated sunlight. From such measurements Claverie *et al.* (1981) found evidence of triplet structure for the $\ell = 1$ modes and quintuplet structure for the $\ell = 2$ modes. This was interpreted as due to rotation; the average splitting was found to be 0.75 μHz, *i.e.* substantially higher than the surface value of the angular velocity. Pallé *et al.* (1988) analyzed more extensive sets of data. To minimize the problems caused by the finite linewidth of the modes (*cf.* section 3) they selected subsets of the timestrings where a given mode appeared to maintain its phase and amplitude. The resulting spectra again showed triplet structure for several $\ell = 1$ modes, with an average splitting of 0.72 μHz. From an autocorrelation analysis the splitting for modes of degree 2 was found to be 0.55 μHz, whereas the splitting for $\ell = 3$ and 4 was estimated to be 0.47 μHz, *i.e.* close to the value expected from the surface rotation. It is interesting that these results are consistent with the values obtained by Duvall & Harvey (1984) from spatially resolved observations (*cf.* section 5.1).

A severe difficulty with the results of Claverie *et al.* (1981) and Pallé *et al.* (1988) is the detection of the $\ell = 1$, $m = 0$ mode which, as discussed in section 3, should be essentially invisible in observations of this nature; the same is true for the $\ell = 2$, $m = \pm 1$ modes that were apparently detected by Claverie *et al.* It was suggested by Isaak (1982) that the observations could be explained if the oscillations had an axis of symmetry differing from the surface rotation axis, possibly due to a very strong magnetic field in the deep solar interior. However, except under very special circumstances such a configuration would lead to a far more complicated spectrum of oscillations (Dicke 1982, Gough 1982a, Gough & Taylor 1984; Gough & Thompson 1990a). An alternative explanation is that the observed fine structure in the peaks is predominantly due to amplitude variation or phase shifts. It is not inconceivable that such effects could mimic rotational splitting, even in the data of Pallé *et al.* who attempted to minimize them. Also it might be noted that in the examples of spectra shown by Pallé *et al.* the evidence for triplet structure is rather marginal; if these cases are disregarded it is possible that the measured splittings correspond to the difference between $m = -1$ and $+1$, in which case the corresponding averaged angular velocity $\langle\Omega\rangle$ (*cf.* equation (2.17)) might even be smaller than the surface value. Observations of low-degree p modes that provide some separation of the different values of m are clearly highly desirable.

Observation of g modes would provide far more sensitive measures of the core rotation. Delache & Scherrer (1983) found evidence for rotationally split $\ell = 1$ modes in differential Doppler velocity measurements; the splitting corresponded to an average rotation frequency $\langle\Omega\rangle/2\pi$ of 1.2 μHz. From a statistical analysis of irradiance measurements, using the asymptotic properties of the the modes, Frölich & Delache (1984a,b) found that $\langle\Omega\rangle/2\pi$ was 1 to 1.2 μHz. Subsequent analysis of similar data (Fröhlich 1988) yielded $\langle\Omega\rangle/2\pi = 1.05$ μHz. Pallé & Roca Cortês (1988) made a statistical analysis of several years' Doppler velocity measurements in integrated light and found that the average of $\langle\Omega\rangle/2\pi$ was between 1.0 and 1.3 μHz; on the other hand Garcia, Pallé & Roca Cortês (1988) found no statistically significant evidence for g modes in more extensive datasets of this nature. Finally Henning & Scherrer (1988) identified several pairs of peaks in spectra of differential Doppler velocity observations as rotationally split $\ell = 1$ modes; the average splitting corresponded to $\langle\Omega\rangle/2\pi = 1.6$ μHz. Although these results are interesting, it is probably fair to say that the significance of the detection of g modes has yet to be definitely established.

It is evident that there is considerable uncertainty concerning the observational situation for the splitting of modes that are sensitive to the rotation of the core. Nevertheless, it is of some interest to consider the consequences of the observed values that have been quoted. For example, it is immediately obvious from Figure 3 that to obtain a splitting of 0.7 μHz for the $\ell = 1$ p modes due to rapid rotation of a small core, a very high angular velocity would be needed. To illustrate this in more detail, I restrict myself to rotation that depends on r alone, and consider piecewise constant rotation laws on the form

$$\Omega(r) = \begin{cases} \Omega_c & \text{for } r < r_c \\ \Omega_s & \text{for } r > r_c ; \end{cases} \tag{5.2}$$

for Ω_s the equatorial plasma rate given in equation (1.1) is used. Furthermore, I assume that the observed average splittings of p modes can be adequately represented by the splitting corresponding to the average of the kernels for modes of the appropriate ℓ within the frequency range 2500 to 3500 μHz, whereas the average splitting of g modes can be similarly represented by modes in the frequency range 50 to 100 μHz. Finally I impose the constraint that the average splitting of the $\ell = 1$ p modes be 0.7 μHz; the required Ω_c can then be evaluated for any assumed radius r_c, and the splittings for other modes can be computed.

In Figure 12a the dashed line illustrates the resulting $\Omega_c(r_c)$. As was already inferred from Figure 3, quite rapid core rotation is required if the core is not extended. If, for example, the core extends over the inner 20 per cent of the solar radius, it would have to rotate at about 9 times the surface rate. Computed p-mode splittings, as functions of r_c, are shown in Figure 12b. Also shown are observed splittings for $\ell = 2$, 3 and 4 from Duvall & Harvey (1984), indicated by the intervals corresponding to the standard errors. For higher values of ℓ, the observations are generally at or below the surface angular velocity (*cf.* Figure 6). It is evident that the computed splittings are in general considerably outside the error limits of the observations. For $\ell = 2$ only a narrow range of r_c around 0.14 R appears to be marginally acceptable, and here Ω_c would have to be around 15 Ω_s. The splittings for $\ell = 3$ and 4 then deviate substantially more than 1 standard deviation from the observed value, however. The situation is further complicated if also the suggested observed splittings for g modes, corresponding to 2 - 3 times the surface angular velocity, are taken into account; for these modes most of the contribution to the splitting comes from the core, as was seen in Figure 3, and hence the average angular velocities $\langle\Omega\rangle$, as sensed by the g modes, are only slightly below Ω_c. This is confirmed by Figure 12a, where the solid curves show computed $\langle\Omega\rangle$ in units of Ω_s, as functions of r_c, for g modes. It is evident that the observations are only consistent with the model if r_c is as large as 0.5 R; but this is incompatible with the p-mode data.

Obviously no definite conclusions can be drawn on the basis of this analysis, given that it is restricted to a very special form of the rotation law. Indeed, there is little doubt that an angular velocity distribution $\Omega(r)$ can be found which agrees with all the observed splittings, including those for g modes; the results obtained here suggest that it would have to be rather peculiar, however. If only the p-mode data are taken into account, it appears likely that reasonably plausible rotation laws can be found which are consistent with the data within their error bars. However, the constraints of the observed splittings for $\ell \geq 4$, which appear to be fairly well-determined, would limit the region of rapid rotation to radii below about 0.2 R, and consequently this region would have to rotate at many times the surface rate. Better data than now available, in particular definite and accurate measurements of

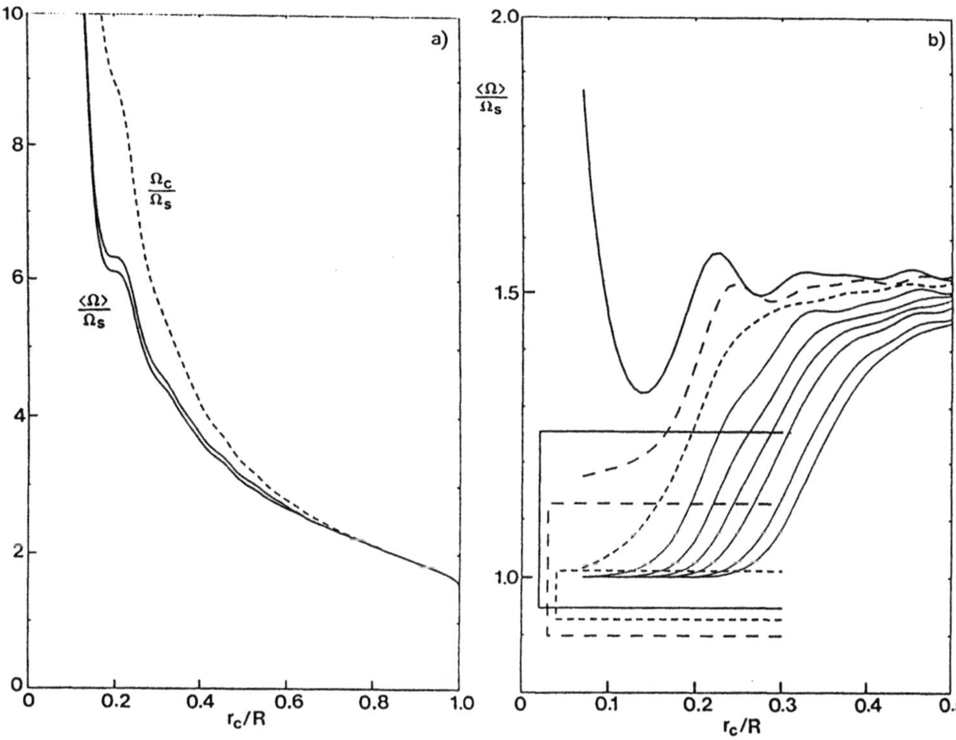

Figure 12. In panel a) the dashed curve shows the angular velocity Ω_c of the core, in units of the surface angular velocity Ω_s, as a function of the core radius r_c, for a rotation law of the form (5.2); Ω_c was determined such that the average splitting for $\ell = 1$ p modes was 0.7 μHz. The solid curves show the corresponding average angular velocities $\langle\Omega\rangle$ (*cf* equation (2.17)), also in units of Ω_s, as sensed by $\ell = 1$ and $\ell = 2$ g modes. Panel b) shows $\langle\Omega\rangle/\Omega_s$ for p modes for this rotation law, again as a function of r_c. The thin continuous curves are for $\ell = 5 - 10$. The heavy curves give splittings for $\ell = 2$ (————), $\ell = 3$ (----------) and $\ell = 4$ (— — — — — — — —). Also shown, with the same line styles, are the one standard deviation intervals for the splittings observed by Duvall & Harvey (1984).

individual splittings for $\ell = 1$ and 2, are needed before a more detailed determination of the rotation of the core can be made.

5.4 Measurements at the solar limb.

Apart from some fairly minor systematic differences the observations discussed so far present a reasonably consistent picture. In particular, with the possible exception of the core, there is no indication that parts of the Sun are rotating at substantially more than the surface rate. However, completely discordant results have been obtained from observations of the solar limb intensity made by the SCLERA group. Analysis of these data indicates that most of the solar interior rotates at more than 4 times the surface rate (*e.g.* Hill, Bos & Goode 1982). Furthermore evidence has been found for a substantial increase in the polar angular velocity, to well over the equatorial value, in the lower part of the convection zone (Hill *et al.* 1986). These results are inconsistent with the rotational splittings discussed in the preceding sections.

The reasons for this discrepancy are not clear. It should be noted, however, that analysis of the limb measurements provides a far less efficient separation between different modes than does the analysis of the two-dimensional velocity or intensity maps. Therefore, the resulting power spectra are extremely complicated, and the tasks of identifying individual modes and measuring the rotational splitting are correspondingly difficult. Given this, and the agreement between the remaining measurements, it appears likely that problems with the mode identification are responsible for the discrepant SCLERA measurements.

A more detailed discussion of these results was presented by Harvey (1988a).

6 Discussion.

It is evident that helioseismology has already provided a considerable amount of information about rotation in the solar interior. Although the results are incomplete for much of the Sun, and the details are still uncertain, it seems likely that the main features are robust. There appears to be an increase with depth of the equatorial angular velocity Ω_{eq}, by about 5 per cent, in the upper few Mm of the solar convection zone; in the rest of the convection zone Ω_{eq} probably decreases slightly. The latitude dependence of Ω is similar to the surface behaviour in much of the convection zone, with a transition, which could be quite rapid, near its base to a substantially smaller variation with latitude. The present data provide relatively little information about the latitude dependence in the radiative interior; however, they appear to be consistent with latitude-independent rotation in this region. More information is available about the variation of Ω_{eq} beneath the convection zone: in much of the radiative interior Ω_{eq} is slightly, but apparently significantly, smaller than its surface value, although the behaviour may be more complicated near the outer edge of the energy generating core. The rotation of the core is still very uncertain; there are indications from several independent sets of observations that it rotates at several times the surface rate, but the evidence cannot as yet be regarded as conclusive.

These results appear to be consistent with observations of the rotation on the solar surface. Thus the slightly higher angular velocity for magnetic features, compared with the surface plasma rate (*cf.* equations (1.1) and (1.2)) could be linked to the increase with depth in the equatorial angular velocity near the surface. Further observations of similar nature, reviewed by Schüssler (1987), also suggest faster rotation in the upper part of the convection zone, whereas there are indications that the rotation in the deeper part of the convection zone could be *slower* than at the surface. Finally Stenflo (1989) found that the rotation of large-scale magnetic patterns varied little with latitude; assuming that such patterns originate near the base of the convection zone, he inferred a distribution of rotation within the convection zone which is similar to that obtained in Figure 10 from inversion of the rotational splittings.

It is of obvious interest to investigate how the observed rotation profile might have come about or is maintained. It is generally believed that solar-type stars rotate rapidly when they arrive on the main sequence but lose angular momentum from the surface through a magnetic stellar wind during their evolution. The distribution of rotation in the interior of the star then depends on the transfer of angular momentum to the convection zone, and hence to the surface where the braking takes place; however it is not unreasonable to expect that a rapidly rotating core would remain. Spruit, Knobloch & Roxburgh (1983) estimated the resulting rotation profile, by assuming that it would be such as to maintain marginal stability towards instabilities that would cause angular momentum transport. Pinsonneault *et al.* (1989) computed detailed evolution sequences of rotating stellar models, taking into account angular momentum loss from the surface and transport of angular momentum through assumed rotationally induced instabilities. These processes are not fully understood, and the calculation therefore depends on a number of unknown parameters; as a result there is considerable latitude in the results. However, in all cases Ω was a decreasing function of radius, and the angular velocity of the core exceeded the surface value by a large factor. In fact, calculations by Demarque & Guenther (1988), assuming such angular velocity profiles, of rotational splittings for $\ell = 7 - 10$ showed that these far exceeded the values observed by Duvall & Harvey (1984). Thus the models appear to be inconsistent with the observations, and additional mechanisms for angular momentum transport are required. Indeed, it was pointed out by Rosner & Weiss (1985) that even a very modest magnetic field in the radiative interior would transport angular momentum so efficiently that rotation would become essentially rigid. A more detailed discussion of the possible mechanisms for angular momentum transport, with special emphasis on magnetic effects, was given by Spruit (1987).

Pinsonneault *et al.* (1989) pointed out that if the Sun were to be rotating approximately rigidly, as appears to be indicated by the helioseismic data, the rotation period will increase rapidly, to more than 200 days, when the Sun evolves into a subgiant. This seems to be in conflict with inferences about subgiant rotation made by Noyes *et al.* (1984) who found that the rotation period was roughly constant, and below 50 days, for subgiants of effective temperature down to about 4900 K. On the other hand the subgiant data were in good agreement with the computations of Pinsonneault *et al.* It is evident that there is no formal conflict between the solar helioseismic results and the subgiant data: there is no assurance that either the Sun, or the subgiants, are representative, and hence there is no compelling reason to expect them to behave similarly. It should also be pointed out that the subgiant rotation periods were determined by means of an indirect method based on chromospheric activity and measured rotation of main-sequence stars; this may not hold equally for subgiants. Nevertheless, there are indications of a very interesting problem here: It is obviously unsatisfactory to be forced to claim that the Sun is special in terms of its rotational properties. Yet, given the observed rotational splittings, it seems highly unlikely that the Sun can hide a sufficient amount of angular momentum to enable it to maintain relatively rapid rotation as it evolves away from the main sequence. Indeed, for the simple model of rotation discussed in section 5.3 (*cf.* equation (5.2)), imposing again an average splitting of 0.7 μHz for the $\ell = 1$ modes and demanding that the model be consistent with the other splitting observations, the angular momentum is less than 1.5 times the value corresponding to uniform rotation; this is insufficient to maintain rapid rotation when the Sun evolves into a subgiant. - Better, and more extensive, data on the rotation of stars away from the main sequence are badly needed.

The inferred angular velocity distribution in the convection zone is in conflict with results of detailed hydrodynamical simulations of convection zone dynamics, which tend to predict rotation that is constant on cylinders, *i.e.* depending only on the distance from the rotation axis (*e.g.* Gilman & Miller 1986; Glatzmaier 1985). Gilman, Morrow & DeLuca (1989) argued that the observed distribution may be explained through angular momentum transport between the convection zone and the radiative interior, through some as yet unspecified mechanism; the details of this model remain to be worked out, however. It should also be noted that simplified models incorporating anisotropic viscosity or inhomogeneous heat transport (Pidatella *et al.* 1986; see also Stix 1987) have produced rotation profiles that are superficially similar to those obtained from the helioseismic measurements.

The centrifugal distortion of the Sun due to rotation perturbs the gravitational field outside the Sun. It has been speculated that this might make a significant contribution to the precession of the orbit of Mercury, possibly invalidating the apparent agreement between the observed value and the predictions of Einstein's theory of general relativity. It is in principle possible to measure the distortion of the visible solar surface, and from this to infer the distortion of the equipotential surfaces, as measured by the quadrupole moment J_2. Such measurements have in the past indicated a substantial oblateness (Dicke & Goldenberg 1974), although more recent measurements have found smaller values (Hill & Stebbins 1975; Dicke, Kuhn & Libbrecht 1987). In any case the determination of the shape of the visible surface is complicated by the variation in solar activity, and hence possibly in surface brightness, between equator and poles. On the other hand, given a helioseismic determination of $\Omega(r,\theta)$ it would be straightforward to calculate J_2 (Gough 1982b). Indeed, the main areas of uncertainty in the current determination of the angular velocity are in the solar core and very near the surface, both of which regions have little effect on J_2. An additional problem is the lack of definite information about the latitude variation of rotation in the radiative interior; however if, as seems plausible, the latitude dependence is modest, J_2 is unlikely to be substantially different from the value $J_2 = 1.7\times10^{-7}$ which results from the equatorial angular velocity determined by Duvall *et al.* (1984) assuming no latitude dependence. In that case there is essentially perfect agreement between the observed orbit of Mercury and the predictions of general relativity (*e.g.* Narlikar & Rana 1985; see their "Note added in proof").

The rotation in the solar convection zone bears directly on those models of the generation of solar magnetic activity that are based on dynamo action spread out through the convection zone. As reviewed, for example, by Gilman & DeLuca (1986), such models require an increase of the angular velocity with depth in the convection zone, to reproduce the observed migration of active regions towards the equator during the solar cycle; this is appears to be inconsistent with the helioseismic measurements. This difficulty can be avoided by considering dynamo action in the region of convective overshoot beneath the convection zone. Here, according to results shown in Figure 10, there is a slight decrease in Ω with depth at low latitudes, and a substantial increase with depth at high latitudes. As discussed by Gilman, Morrow & DeLuca (1989), who made a careful analysis of the implications on dynamo models of the helioseismic measurements of the angular velocity, it may be possible under these circumstances to explain the observed surface migration of magnetic activity.

As a further contribution to understanding the generation of solar activity, and the dynamics of the convection zone, measurements of other large-scale velocity fields would be very valuable. In particular one might be able to detect the flow associated with giant convective cells, which are predicted by numerical models of the convection zone. Gough & Toomre (1983) considered the effects of such large-scale convective eddies on the oscillations in a simplified models. Some evidence for day-to-day variations in the horizontal velocities inferred by inverting frequency splittings of high-degree modes was found by Hill, Gough & Toomre (1984a); this was tentatively attributed to the result of advection of giant cells past the region of the solar surface where the oscillations were observed. Hill, Gough & Toomre (1984b) carried out a sensitivity analysis, which provided some support for the reality of the results. Further evidence for such flows was obtained by Hill (1988a, 1989a). However, it should be pointed out that the analysis could be quite sensitive to beating between unresolved modes, or possibly to effects of atmospheric seeing. Thus the detection of convective velocity fields must so far be regarded as preliminary.

Measurements of this nature are probably at the limit of the capabilities of present observations. Indeed, despite their impressive achievements, helioseismic investigations of solar internal rotation are still in their infancy. Coming developments in the observation of rotational splittings should permit a far more accurate determination of $\Omega(r,\theta)$ throughout the Sun.

Observations of solar oscillations, and analyses of the data, are longterm projects which are continually being pursued by several independent groups. As a result there is a steady increase in the material available for inverse analyses. In the near future one may hope in particular to obtain improved data for high-degree modes; this should enable a better resolution of the equatorial angular velocity near the surface and, one may hope, information about the variation with latitude in this region. Additional data for modes of moderate degree are also becoming available, which may allow us to limit further the extent of the region near the base of the convection zone where the transition to almost latitude-independent rotation takes place.

In the slightly longer term, dramatic improvements in the observational situation can be expected as new instruments become operational during the coming decade. Two networks of observing stations using integrated light, Doppler velocity techniques have already begun observations (Aindow *et al.* 1988; Fossat 1988), and in both cases further stations will be added. The GONG project (Hill 1989b; see also Harvey 1988b, Hill 1988b, Pintar 1988) will establish six stations for Doppler velocity observations with a spatial resolution which allows observation of modes of degree up to about 200. This network is expected to become operational in 1992. Finally, the satellite SOHO (Domingo 1988), to be launched in 1995, will carry instruments for unresolved (Damé 1988) and highly resolved (Scherrer, Hoeksema & Bogart 1988) Doppler velocity measurements, as well as instruments for measuring solar irradiance and intensity (Fröhlich *et al.* 1988). The data from these instruments will allow resolution of the depth and latitude variation of rotation in far greater detail than is now possible; undoubtedly this will cause the results presented in this review to appear extremely crude. Nevertheless it may not be unreasonable to hope that their main features will still be found to be valid.

Acknowledgements: I am very grateful to J. R. Kuhn, K. G. Libbrecht, R. W. Noyes, M. Stix, J. M. Sørensen and S. Tomczyk for useful correspondence or conversations in the course of the preparation of this review.

References.

Aindow, A., Elsworth, Y. P., Isaak, G. R., McLeod, C. P., New, R. & van der Raay, H. B., 1988. *Seismology of the Sun & Sun-like Stars*, p. 157, eds Domingo, V. & Rolfe, E. J., ESA SP-286.

Backus, G. & Gilbert, F., 1967. *Geophys. J.*, **13**, 169.

Backus, G. & Gilbert, F., 1970. *Phil. Trans. Roy. Soc. London*, **A266**, 123.

Brown, T. M., 1985. *Nature*, **317**, 591.

Brown, T. M., 1986. *Seismology of the Sun and the distant Stars*, p. 199, ed. Gough, D. O., D. Reidel Publ. Co., Dordrecht.

Brown, T. M., 1988. *Proc. IAU Symposium No 123*, p. 453, eds Christensen-Dalsgaard, J. & Frandsen, S., D. Reidel Publ. Co., Dordrecht.
Brown, T. M. & Morrow, C. A. 1987. *Astrophys. J.*, **314**, L21.
Brown, T. M., Christensen-Dalsgaard, J., Dziembowski, W. A., Goode, P., Gough, D. O., Morrow, C. A., 1989. *Astrophys. J.*, **343**, 526.
Christensen-Dalsgaard, J., 1986. *Cool Stars, Stellar Systems and the Sun. Lecture Notes in Physics*, **254**, p. 145, eds Zeilik, M. & Gibson, D. M., Springer.
Christensen-Dalsgaard, J. & Berthomieu, G., 1990 In *Solar interior and atmosphere*, eds Cox, A. N., Livingston, W. C. & Matthews, M., Space Science Series, University of Arizona Press, submitted.
Christensen-Dalsgaard, J. & Gough, D. O., 1982. *Mon. Not. R. astr. Soc.*, **198**, 141.
Christensen-Dalsgaard, J. & Schou, J., 1988. *Seismology of the Sun & Sun-like Stars*, p. 149, eds Domingo, V. & Rolfe, E. J., ESA SP-286.
Christensen-Dalsgaard, J., Gough, D. O. & Toomre, J., 1985. *Science*, **229**, 923.
Christensen-Dalsgaard, J., Schou, J. & Thompson, M. J., 1990. *Mon. Not. R. astr. Soc.*, in the press.
Claverie, A., Isaak, G. R., McLeod, C. P. & van der Raay, H. B., 1981. *Nature*, **293**, 443.
Cowling, T. G. & Newing, R. A., 1949. *Astrophys. J.*, **109**, 149.
Craig, I. J. D. & Brown, J. C., 1986. *Inverse problems in astronomy: a guide to inversion strategies for remotely sensed data*, Adam Hilger, Bristol.
Cuypers, J. 1980. *Astron. Astrophys.*, **89**, 207.
Damé, L., 1988. *Seismology of the Sun & Sun-like Stars*, p. 367, eds Domingo, V. & Rolfe, E. J., ESA SP-286.
Deepak, A., 1977. *Inversion Methods in Atmospheric Remote Sounding*, Academic, New York.
Delache, P. & Scherrer, P. H., 1983. *Nature*, **306**, 651.
Demarque, P. & Guenther, D. B., 1988. *Seismology of the Sun & Sun-like Stars*, p. 99, eds Domingo, V. & Rolfe, E. J., ESA SP-286.
Deubner, F.-L. & Gough, D. O., 1984. *Ann. Rev. Astron. Astrophys.*, **22**, 593.
Deubner, F.-L., Ulrich, R. K. & Rhodes, E. J., 1979. *Astron. Astrophys.*, **72**, 177.
Dicke, R. H., 1982. *Nature*, **300**, 693.
Dicke, R. H. & Goldenberg, H. M., 1974. *Astrophys. J. Suppl.*, **27**, 131.
Dicke, R. H., Kuhn, J. R. & Libbrecht, K. G., 1987. *Astrophys. J.*, **318**, 451.
Domingo, V., 1988. *Seismology of the Sun & Sun-like Stars*, p. 363, eds Domingo, V. & Rolfe, E. J., ESA SP-286.
Durney, B. R., Hill, F. & Goode, P. R., 1988. *Astrophys. J.*, **326**, 486.
Duvall, T. L. & Harvey, J. W., 1984. *Nature*, **310**, 19.
Duvall, T. L., Dziembowski, W. A., Goode, P. R., Gough, D. O., Harvey, J. W. & Leibacher, J. W., 1984. *Nature*, **310**, 22.
Duvall, T. L., Harvey, J. W. & Pomerantz, M. A., 1986. *Nature*, **321**, 500.
Dziembowski, W. A., 1977. *Acta Astron.*, **27**, 203.
Dziembowski, W. & Goode, P. R., 1984. *Mem. Soc. astr. Ital.*, **55**, 185.
Dziembowski, W. A., Goode, P. R. & Libbrecht, K. G., 1989. *Astrophys. J.*, **337**, L53.
Fossat, E., 1988. *Seismology of the Sun & Sun-like Stars*, p. 161, eds Domingo, V. & Rolfe, E. J., ESA SP-286.
Fröhlich, C., 1988. *Proc. IAU Symposium No 123*, p. 83, eds Christensen-Dalsgaard, J. & Frandsen, S., Reidel, Dordrecht.
Fröhlich, C. & Delache, Ph., 1984a. *Mem. Soc. astr. Ital.*, **55**, 99.
Fröhlich, C. & Delache, Ph., 1984b. *Solar seismology from space*, p. 183, eds Ulrich, R. K., Harvey, J., Rhodes, E. J. & Toomre, J., NASA, JPL Publ. 84-84.
Fröhlich, C., Andersen, B. N., Berthomieu, G., Crommelynck, D., Delache, Ph., Domingo, V., Jimenez, A., Jones, A. R., Roca Cortés, T., Wehrli, Ch., 1988. *Seismology of the Sun & Sun-like Stars*, p. 371, eds Domingo, V. & Rolfe, E. J., ESA SP-286.
Garcia, C., Pallé, P. L. & Roca Cortés, T., 1988. *Seismology of the Sun & Sun-like Stars*, p. 353, eds Domingo, V. & Rolfe, E. J., ESA SP-286.
Gilman, P. A. & DeLuca, E. E., 1986. *Cool Stars, Stellar Systems and the Sun. Lecture Notes in Physics*, **254**, p. 163, eds Zeilik, M. & Gibson, D. M., Springer.

Gilman, P. A. & Miller, J. 1986. *Ap. J. Supp.*, **61**, 585.
Gilman, P. A., Morrow, C. A. & DeLuca, E. E., 1989. *Astrophys. J.*, **338**,, 528.
Glatzmaier, G., 1985. *Ap. J.*, **291**, 300.
Gough, D. O., 1978. *Proc. Workshop on solar rotation*, p. 255, eds Belvedere, G. & Paterno, L., University of Catania Press.
Gough, D. O., 1981. *Mon. Not. R. astr. Soc.*, **196**, 731.
Gough, D. O., 1982a. *Nature*, **298**, 350.
Gough, D. O., 1982b. *Nature*, **298**, 334.
Gough, D. O., 1984. *Phil. Trans. R. Soc. London*, A **313**, 27.
Gough, D. O., 1985a. *Solar Phys.*, **100**, 65.
Gough, D. O., 1985b. *Future missions in solar, heliospheric and space plasma physics*, p. 183, eds Rolfe, E. & Battrick, B., ESA SP-235, ESTEC, Noordwijk.
Gough, D. O., 1990. In *Astrophysical Fluid Dynamics*, eds Zahn, J.-P. & Zinn-Justin, J., Elsevier Science Publisher B.V., in the press.
Gough, D. O. & Taylor, P. P., 1984. *Mem. Soc. astr. Ital.*, **55**, 215.
Gough, D. O. & Thompson, M. J., 1990a. *Mon. Not. R. astr. Soc.*, in the press.
Gough, D. O. & Thompson, M. J., 1990b. In *Solar interior and atmosphere*, eds Cox, A. N., Livingston, W. C. & Matthews, M., Space Science Series, University of Arizona Press, submitted.
Gough, D. O. & Toomre, J., 1983. *Solar Phys.*, **82**, 401.
Hansen, C. J., Cox, J. P. & van Horn, H. M., 1977. *Astrophys. J.*, **217**, 151.
Harvey, J., 1988a. *Seismology of the Sun & Sun-like Stars*, p. 55, eds Domingo, V. & Rolfe, E. J., ESA SP-286.
Harvey, J., 1988b. *Seismology of the Sun & Sun-like Stars*, p. 203, eds Domingo, V. & Rolfe, E. J., ESA SP-286.
Henning, H. M. & Scherrer, P. H., 1988. *Seismology of the Sun & Sun-like Stars*, p. 419, eds Domingo, V. & Rolfe, E. J., ESA SP-286.
Hill, F., 1987. *The internal solar angular velocity*, p. 45, eds Durney, B. R. & Sofia, S., Reidel, Dordrecht.
Hill, F., 1988a. *Astrophys. J.*, **333**, 996.
Hill, F., 1988b. *Seismology of the Sun & Sun-like Stars*, p. 209, eds Domingo, V. & Rolfe, E. J., ESA SP-286.
Hill, F., 1989a. *Astrophys. J.*, **343**, L69.
Hill, F., 1989b. *Inside the Sun, Proc. IAU Colloquium No. 121*, eds Berthomieu G. & Cribier M., in the press.
Hill, F., Gough, D. O. & Toomre, J., 1984a. *Mem. Soc. astr. Ital.*, **55**, 153.
Hill, F., Gough, D. O. & Toomre, J., 1984b. *Solar seismology from space*, p. 95, eds Ulrich, R. K., Harvey, J., Rhodes, E. J. & Toomre, J., NASA, JPL Publ. 84-84.
Hill, F., Gough, D. O., Toomre, J. & Haber, D. A., 1988. *Proc. IAU Symposium No 123*, p. 45, eds Christensen-Dalsgaard, J. & Frandsen, S., D. Reidel Publ. Co., Dordrecht.
Hill, H. A., 1978. *The New Solar Physics*, chapter 5 ed. Eddy, J. A., Westview Press, Boulder CO.
Hill, H. A. & Stebbins, R. T., 1975. *Astrophys. J.*, **200**, 471.
Hill, H. A., Bos, R. J. & Goode, P. R., 1982. *Phys. Rev. Lett.*, **49**, 1794.
Hill, H. A., Rabaey, G. F., Yakowitz, D. S., & Rosenwald, R. D., 1986. *Astrophys. J.*, **310**, 444.
Isaak, G. R., 1982. *Nature*, **296**, 130.
Jefferies, S. M., Pomerantz, M. A., Duvall, T. L., Harvey, J. W. & Jaksha, D. B., 1988. *Seismology of the Sun & Sun-like Stars*, p. 279, eds Domingo, V. & Rolfe, E. J., ESA SP-286.
Korzennik, S. G., Cacciani, A., Rhodes, E. J., Tomczyk, S. & Ulrich, R. K., 1988. *Seismology of the Sun & Sun-like Stars*, p. 117, eds Domingo, V. & Rolfe, E. J., ESA SP-286.
Kosovichev, A. G., 1988. *Pis'ma Astron. Zh.*, **14**, 344 (English translation: *Sov. Astron. Lett.*, **14**, 145).
Kosovichev, A. G. & Parchevskii, K. V., 1988. *Pis'ma Astron. Zh.*, **14**, 473 (English translation: *Sov. Astron. Lett.*, **14**, 201).
Ledoux, P., 1949. *Mem. Soc. R. Sci. Liège, 4th ser.*, **9**, 263.
Libbrecht, K. G., 1986. *Nature*, **319**, 753.

Libbrecht, K. G., 1988a. *Astrophys. J.*, **334**, 510.
Libbrecht, K. G., 1988b. *Seismology of the Sun & Sun-like Stars*, p. 131, eds Domingo, V. & Rolfe, E. J., ESA SP-286.
Libbrecht, K. G., 1989. *Astrophys. J.*, **336**, 1092.
Loumos, G. L. & Deeming, T. J., 1978. *Astrophys. Space Science*, **56**, 285.
Morrow, C.A. 1988. *Seismology of the Sun & Sun-like Stars*, p. 137, eds Domingo, V. & Rolfe, E. J., ESA SP-286.
Narlikar, J. V. & Rana, N. C., 1985. *Mon. Not. R. astr. Soc.*, **213**, 657.
Noyes, R. W., Hartman, L. W., Baliunas, S. L., Duncan, D. K. & Vaughan, A. H., 1984. *Astrophys. J.*, **279**, 763.
Pallé, P. L. & Roca Cortés, T., 1988. *Proc. IAU Symposium No 123*, p. 79, eds Christensen-Dalsgaard, J. & Frandsen, S., D. Reidel Publ. Co., Dordrecht.
Pallé, P. L., Pérez Hernandez, F., Régulo, C. & Roca Cortés, T., 1988. *Seismology of the Sun & Sun-like Stars*, p. 125, eds Domingo, V. & Rolfe, E. J., ESA SP-286.
Parker, R. L., 1977. *Ann. Rev. Earth Planet. Sci.*, **5**, 35.
Pidatella, R. M., Stix, M., Belvedere, G. & Paterno, L., 1986. *Astron. Astrophys.*, **156**, 22.
Phillips, D. L., 1962. *J. Assoc. Comput. Mach.*, **9**, 84.
Pinsonneault, M. H., Kawaler, S. D., Sofia, S. & Demarque, P., 1989. *Astrophys. J.*, **338**, 424.
Pintar, J. A., 1988. *Seismology of the Sun & Sun-like Stars*, p. 217, eds Domingo, V. & Rolfe, E. J., ESA SP-286.
Rhodes, E. J., Cacciani, A. & Korzennik, S. G., 1988. *Seismology of the Sun & Sun-like Stars*, p. 81, eds Domingo, V. & Rolfe, E. J., ESA SP-286.
Rhodes, E. J., Cacciani, A., Korzennik, S. G., Tomczyk, S., Ulrich, R. K. & Woodard, M. F., 1988. *Seismology of the Sun & Sun-like Stars*, p. 73, eds Domingo, V. & Rolfe, E. J., ESA SP-286.
Rosner, R. & Weiss, N. O., 1985. *Nature*, **317**, 790.
Schüssler, M., 1987. *The internal solar angular velocity*, p. 303, eds Durney, B. R. & Sofia, S., Reidel, Dordrecht.
Scherrer, P. H., Hoeksema, J. T. & Bogart, R. S., 1988. *Seismology of the Sun & Sun-like Stars*, p. 375, eds Domingo, V. & Rolfe, E. J., ESA SP-286.
Snodgrass, H. B., 1983. *Astrophys. J.*, **270**, 288.
Snodgrass, H. B., 1984. *Solar Phys.* **94**, 13.
Spruit, H. C., 1987. *The internal solar angular velocity*, p. 185, eds Durney, B. R. & Sofia, S., Reidel, Dordrecht.
Spruit, H. C., Knobloch, E. & Roxburgh, I. W., 1983. *Nature*, **304**, 520.
Stenflo, J. O., 1989. *Astron. Astrophys.*, **210**, 403.
Stix, M., 1987. *The internal solar angular velocity*, p. 329, eds Durney, B. R. & Sofia, S., Reidel, Dordrecht.
Sørensen, J. M., 1988. *Seismology of the Sun & Sun-like Stars*, p. 41, eds Domingo, V. & Rolfe, E. J., ESA SP-286.
Tarantola, A., 1987. *Inverse Problem Theory*, Elsevier, Amsterdam.
Thompson, M. J., 1990. *Solar Phys.*, in the press.
Tomczyk, S., 1988. *A measurement of the rotational frequency splitting of the solar five-minute oscillations*, PhD dissertation, University of California, Los Angeles.
Tomczyk, S., Cacciani, A., Korzennik, S. G., Rhodes, E. J. & Ulrich, R. K., 1988. *Seismology of the Sun & Sun-like Stars*, p. 141, eds Domingo, V. & Rolfe, E. J., ESA SP-286.
van der Raay, H. B., Pallé, P. L. & Roca Cortés, T., 1986. *Seismology of the Sun and the distant stars*, p. 215, ed. Gough, D. O., Reidel, Dordrecht.

Fluctuations of the Interstellar Medium

B.M. Deiss

Institut für Theoretische Physik, J.W. Goethe Universität,
Robert-Mayer-Str. 10, D-6000 Frankfurt, Fed. Rep. of Germany

1. Introduction

One of the outstanding features of the interstellar medium is its extreme inhomogeneity. Strong small-scale variations of the column density and mean radial velocity, doppler-broadened spectral lines partly consisting of several distinct (velocity-) components all indicate that the ISM is in a highly disordered state on small scales. The smallest structures currently observed in molecular clouds are of the order of a tenth of a parsec. This lower limit seems to be determined only by the resolution of the observational instruments. Thus in considering the dynamics of the ISM on a given scale one nearly always finds fluctuations within the scale.

The interaction of such 'microscopic' fluctuations of the physical parameters may have a crucial 'macroscopic' effect on the large-scale dynamics of a physical system. For instance, the large-scale flow of the ISM along the spiral arms can be substantially influenced by small-scale fluctuations of the gravitational potential arising from density inhomogeneities in both the stellar and the gaseous component. The collapse of a molecular cloud may serve as another example demonstrating possible effects of small-scale fluctuations: If the cloud is highly fragmented into dense substructures and protostellar condensations, dynamical friction between the infalling gas and the 'protostellar system' leads to a delay of the further collapse. In addition, a turbulent small-scale velocity field generates a turbulent pressure supporting the cloud against collapse.

Due to the stochastic nature of the observed fluctuations a statistical approach seems to be appropriate in order to investigate the dynamics of the interstellar medium. Such a mean-field description consists on the one hand in the deduction of equations governing the large-scale dynamics and on the other hand in the selfconsistent determination of the small-scale fluctuations. However, the latter step is not in general feasible. A special approximation scheme consists in the assumption of a physical mechanism inducing fluctuations of the ISM, as it were, from outside the system considered. One can then study the influence of the excited fluctuations on the large-scale dynamics of the gas. A possible feedback on the excitation mechanism can be investigated in a separate step. Here we consider the fluctuations being induced by the inhomogeneous gravitational field of the system of stars (Kegel and Völk, 1983).

2. Large-scale dynamics

We are interested in equations governing the large-scale temporal and spatial evolution of the mean physical parameters

$$\rho_O = \langle\rho\rangle, \; \mathbf{V}_O = \langle\mathbf{V}\rangle, \; p_O = \langle p\rangle \text{ etc.,}$$

where ρ, $\mathbf{V}$ an p denote mass density, velocity and thermal pressure of the gas, respectively. $\langle\rangle$ symbolizes the averaging procedure. The average volume has to be large compared to the length-scales of the fluctuations, but small compared to those scales over which the mean background parameters vary considerably. The averaging procedure has been described in detail by Just et al. (1986).

As an example, we consider a partially ionized gas gravitationally coupled to the system of stars. The gas is treated as optically thin with heating and cooling rates (Γ,Λ) taken into account. The interaction with the magnetic field is, for the sake of simplicity, described in terms of an ideal MHD-approximation. We are looking for the dynamics of the mean momentum and mean thermal energy density ε_O^{th} of the gas and the mean distribution function f_O of the system of stars.

Splitting each physical quantity ψ into a mean and a fluctuating part

$$\psi = \psi_O + \psi_1$$

and averaging the momentum and thermal energy equation of the gas and the Vlasov equation yields the following equations

$$\frac{\partial}{\partial t}(\rho_O\mathbf{V}_O) + \mathbf{Div}(\rho_O\mathbf{V}_O\otimes\mathbf{V}_O) + \nabla p_O - \mathbf{Div}\,\hat{S}_O - \frac{1}{4\pi}(\nabla\times\mathbf{B}_O)\times\mathbf{B}_O + \rho_O\nabla(\Phi_{gO} + \Phi_{sO}) =$$

$$- \frac{\partial}{\partial t}\langle\rho_1\mathbf{V}_1\rangle \;-\; \mathbf{Div}\langle\rho\mathbf{V}\otimes\mathbf{V} - \rho_O\mathbf{V}_O\otimes\mathbf{V}_O\rangle \qquad \textbf{(1a,b)}$$

$$+\frac{1}{4\pi}\left[\mathbf{Div}\langle \mathbf{B}_1\otimes\mathbf{B}_1\rangle-\frac{1}{2}\nabla\langle B_1^2\rangle\right] \tag{1c}$$

$$-\frac{1}{4\pi G}\left[\mathbf{Div}\langle\nabla\Phi_{g1}\otimes\nabla\Phi_{g1}\rangle-\frac{1}{2}\nabla\langle|\nabla\Phi_{g1}|^2\rangle\right] \quad - \quad \langle\rho_{g1}\nabla\Phi_{s1}\rangle\ , \tag{1d,e}$$

$$\frac{\partial}{\partial t}\varepsilon_O^{th}+\nabla(\varepsilon_O^{th}\mathbf{V}_O)+p_O(\nabla\mathbf{V}_O)-\sum_{\alpha\beta}S_{\alpha\beta,O}\frac{\partial}{\partial x_\alpha}V_{\beta,O}-(\Gamma_O-\Lambda_O)=$$

$$-\nabla\langle\varepsilon_1^{th}\mathbf{V}_1\rangle+\langle -p_1(\nabla\mathbf{V}_1)\rangle+\langle\sum_{\alpha\beta}S_{\alpha\beta,1}\frac{\partial}{\partial x_\alpha}V_{\beta,1}\rangle\ , \tag{2a,b,c}$$

and

$$\left[\frac{\partial}{\partial t}+\mathbf{v}\cdot\nabla-\nabla(\Phi_{gO}+\Phi_{sO})\frac{\partial}{\partial\mathbf{v}}\right]f_O = \tag{3a}$$

$$\langle\nabla(\Phi_{gO}+\Phi_{s1})\frac{\partial}{\partial\mathbf{v}}f_1\rangle \tag{3b}$$

(see Just et al., 1986, Kegel, 1987, Deiss, 1989), where **Div** is the usual vector operator ∇, but operating on second rank tensors.

Equations (1) and (2) ensue from the Euler equation and from the first law of thermodynamics, respectively. Eq. (3) is the averaged Vlasov equation. $\hat{S}$, **B**, Φ_g and Φ_s denote the viscous stress tensor, magnetic induction, and gravitational potential of the gas and of the stellar system respectively.

The left hand sides of (1), (2) and (3) can formally be regarded as the explicit basic equations with the exact physical quantities substituted by their mean values. But, as the right hand sides show, additional forces and energy contributions arise from small-scale fluctuations acting upon the large-scale dynamics of the system. These rhs. terms represent:

(1a) : temporal variation of the mean momentum density contained in the fluctuations;

(1b) : spatial variation of the (kinetic) turbulent pressure tensor;

(1c) : spatial variation of the magnetic turbulent pressure tensor arising from fluctuations of the magnetic field;

(1d) : spatial variation of a 'gravitational turbulent pressure tensor' arising from mass density fluctuations in the gas;

(1e) : dynamical friction between the ISM and the system of stars;

(2a) : spatial variation of the flux of thermal energy contained in the fluctuations;

(2b) : mechanical heating by dissipation of the kinetic energy of fluctuations due to small-scale compressive and expansive processes;

(2c) : mechanical (viscous) heating due to small-scale viscous frictional processes;

(3b) : a 'collision' term arising from the interaction of fluctuations of the distribution function of the stars and of the coupled gravitational field of ISM and stellar system.

This rather simple example of an astrophysical system demonstrates that - caused by nonlinear interactions of small-scale fluctuations - a number of additional physical effects arise, which may influence the large-scale dynamics considerably, and which one would not find by just substituting the physical quantities (e.g. density, velocity, etc.) by their mean values.

But how do the additional nonlinear correlation terms depend on the mean physical parameters ? The importance of this question becomes evident when for example considering the spatial variation of the turbulent pressure. When the turbulent pressure gradient ∇P_{turb} is taken into account in the Euler equation, it is frequently assumed to be

$$\nabla P_{turb} = \langle V_r^2 \rangle \nabla \rho , \tag{4}$$

i.e. to be analogous to the thermal pressure gradient $\nabla p = c_s^2 \nabla \rho$. V_r and c_s denote the one-dimensional turbulent velocity dispersion and the velocity of sound of the gas, respectively. From (1b), however, we find

$$\mathbf{Div} \langle \rho \mathbf{V} \otimes \mathbf{V} - \rho_O \mathbf{V}_O \otimes \mathbf{V}_O \rangle = \mathbf{Div} \left[\langle \mathbf{V}_1 \otimes \mathbf{V}_1 \rangle \rho_O \right]$$

$$+ \mathbf{Div} \left[\langle \rho_1 \mathbf{V}_1 \rangle \otimes \mathbf{V}_O + \mathbf{V}_O \otimes \langle \rho_1 \mathbf{V}_1 \rangle \right] + \mathbf{Div} \langle \rho_1 \mathbf{V}_1 \otimes \mathbf{V}_1 \rangle . \tag{5}$$

To arrive at expression (4), various simplifying assumptions are necessary: (i) the triple correlation term is negligible compared to the other correlation terms; (ii) the velocity fluctuation field is isotropic, hence the second term on the rhs. of (5) vanishes; (iii) velocity fluctuations in different directions are independent of each other and the velocity fluctuation field is isotropic. In this case the tensor in the first term on the rhs. of (5) becomes diagonal with each element equal to V_r.

With these assumptions we obtain

$$\mathrm{Div}\langle \rho \mathbf{V}\otimes\mathbf{V}-\rho_O\mathbf{V}_O\otimes\mathbf{V}_O\rangle = \nabla\left[\langle V_r^2\rangle \rho_O\right]. \tag{6}$$

To arrive ultimately at (4) we have further to assume that $\langle V_r^2\rangle$ exhibits no spatial gradients, which implies that $\langle V_r^2\rangle$ is independent of any mean physical parameter, such as mean density, mean temperature etc. This seems to be rather unlikely.

Let us consider the scaling relations, frequently discussed in the literature (see e.g. Scalo, 1987), between radius and mean density of a cloud $\rho_O \propto R^p$ on the one hand and between radius and velocity dispersion $\langle V_r^2\rangle \propto R^q$ on the other hand, where $p\approx -1$ and $q\approx 1$. Should these relations be proven one would conjecture from the case $p=-1$ and $q=1$ that

$$\langle V_r^2\rangle \propto \rho_O^{-1}. \tag{7}$$

regardless of the physical mechanism generating the turbulent motions within the clouds. Relation (7), if correct, implies that the turbulent pressure is independent of the gas density and that its gradient (6) is determined by the gradients of some other mean physical parameters. However, it should be mentioned that there are serious objections against the cited scaling relations concerning their deduction from observational data (Kegel, 1989, Issa et al., 1989). Nevertheless, the above discussion demonstrates the importance of calculating the equation of state of the turbulent pressure by means of a 'microscopic' fluctuation theory instead of making heuristic assumptions. This holds for each of the correlation terms in (1) - (3).

In a rigorous fluctuation theory one has to determine the fluctuations from equations obtained by subtracting the averaged equations from the exact ones. The resulting equations, however, also contain higher order correlation terms of the fluctuations, thus leading to the well-known closure problem. In the lowest order one can linearize these equations by neglecting the higher correlation terms and by assuming the mean quantities to be quasihomogeneous and quasistationary within the averaging volume. From the resulting set of linear differential equations, one obtains linear relations between the amplitudes of the various fluctuating quantities. Then the nonlinear correlation terms in the averaged equations can be evaluated by inserting the linear relations (quasi-linear approximation). So far, however, there is still one free parameter left determining the absolute values of the linear amplitudes. Considering some mechanism to excite the fluctuations from 'outside' the system, the absolute (induced) amplitudes are given. A special mathematical technique consists in the formulation of the excitation process as an initial value problem and in the investigation of the time asymptotic limit of the correlation terms. This technique has been used in the calculation of the correlation terms, which are described in the following section. It should be mentioned that the consideration of the time asymptotic limit restricts one to the consideration of stable modes only.

3. Fluctuations induced by the system of stars

Due to the point-like structure of the stars, the gravitational potential of the stellar system is extremely inhomogeneous on small scales. Kegel and Völk (1983) showed that, being based on the stochastic motion of the randomly distributed stars, velocity and density fluctuations are generated in the interstellar gas, which is gravitationally coupled to the system of stars.

3.1. *The effect on the interstellar gas*

In the paper of Kegel and Völk (1983) the thermal pressure of the interstellar gas was described by a simple polytropic law. As an extension to that work, Deiss and Kegel (1986) treated the gas as optically thin in the relevant cooling spectral lines and took dissipative effects due to heating and cooling processes into account. They obtained mean velocity fluctuations several times the magnitude of the velocity of sound.

Dissipation of the kinetic energy of the fluctuations heats the interstellar gas. This process is represented by the terms (2b) and (2c). Deiss (1989) and Deiss et al. (1989) deduced an analytical expression for the resulting mechanical heating rates denoted by Γ_f.

Just (1987) investigated the dynamical friction $\langle \rho_{g1} \nabla \Phi_{s1} \rangle$ between interstellar gas and the system of stars arising in the case of a systematic motion between both components and deduced a corresponding analytical expression (see also Deiss et al., 1989).

By way of example we consider a molecular cloud with number density $n_0 = 10^4/\mathrm{cm}^3$, temperature $T = 40$ K, and cooling time $\tau_{cool} = 0.5 \times 10^2$ yrs and a protostellar system with mass density $\rho_{s0} = 0.05\ \rho_{g0}$ and Gaussian velocity distribution with a velocity dispersion $\sigma = 0.9$ km/s. The induced turbulent one-dimensional velocity dispersion of the interstellar gas is then $\sqrt{\langle V_r^2 \rangle} = 0.6$ km/s . For a Gaussian line profile this corresponds to a FWHM of 1.4 km/s , which lies in the range of line widths observed in molecular clouds having sizes of a few parsec.

The corresponding mechanical heating rate amounts to $\Gamma_f = 1.2 \times 10^{-24}$ erg/cm^{3}s (Deiss et al., 1988). This is of the order of the cosmic ray heating rate $\Gamma_{CR} \approx 4 \times 10^{-28} n(H_2)$ (Black, 1987).

If the cloud is in a collapsing state, there is a systematic motion of the infalling gas and the system of protostars within the cloud, hence a dynamical friction acts upon the motion of the gas. The corresponding timescale τ_g^{dec} for deceleration of the gas is $\tau_g^{dec} = 6 \times 10^5$ yrs , which is of the order of the free-fall time of the molecular cloud (Spitzer, 1978). This indicates that the collapse of the molecular cloud is delayed. Thus dynamical friction with protostars embedded within the cloud may have an important effect on the dynamics of the cloud gas.

3.2. *The feedback on the stellar system*

The excitation of fluctuations of the interstellar gas by the system of stars causes a feedback on the distribution function of the stars themselves, following from reasons of momentum and energy conservation. This feedback is represented by the correlation term (3b). In quasilinear approximation Deiss (1989) derived an analytical expression of the following structure

$$\langle \nabla(\Phi_{g0} + \Phi_{s1}) \frac{\partial}{\partial \mathbf{v}} f_1 \rangle =$$

$$\frac{\partial}{\partial \mathbf{v}} \cdot \int \left[\hat{D}^{(2)}(\mathbf{k}) + \hat{D}^{collec.}(\mathbf{k},\mathbf{v}) \right] \left\{ F_0(u) \frac{\partial}{\partial \mathbf{v}} f_0(\mathbf{v}) - f_0(\mathbf{v}) \frac{\partial}{\partial \mathbf{v}} F_0(u) \right\} d^3k \tag{8a}$$

$$- M \rho_{g0} \frac{\partial}{\partial \mathbf{v}} \cdot \left[\mathbf{R}(\mathbf{v}) f_0(\mathbf{v}) \right], \tag{8b}$$

where $\mathbf{k}$ is the wave vector; $F_0(u)$ the one-dimensional distribution function in the direction of the wave vector corresponding to f_0; u the component of $\mathbf{v}$ parallel to the wave vector; M the mass of a single star. $\hat{D}^{(2)}(\mathbf{k})$ and $\hat{D}^{collec.}(\mathbf{k},\mathbf{v})$ are second rank tensors describing dynamical friction and diffusion processes within the stellar system, and $\mathbf{R}(\mathbf{v})$ is an vectorial friction coefficient describing the dynamical friction with the interstellar gas.

The tensor $\hat{D}^{(2)}(\mathbf{k})$ in (8a) represents the pure two-body interaction of the point-like stars and is identical with the corresponding expression one obtains considering the usual small-angle scattering approximation (e.g. Binney and Tremaine, 1987). Each single star induces collective reactions both in the stellar system and the interstellar gas. The tensor $\hat{D}^{collec.}(\mathbf{k},\mathbf{v})$ represents both the interaction of single stars with collective modes generated by all other stars and the mutual interactions of collective modes. The gravitationally coupled interstellar gas has an additional effect on the distribution function of the stars - the action of a dynamical friction (8b) of the individual stars with the ambient gas.

Expression (8b) corresponds to the dynamical friction term of the gas (1e), while (8a) represents the momentum and energy exchanges within the stellar system. For a Gaussian velocity distribution f_O the expression in the curved brackets in (8a) vanishes. In a large-scale homogeneous stellar system without any interstellar gas this situation indicates an equilibrium state of the system. Taking interstellar gas into account, the collective modes of the stellar system in $\hat{D}^{collec.}(\mathbf{k},\mathbf{v})$ are enhanced. In general, this implies a shortening of the relaxation time scale of the system. But, since the stars suffer an additional friction with the interstellar gas (8b), which is a thermodynamically open system due to radiative energy losses, a Gaussian velocity distribution cannot, strictly speaking, represent an equilibrium state. However, the importance of the additional frictional effect depends on the ratio of the friction time scale to the diffusion time scale.

Again, we consider as an example, a molecular cloud with embedded protostars within the cloud, assuming the same parameters as given in section 3.1. Deiss (1989) showed that the amplitudes of the collective reaction modes in $\hat{D}^{collec.}(\mathbf{k},\mathbf{v})$ decrease rapidly with increasing wave number. Since expression (8) is obtained in the time asymptotic limit, only linearly stable modes can be considered. That is, the range of wave numbers contributing to (8) has a lower limit determined by the critical wave number resulting from a linear Jeans analysis. For the molecular cloud considered, there is only a small range of wave numbers near the marginally stable mode, where collective modes become relevant as compared to the two-body encounters represented by $\hat{D}^{(2)}(\mathbf{k})$. The corresponding time scale is much longer than the time scale resulting from the dynamical friction (8b).

In fig. 1 we plotted the deceleration $\partial v/\partial t$ of a single star due to this frictional force as a function of the velocity v of the star. It is evident that the dynamical friction is strongest for a star having a slightly supersonic velocity. For higher velocities it falls off proportional to v^{-2}. Towards the subsonic region there is a steep decrease in the magnitude of the dynamical friction.

Also plotted in fig. 1 is the function

$$\frac{\partial v}{\partial t} = \frac{v}{\tau_s^{dec}}, \qquad (9)$$

where τ_s^{dec} denotes the deceleration time scale of the star. Considering a Gaussian initial distribution function of the protostellar system with a velocity dispersion of ~ 1 km/s, it ensues from fig. 1 that the distribution function is highly deformed within less than 10^6 yrs. The resulting velocity distribution function will then exhibit a steep slope near the velocity of sound.

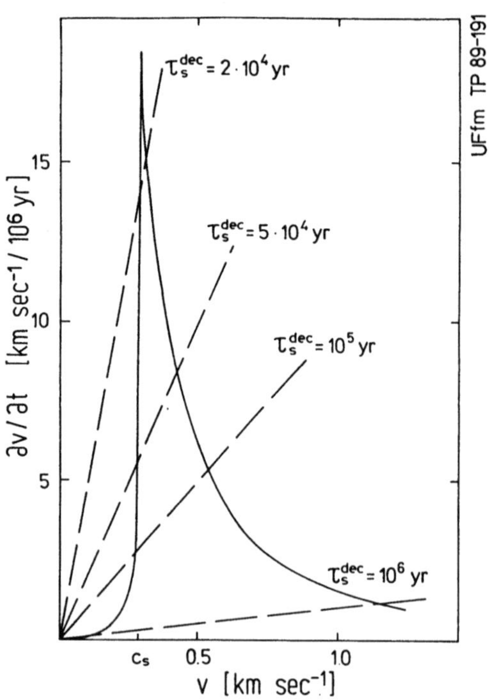

Fig. 1: (Solid line): Deceleration of a single star due to the dynamical friction (8b) with the ambient gas plotted as a function of the velocity v of the star. The assumed parameters are the following: number density of the molecular gas $n_O = 10^4 cm^{-3}$, temperature T = 40 K, cooling time τ_{cool} = 50 yrs, and mass of the star $M = 1\ M_\odot$. (Dashed lines): Deceleration following from a force proportional to the velocity v (see (9)), with the proportionality constant being the inverse of the deceleration time τ_s^{dec}. The dashed lines, plotted for four different deceleration time scales, indicate the range of the corresponding deceleration time scale of the dynamical friction.

4. Summary and conclusion

The interstellar medium is extremely inhomogeneous revealing strong small-scale fluctuations of the observed physical parameters. The interaction of these fluctuations may have a considerable effect on the large-scale dynamics of the ISM. We deduced exact equations governing the large-scale motion and large-scale thermal energy balance of the gas. These equations contain correlation terms representing the various interactions of small-scale fluctuations. We stressed the importance of calculating the correlation terms by means of a fluctuation theory instead of making some heuristic assumptions about these terms (e.g. the turbulent pressure tensor).

We considered a special mechanism for exciting fluctuations in the ISM, this being based on the inhomogeneous gravitational potential of the randomly distributed (point-like) stars (Kegel and Völk, 1983). In quasilinear approximation we obtained analytical expressions for the rms. of the induced velocity fluctuations of the gas; for the mechanical heating rate due to dissipation of the kinetic energy of the fluctuations; for the dynamical friction between the ISM and the stellar system arising in the case of a systematic motion between both components; for the feedback on the distribution function of the stellar system

In the present paper we only cited the references where the analytical expressions can be found, but gave some numerical results. As an example, we considered a typical molecular cloud of a few parsec diameter with protostars embedded within the cloud. The magnitude of the induced velocity dispersion of the gas lies in the range of what one deduces from observed line widths and the resulting mechanical heating rate is of the order of the cosmic ray heating rate. The time scale for dynamical friction between the ISM and the protostellar system τ_g^{dec} is of the order of the free-fall time of the cloud, indicating that a possible collapse of the cloud may be delayed considerably.

In addition, we considered the feedback on the exciting protostellar system and found that the velocity distribution function is deformed considerably within a time interval shorter than the free-fall time of the cloud. The deformation results from a dynamical friction of each individual star with its surrounding gas. Stars with velocities of up to four times the velocity of sound are slowed down to subsonic velocities. The resulting velocity distribution function will then exhibit a steep slope near the velocity of sound. This has some implications for the time scale for the dynamical friction of the gas τ_g^{dec}, which is obtained by assuming a Gaussian velocity distribution function of the stars. Since the magnitude of dynamical friction of the gas depends on the steepness of the velocity distribution function of the stellar system (Just, 1987, Deiss et al., 1989), the corresponding deceleration timescale τ_g^{dec} may be much smaller as stated above, implying a strengthend delay of the cloud collapse.

We conclude that, after star formation has started, the dynamics of the cloud gas may considerably be affected by the gravitational interaction with the system of protostars within the cloud. This has implications both on the star formation rate and on the initial mass function.

Acknowledgement: This talk is based on work performed in collaboration with A. Just and W.H. Kegel. Part of this work has been supported by the Deutsche Forschungsgemeinschaft (Schwerpunkt "Theorie kosmischer Plasmen").

References

1. J.H. Black: In *Interstellar Processes*, eds. D.J. Hollenbach, H.A. Thronson Jr. (D. Reidel Publ. Comp., Dordrecht, 1987) p. 731
2. B.M. Deiss: Thesis, Frankfurt, 1989
3. B.M. Deiss and W.H. Kegel: Astron. Astrophys. 161, 23 (1986)
4. B.M. Deiss, A. Just and W.H. Kegel: *Astron. Gesell. abstr. ser.* 1, 25 (1988)
5. B.M. Deiss, A. Just and W.H. Kegel (1989): in preparation
6. M. Issa, I. MacLaren, A.W. Wolfendale: submitted to Astrophys. J. (1989)
7. A. Just: Thesis, Frankfurt, 1987
8. A. Just, W.H. Kegel and B.M. Deiss: Astron. Astrophys. 164, 337 (1986)
9. W.H. Kegel: In *Physical Processes in Interstellar Clouds*, eds. G.E. Morfill, M. Scholer (D. Reidel Publ. Comp., Dordrecht, 1987) p. 115
10. W.H. Kegel and H.J. Völk: Astron. Astrophys. 119, 101 (1983)
11. W.H. Kegel: Astron. Astrophys., in press (1989)
12. J.M. Scalo: In *Interstellar Processes*, eds. D.J. Hollenbach, H.A. Thronson Jr. (D. Reidel Publ. Comp., Dordrecht, 1987) p. 349

Acceleration of Cosmic Rays in Supernova Remnants

E.A. Dorfi

Universitätssternwarte, Institut für Astronomie der Universität Wien, Türkenschanzstr. 17, A-1180 Wien, Austria

1. Introduction

Cosmic rays have been an interesting and challenging problem in astrophysics since their discovery by Victor Hess (1912). The up-to-date information is collected in the proceedings (and references therein) of the biannual *International Cosmic Ray Conference*, the most recent ones in Bangalore (1983), La Jolla (1985) and Moscow (1987). Furthermore, this introduction is not intended as a balanced review on the field and we will concentrate on certain aspects of the acceleration of cosmic rays in supernova remnants (SNR's). Therefore we restrict the following discussion on the socalled galactic component of cosmic rays in the range between 1 GeV/nuc and 10^6 GeV/nuc, where most of the cosmic ray energy is concentrated (e.g. Axford 1981), and in particular to the acceleration of particles in SNR's. The importance of the acceleration of cosmic rays and their influence on other astrophysical events has been discussed e.g. by Völk (1987). Solar modulation affects significantly the interpretation of the observations for the lower particle energies between 1 GeV and 10 GeV (Parker 1958). Particles with energies as large as 10^{20} eV/nuc have been detected and this raises the question which process can accelerate single particles to such energies. The cosmic ray energy density is about $\epsilon_c = 1\,\mathrm{eV\,cm^{-3}}$ (at least in the solar neighbourhood) and since the charged particles are tied to the magnetic field which permeates the interstellar medium, we have an intimate relation between these energetic particles and the other interstellar components. The gyroradius of a high energy proton is given by $r_g \simeq 10^{-6} E[\mathrm{GeV}]/B[\mu G]\,\mathrm{pc}$. Note that this energy density is comparable to that of the magnetic field, the interstellar photon field, the turbulent and ordered motions of the interstellar gas. The interstellar gas is mainly heated by shock waves and this equipartition of energies suggests a connection between shock waves and the acceleration process of particles. According to this energy argument cosmic rays play an important role for the dynamics of the interstellar medium.

A further property of galactic cosmic rays is their high degree of isotropy which decreases for higher energies ($T > 10^6\,\mathrm{GeV}$) suggesting an increasing contribution of extragalactic particles. The chemical abundances correspond essentially to the normal solar composition if they are corrected for spallation in the interstellar medium. From the radioactive decay of Be^{10} in the secondary component one can estimate a typical residence timescale of $\tau \simeq 2 \cdot 10^7$ years in our galaxy. The observed grammage is about $X = 5\,\mathrm{g\,cm^{-2}}$ and $X = c\bar{\rho}\tau$ suggests a mean density of $\bar{\rho} \simeq 0.16$ H-atoms $\mathrm{cm^{-3}}$. Since the mean density in our galaxy within a vertical scale hight of ± 100 pc is about 0.4 H-atoms $\mathrm{cm^{-3}}$ (e.g. Burton and Liszt, 1981) we can estimate the cosmic ray storage (or confinement) volume to be about ± 250 pc times the galactic disc radius of 15 kpc, i.e. $V_{gal} \simeq 10^{67}\,\mathrm{cm^{-3}}$. Therefore the power necessary to maintain the observed galactic cosmic ray energy density is roughly given by integrating the observed energy density over the galactic volume

$$L_{CR,gal} \simeq \int_{V_{gal}} \frac{\epsilon_c}{\tau} dV \simeq 3 \cdot 10^{40}\,\mathrm{erg\,s^{-1}}. \tag{1}$$

The differential spectrum of primary galactic cosmic rays is given by a power-law $N(E) \propto E^{-2.65}dE$ with a break at a few 10^6 GeV (e.g. Simpson 1983). The secondary particles are produced by nuclear interaction with the interstellar medium but their spectrum is steeper and this difference in the spectral index results in an energy-dependent confinement time τ. Particles with higher energies can escape more easily from our galaxy. Taking this fact into account we have to increase $L_{CR,gal}$. This order of magnitude estimate is compared with the canonical supernova explosion energy of $E_{SN} = 10^{51}$ erg and the typical supernova rate of 1/30 per year yielding $L_{SN,gal} \simeq 10^{42}$ erg s^{-1} and consequently a production efficiency of about 3 percent. On this level of approximations we have also to consider that the supernova energy may be overestimated. Putting together all arguments we can conclude that a supernova should transfer about 10 or 30 percent of its explosion energy into cosmic rays if the particles are accelerated in the evolving SNR (e.g. Drury et al., 1989). However, a sucessful theory of particle acceleration in SNR's should give more precise numbers of the amount of energy transformed into cosmic rays.

We concentrate on the process of diffusive shock acceleration (Axford et al. 1977, Krymsky 1977, Bell 1978a,b, Blandford and Ostriker 1978) to produce and reaccelerate the suprathermal particles called cosmic rays. These particles have high enough momenta to see the shock wave as a discontinuity and they do not interact with the electromagnetic turbulence in the shock. These particles are not heated by two-body collisions (or collective electromagnetic effects) but can gain energy being scattered between the downstream and upstream region (and vice versa) on magnetic irregularities. In the case of a plane shock the steady solutions for test particles are a power-law distribution depending on the compression ratio of the shock. For a strong shock in a non-relativistic plasma with $\gamma_g = 5/3$ we get $f(p) \propto p^{-q}$, $q = 4$, remarkably close to the index 4.3 inferred for the source spectrum of galactic cosmic rays. However, one should keep in mind that the obtained spectrum can be affected by adiabatic effects, non-linearities, time-dependence and energy losses (like ionization and synchrotron radiation for electrons). Additional complications arise from the irregularities in the background plasma which scatter the particles. They have to be considered simultaneously with the acceleration process since these high energy particles can stream along the field pushed by a cosmic ray gradient. In this case hydromagnetic waves are generated by a resonant instability (Lerche 1967) which scatter the particles and tend to reduce this anisotropy. The mean cosmic ray velocity approaches then the Alfvén velocity. For more details and a rigorous discussion of the theory of diffuse shock acceleration we refer e.g. to some reviews and references therein (Drury 1983, Blandford and Eichler 1987, Völk 1987).

2. Equations, initial conditions, method of solution

Assuming spherical symmetry the evolution of a SNR is described by the equations of one-dimensional gas dynamics. The equation of motion contains an additional cosmic ray pressure term P_c. For simplicity in this paper we neglect cooling of the gas although the late evolution of a SNR is also modified by radiative effects. The cosmic ray energy E_c is given by integrating the particle distribution function over the particle momentum and with appropriate definitions (e.g. Drury 1983) we obtain a hydrodynamical equation for E_c

$$\frac{\partial E_c}{\partial t} + \nabla \cdot (E_c \mathbf{u}) - P_c \nabla \cdot \mathbf{u} = \nabla \cdot (\bar{\kappa} \nabla E_c), \tag{2}$$

where $\mathbf{u}$ denotes the gas velocity and $\bar{\kappa}$ the averaged, mean effective diffusion coefficient of cosmic rays. In this section we assume $\bar{\kappa} = const.$ throughout the SNR evolution and further aspects of $\bar{\kappa}$ are given in the next section. The system is closed by an adiabatic equation of state for the cosmic rays and γ_c is taken constant or calculated according to the approximations discussed in section 3.2

$$P_c = (\gamma_c - 1)E_c. \tag{3}$$

The value of γ_c is clearly bounded between the limits of relativistic and non-relativistic particles, $4/3 \leq \gamma_c \leq 5/3$ and depends on the detailed knowledge of the particle distribution function.

The supernova energy E_{SN} of 10^{51}erg is deposited purely as thermal energy within a radius of 10^{12}cm. A constant density up to 10^{14}cm and then an exponential decrease towards the external density of ρ_{ext} simulates the density structure of the progenitor star. The external material is at rest and the pressure corresponds to the warm interstellar medium of $T_{ext} = 8000$ K and $\rho_{ext} = 5 \cdot 10^{-25}\mathrm{g\,cm}^{-3}$. The cosmic ray pressure is set equal to the external gas pressure, *i.e.* $P_{c,ext} = P_{g,ext}$. The ejected mass of $5\,M_\odot$ fixes the mean density of the progenitor star. A non-uniform external medium generated e.g. by a stellar wind before the supernova explosion is not considered in this paper (cf. Völk and Biermann, 1988).

The initial conditions define a sweep-up radius R_{sw} where the ejected mass equals the swept-up mass in the external medium and R_{sw} is given by $1.68 \cdot 10^{19}$cm. Under the assumption that most of the supernova energy is kinetic energy we get a typical ejection velocity v_{ej} of $4500\,\mathrm{km\,s}^{-1}$. Therefore we calculate a typical time-scale of $t_{sw} = R_{sw}/v_{ej}$ and with our initial values $t_{sw} = 3.75 \cdot 10^{10}$s corresponding to about 1200 years.

The equations are discretized on a staggerd mesh according to the formulae discussed in Winkler and Norman (1986) yielding a first-order donor-cell, conservative and implicit system of non-linear finite difference equations. The system is enlarged by the socalled grid-equation (Dorfi and Drury, 1987) which continuously redistributes the grid points during the calculation to obtain a sufficient resolution in all physical quantities and to trace the flow discontinuites. An artificial viscosity in the tensor formulation of Tscharnuter and Winkler (1979) is included to handle shock fronts. The typical thickness of a shock front is usually less than 10^{-4} of the local radius and less than the scale-length defined by the diffusion coefficient of the cosmic rays and the gas velocity, i.e. $l_d = \bar{\kappa}/u$. This implicit discretization results in a system of five non-linear algebraic equations which is solved by conventional Newton-Raphson techniques modified for the special structure of the associated Jacobian.

As a first step we choose constant parameters for $\gamma_c = 4/3$ and $\bar{\kappa} = 10^{25}\mathrm{cm}^2\,\mathrm{s}^{-1}$ to illustrate the evolution of a SNR during the Sedov-phase including cosmic rays. The results are depicted in Figure 1 and are contrasted with the next section where we calculate the time-dependence of γ_c and $\bar{\kappa}$ with respect to several approximations of the particle spectrum. For the adopted parameters the reverse shock reaches its maximum radius of $1.5 \cdot 10^{19}$cm at $t = 8.74 \cdot 10^{10}$s or $2.33\,t_{sw}$. Then the reverse shock propagates inwards and the interior of the remnant is heated up and refilled with energetic particles. The first curve plots the variables at $t = 3.04 \cdot 10^{11}$s and the forward shock is at $r_s = 3.92 \cdot 10^{19}$cm with a Mach-number of 46.6. The maximum cosmic ray pressure is $P_{c,max} = 11.8\,P_{c,ext}$. In contrast to the steep gradients behind the shock waves during the free expansion phase we see the Sedov-phase with more moderate variations of P_c inside the remnant. The interior variations left over in all physical variables at $t = 6.17 \cdot 10^{11}$s and $t = 1.53 \cdot 10^{12}$s are damped out, the Mach-numbers are given by 45.5 and 16.2, respectively. The cosmic ray enhancement ranges between $P_{c,max} = 11.8\,P_{c,ext}$ and $P_{c,max} = 10.4\,P_{c,ext}$, and P_c decreases about one order of magnitude towards the interior. In the later evolution phases at $t = 5.96 \cdot 10^{12}$s a small precursor develops in the interstellar medium infront of the shock at $r_s = 1.34 \cdot 10^{20}$cm with a Mach-number of 7.9. The subsequent Sedov-expansion is shown at times $t = 1.56 \cdot 10^{13}$s, $t = 4.07 \cdot 10^{13}$s and $t = 6.81 \cdot 10^{13}$s, where the shock is located at $r_s = 2.01 \cdot 10^{20}$cm, $r_s = 3.06 \cdot 10^{20}$cm and

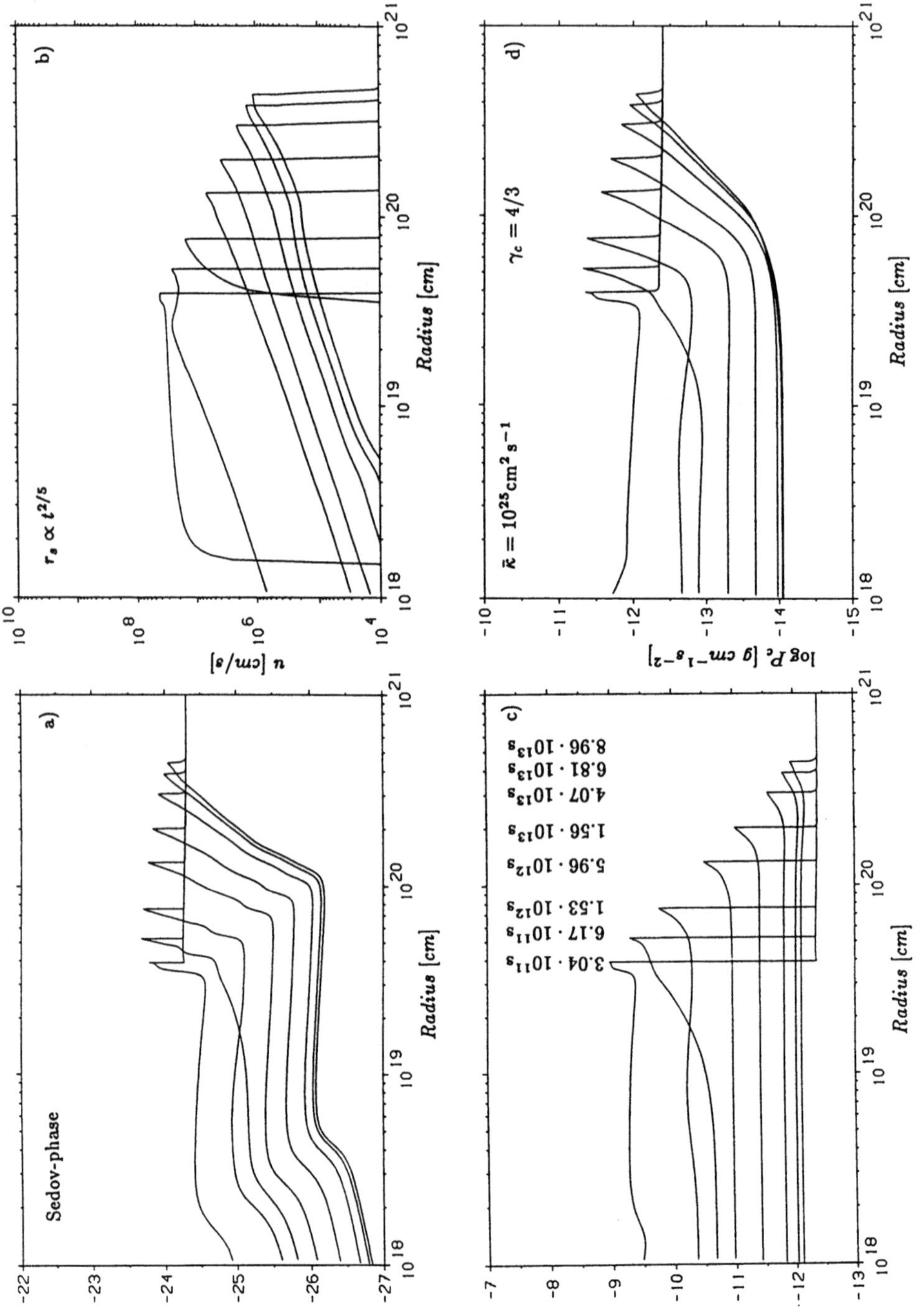

Figure 1: The temporal evolution during the Sedov-phase between $t = 3.03 \cdot 10^{11}$s and $t = 8.96 \cdot 10^{13}$s for $\bar{\kappa} = 10^{25}\mathrm{cm}^2\,\mathrm{s}^{-1}$. a) Density: For times $t \geq 1.53 \cdot 10^{12}$s we see a self-similar expansion of the remnant. b) Velocity: At later times $t \geq 1.53 \cdot 10^{12}$s a small cosmic ray precursor develops infront of the forward shock. c) The thermal pressure corresponds to the usual Sedov-phase and drops rapidly behind the shock wave. d) The particle pressure exhibits a region of enhanced cosmic ray pressure behind the shock wave.

$r_s = 3.87 \cdot 10^{20}$cm, respectively while the Mach-numbers decrease from 4.4 to 2.4 and 1.8. The maximum particle pressure drops from $P_{c,max} = 5.1\, P_{c,ext}$ to $P_{c,max} = 3.6\, P_{c,ext}$ and $P_{c,max} = 2.8\, P_{c,ext}$. The last calculated model for $\bar{\kappa} = 10^{25}\mathrm{cm}^2\,\mathrm{s}^{-1}$ is placed at $t = 8.96 \cdot 10^{13}$s or about $2400\, t_{sw}$, $r_s = 4.41 \cdot 10^{20}$cm and with a Mach-number of 1.6. The value of $P_{c,max} = 2.3\, P_{c,ext}$ is still above the external cosmic ray pressure and the interior P_c is lower by 2 orders of magnitude relative to $P_{c,max}$.

In contrast to the models with larger values of $\bar{\kappa}$ we always observe during the Sedov-phase a region of enhanced cosmic ray pressure in the vicinity of the forward shock, but the remnant is never dominated by cosmic rays. However, the diffusive coupling of the interior of the SNR plays an important role during the entire remnant evolution and determines the value of $P_{c,max}$ in the vicinity of the shock and consequently the thickness of the region of enhanced cosmic ray energy. This shell corresponds to about 60% of the SNR volume and stays fairly constant at about 0.75 in units of the shock radius r_s during the Sedov-phase.

3. Time-dependent effects

3.1 Plane-parallel flows

We study the temporal development of a hydrodynamical shock wave containing cosmic rays and thermal plasma in linear geometry. The flow is reflected at a rigid wall on the left side and a shock wave starts propagating to the right. Figure 2 depicts the variables at different times given in units of the diffusion time $t_d = \bar{\kappa}/u_s^2$ where u_s denotes the shock speed. The Mach-number defined through

$$M^2 = \frac{\rho u^2}{\gamma_c P_c + \gamma_g P_g} \tag{4}$$

corresponds to $M = 11.2$. The quantity $N = 0.5$ gives the amount of cosmic ray pressure relative to the total pressure in the upstream region, i.e.

$$N = \frac{P_c}{P_c + P_g}. \tag{5}$$

The value of the adiabatic coefficient of cosmic rays is set to $\gamma_c = 3/2$.

Following the curves from left to right we see the formation of a strong gas shock, most of the converted kinetic energy of the incoming flow heats the thermal gas. The compression ratio is almost 4 according to $\gamma_g = 5/3$ (cf. Fig. 2a). Further in time we see the increase of the cosmic ray pressure (Fig. 2d) due to the ongoing particle acceleration and a decrease of the gas pressure (Fig. 2c). The overall compression ratio increases temporarily, for $t = 5.9\, t_d$ we have a value of about 9. The gas is compressed adiabatically in the pronounced precursor by a factor of about 3 followed by a gas subshock with a compression ratio of 2.7. However, this transient phenomenon is convected downstream and the shock changes to a purely cosmic ray dominated one, where almost all of the incoming kinetic energy is deposited in energetic particles, the gas remains cool (Fig. 2c) and is only compressed adiabatically. In addition we see a flow towards the shock wave in the downstream region caused by the diffusive reduction of the cosmic ray gradient which is compensated by motions pushed by the leftover gradient of the thermal plasma. The small positive velocity (Fig. 1b) illustrates this fact of maintaining pressure equilibrium in the downstream region. The late evolution at $t = 10.2\, t_d$ and $t = 14.2\, t_d$ represents the time asymptotic solutions which can be compared to the analytical solutions of Drury and Völk (1981). The compression ratio is reduced to 6 in agreement with $\gamma_c = 3/2$.

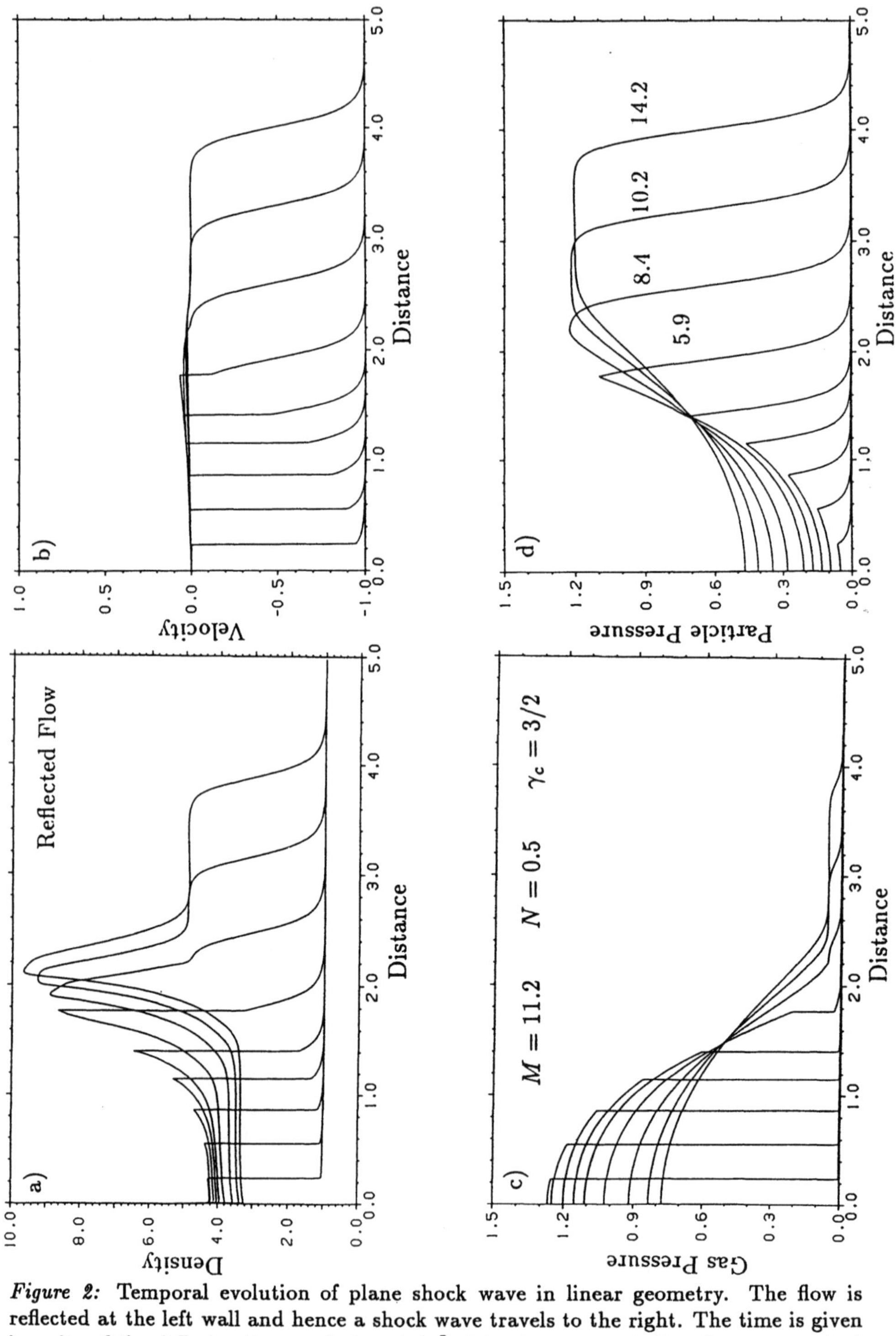

Figure 2: Temporal evolution of plane shock wave in linear geometry. The flow is reflected at the left wall and hence a shock wave travels to the right. The time is given in units of the diffusion time-scale $t_d = \bar{\kappa}/u_s^2$. The time-asymptotic values are reached after about 8 diffusion time-scales.

3.2 Spherical flows with time-dependent values of γ_c and $\bar{\kappa}$

Variations in the compression ratio of the shock fronts accompanied with more or less smooth transitions of the background plasma yield a time variation of γ_c and hence taking γ_c constant is only a rough approximation to the adiabatic index of the cosmic rays in an evolving SNR. In contrast to the models presented in section 3.1 we calculate now the adiabatic exponent of the cosmic rays γ_c and the mean diffusion coefficient $\bar{\kappa}$ according to approximations introduced by Drury et al. (1989). Under their assumptions the value of γ_c at the shock is given by a simple expression

$$\gamma_c = \frac{4E_c + E_n}{3E_c}, \tag{6}$$

where E_n denotes the amount of the non-relativistic energy density of cosmic rays; E_n is then calculated on the basis of the test particle theory (no reaction of the energetic particles on the flow). They also derive an expression for the mean diffusion coefficient

$$\bar{\kappa} = \frac{\kappa(p_{max})}{\max\left(3, \ln(p_{max}/mc)\right)} \tag{7}$$

and $\kappa(p_{max})$ denotes the momentum-dependent diffusion coefficient for particles with the highest momentum p_{max} reached during the SNR evolution. Therefore one has to integrate an equation for p_{max} and has to assume that $\kappa(p)$ is given by the Bohm limit for strong wave exitation and we take a mean magnetic field of $5\,\mu G$ to get the gyroradius. Again we refer to Drury et al. (1989) for further details. Several constraints for $\kappa(p)$ from theory and observations are compiled by Axford (1981).

Some preliminary results including the aforementioned time-dependence of γ_c and $\bar{\kappa}$ are presented in Figures 3 and 4. The gas density and the gas velocity in the shock frame are shown in Fig. 3a,b. As mentioned in 3.1 we observe the transition from gas dominated to cosmic ray dominated shocks before the Sedov-phase begins. This process occurs therefore also in spherical calculations if the cosmic ray pressure exceeds a certain level and it is almost independent on the detailed knowledge of the particle distribution function. At $t = 7.2 \cdot 10^{10}$s we have two density peaks due to forward and reverse shock and later at $t = 2.1 \cdot 10^{11}$s the forward shock contains a developed precursor followed by a gas subshock. The pronounced precursors are seen in the velocity (Fig. 3b) and correspond to times between $t = 2.3 \cdot 10^{12}$s, $t = 1.5 \cdot 10^{13}$s and $t = 5.9 \cdot 10^{13}$s. The overall compression ratio varies between 12.2, 4.8 and 2.8, the adiabatic compression in the precursor between 3.1, 2.3 and 1.5. The Mach-numbers are in the range of 29.4, 5.4 and 2.9.

Cosmic ray shocks are unstable with respect to small density fluctuations in the upstream region advected through the shock (Dorfi and Drury 1985). A more detailed analysis (Drury and Falle 1986) shows how these density fluctuations are amplified and several numerical examples are calculated for a given density fluctuation of the order of $\delta\rho/\rho = 0.1$. A variation of the mean diffusion coefficient $\bar{\kappa} \propto \rho^{-1}$ suppresses this instability and hence we modify $\bar{\kappa}$ in unstable regions in this way to avoid a rapid growth of small fluctuations (cf. also Dorfi, 1990), in particular in the precursor region of the reverse shock propagating inwards.

In addition to these transient phenomena we plot in Figure 4 the different energy contributions in units of the initial explosion energy $E_{SN} = 10^{51}$ erg between 10^{10}s and $8 \cdot 10^{13}$s. From $3 \cdot 10^{10}$s we see a rapid increase of the cosmic ray energy E_{CR} and at $t = 1.3 \cdot 10^{11}$s the reverse shock changes the kinetic, thermal and cosmic ray energy by only small amounts. Most of the supernova energy is transferred to high energy particles and after about 10^{12}s the total cosmic ray energy E_{CR} stays almost at a level of about

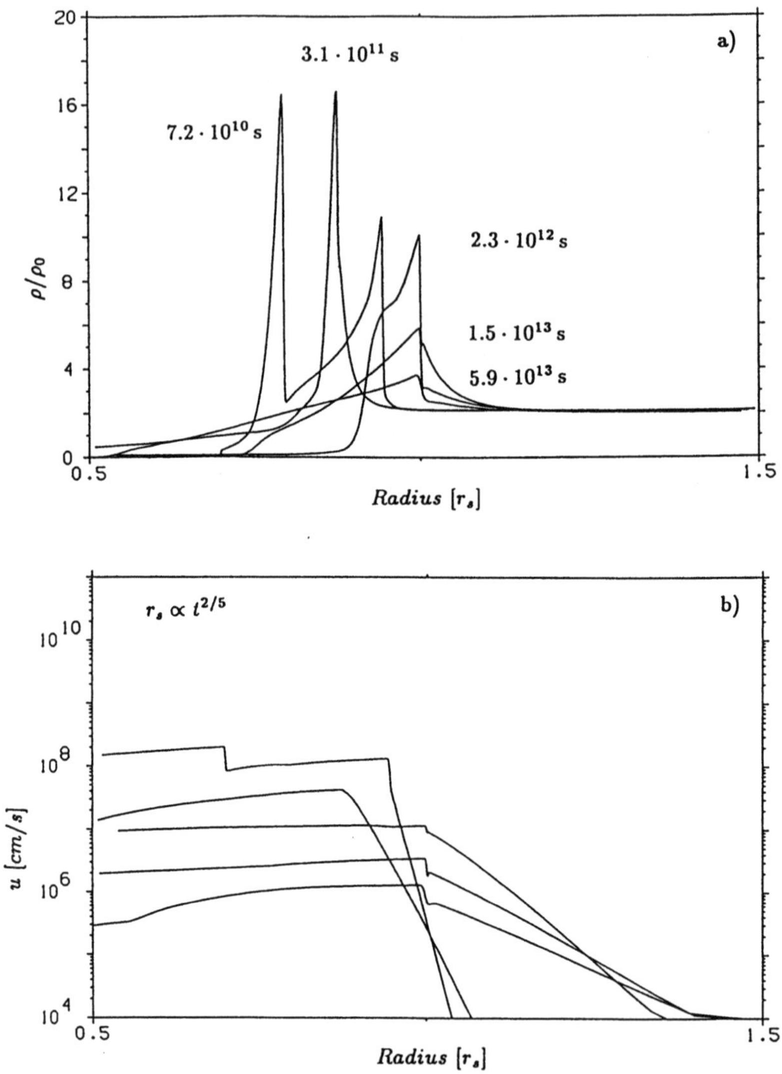

Figure 3: Density and velocity during the Sedov-phase in the shock frame. a) The strong density enhancements are caused by the transition from a gas dominated to a cosmic ray dominated shock wave. b) The cosmic ray precursor infront of the forward shock decelerates the incoming plasma and leads to almost adiabatic compressions of the thermal plasma.

70%. SNR models with constant values of γ_c and $\bar{\kappa}$ do not give such high cosmic ray effects (cf. Dorfi, 1990) although the injection of energetic particles at the shock front have been included in some models.

4. Concluding remarks

From the presented calculations we clearly see the importance and radial dependence of time-dependent phenomena during the SNR evolution. First on the (macroscopic) hydrodynamic level the cosmic ray transition leads to large density enhancements and they

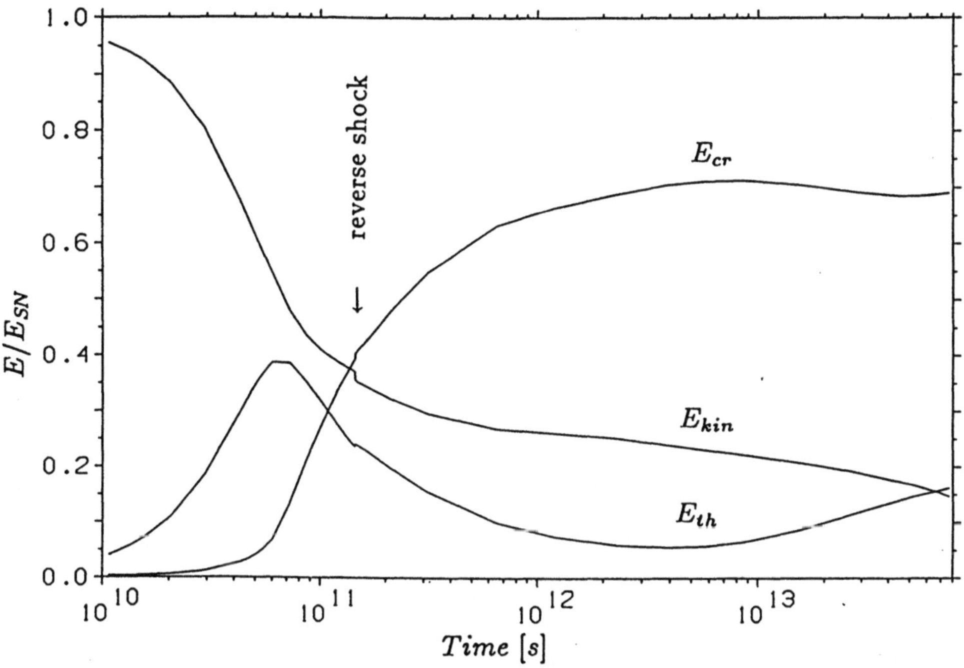

Figure 4: The overall energy partition in units of the initial supernova explosion energy E_{SN} during the time interval 10^{10}s and $8 \cdot 10^{13}$s. At time $t = 1.3 \cdot 10^{11}$s the reverse shock is seen propagating inwards.

become more important if cooling were also be considered. Values of $\bar{\kappa} = 10^{25}\mathrm{cm}^2\,\mathrm{s}^{-1}$ and $\gamma_c = 4/3$ allow up to about 30% of E_{SN} to be converted into cosmic rays at $3000\,t_{sw}$. In this case at about $20\,t_{sw}$ the overall cosmic ray energy exceeds 1% of E_{SN}. During the entire Sedov-phase a region of enhanced cosmic ray pressure is located around the forward shock. The particle pressure decreases from $P_{c,max}$ towards the interior and levels the external value at an approximate radial dimension of 0.75 in units of the radius of the forward shock and this enhanced cosmic ray pressure contains about 60% of the SNR volume.

A careful treatment of γ_c and $\bar{\kappa}$ is necessary to simulate more realistic SNR's including cosmic ray acceleration because these preliminary results indicate a very efficient conversion of the supernova explosion energy into high energy particles. However, these high cosmic ray energy densities accompanied with low plasma temperatures have to be compared to observational constraints, in particular to the X-ray luminosities of young remnants (in case of Tycho's remnant see e.g. Heavens, 1984). For the cosmic ray dominated remnants we get an almost homogeneous cosmic ray pressure inside the remnant which can exceed the external cosmic ray pressure by several orders of magnitude. In conclusion we find that the bulk of the galactic cosmic rays can be produced by SNR's but on the other hand we need further calculations including more realistic physics to quantify this statement.

Acknowledgements

Part of this work was done while I was Alexander-von-Humboldt fellow at the Max-Planck-Institut für Astrophysik, (Garching). I would also like to thank H.J. Völk (Heidelberg) and L.O'C. Drury (Dublin) for many enlightning and fruitful discussions on the problem of acceleration of cosmic rays. The numerical calculations have been carried out at the Rechenzentrum of the MPG (Garching) on the CRAY-XMP/24 and have been made possible by support of the Max-Planck-Institut für Kernphysik, Heidelberg.

References

Axford, W.I., Leer, E., Skadron, G.: 1977, Proc. 15th Int. Cosmic Ray Conf., Plovdiv, **11**, 132

Axford, W.I.: 1981, Proc. Int. School and Workshop on Plasma Astrophysics, Varenna, ESA SP–161, 425

Bell, A.R.: 1978a, Mon. Not. Roy. Astron. Soc. **182**, 147

Bell, A.R.: 1978b, Mon. Not. Roy. Astron. Soc. **182**, 443

Blandford, R.D., Ostriker, J.P.: 1978, Astrophys. J. **221**, L29–32

Blandford, R.D., Eichler, D.: 1987, Phys. Rep. **154**, 1

Burton, W.B., Liszt, H.S.: 1981, in: 'IAU Symp. No. 94, Origin of Cosmic Rays', eds. G. Setti, G. Spada, A.W. Wolfendale, D. Reidel, Dordrecht, p. 231

Dorfi, E.A.: 1990, Astron. Astrophys. **234**, 419

Dorfi, E.A., Drury L.O'C.: 1985, Proc. 19th Int. Cosmic Ray Conf., La Jolla, **3**, 121

Dorfi, E.A., Drury L.O'C.: 1987, J. Comput. Phys. **69**, 175

Drury, L.O'C.: 1983, Rep. Prog. Phys. **46**, 973

Drury, L.O'C., Falle, S.A.E.G.: 1986, Mon. Not. Roy. Astron. Soc. **223**, 353

Drury, L.O'C., Völk, H.J.: 1981, Astrophys. J. **248**, 344

Drury, L.O'C., Markiewicz, W., Völk, H.J.: 1989, Astron. Astrophys. **225**, 179

Heavens, A.F.: 1984, Mon. Not. Roy. Astron. Soc. **211**, 195

Hess, V.F.: 1912, Physik. Zeitschr. **13**, 1804

Krymsky, G.F.: 1977, Dokl. Nauk. SSR **234**, 1306, (Engl. trans.: Sov. Phys. Dokl. **23**, 327)

Lerche, I.: 1967, Astrophys. J. **147**, 689

Parker, E.N.: 1958, Phys. Rev. **110**, 1445

Tscharnuter, W.M., Winkler, K.-H.A.: 1979, Comp. Phys. Comm. **18**, 171

Simpson, J.A.: 1983, Ann. Rev. Nucl. Part. Sci. **33**, 323

Völk, H.J.: 1987, Proc. 20th Int. Cosmic Ray Conf., Moscow, **7**, 157

Völk, H.J., Biermann, P.L.: 1988, Astrophys. J. **333**, L65–68

Winkler, K.-H.A., Norman, M.L.: 1986, in 'Astrophysical Radiation Hydrodynamics', eds. K.-H.A. Winkler and M.L. Norman, NATO ASI Series C: Mathematical and Physical Sciences Vol. 188, D. Reidel, Dordrecht

Index of Contributors

MIX
Papier aus verantwortungsvollen Quellen
Paper from responsible sources
FSC® C105338

If you have any concerns about our products,
you can contact us on
ProductSafety@springernature.com

In case Publisher is established outside the EU,
the EU authorized representative is:
Springer Nature Customer Service Center GmbH
Europaplatz 3, 69115 Heidelberg, Germany

Printed by Libri Plureos GmbH
in Hamburg, Germany